AF443062

MATERIALS ANALYSIS USING A NUCLEAR MICROPROBE

MATERIALS ANALYSIS USING A NUCLEAR MICROPROBE

MARK B.H. BREESE
University of Oxford
Oxford, United Kingdom

DAVID N. JAMIESON
University of Melbourne
Melbourne, Australia

PHILIP J.C. KING
University of Oxford
Oxford, United Kingdom

A Wiley-Interscience Publication

JOHN WILEY & SONS, INC.

New York   •   Chichester   •   Brisbane   •   Toronto   •   Singapore

Library of Congress Cataloging in Publication Data:
Breese, Mark B.H., 1966–
    Materials analysis using a nuclear microprobe / Mark B.H. Breese,
  David N. Jamieson, and Philip J.C. King.
      p.   cm.
    Includes index.
    ISBN 0-471-10608-9 (alk. paper)
    1. Ion bombardment.   2. Materials–Effect of radiation on.
  3. Materials–Analysis.   I. Jamieson, David N.   II. King, Philip J.C.
  III. Title.
  QC702.7.B65B74 1996
  502′.8′2–dc20   95-11166

Printed in the United States of America

10 9 8 7 6 5 4 3 2 1

Contents

3 Microprobe Ion Optics 81

4 Analytical Techniques 139

5 Spatially Resolved Ion Channeling Techniques 201

6 Ion Beam Induced Charge Microscopy 247

7 Microelectronics Analysis 286

8 Crystal Defect Imaging with a Nuclear Microprobe   330

9 Other Materials Analysis and Modification   389

Appendices

Index   423

Foreword

The fundamentals of ion–solid interactions have been extensively studied for decades and are well known. Although the initial motivation for these studies was basic science, the important applications of ion beams for materials modification and ion beam-related analysis techniques that have emerged turned the use of ion beams to great practical importance. The need for these ion beam-related techniques has been mainly triggered by the development of semiconductor devices that require ion implantation and diagnostics, methods for the evaluation of crystal quality, impurity content, and thin-film quality. As a result, both the fields of "ion beam modification of materials" (IBMM) and "ion beam analysis" (IBA) are nowadays well established. Besides ion implantation, a wide variety of spectroscopies that make use of unfocused light ion beams has evolved. These include Rutherford backscattering spectrometry, particle induced X–ray emission, nuclear reaction analysis, and ion channeling. Several reviews and textbooks have been published recently describing the basics and applications of these techniques.

The field of electron microscopy and the use of finely focused electron beams has developed in parallel and has found many applications in microanalysis of materials. These include scanning electron microscopy, transmission electron microscopy, various diffraction techniques, energy dispersive X–ray analysis, electron beam induced current, cathodoluminescence and electron spectroscopies, as well as applications in writing ultrafine features as needed for submicron lithography.

The combination of both these fields, namely the use of finely focused beams of MeV light ions to turn ion beam spectrometry into ion beam microscopy, is the topic of this book. This merging of two very powerful research fields offers many advantages owing to the different interactions of electrons and MeV ions with matter, such as the differences in their rates of energy loss, the differences in electronic excitation, and the differences in the resulting particle trajectories. However, focusing of MeV ion beams down to submicron spots, as is commonly done for electrons, poses many new technical difficulties, which have been overcome only recently. Furthermore, because of the more complex stopping processes of ions in matter and information obtainable from the kinematics involved in the ion–atom collisions, vast amounts of information is contained even in unfocused MeV ion beam analysis experiments. This information is greatly increased when combined with the raster scanning of a microfocused MeV ion beam across the sample. In such experiments, information about the beam position, and on the crystal structure from channeling experiments are added into the already existing depth, impurity, and density information of the common IBA experiments. Hence a vast amount of data needs to be collected and stored in real-time. In most microanalysis experiments, these data sets have to be sorted out off-line to yield the desired information on the spatially resolved structure of the sample being investigated. Hence, the additional complexity for data acquisition and analysis adds another major technical challenge to microprobe analysis.

Despite these complications, the need for a tool capable of probing solids with a spatial resolution of microns and with the analyzing power of the commonly used IBA techniques has recently grown due to the development of microelectronic devices and other technologies involving micron-sized structures. Some forty laboratories are now active worldwide in experiments involving nuclear microprobes. It is anticipated that in light of the large potential of this field, this number will grow as more laboratories, previously involved in unfocused IBA work or even in low-energy nuclear physics experiments, will undergo modifications to incorporate microprobe capabilities.

This book, written by three leading scientists in the field, M.B.H. Breese, D.N. Jamieson, and P.J.C. King, brings together the know-how gained over many years in two major microprobe facilities, namely at Oxford University, U.K., and at the University of Melbourne, Australia. It is meant to give readers from the wider ion beam community an overview of the basic aspects of the technical requirements and detailed descriptions of the various uses of nuclear microprobes for materials analysis.

This book brings the topics of MeV ion optics, crystallography, and solid-state physics together into a cohesive framework and is intended to be of interest to researchers in the fields of nuclear microscopy and accelerator technology and also to a wide range of scientists involved in materials and microelectronics analysis and fabrication. The book provides both the basic theory necessary for an introduction to the research fields discussed, advances in those fields, and to the current state of the art in each subject considered. The fundamental processes involved in nuclear microscopy and ion–solid interactions are described in an accessible manner to the diverse range of readers who may be unfamiliar with this field. All the experimental aspects associated with the new analytical developments are fully described. Because this book makes the expertise of the authors available to the reader, it will certainly prove to be invaluable to both established researchers as well as newcomers to this field.

My personal encounter with the field of nuclear microprobes came about through my collaborations with two of the authors when at the University of Melbourne. Work with the excellent facilities there and interactions with the researchers have made me appreciate the great potential of this new field and the many possible analytical applications. Much of this was not known to me from my previous experience of IBA and IBMM. I hope that the new horizons which microprobe analysis opens up, as I have witnessed and as described in this book, will be valued by the worldwide ion beam community and thus will give a large impetus to an important, promising, and rapidly evolving field of research.

RAFI KALISH

Haifa, Israel

Historical Background

The use of focused MeV ion beams for elemental analysis arose in the 1960s when Van de Graaff accelerators, found in many physics laboratories across the world, started to become too small to produce the higher energies needed for nuclear physics research.

Even before that time, users of MeV ion beams had realized that elastically scattered particles or emitted gamma rays could give information about the elemental or isotopic composition of the target. By the mid-1960s it had also been realized that characteristic X–rays generated by MeV ions were accompanied by far less brehmsstrahlung background than those generated in an electron microprobe, and that this might give the ion beam much better detection limits than an electron beam.

By that time there were two groups, one at Lucas Heights in Australia and the other at Harwell in Britain, doing positional analysis with nuclear reactions, and a second group at Harwell using a crystal spectrometer for what has since become known as PIXE analysis with X–rays. However the low current density in the collimated beams made measurements slow and severely limited the attainable positional resolution. The logical step forward was taken at Harwell of trying to improve the beam-line focusing. The aim was to increase the current density at the collimator near the target or even to produce a demagnified image of an object aperture placed just after the energy defining slits of the Van de Graaff accelerator.

At just that time, studies relating to electron microscopy at MeV energies, including theoretical work on magnetic quadrupole lenses, had been published. These implied that with care a significant improvement in the target current density of the Harwell system should be achievable. A simple trial was set up using four existing magnetic quadrupoles mounted close together with their yokes, which were circular, resting in an aluminum V-groove. Initially it was intended to measure the shape of the resulting focused beam with an orthogonal pair of oscillating wires, but this arrangement rapidly gave way to a thin glass cover-slip viewed by a ×100 microscope. The currents through the quadrupoles were chosen to give the 'Russian Quadruplet' arrangement which had been recommended for its low spherical aberration.

Initial attempts at focusing the proton beam coming through a 0.1 mm diameter hole drilled in a tantalum sheet gave the type of unstable effect frequently seen on beam-line viewing quartzes, and no combination of lens currents gave an image approaching the hoped-for 20 μm diameter spot. The only readily accessible degrees of freedom of the lenses were the relative rotations of the four magnets which could be easily if somewhat jerkily carried out using a screwdriver as a lever because of the way in which the lenses were mounted. As subsequent theory explained, the precise rotation of just one lens stopped the wild sensitivity of the image to lens excitations and allowed a spot of about 20 μm diameter to be produced.

The Harwell focusing system was then rebuilt using better quadrupole lenses to give a spot diameter of about 3 μm. A versatile target chamber equipped with detectors and their electronics allowed the nuclear reaction microanalysis of materials to continue with a spatial resolution of a few microns. Even with the improved current density, X–ray analysis with a crystal spectrometer was not viable. It was only in 1972 that this area was revived using the much more efficient solid state detector and has become a major feature of nuclear microprobe analysis in biomedical and geological research.

As this book describes, the use of nuclear microprobes has by now expanded greatly from these humble beginnings. The spatial resolution has improved by an order of magnitude and the data collection has become more and more sophisticated. But, perhaps of more importance, the range of ways in which these focused beams are now being used, particularly in materials science, has greatly expanded to give to structural, crystallographic and electrical information along with elemental analysis.

DR. JOHN COOKSON

Preface

Objectives

The use of a nuclear microprobe to image and analyze three dimensional microstructures, such as integrated circuits and crystal defects, is the main focus of this book. The ability of MeV ions to penetrate through surface layers on a sample with little scattering, together with the ability to focus the ion beam to a probe size smaller than one micrometer is what gives the nuclear microprobe its analytical power.

Successful operation of a nuclear microprobe requires an unusual range of skills drawn from the fields of ion optics, nuclear, atomic and solid state physics. This book presents the necessary concepts with the aim of guiding readers with diverse backgrounds through both the theory and practice of nuclear microscopy.

The main objective of the book is to be a self-contained handbook to anyone who wishes to construct, operate, optimize or otherwise become involved in a system for nuclear microprobe analysis of materials.

Outline

Chapter 1 reviews the fundamentals of ion–solid interactions which are relevant to the MeV ion analytical techniques used in this book. The different effects of the electronic and nuclear components of the ion energy loss are described. Ion channeling and the basic mechanisms of ion induced defect formation in semiconductors, which are important features of nuclear microprobe analysis are also introduced. Finally the passage of MeV light ions and keV electrons through matter is compared to provide an understanding of the relative merits of each type of charged particle for microscopy.

Chapters 2 and 3 describe the hardware of the nuclear microprobe and the ion optics needed to understand and optimize its performance. Chapter 2 is a general discussion of a nuclear microprobe system and covers all the important components. It includes a discussion of accelerators, probe-forming lens systems, sample chambers, detectors and data acquisition systems. The topic of ion optics is introduced in Chapter 2, but Chapter 3 goes into this important subject in more detail. The causes, effects and characterization of various types of aberrations which limit the beam spot size of a nuclear microprobe are all discussed from a practical point of view. Particular attention is given to the diagnosis and elimination of parasitic multipole aberrations which would otherwise prevent the focusing of high resolution probes.

Chapter 4 first describes several of the MeV ion beam analytical techniques which have been widely used with unfocused MeV ion beams, but are also important for focused beams. These are particle induced X–ray emission (PIXE), backscattering spectrometry, nuclear reaction analysis (NRA), elastic recoil detection analysis (ERDA) and ion beam induced luminescence (IBIL). All the other analytical techniques described in Chapters 4, 5 and 6 are solely for use with the nuclear microprobe since they rely on the generation of spatially resolved information. The microprobe methods detailed in Chapter 4 are ion induced electron imaging, scanning transmission ion microscopy (STIM) and ion microtomography (IMT).

Chapter 5 describes focused ion beam channeling techniques which can generate images showing spatial variations in crystallographic quality. A brief review of those aspects of ion channeling relevant to these two methods is given and the dechanneling effects associated with different types of dislocations and defects are discussed.

Chapter 6 describes the theoretical and experimental aspects of ion beam induced charge (IBIC) microscopy for imaging the distribution of *pn* junctions in microelectronic devices and dislocations in semiconductors. The contrast mechanisms for IBIC are described, preceded by some basic relevant semiconductor theory. It is then shown how the ion induced charge pulse height can be calculated in terms of the semiconductor diffusion length, surface and depletion layer thickness, ion channeling effects and ion induced damage, so that the beam type and energy for the maximum signal to noise ratio can be optimized.

Chapters 7, 8 and 9 describe the use of these analytical techniques for the analysis of a wide variety of mainly crystalline materials, such as microelectronic devices, $Si_{1-x}Ge_x/Si$ epilayers and diamond. Chapter 7 first describes analysis of device active areas using IBIC microscopy and then describes the analysis of the physical device structure using PIXE, backscattering spectrometry and STIM. Chapter 8 describes the use of different microprobe techniques to image and characterize misfit dislocations present in epitaxial $Si_{1-x}Ge_x/Si$ layers and stacking faults. The main alternative analytical methods used for crystal defect imaging are reviewed to show the complimentary information provided by nuclear microscopy. Chapter 9 describes work on imaging different kinds of defects such as point defects, growth defects and dislocations present in superconductor crystals, laser annealed diamonds and in microelectronic devices.

MARK B.H. BREESE
DAVID N. JAMIESON
PHILIP J.C. KING

Acknowledgments

We would like to acknowledge the great contribution made to the field of nuclear microscopy by Geoff Grime, George Legge, and Frank Watt who at various times have been our advisers, mentors, and colleagues. Much of the work presented in this book would not have been possible without their considerable efforts in establishing the nuclear microprobe facilities in Oxford and Melbourne.

The wisdom and foresight of John Cookson, who developed the first nuclear microprobe at Harwell, is greatly appreciated.

There are many friends and colleagues at the University of Oxford who have helped immeasurably in supporting and advising us with the research work presented in this book. In this respect, we particularly wish to acknowledge Drs. Roger Booker, Mike Goringe, Peter Wilshaw, and Linda Romano from the Department of Materials who had sufficient confidence in the value of this work to give invaluable help. We also thank the people who have contributed their excellent work for inclusion in this book, namely Chris Marsh, Michelle de Coteau, and Angus Wilkinson from the Department of Materials. Mark and Philip also wish to thank the Royal Commission for the Exhibition of 1851 for providing Fellowships, during which much of the work in Oxford was carried out.

In Melbourne, there were also many people who helped us with material for this book. We thank our collaborators from outside laboratories, as well as our colleagues and the honors, summer, and graduate students in the Microanalytical Research Centre. A special thanks goes to Sean Dooley, Andrew Bettiol, and Lachlan Witham who allowed us to use some of their work. None of the original research would have been possible without the financial support of the Australian Research Council, the CSIRO/University collaborative research fund, and the Australian Telecommunications Research Board.

Our research work would not have been possible without the expert assistance of Mike Marsh and Mike Dawson (in Oxford) and Roland Szymanski and Steve Gregory, and their staff (in Melbourne) who constructed, maintained, and operated our facilities to a very high standard.

For help in the preparation of this book, we wish to thank Raik Jarjis, John Cookson, Alexander Dymnikov, Andrew 'Flynn' Saint, and Chris Ryan for their critical comments on the early drafts; and the staff at John Wiley and Sons, New York, who professionally managed all aspects of the production.

To the staff who operate the excellent facilities in Lygon St, Melbourne and the Royal Oak and Eagle and Child public houses in Oxford we give our heart-felt thanks for providing a convivial atmosphere for discussions, scientific and otherwise.

M.B.H. BREESE
D.N. JAMIESON
P.J.C. KING

Symbols Used in the Text

Some of the commonly used symbols in this book are defined here. Some equations are simplified by using non S.I. units such as: $e^2 \equiv q^2/(4\pi\epsilon_o) \equiv 14.4$ eV Å, where q is the electron charge.

Ion Optical Parameters

B	Incident ion beam brightness
B_Q	Quadrupole lens pole tip magnetic field
D	Drift space length
E_o	Incident ion energy
f_x, f_y	Focal lengths in the xoz and yoz planes
L_Q	Quadrupole lens effective length
M_1	Incident ion mass
M_x, M_y	Magnifications in the xoz and yoz planes
p	Relativistic ion momentum
P	Grid period
r_Q	Quadrupole lens bore radius
v_1	Ion velocity
V_Q	Quadrupole lens pole tip potential
x	Ion displacement in the xoz plane
y	Ion displacement in the yoz plane
Z	Incident ion charge state
Z_1	Incident ion atomic number
α	Grid rotation angle
β	Quadrupole lens excitation (m.k.s. units)

ϕ	Ion trajectory in the xoz plane
θ	Ion trajectory in the yoz plane

Ion–Solid Interactions

a_{tf}	Thomas–Fermi screening radius
a_o	Bohr radius
$K(\theta_S)$	Kinematic factor
Ω_B	Bohr energy straggle
M	Atomic weight
M_2	Sample nuclear mass
m_e	Electron mass
N_s	Areal density
N_A	Avogadro's number
R_i	Ion range in matter
R_e	Electron range in matter
ρ	Mass density
θ_S	Scattering angle
$\sigma(\theta_S)$	Ion scattering cross-section
v_o	Bohr velocity
Z_2	Sample atomic number

Ion Energy Loss

dE/dz	Rate of electronic of nuclear energy loss per micron
ϵ	Stopping cross-section
LET	Linear energy transfer

Semiconductors

E_{eh}	Energy required to create electron-hole pair
L	Minority carrier diffusion length
z_d	Depletion layer thickness

Channeling

a	Lattice parameter
D_s	Dose for swelling induced dechanneling
χ_{min}	Minimum backscattered ion, or X–ray yield in channeling alignment
ψ	Beam angle to channeling axis or plane
ψ_a	Axial channeling critical angle
ψ_p	Planar channeling critical angle
ϱ	Transverse r.m.s. thermal vibration amplitude

1

ION–SOLID INTERACTIONS

An understanding of the way in which MeV light ions lose energy through amorphous and crystalline materials is essential for describing their use for analyzing the elemental, crystallographic and electronic properties of materials. Because 1 to 10 MeV ^{1}H ions (protons) and ^{4}He ions (α-particles) are the ions most commonly used for the analytical methods of this book, this chapter concentrates on their interactions with solids. In particular, the most relevant features are the ways in which MeV ions lose their energy in collisions and these energy loss mechanisms are first described.

The energy loss and scattering of ions in collisions with the atomic electrons and nuclei of every element has been thoroughly investigated experimentally and theoretically for 80 years, since the observation by Geiger and Marsden of α-particles backscattering from gold foils [1], Rutherford's derivation of ion scattering cross-sections [2], and the development of the Bohr model of the atom [3,4]. As a result of much theoretical and experimental work, there is agreement to within 2% to 10% between calculated and measured rates of ion energy loss through most elements over a large range of energy. More accurate agreement between measured and theoretical values is difficult because of the many different types of interactions that can occur between the ions and the atomic nuclei and electrons. There is still considerable research in this field, with many papers being published in journals such as *Nuclear Instruments and Methods in Physics Research*, and many conference proceedings, as listed in Appendix 1. The most comprehensive and widely used values for the amount of energy lost by ions are semiempirical compilations [5–9] and mathematical expressions [10,11]. Reviews of the energy loss of ions through matter include Refs. 12 to 14 and Refs. 15 to 17 contain more detailed information on much of the work discussed in this chapter. As a result of this large body of work on measuring the energy loss, scattering yields and ionization probability of various elements, the particle induced X–ray emission (PIXE), backscattering spectrometry, nuclear reaction analysis (NRA), and elastic recoil detection analysis (ERDA) techniques have now reversed this process and are used to determine the elemental composition of materials.

Interest in the mechanisms of ion energy loss and scattering was given a further impetus in the 1960s with the observation of an orientation dependence of both the energy loss and the scattering cross-sections in crystalline materials [18,19]. This process is called ion channeling and is introduced in Section 1.4. After the initial development of theoretical models characterizing ion channeling and much experimental work on measuring the different rates of ion energy loss and scattering of channeled ions, it has also become a very valuable ion beam technique for crystallographic analysis [20,21].

There are computer codes available for Monte Carlo simulations of the passage of individual ions through any combination of elements. These codes simulate the electronic and nuclear energy loss, ion range, defect profiles and energy straggling through amorphous materials [22], and through crystalline materials [23–26]. In this book, models for the energy loss and straggling effects of ions through amorphous materials are based on Refs. 5 to 7 and the computer code TRIM [22].

Most of the analytical techniques and applications described in this book are used for the study of the crystallographic and electrical structure of semiconductor materials. The basic mechanisms underlying the analytical techniques described in Chapters 4 to 6 are discussed here using amorphous or crystalline silicon for most examples. The effects of ion induced damage to the material under analysis are described in Section 1.5. This is particularly important for many of the new MeV ion beam analytical techniques described in this book, because it affects their sensitivity for detecting variations of crystalline perfection.

Electron beam interactions with solids have long been used for microscopy and are well established; therefore, this book contrasts the different characteristics of MeV light ions and keV electrons, where relevant. Many of the different processes that can occur when ions and electrons travel through matter are similar, and their energy loss mechanisms are compared in Section 1.6.

1.1. ELECTRONIC ENERGY LOSS

Ions lose kinetic energy during their passage through matter by colliding either with the clouds of atomic electrons or with the atomic nuclei. The resultant energy transfer causes the ions to slow until they come to rest at some depth in the material. The radius, r, of the atomic nucleus is given by $r \sim r_o A^{1/3}$ where $r_o \approx 1.4 \times 10^{-5}$ Å. For silicon $r \sim 4 \times 10^{-5}$ Å, compared with the distance between the nuclei in the silicon lattice of several angstroms. Since the atomic nuclei are so small, a collision between an ion and a nucleus occurs infrequently, so most of the energy of the MeV ion is lost in collisions with the atomic electrons. It is consequently this process that determines the distance ions travel before they come to rest, which is their range in matter.

The energy loss of MeV ions through matter has been treated using many definitions. Strictly speaking, the ion energy losses to both the atomic electrons and to the atomic nuclei should always be taken into account. However, the nuclear energy loss of MeV light ions is small, so consideration of only the electronic energy loss adequately describes the distance which MeV ions travel through matter.

The units most commonly used in this book are the average rate of electronic energy loss per micron of material traversed by the ion, dE/dz, in units of keV/μm. This simple definition is useful for nuclear microscopy because the lateral dimensions of the beam spot size and the scanned area are also typically described in microns. This definition is limited, however, in that it does not take into account the atomic density of the material, N, which is shown as a function of atomic number in Figure 1.1. There is such a large variation in atomic density across the periodic table that it is essential to take this into account when comparing the energy losses of different ions in different elements or for cases such as carbon where the same element exists in several allotropes with different density.

Much of the original experimental work on measuring ion energy loss by passing ions through thin foils of a known atomic density used the following definition of the stopping cross-section, ϵ, which avoids the problem:

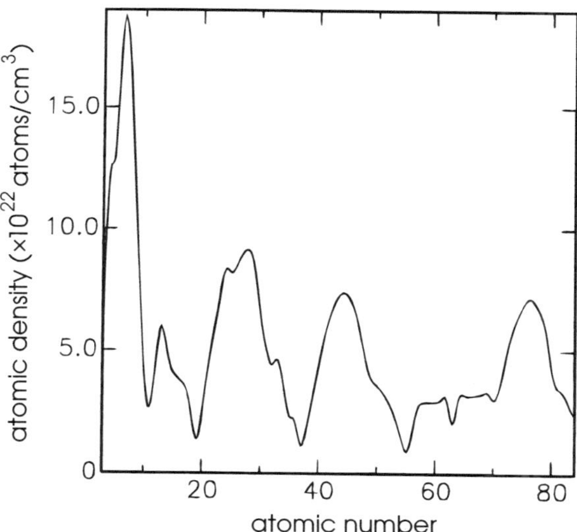

Figure 1.1. Variation in atomic density as a function of atomic number.

$$\epsilon = \frac{1}{N}\frac{dE}{dz} \text{ eV cm}^2 \quad \text{or} \quad \epsilon = \frac{10^{15}}{N}\frac{dE}{dz} \text{ eV}/(10^{15} \text{ atoms}\cdot\text{cm}^{-2}) \qquad (1.1)$$

Here $N = \rho N_A/M$, where ρ is the mass density (g/cm^3), N_A is Avogadro's number (6.025×10^{23}/mole), and M is the atomic weight (g/mole). For silicon, which has a density of $\rho = 2.3$ g/cm^3, and $M = 28$ g/mole, then $N \sim 5 \times 10^{22}$ atoms/cm^3. For a 3 MeV ^{4}He ion which loses energy at a rate of 200 keV/μm through silicon, $\epsilon = 2 \times 10^9$ eV/cm/(5×10^{22} atoms $\cdot$ cm^{-3}) $= 4 \times 10^{-14}$ eV cm^2 or 40 eV/(10^{15} atoms $\cdot$ cm^{-2}).

Another useful definition is to express the rate of energy loss in terms of the areal density of the material, which has units of milligrams per square centimeter. A 1 μm thick silicon layer has an areal density of 0.23 mg/cm^2, so an energy loss of 200 keV/μm is equivalent to 860 keV/(mg $\cdot$ cm^{-2}). Another unit of energy loss is the linear energy transfer (LET; in MeV/(mg $\cdot$ cm^{-2})). This unit is best suited to describing the rate of electronic energy loss of heavy, high-energy ions, which are used to generate soft upsets in microelectronic devices as described in Chapter 6. In silicon, for example, an LET of 100 means that the rate of ion energy loss is 100 MeV/(mg $\cdot$ cm^{-2}).

1.1.1. Electronic Energy Loss Regimes

The ways in which MeV light ions lose energy through matter owing to transfer of kinetic energy to the atomic electrons can be conveniently described using low- and high-energy regimes. An estimate of the ion energy that separates these regimes is obtained by equating the ion velocity to the Bohr velocity, v_o, of an electron in the innermost atomic shell of a hydrogen atom, $v_o = e^2/\hbar = 2.2 \times 10^6$ m/s. This velocity corresponds to a ^{1}H ion energy of 25 keV and a ^{4}He ion energy of 100 keV, so that the energy losses of MeV ^{1}H and ^{4}He ions are described using the high-energy regime. Here, because the ion velocity v_1 is greater than the electron velocities in their atomic shells, the atoms appear static to the ions. The ion's velocity is so great that it becomes stripped of any electrons and can be treated as a fully ionized particle with a positive charge of $Z_1 e$. The energy loss in this regime is virtually independent of the chemical nature of the material traversed. This makes the high-energy regime relatively straightforward to model and facilitates quantitative analysis using MeV ion-beam analytical techniques. Figure 1.2 shows the average rate of electronic energy loss as a function of energy of ^{1}H and ^{4}He ions in amorphous silicon. The rate of energy loss decreases with increasing ion energy in the high-energy regime because the ions pass through the orbiting electron clouds faster and have less chance of colliding with them.

In the low-energy regime, the ion velocity is slow compared with that of the inner shell electrons of the atomic nuclei, so they no longer appear static to the ion. Ions incident on a solid may not become fully stripped of their electrons [27–29] in the low-energy regime, so the average positive ion charge is less

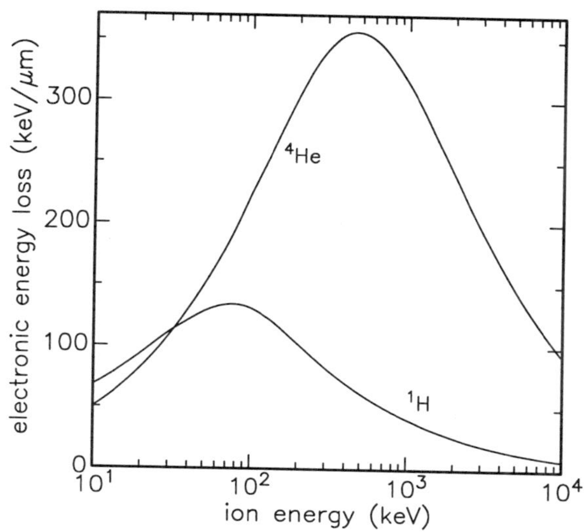

Figure 1.2. Average rate of electronic energy loss for ^{1}H ions and ^{4}He ions in amorphous silicon as a function of ion energy.

than $Z_1 e$. The ion energy loss in the low-energy regime was derived by Lindhard et al. [30,31] and also by Firsov [32] and in both cases it was shown that the number of atomic electrons involved in excitation processes decreases because the inner shell electrons have a declining role in the ion energy loss with lower ion energies. The energy loss increases with increasing ion energy rather than decreasing as in the high-energy regime. The rate of energy loss and the types of interaction of the ions with the atomic nuclei become dependent on the chemical nature of the material in the low-energy regime. Effects in this energy regime are harder to model than in the high-energy regime, so ion beam analytical techniques involving low-energy ions tend to be less quantitative than those using high-energy ions. The maximum rate of ion energy loss occurs at the Thomas–Fermi ion velocity given by $v_1 = v_o Z_1^{2/3}$, which is approximately 25 keV for ^{1}H ions and 250 keV for ^{4}He ions. For heavier ions the relevant energies are much higher, for example, approximately 23 MeV for silicon ions.

The derivation of the rate of ion energy loss in the high-energy regime was first carried out by Bohr [3,4] using a central-force field model of ion scattering in a cloud of free electrons with a subsequent momentum transfer to the atomic electrons. This classic derivation is given in many textbooks [15,17,33]. Later calculations by Bethe and Bloch [34,35] characterized the energy loss in terms of close collisions with large momentum transfer when the ion is within the electron shells and distant collisions with small momentum transfer when the ion is outside the electron shells. These two components were shown to be the same size in the high-energy regime, and they derived an expression which can be summarized as

$$\frac{dE}{dz} = \frac{4\pi Z_1^2 e^4 N Z_2}{m_e v_1^2} \ln \frac{2m_e v_1^2}{I} \tag{1.2}$$

where Z_1 and Z_2 are the atomic numbers of the incident ion and sample nucleus, respectively, v_1 is the incident ion velocity and m_e is the electron mass. The average electron excitation energy $I \simeq 10Z_2$ eV [36], ignoring variations from the electron shell structure and the electron binding energy. The complete electronic energy loss formula derived by Bethe and Bloch has corrections for relativistic terms at high ion energy and corrections for the effect of strongly bound inner electrons. Figure 1.3 shows a gradual increase in the stopping cross-section for 2 MeV ^{4}He ions with increasing atomic number. The stopping cross-section rather than the rate of energy loss is used here to take into account the effect of the variation in atomic density across the periodic table shown in Figure 1.1. For material containing different elements, the total stopping power can be calculated by summing the individual stopping powers of the components according to their stoichiometric fractions, as described in Ref. 15. The kinetic energy of the ion is $M_1 v_1^2 / 2$, and from Eq. (1.2) the electronic energy loss is proportional to Z_1^2 for the same ion velocity v_1; so, for example a 1 MeV ^{1}H ion has approximately a quarter of the rate of energy loss of a 4 MeV ^{4}He ion.

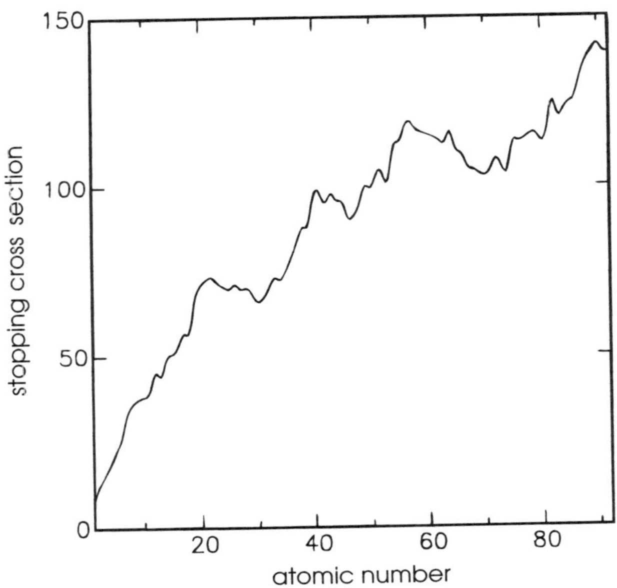

Figure 1.3. Stopping cross-section in units of eV/(10^{15} atoms $\cdot$ cm^{-2}) for 2 MeV ^{4}He ions as a function of atomic number.

For the purposes of ion beam analysis, high rates of signal production are important. Since collisions between the ion and the atomic electrons are much more probable than with the atomic nuclei, there tends to be a greater rate of signal production from collisions with the atomic electrons than by collisions with the nuclei. A collision with an atomic electron can cause any of several different electronic processes to occur. An inner shell electron can be knocked out, leading to ionization. Subsequent decay of an outer shell electron can give rise to emission of an X–ray which forms the basis of PIXE analysis. If the atom is ionized within a few angstroms of the material surface then the electron can escape and contribute to ion induced electron imaging. Scanning transmission ion microscopy (STIM) and ion microtomography (IMT) also utilize the electronic energy loss of the ions as the imaging mechanism since they generate images showing variations in the areal density of the material. If the ionization event occurs within the bulk of a crystalline material then the electrons and holes formed by the positively charged lattice vacancy can drift or diffuse through the crystal lattice. The measurement of these ionized electrons and holes is the basis of the ion beam induced charge (IBIC) microscopy. Radiative recombination processes can lead to emission of a photon and this is the basis of ion beam induced luminescence (IBIL) microscopy. These processes are further described in Chapters 4 and 6.

1.1.2. Ion Range

MeV ions continually lose their kinetic energy in collisions with the atomic electrons until they come to rest at some depth within the material. The deepest part of the range of MeV light ions is not important for NRA, ERDA, and electron imaging, because these signals usually originate from very close to the surface. However, for most other nuclear microprobe techniques, the long range of MeV light ions is an important asset allowing them to access features of interest below the material surface.

Figure 1.4 shows the average rate of electronic energy loss of 3 MeV ^{1}H ions and 3 MeV ^{4}He ions with distance traveled in amorphous silicon. The rate of ion energy loss increases initially as the ion penetrates further, which is in accordance with Figure 1.2, where the rate of ion energy loss is initially to the right of the maximum value. As the ion penetrates further, the rate of energy loss reaches a maximum, which is seen close to the end-of-range for MeV light ions in Figure 1.4.

The average range, R_i, which MeV light ions travel through matter before coming to rest can be evaluated as

$$R_i = \int_0^{E_o} \left(\frac{dE}{dz} \right)^{-1} dE \tag{1.3}$$

where E_o is the incident ion energy. Figure 1.5 shows the ranges of ^{1}H ions

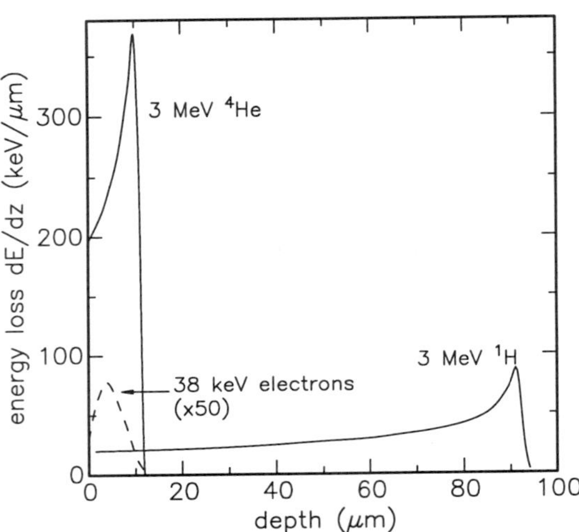

Figure 1.4. Average rate of electronic energy loss for 3 MeV ^{1}H ions and 3 MeV ^{4}He ions in amorphous silicon as a function of particle penetration. The energy curve for electrons is discussed in Section 1.6.

and ^{4}He ions in silicon as a function of ion energy calculated by Eq. (1.3). A ^{4}He ion travels a much shorter distance than a ^{1}H ion of the same energy because of its higher rate of energy loss. The average range of 3 MeV ^{1}H ions and ^{4}He ions in several different elements is shown in Table 1.1. The ^{1}H ion range is roughly seven times greater than that of the same energy ^{4}He ion,

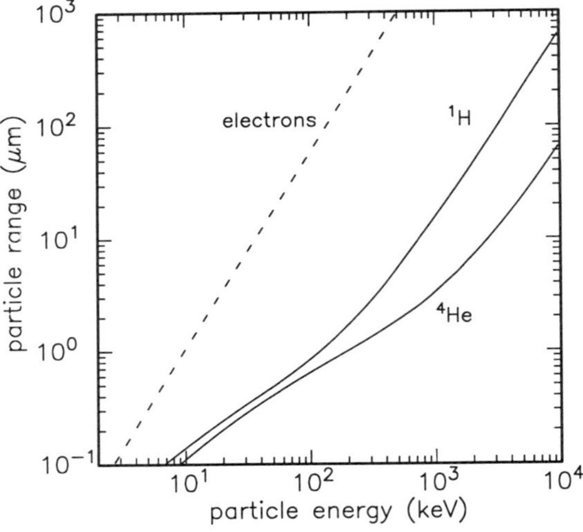

Figure 1.5. Average range of ^{1}H ions, ^{4}He ions, and electrons in amorphous silicon as a function of particle energy. The range curve for 38 keV electrons is discussed in Section 1.6.

TABLE 1.1 Ion Ranges

Element	Atomic Number	Density (g/cm^3)	Ion Range (μm)	
			3 MeV ^{1}H Ion	3 MeV ^{4}He Ion
C	6	2.2	74	9
Si	14	2.3	90	11.6
K	19	0.86	261	32.6
Fe	26	7.8	35	5.0
Ga	31	5.9	55	8.4
Ag	47	10.5	36	5.3
Ba	56	3.5	110	15.2
W	74	19.3	27	4.2
Au	79	19.3	27	4.4
Pb	82	11.3	46	7.4
U	92	19.0	28	4.6

and there is a wide variation in the range in different elements owing to the widely differing density. The longer ^{1}H ion range is an important asset for STIM analysis, which is described in Section 4.7, because heavier ions often cannot be used owing to their insufficient penetration through materials thicker than 10 μm. The variation in the range of monoenergetic light ions due to statistical fluctuations in the number of collisions they undergo with the atomic electrons is described in Section 1.3.

In contrast with MeV ions, at very low energies of only a few electron volts, ions do not have enough energy to penetrate even the outermost monolayer of the material; so they tend to be deposited on the material surface. This is the basis of ion beam deposition methods described, for example, in the Ion Implantation conference series listed in Appendix 1. At very high incident ion energies the rate of energy loss continues to decrease with increasing energy. For example, a 1000 MeV ^{1}H ion has a rate of energy loss of less than 4 keV/μm and a range of about 1 m in silicon.

1.2. NUCLEAR ENERGY LOSS

Although the energy lost by the vast majority of MeV ^{1}H and ^{4}He ions in collisions with the atomic nuclei is less than 10 keV, nuclear collisions are extremely important because they provide the analytical signals measured with backscattering spectrometry and NRA. Furthermore, the transfer of even modest amount of energy to atomic nuclei of the sample causes ion induced defects owing to displacements of the atomic nuclei from their original sites. The displacement of light nuclei by heavy ions is also the basis of ERDA described in Section 4.4. However, these displacements adversely effect the sensitivities of other MeV ion beam techniques such as IBIC, IBIL, and channeling contrast microscopy (CCM), because they limit the ion dose that these methods use to make an accurate measurement. Ion induced defects resulting from collisions with atomic nuclei are described in more detail in Section 1.5. A considerable fraction of the energy of heavy ions and keV light ions is transferred to the sample nuclei; a description of nuclear energy loss as an important mechanism in the stopping process of ions through matter was first developed by Bohr [37] and refined by Lindhard [31].

Figure 1.6 shows the average nuclear energy loss along the paths of 3 MeV ^{1}H ions and 3 MeV ^{4}He ions in amorphous silicon. The maximum rate of nuclear energy loss occurs at an energy between 100 eV and 10 keV for light ions and for heavy ions respectively. The rate of nuclear energy loss thus rises toward the end of the ion range. Because the total nuclear energy loss of a MeV ^{1}H ion is less than that of a MeV ^{4}He ion, the former create less damage to the material being analyzed.

Although the average rate of nuclear energy loss of MeV light ions is small compared with the electronic energy loss rate, any individual ion may lose a large amount of energy in a very violent collision with the atomic nucleus, which results in the ion being scattered through a large angle, or even backscattered out of the sample. Nuclear energy loss involves energy transfer from the ion to the atomic nucleus by an interaction between the two positive nuclear charges; such scattering of ions from atomic nuclei was first characterized by Rutherford [2]. In the absence of nuclear penetration, the interaction between

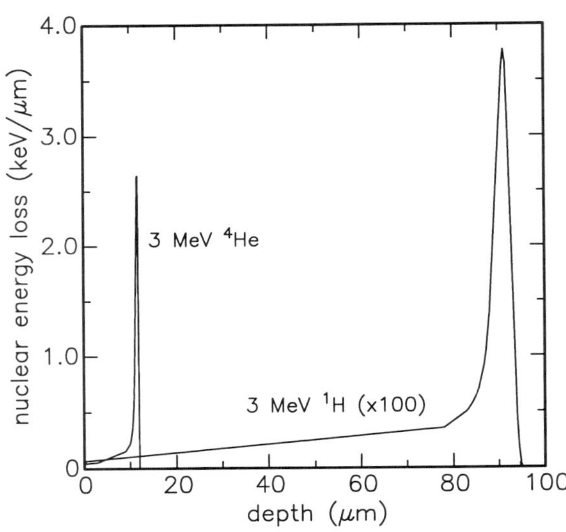

Figure 1.6. Average nuclear energy loss for 3 MeV ^{1}H ions and 3 MeV ^{4}He ions in amorphous silicon as a function of ion penetration.

the ion and the nucleus can be well represented as a repulsive Coulomb potential $V(r)$ between the positive ion charge $Z_1 e$ and the positive charge $Z_2 e$ on the sample nucleus at a separation distance r as

$$V(r) = \frac{Z_1 Z_2 e^2}{r} \chi \tag{1.4}$$

where χ is a screening function. Rutherford assumed that the ion velocity in the high-energy regime was large enough to fully penetrate inside the innermost atomic electron shells. In this case, the atomic nucleus is not shielded by the inner electrons; so the screening function in Eq. (1.4) can be ignored and the collision treated as a pure Coulomb interaction between two bare positive charges. If the ion trajectory does not completely penetrate the inner electron shells, the charge of the atomic nucleus is screened from the ion, which leads to a modification of the unscreened Coulomb potential and hence the screening function χ in Eq. (1.4). Several different models of the screening function have been used [30,38], and, in each case, a screening radius is defined to characterize the variation of the screening potential away from the nucleus. This distance is usually taken to be the Thomas–Fermi screening radius a_{tf} [39], given by

$$a_{tf} = \frac{0.885 a_o}{(Z_1^{1/2} + Z_2^{1/2})^{2/3}} \tag{1.5}$$

where a_o is the Bohr radius equal to 0.53 Å, and a_{tf} typically has a value of 0.1 to 0.2 Å. An estimate of the lower ion energy limit, $E_{\min}$, where it becomes necessary to use a screened potential instead of a pure Coulomb potential to correctly model the kinematics of the ion–nucleus collision can be found from the energy that allows the ion to approach within a radius r equal to the radius of the K electron shell, a distance of approximately a_o/Z_2. Figure 1.7 shows the variation of this lower energy limit with atomic number for ^{1}H ions. With ^{1}H ions the lower limit varies from less than 10 keV for light elements to nearly 200 keV for heavy elements, and with ^{4}He ions this minimum energy is twice as large. MeV light ions are thus well above this limit, except very near to the end of their range, so that their collisions with the atomic nuclei can be treated using an unscreened Coulomb potential.

An ion can approach closer to the atomic nucleus with increasing energy, and eventually it can approach to within a distance comparable with the nuclear radius. When this occurs, the interaction again cannot be represented using a pure Coulomb potential since nuclear penetration occurs and the forces operating within the nucleus itself affect the collision. This results in new scattering effects, which can be accompanied by structural changes to the atomic nucleus with emission of reaction products that are measured using NRA, as

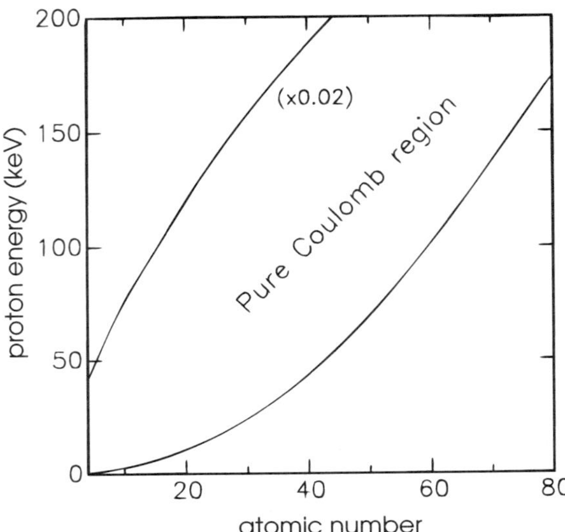

Figure 1.7. Illustration of the region for treating the ^{1}H ion–nucleus interaction as an unscreened Coulomb potential. The upper limit is shown divided by 50.

described in Section 4.3. The maximum energy for treating the ion–nucleus collision classically can be taken as $E_{\max} \sim (Z_1 Z_2 e^2)/r$ where $r = r_o A^{1/3}$. Figure 1.7 shows the variation of this maximum ^{1}H ion energy with atomic number, and for ^{4}He ions the energies are again twice as great. These values are only approximate, but, according to Figure 1.7, this energy is approximately 4 MeV for ^{1}H ions and 8 MeV for ^{4}He ions in silicon. In practice, these effects can occur on light element samples at considerably lower energies, such as the elastic resonance for 3.04 MeV ^{4}He ions on ^{16}O, measured at backward angles.

1.2.1. Classical Scattering Theory

Important parameters to model the collision between the ion and the atomic nucleus are the probability of ions being scattered through an angle θ_S, called the scattering cross-section $\sigma(\theta_S)$, and the fraction of the ion energy which remains with the ion after the collision, called the kinematic factor $K(\theta_S)$. The energy of the recoiling nucleus is $[1 - K(\theta_S)]E$. The kinematic factor and scattering cross-section must be accurately known to interpret the measured spectrum of ion energies in backscattering spectrometry. They are tabulated in Refs. 15 to 17, and only the general trends are outlined here.

When an ion of mass M_1 collides with an atomic nucleus of mass M_2, energy is transferred to the nucleus. In a pure Coulomb collision, the scattered ion retains all its energy except that lost making the nucleus recoil, and the kinematics of this elastic collision are defined by the conservation of energy and momentum. The kinematic factor can be derived in a non-relativistic form in the laboratory reference frame [15] as

$$K(\theta_S) = \left\{ \frac{[1 - (M_1/M_2)^2 \sin^2 \theta_S]^{1/2} + (M_1/M_2)\cos \theta_S}{1 + M_1/M_2} \right\}^2 \tag{1.6}$$

The kinematic factor thus depends only on the scattering angle and the ratio of the masses of the ion and the atomic nucleus: it does not depend on the ion energy. Its variation with atomic mass according to Eq. (1.6) is shown in Figure 1.8a for ^{4}He ions at scattering angles of $\theta_S = 30°, 90°$, and $180°$. Figure 1.8b shows the variation of the kinematic factor with the angle of scattering for ^{4}He ions incident on three different atomic masses. A close approach results in a large scattering angle. More energy is imparted to the atomic nucleus with increasing ion scattering angle so the kinematic factor is lower. Most energy is transferred in a head-on collision, which results in the ion being scattered through $180°$. This can only occur, however, when the incident ion is lighter than the nucleus.

The elastic scattering cross-section is usually given in units of barns (10^{-24} cm^2), which is roughly the size of the atomic nucleus. It can be derived, in the laboratory reference frame to second order, by treating the collision as a two-body scattering problem as

$$\sigma(\theta_S) \simeq \left(\frac{Z_1 Z_2 e^2}{4E} \right)^2 \left[\sin^{-4} \frac{\theta_S}{2} - 2 \left(\frac{M_1}{M_2} \right)^2 \right] \tag{1.7}$$

Equation (1.7) provides the Rutherford scattering cross-sections, which are used in Rutherford backscattering spectrometry (RBS). Figure 1.8c shows the Rutherford scattering cross-section as a function of atomic number for 2 MeV ^{1}H ions for three scattering angles, according to Eq. (1.7). Figure 1.8d shows the variation of the scattering cross-section for 2 MeV ^{1}H ions at $\theta_S = 179.5°$ from silicon with scattering angle, from Eq. (1.7). Increasing from $\theta_S = 0°$, the number of scattered ions decreases rapidly toward $\theta_S = 90°$ and then slowly decreases. Nuclear microprobes most commonly use MeV ^{1}H ions for elemental analysis, and these often give rise to non-Rutherford scattering cross-sections with light elements present in the material. This is why the term backscattering spectrometry, and not RBS, is used in this book, as it is generally more appropriate. Non-Rutherford scattering cross-sections are described further in Section 4.3.

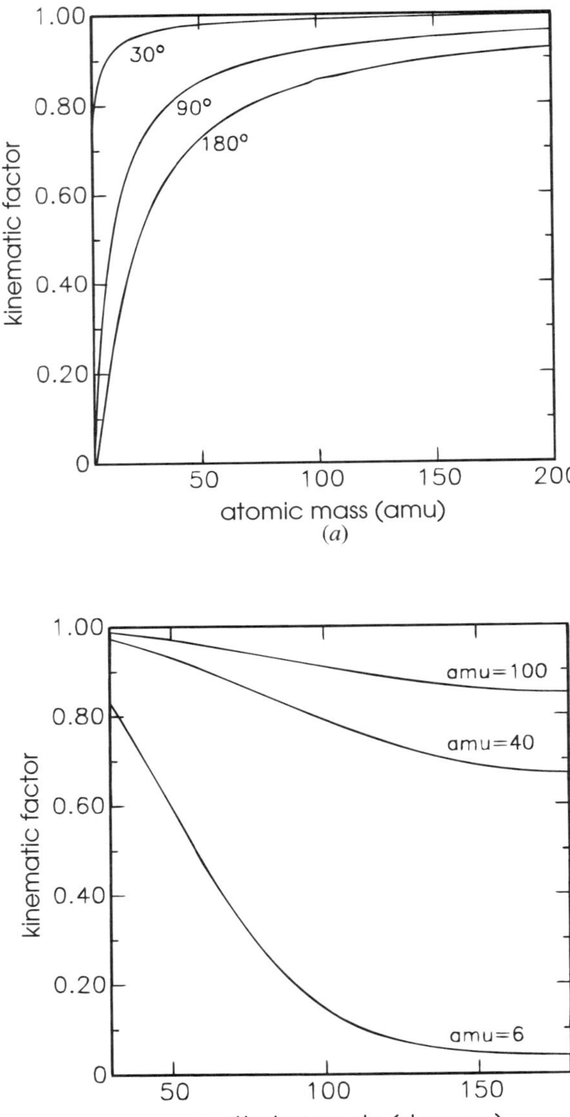

Figure 1.8. (a) Kinematic factor for MeV ^{4}He ions as a function of sample atomic mass for three scattering angles. (b) Kinematic factor for MeV ^{4}He ions for three different atomic masses (a.m.u.) as a function of scattering angle. (c) Scattering cross-section as a function of atomic number for 2 MeV ^{1}H ions for three fixed scattering angles, according to Eq. (1.7). (d) Scattering cross-section as a function of scattering angle for 2 MeV ^{1}H ions in silicon, according to Eq. (1.7). In reality, the Rutherford scattering cross-section breaks down below an atomic number of approximately 20.

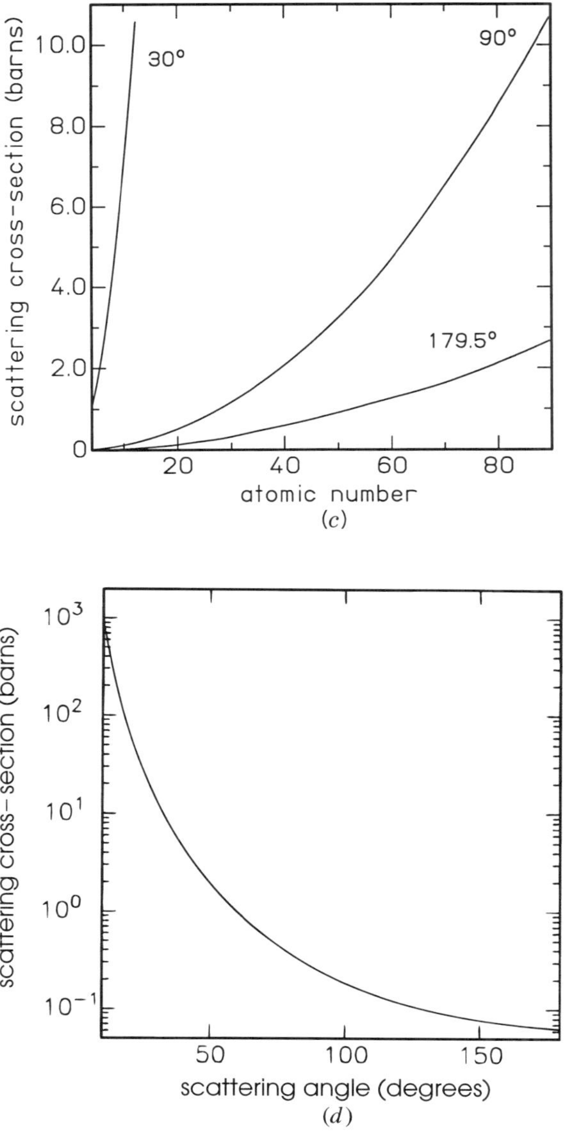

Figure 1.8. (*Continued*)

1.3. ION STRAGGLING

The average rate of electronic energy loss, dE/dz, was used in Section 1.1 to determine the average ranges of ^{1}H ions and ^{4}He ions. In practice, MeV ions lose their energy in discrete collisions with individual atomic electrons. This process is subject to fluctuations in the number and geometry of collisions. This gives rise to a distribution in the energies of initially monoenergetic ions after traversing a depth of material and this energy straggle limits the precision with which measurements based on ion energy losses can be made. This same effect also results in a variation in the distance to which individual ions penetrate the material, which is called range or longitudinal straggling. Variations in the number and geometry of the ion–electron collisions also results in a distribution of the transverse momentum acquired by the ions. This alters the trajectory angle of the ions through the material, so that they stop at different lateral distances away from the beam axis; that is, there is a lateral straggle or spread.

Figure 1.9 shows Monte Carlo computer simulations of the trajectories of one thousand 3 MeV ^{1}H and 3 MeV ^{4}He ions in amorphous silicon. The increase in the lateral spread with increasing ^{1}H ion penetration can be seen, with no backscattered ions in this example. Both of these light ions undergo little lateral spread in the first few microns of their trajectory. The shorter ^{4}He ion range and lower lateral straggling at end-of-range, compared with ^{1}H ions can be seen. Figure 1.10 shows the lateral straggle away from the ion axis and the longitudinal straggle about the average range as a function of ^{1}H and ^{4}He ion energy, based on values given in Ref. 6. ^{1}H ions have the greater variation in the end-of-range distribution, because they travel much farther than ^{4}He ions of

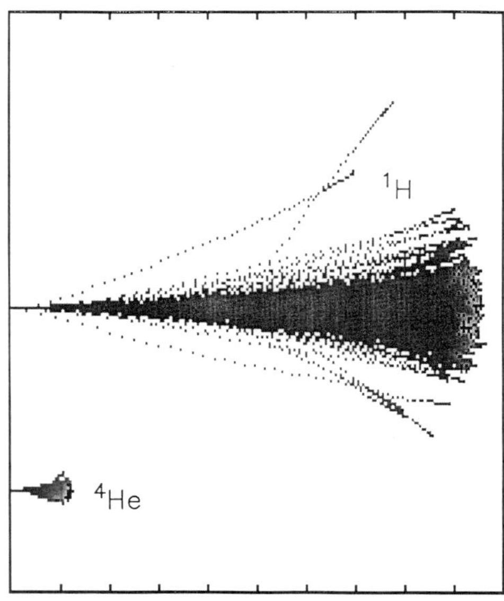

Figure 1.9. TRIM Monte Carlo computer simulations of the passage of 1,000 3 MeV ^{1}H ions and 3 MeV ^{4}He ions in amorphous silicon. The box size is 100 μm horizontally by 8 μm vertically.

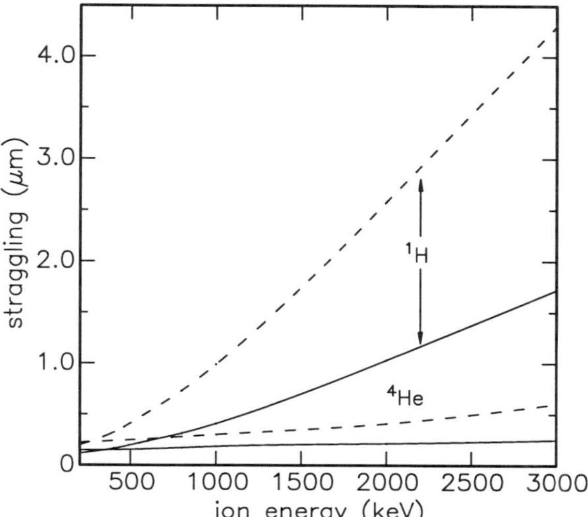

Figure 1.10. Longitudinal straggle (solid lines) about the average range and lateral straggle (dashed lines) about the ion axis for ^{1}H and ^{4}He ions in amorphous silicon as a function of energy.

the same energy. The end-of-range distributions for 3 MeV ions are similarly wider than for the same type of MeV ions.

These straggling effects limit the depth and lateral resolution attainable with MeV ion beam analytical and imaging techniques. Energy straggling limits the mass and depth resolution attainable with backscattering spectrometry and the depth resolution with NRA. Similarly with STIM and IMT, the minimum resolvable areal density is limited by the ion energy straggle. Lateral straggling, rather than the beam spot size on the material surface, ultimately defines the minimum spatial resolution attainable with the nuclear microprobe in thick layers. However, a spatial resolution of approximately 100 nm can be maintained for several microns through the sample. The lateral spread of a MeV ion beam is considerably less than for keV electrons, which is why a nuclear microprobe is a uniquely valuable tool for generating high spatial resolution images of buried layers.

The distribution of energy lost by individual MeV ions about an average value at a given depth in the material can be adequately described by a Gaussian distribution as

$$P(E)dE = \frac{1}{\sqrt{2\pi}\Omega_B} \exp\left(\frac{-E^2}{2\Omega_B^2}\right) dE \qquad (1.8)$$

where Ω_B^2 is the variance of the energy straggle, called the Bohr energy straggle. In the Bohr model of energy straggle, collisions between fully stripped ions and the atomic electrons are assumed to be the dominant contribution to the energy straggle, which is the case in the high-energy regime defined in Section 1.1. In the low-energy regime, the ions are not fully ionized and Bohr's original theory was subsequently modified to incorporate this [details can be found in Refs. 40,41]. In the high-energy regime, the variance of the energy straggle after traversing a distance z through matter is

$$\Omega_B^2 = 4\pi Z_1^2 e^4 N Z_2 z \tag{1.9}$$

This predicts that Ω_B^2 increases with the areal electron density, NZ_2z, traversed by the ions but does not depend on the ion energy. Figure 1.11 shows the value of Ω_B^2 as a function of atomic number of a 1 μm thick layer traversed by MeV ^{1}H ions. The large fluctuations arise mainly from the variations in density from element to element illustrated in Figure 1.1. Chu et al. [15] derived a simplified approximation for the energy straggle of ions that lose an energy of ΔE as

$$\Omega_B \approx n \sqrt{\overline{E}\Delta E} \times 0.01 \tag{1.10}$$

where $\overline{E}$ is the average ion energy within the material, and $n = 1$ for ^{4}He ions and $n = 2$ for ^{1}H ions. For a 3 MeV ^{4}He ion which loses 200 keV through 1 μm of silicon, the energy straggle is thus $\Omega_B \sim 8$ keV. Similarly for 3 MeV ^{1}H ions through the same 1 μm layer, $\Omega_B \sim 5$ keV.

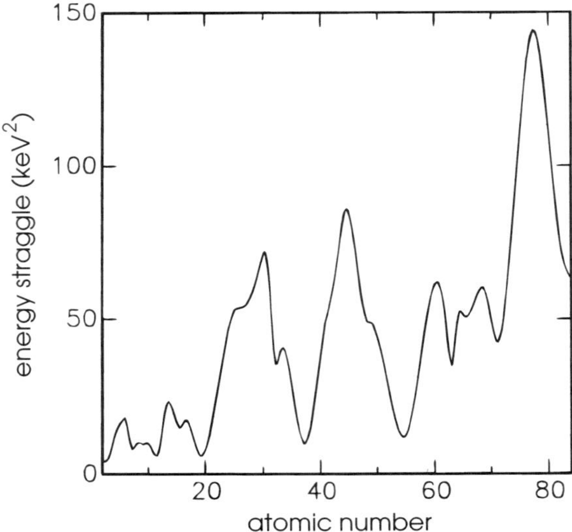

Figure 1.11. Energy straggle for MeV ^{1}H ions passing through a 1 μm thick layer of different atomic number.

1.4. EFFECTS OF CRYSTALLINITY ON THE PASSAGE OF IONS

So far, it has been assumed that the materials through which the MeV light ions pass were amorphous. Many applications in this book consider the analysis of crystalline materials. Both an ion's rate of energy loss and its probability of scattering can be greatly influenced by this crystallinity because of an effect called ion channeling. A brief explanation is given in Section 1.4.1 of relevant features of crystal structures, and then ion channeling is described in the remaining parts of Section 1.4.

1.4.1. Silicon Crystal Structure

A regular atomic arrangement is the defining characteristic of crystalline materials and detailed descriptions of crystal structures can be found in textbooks on solid state physics or crystallography [42,43]. Many materials have crystal lattices that are cubic in form, either with a site at the cube center (body-centered cubic, bcc, lattice) or with a site in the middle of the each cube face (face-centered cubic, fcc, lattice). Some crystalline materials for which ion-beam studies are presented in this book are silicon, silicon-germanium alloys, and diamond. All these materials have the same basic crystal structure, shown in Figure 1.12.

The structure can be thought of as two face-centered cubic lattices, one displaced by one quarter of the cube body diagonal from the other. The lattice parameters (length of side of one face-centered cube), a are, respectively, 5.431 Å and 3.567 Å for silicon and diamond. An alloy of silicon with germanium has the same crystal structure as that of pure silicon, with germa-

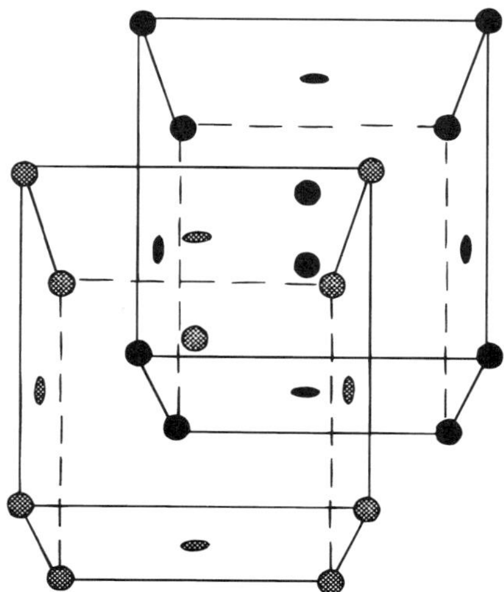

Figure 1.12. The structure of silicon and diamond crystals showing the two interpenetrating face centered cubic (fcc) lattices.

nium atoms distributed at random among the silicon atoms. However, the lattice parameter of pure germanium is 5.646 Å, which is 4% larger than that of silicon. This means that a $Si_{1-x}Ge_x$ alloy, where x is the germanium molar fraction, has a larger lattice parameter than pure silicon and is given by $a_{SiGe} = a_{Si}(1 + 0.04x)$.

It is often necessary to specify particular directions and planes within the crystal structure; this is done using a triad of numbers enclosed within brackets, Miller indices. Directions are indicated by $\langle \ \rangle$ for a general set of equivalent directions (e.g., $\langle 100 \rangle$), and by [ ] for a specific direction (e.g., [010] is one particular $\langle 100 \rangle$ direction). In crystals with a cubic lattice, $\langle 100 \rangle$ directions are the cube edges, $\langle 110 \rangle$ directions are the cube face diagonals, and $\langle 111 \rangle$ directions are the cube body diagonals, as shown in Figure 1.13. Planes are indicated by $\{ \ \}$ for a general set of planes (e.g., $\{100\}$, which is the set of cube faces) and by ( ) for a particular plane (e.g., (001) is one particular cube face). In cubic crystals, the [abc] direction is normal to the (abc) plane. Negative indices are indicated by a bar above them, for example [$\bar{1}$10].

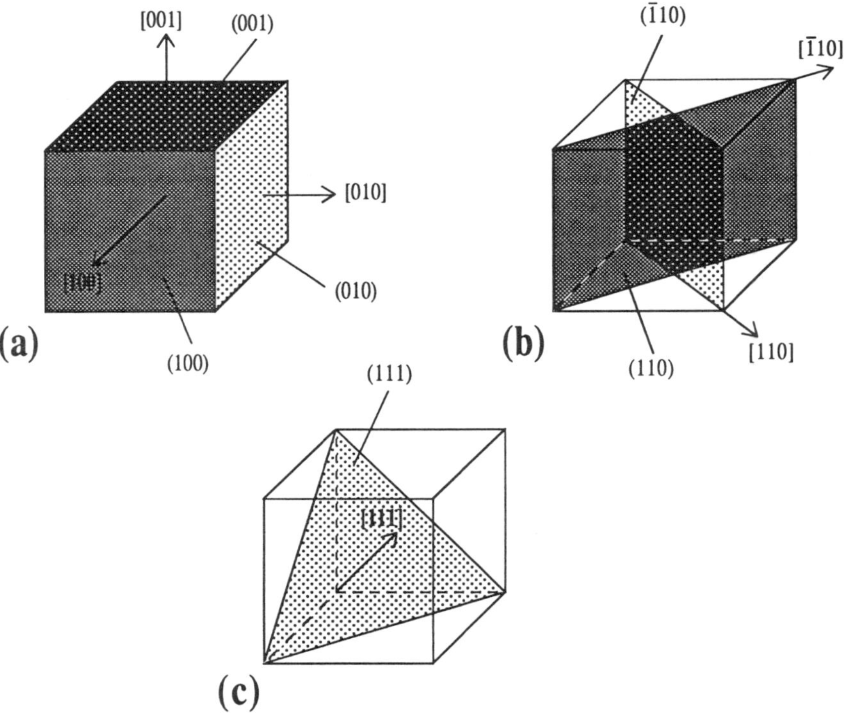

Figure 1.13. Diagram showing major planes and directions in a cubic crystal. (a) Some of the $\langle 100 \rangle$ directions and $\{100\}$ planes. (b) Some of the $\langle 110 \rangle$ directions and $\{110\}$ planes. (c) A $\langle 111 \rangle$ direction and a $\{111\}$ plane.

The calculation of certain parameters associated with channeling (e.g., Eqs. (1.12), (1.13) and (1.19), (1.20) below) involves knowledge of the distances between atoms along crystal directions and the spacing between atomic planes. Table 1.2 gives expressions for these quantities for the major directions and planes of crystals with the silicon and diamond structure. There are two values for the inter-atomic spacing along $\langle 111 \rangle$ directions and for the spacing of $\{111\}$ planes, and the average of these two values is used for the calculation of channeling parameters.

Ion channeling analysis typically involves the angular location of different crystallographic directions and planes. This process can be facilitated by the use of a stereographic projection of the crystal, which is a depiction of the angular locations of planes and directions in two dimensions. For cubic crystals, the angle, α, between two sets of planes with indices (h_1, k_1, l_1) and (h_2, k_2, l_2) is given by

$$\cos \alpha = \frac{h_1 h_2 + k_1 k_2 + l_1 l_2}{(h_1^2 + k_1^2 + l_1^2)^{1/2}(h_2^2 + k_2^2 + l_2^2)^{1/2}} \tag{1.11}$$

Stereographic projections for face centered cubic crystals are given in Appendix

TABLE 1.2. The Interatomic Distances and Interplanar Separations for the Major Directions and Planes in Crystals with the Silicon and Diamond Structures. a is the Lattice Parameter

Atomic Separations Along Crystal Directions	
Direction	Interatomic Distance
$\langle 100 \rangle$	a
$\langle 110 \rangle$	$\dfrac{a}{\sqrt{2}}$
$\langle 111 \rangle$	$\dfrac{\sqrt{3}a}{4}, \dfrac{3\sqrt{3}a}{4}$

Plane	Interplanar Separation
$\{100\}$	$\dfrac{a}{4}$
$\{110\}$	$\dfrac{a}{2\sqrt{2}}$
$\{111\}$	$\dfrac{a}{4\sqrt{3}}, \dfrac{3a}{4\sqrt{3}}$

2, and other examples of stereographic projections are given in Figure 1.16 below and in Chapter 5.

1.4.2. Ion Channeling

The regular arrangement of atoms in a crystalline material can have a very great effect on the passage of ions through the material. Alignment of a major direction of the crystal lattice (i.e., an axis), or of a direction contained within a set of the lattice planes, with the ion beam can lead to a large reduction in the ions' rate of energy loss and scattering probability. Channeling effects were first observed during the large amount of effort into keV ion-implantation technology in the 1950s. MeV ion channeling was experimentally observed only in the 1960s [18,19], when it was also accidentally discovered by computer simulation [44], although it had been predicted much earlier [45]. Detailed descriptions of the channeling process can be found in Refs. 20, 21, and 46 and useful information for calculation of channeling parameters in Ref. 16.

Channeled ions are steered by the rows or planes of the lattice, so that they travel in regions of lower atomic electron density. Their energy loss rate is therefore significantly reduced; MeV ^{1}H ions channeled in the {111} and {110} planes of silicon suffer an average energy loss rate of the order of 0.45 and 0.60 of the nonchanneled rate, respectively [47]. The ions do not closely approach the atoms, but rather suffer a gentle steering effect caused by glancing collisions with many atoms, as shown schematically in Figure 1.14. The atomic nuclei and inner-shell electrons are shielded from the beam, which results in a reduction in the production cross-sections for various analytical signals. The change in the measured yield of signals from close encounter processes between channeled and nonchanneled alignments of the crystal gives information on crystal quality, epitaxy, and the lattice position of interstitial elements [20,21]. Most uses of ion channeling have involved measurement of the change in yield of backscattered ions to give depth-resolved channeling information, but PIXE and NRA have also been used to identify interstitial impurity elements using ion channeling [21].

The nuclear microprobe has an important capability in that it can produce spatially resolved ion channeling information, which enables local variations

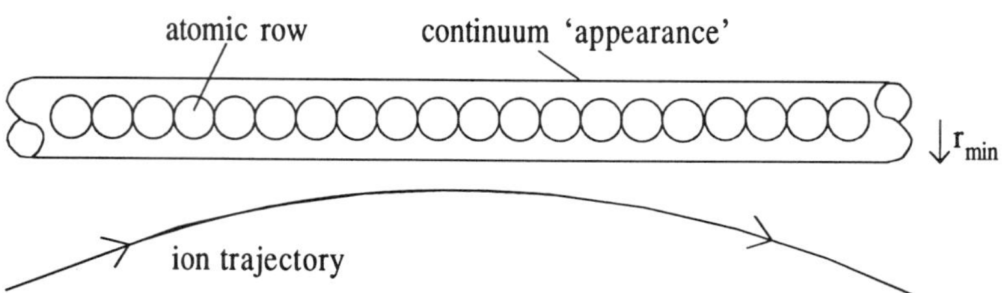

Figure 1.14. Schematic showing the trajectory of a channeled ion near an atomic row. The ion is gently steered by the row as long as it does not approach the row closer than r_{min}.

in crystal quality to be imaged. Variations across the material surface in the yields of backscattered ions or ion induced X–rays can be mapped with the beam at channeling alignment, which is called channeling contrast microscopy (CCM). Local changes in the energy loss of ions transmitted through thinned crystals can be used to produce images of crystal defects such as dislocations and stacking faults, and this method is called channeling scanning transmission ion microscopy (CSTIM) or transmission ion channeling. These channeling techniques and some considerations that the channeling process imposes on the focused beam divergence produced by a nuclear microprobe are discussed in Chapter 5. Applications of CSTIM for the analysis of defects in crystalline materials are described in Chapter 8, and examples of CCM for a variety of applications are described in Chapter 9.

1.4.3. Channeling Theory

When an ion beam is aligned with a channeling direction of a crystal, not every ion becomes channeled. This is because some ions are scattered through too large an angle by the atoms in the first monolayer or so of the crystal for channeling to occur. For axial channeling, the number of atomic rows per unit surface area of the crystal is Nd, where N is the the number of atoms per unit volume and d is the atom spacing along the rows. The area associated with each row is therefore $1/(Nd)$. If ions approaching within a radius r_{min} of a row are not channeled, then the fraction of ions not channeled when the beam is aligned with a crystal axis is just

$$\chi_{min} = \frac{\pi r_{min}^2}{1/(Nd)} = \pi r_{min}^2 Nd \qquad (1.12)$$

The transverse root-mean-square thermal vibration amplitude of the sample atoms, ρ, can be used for r_{min}. The fraction of ions not initially channeled can be measured in a backscattering experiment, when it is called the minimum yield, χ_{min}. This parameter is widely used to characterize the amount to which ions are able to channel in a crystalline sample and hence its crystalline quality. Computer simulations and experiment show that the axial minimum yield actually is approximately a factor of three greater than predicted by this simple model. For planar channeling, a geometrical argument similar to that above gives

$$\chi_{min} = \frac{2}{d_p} \frac{\rho}{\sqrt{2}} \qquad (1.13)$$

where d_p is the interplanar spacing. This again underestimates the planar minimum yield. χ_{min} values for $\langle 110 \rangle$ and $\langle 111 \rangle$ axial directions of silicon are approximately 3%, and for $\{111\}$, $\{110\}$, and $\{100\}$ planar directions approximately 30% to 40%, for 3 MeV ^{1}H ions [48].

Ions that are channeled at the crystal surface are shepherded by the bounding rows or planes, as shown in Figure 1.14. The gentle steering experienced by channeled ions arises from small angle deflections produced by interactions with successive atoms in the channel walls. The atomic rows or planes can be considered as strings or sheets of charge and are described by a continuum potential that is formed by averaging the individual atomic potentials. For the axial case, the continuum potential is

$$U_a(r) = \frac{1}{d} \int_{-\infty}^{\infty} V(\sqrt{z^2 + r^2})\, dz = \frac{Z_1 Z_2 e^2}{d} \ln\left[\left(\frac{Ca_{tf}}{r}\right)^2 + 1\right] \tag{1.14}$$

where

$$V(r') = \frac{Z_1 Z_2 e^2}{r'} \left(1 - \frac{r'}{\sqrt{r'^2 + C^2 a_{tf}^2}}\right) \tag{1.15}$$

which is a form of the Coulomb potential discussed in Section 1.2 and has been modified with a screening function as commonly used for analytical treatments of channeling. r' is the spherical polar coordinate given by $r' = (z^2 + r^2)^{1/2}$, r is the transverse distance from the atomic row, C is a constant equal to $\sqrt{3}$, and a_{tf} is the Thomas–Fermi screening radius given by Eq. (1.5).

The small angle scattering that each ion suffers as a result of its interaction with the first few atomic layers means that even when the beam is perfectly aligned with a crystal channeling direction, the resulting channeled ions have small components of their motion perpendicular to the bounding rows or planes. This transverse energy, $E_\perp$ is given for the axial case by

$$E_\perp = U_a(r) + E\psi^2 \tag{1.16}$$

where E is an ion's total energy and ψ is its angle to the channeling direction (ψ is small so that $\sin \psi \simeq \psi$).

If an ion approaches one of the channel walls too closely, it begins to see the discrete nature of the wall rather than the continuum potential, and channeling breaks down. This closest approach distance, put equal to ρ for axial channeling, means that there is a maximum transverse energy that an ion can have for it to remain channeled, given by $U_a(\rho)$. Alternatively, this condition for channeling can be expressed as the maximum angle, ψ_a, that an ion can make with the axial channel direction when it crosses the channel center, where

$$E\psi_a^2 = U_a(\rho) \tag{1.17}$$

For axial channeling, the critical channeling angle is given from Eqs. (1.14) and (1.17) in radians by

$$\psi_a = \left(\frac{Z_1 Z_2 e^2}{Ed} \right)^{1/2} \left| \ln \left[\left(\frac{Ca_{tf}^2}{\varrho} \right) + 1 \right] \right|^{1/2}, \tag{1.18}$$

with a good estimate of the axial channeling critical angle given by

$$\psi_a \simeq \sqrt{\frac{2Z_1 Z_2 e^2}{Ed}} \tag{1.19}$$

For planar channeling, similar arguments to the above give

$$\psi_p \simeq \sqrt{\frac{2\pi Z_1 Z_2 e^2 a_{tf} N d_p}{E}} \tag{1.20}$$

The channeling critical angle gives an estimate of how close in angle an ion beam must be to a channeling direction for a significant fraction of the beam to be channeled. Critical angles are usually measured experimentally from the half-width at half-maximum of backscattered ion yield as a function of angle to the channeling direction. For example, the critical channeling angles for the $\langle 110 \rangle$, $\{111\}$, $\{110\}$, and $\{100\}$ channeling directions of silicon are $0.26°$, $0.092°$, $0.087°$, and $0.070°$ respectively for 3 MeV ^{1}H ions [48].

Figure 1.15 shows an example of the measured variation in the number of

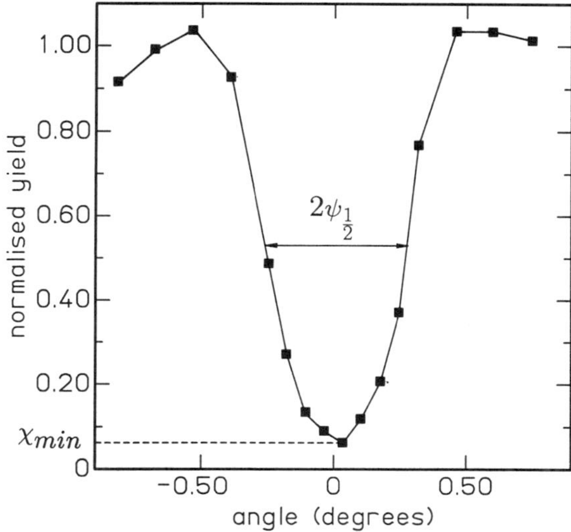

Figure 1.15. Measured variation with tilt angle from the substrate [001] axis of the yield of backscattered 2 MeV ^{1}H ions from the silicon atoms in the alloy layer of a $Si_{0.85}Ge_{0.15}/Si$ sample. The minimum yield in channeled alignment is χ_{min} and the half-width-at-half-maximum is $\psi_{\frac{1}{2}}$.

backscattered ^{1}H ions (i.e., the yield) versus tilt angle of the beam from the substrate [001] axis for 2 MeV ^{1}H ions scattering from silicon atoms in the silicon–germanium layer of a $Si_{0.85}Ge_{0.15}/Si$ crystal. The half-width at half-maximum of the yield curve is $\psi_{\frac{1}{2}} = 0.27°$, and the minimum yield is $\chi_{min} = 6\%$. The minimum yield is slightly greater than that expected from a perfect crystal owing to the presence of misfit dislocations, as described in Chapter 8.

Figure 1.16a shows a contour map of the measured channeling probability for 1 MeV ^{4}He ions backscattering from silicon for incident angles about the silicon $\langle 001 \rangle$ axis. Figure 1.16b shows the corresponding planes near the $\langle 001 \rangle$ axis in silicon. More comprehensive stereographic projections for planes about the $\langle 001 \rangle$, $\langle 011 \rangle$, and $\langle 111 \rangle$ axes are given in Appendix 2. The closely spaced lines in Figure 1.16a are where the measured yield changes rapidly from an adjacent channeling axis or plane. The use of these "ion channeling patterns" is further described in Chapter 5 as a means of locating a particular axis or plane for analysis.

From Eqs. (1.18) and (1.19), the channeling critical angles increase if heavier ions are used or if the beam energy is reduced. It is, therefore, very difficult to avoid the channeling effect for low-energy heavy ions, which is a great problem for fabrication by ion implantation of shallow implants using low-energy ions. Implications of such effects are discussed in the proceedings of the Ion Implantation Conferences listed in Appendix 1.

That there is a critical angle for channeling means that the trajectory of channeled ions remains nearly parallel (to within an angle of ψ_a or ψ_p) to the bounding rows or planes of the channel. Close interaction with the atomic nuclei, which may cause the ions to be scattered through larger angles, is prevented. The angular distribution of channeled ions transmitted through a thinned crystal is therefore modified from that displayed by non-channeled ions [47], where the distribution is approximately Gaussian about the beam direction. The channeled ion distribution, however, is highly peaked about the channeling direction, with the majority of channeled ions being transmitted within an angle of the order of the channeling critical angle.

The closest approach that a channeled ion can make to the bounding rows or planes of atoms is controlled by its transverse energy, $E_\perp$. Ions with small $E_\perp$ remain close to the channel center, whereas those with larger $E_\perp$ can approach the bounding walls of the channel more closely. All channeled ions therefore have access to the center of the channel, but relatively few to the regions close to the channel walls. This results in the distribution of channeled ions across the channel being highly nonuniform, which is called flux peaking. The chance of a channeled ion interacting with an interstitial atom can, therefore, be several times greater than that of a nonchanneled ion, which is a characteristic feature of experiments for determining the lattice site location of impurity atoms in crystals using ion channeling [21].

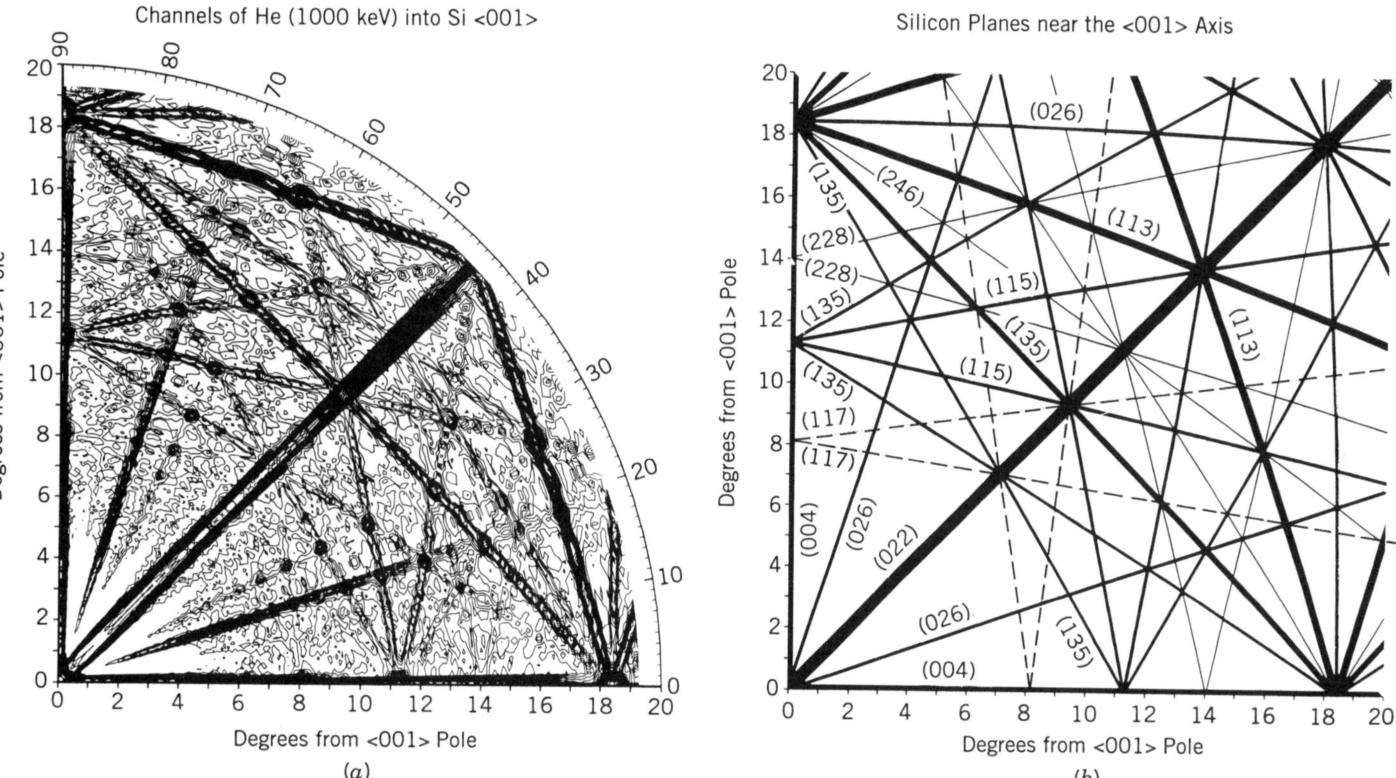

Figure 1.16. (a) Channeling probability of ^{4}He ions near the silicon $\langle 001 \rangle$ axis, with contour lines at intervals of 2% channeling probability. The ordinate and abscissa both indicate polar angles, which are the latitude angles from the $\langle 001 \rangle$ pole, whereas the azimuthal angles indicate the longitude angles from the (004) plane. (b) The major planes that can be seen in (a). Negative signs have been omitted from the plane indices. Reprinted with permission from *Ion Channeling Near the Silicon $\langle 001 \rangle$ Axis*, J.F. Ziegler and R.F. Lever, IBM Corporation.

1.4.4. Dechanneling

Channeled ions eventually leave their privileged trajectories and revert to ordinary, nonchanneled paths through the crystal, which have essentially random directions. This process, called dechanneling, is caused by the interaction of the ions with the electrons in the channel, and with the nuclei of the channel walls, which are displaced from their equilibrium positions by the thermal vibrations of the lattice. As an ion progresses along its path, collisions with the electrons and displaced nuclei have the effect of slowly increasing its transverse energy, until eventually the ion can approach the channel wall closer than the critical distance and dechanneling occurs.

For planar channeling, the dominant contribution to the ions' increase in transverse energy comes from collisions with the valence electrons in the channel [49]. The fraction of ions remaining channeled at a depth z in the crystal decreases exponentially as z increases. Planar dechanneling can therefore be described by a half thickness, $z_{1/2}$, which is the depth into the crystal at which 50% of the initially channeled ions have been dechanneled. The value of $z_{1/2}$ is found to be relatively insensitive to the crystal temperature, and to increase linearly with beam energy. The value of $z_{1/2}$ also depends on the particular planar direction chosen; for example the values of $z_{1/2}$ for the {111}, {110}, and {100} planar channeling directions in silicon are approximately 5.0 μm, 4.5 μm, and 2.3 μm, respectively for 3 MeV ^{1}H ions [48]. Axial dechanneling is more strongly affected by the crystal temperature; the main factor increasing the ions' transverse energy is scattering by the displaced atomic nuclei in the axis walls. In this case, the fraction of the ions dechanneled at depth z is proportional to $z\rho^2/E$ [50].

While ions will be naturally dechanneled even in a perfect crystal, the presence of crystal defects such as dislocations and stacking faults which disrupt the regular arrangement of the atomic rows and planes leads to a greatly increased probability of dechanneling. This enhanced dechanneling is exploited by the CSTIM technique to produce images of crystal defects. When an ion is dechanneled, it returns to the higher, nonchanneled energy loss rate. The average energy loss of ions transmitted through a thin crystal is, therefore, greater where there is a defect than where the crystal is defect-free. In CSTIM, the average energy loss of transmitted ions is mapped as a focused ion beam is scanned over the crystal surface, and this produces images showing where dechanneling is occurring. In Chapters 5 and 8, we show how images of dislocations and stacking faults can be produced by this method.

1.4.5. Blocked Trajectories

It has been described above how ions can be steered through a crystal by successions of correlated collisions with the rows or planes of the lattice. For planar channeling, such ions oscillate between the bounding planes. For axial channeling, the motion is more complex; ions with a very low transverse energy can stay within a single axial channel, whereas channeled ions with a larger transverse energy are free to wander between neighbouring channels. For all channeled ions, however, the transverse energy must remain below that required to bring them closer than a distance of r_{min} to the atomic rows or planes. The effects of this are a reduction in the yields from close encounter processes and the energy loss rate.

However, with an ion beam aligned with a channeling direction, some ions can be transmitted with a higher than normal energy loss [51]. This fraction of the transmitted ions is very small when the beam is at exact alignment, but increases as the sample is tilted so that the beam makes a small angle to the channeling direction and reaches maximum when the beam is at approximately 1.5 to 2 times the channeling critical angle from alignment [47,52]. At this angle, the yield of backscattered ions is increased to above that expected for nonchanneled ions. The ions responsible for these effects are those that have a transverse energy sufficient to take them just through the channeling rows or planes. For part of their path through the crystal, they suffer the steering action produced by the atomic rows in a similar fashion to channeled ions. However, they penetrate into the channel wall and because of the steering they have experienced, they end up traveling for part of their path very close to the center of the atomic row or plane. Such ions, which have trajectories described as *blocked* [51], spend a longer than normal time within these regions of high atomic density. As a result, they suffer a greater than normal energy loss rate and a higher than normal probability of scattering. Therefore, such trajectories are very unstable, and the blocked ions are lost from these paths much more rapidly than channeled ions are dechanneled. Consideration of ions with blocked trajectories is necessary for understanding CSTIM image contrast of stacking faults, as described in Section 8.6.

1.5. ION INDUCED DAMAGE IN SEMICONDUCTORS

Ion induced damage is one of the main drawbacks of MeV ion analytical techniques and should always be considered in nuclear microprobe analysis, because the high current density in the focused beam produces a higher ion induced defect density than occurs with an unfocused beam. References 33 and 53 consider high-energy charged particle damage in semiconductors and insulators in greater detail and a list of references for damage in semiconductors is given in Ref. 33 (p. 12). The proceedings of the Ion Beam Modification of Materials and of the Ion Implantation conferences, which are listed in Appendix 1, also have many discussions of the effects of ion irradiation of metals, semiconductors and insulators.

The nuclear energy loss of the ion results in energy being transferred to the atomic nucleus involved in the collision. An atomic nucleus can be displaced if it receives 20 eV during the collision with the ion, which is enough to break its lattice bonds and displace it sufficiently far away that it does not fall back into the vacant lattice site. Simple lattice defects can occur when an atomic nucleus receives just enough energy from the ion to escape from its lattice site, as it then does not have sufficient energy to displace any other atomic nuclei in subsequent collisions. In this simple low kinetic energy transfer collision, the resulting vacancy/interstitial pair consisting of the displaced lattice atom and the resultant empty lattice site, is called a Frenkel defect. If the struck lattice atom is displaced with enough kinetic energy, then it can travel away from its original lattice site and may collide with other lattice atoms, causing displacements of more lattice nuclei which in turn may displace others. The high kinetic energy of the originally displaced lattice atom is thus dissipated in producing a cascade of vacant lattice sites and interstitial displaced nuclei. For example, Figure 1.17a [55] shows a transmission electron microscopy (TEM) image of defect clusters produced by irradiation of gallium arsenide with high-energy xenon ions. The dark spots are defect clusters produced by individual xenon ions generating a high defect density of displaced atomic nuclei and localized amorphization of the crystal lattice along their trajectory. Figure 1.17b shows a high-resolution TEM image of this same material, where the defect clusters appear as white regions approximately 10 nm across.

Different combinations of defects are created by different ions at different energies. ^{1}H ions tend to produce a mixture of simple defects and complex defect clusters, whereas heavier ions tend to produce a higher proportion of defect clusters as they can impart higher kinetic energy to the displaced lattice

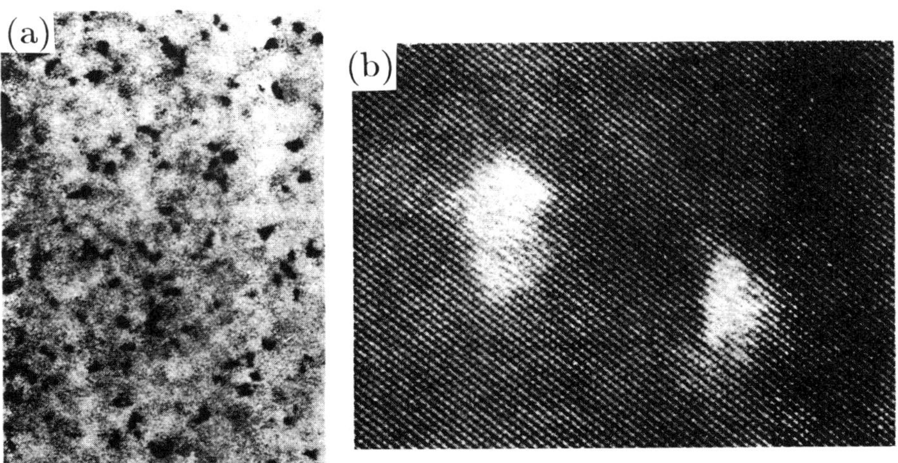

Figure 1.17. (a) Plan view TEM image of defect clusters produced by irradiation of 6×10^{11} Xe ions/cm^2 of GaAs at 300 K. (b) High-resolution TEM image of this same material, where the defect clusters appear as white regions about 10 nm across. Reprinted from Ref. 55 with kind permission from Elsevier Science B.V., Amsterdam, The Netherlands.

nuclei. Figure 1.18 shows a simple representation of the defect depth distribution caused by 3 MeV ^{1}H ions and 3 MeV ^{4}He ions in amorphous silicon. The assumption is made that the number of simple Frenkel defects created in both cases is equal to the nuclear energy loss per micron shown in Figure 1.6 divided by the sum of the binding energy and displacement energy. MeV ^{4}He ions create many more defects than MeV ^{1}H ions, and, furthermore, they are created within a much shorter range, resulting in a much higher defect density.

This description presents a very simple view of ion induced damage. In practice there are many factors that can affect the type of defects formed in semiconductors. The precise effects of ion irradiation are therefore difficult to characterize accurately. There may be a mixture of ion–electron clusters, trapped ions or electrons, new molecular species, vacancy–interstitial clusters, and trapped simple defects. [Different types of vacancies and interstitial complexes are discussed in 53.] Factors that determine the final defect state and its effect on the material properties include the type and energy of the ion, the temperature, and the charge state of the defect. In crystalline materials, the thermal energy of the crystal lattice enables some of the ion induced defects to move even at room temperature, which may result in defect annealing and the formation of extended defects. At a low ion dose, the semiconductor impurities and original lattice imperfections may also affect the nature of the ion induced defects once thermal equilibrium has been reached.

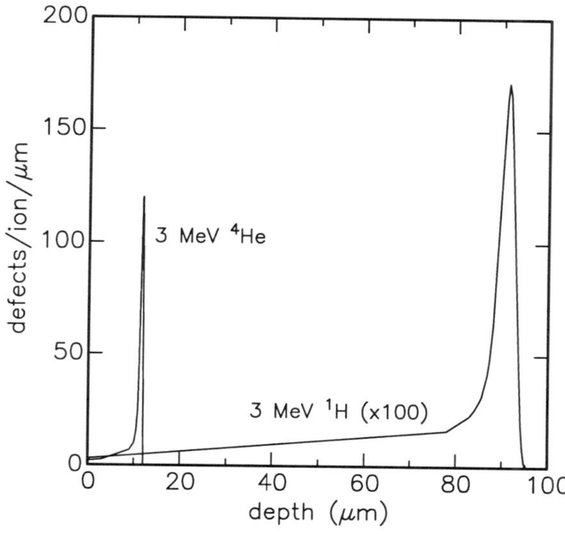

Figure 1.18. Simple comparison of the rate of generation of Frenkel defects in amorphous silicon as a function of ion penetration for 3 MeV ^{1}H ions and 3 MeV ^{4}He ions.

1.5.1. Effect on Measured Electronic and Structural Properties of Semiconductors

A full discussion of charge transport in semiconductors is given in Ref. 54 and only described briefly here. The electronic properties most seriously affected by ion induced damage are the diffusion length of minority charge carriers because the ion induced defects act as trapping and recombination centers, the majority carrier mobility because the defects act as scattering centers, and the majority carrier concentration because the defects exhibit donor/acceptor characteristics.

To illustrate this, consider a low rate of ion induced defect generation of 1 defect/μm/ion in Figure 1.18, such that a dose of 10 ions/μm^2 (10^9 ions/cm^2) creates 10^{13} defects/cm^3. This represents only a very small fraction of nuclei that are displaced from their lattice sites compared with an atomic density of $\sim 10^{22}$ atoms/cm^3. Silicon, with a carrier lifetime of 1 μs and a resistivity of 10 Ω cm, has approximately 10^{14}/cm^3 recombination centers and a carrier concentration of 10^{15}/cm^3. If 10^{13} defects/cm^3 are generated by ion irradiation, the carrier lifetime can be expected to decrease by approximately 10%. A defect density of 10^{14}/cm^3 may similarly be expected to alter the carrier density by 10%. The sensitivity of the carrier mobility to ion induced defects depends on the temperature, but 10^{14} defects/cm^3 will generally begin to affect the mobility. The diffusion length is thus generally the electronic property most easily altered by ion induced damage, and this is further described in Chapter 6 since it is relevant for IBIC microscopy.

Whereas IBIC is sensitive to variations in the electrical properties of the semiconductor material, ion channeling is sensitive to imperfections in the physical structure of the crystal lattice. The minimum yield measured with ion channeling increases with ion induced damage, because the atomic nuclei displaced by the ions into interstitial locations cause subsequent ions to have a greater dechanneling probability. The lowest minimum yield measurable with ion channeling is approximately 1%, and the presence of a low concentration of interstitial defects may not cause a significant increase over this value. The cross-section for dechanneling from a single interstitial nucleus can be shown to be approximately 10^{-19} cm^2 [21], so a single interstitial nucleus is not able to prevent channeling by closing off the channel area of typically $\sim 10^{-15}$ cm^2 to subsequent ions. A dose of 10^9 ions/μm^2 (10^{17} ions/cm^2) might be expected to increase the minimum yield to 10% by displacing enough nuclei from their lattice sites to dechannel 10% of the initially channeled ions.

In practice, an increase in the channeling minimum yield may be detectable in some materials after doses of approximately 10^{16} MeV ^{4}He ions/cm^2, and damage related effects are discussed further in Section 1.5.3. However, an ion dose greater by many orders of magnitude is obviously needed to generate a change in the physical crystal structure detectable with ion channeling than to generate a change measurable with IBIC. This makes techniques for imaging structural defects based on ion channeling such as CCM and CSTIM considerably less sensitive to ion induced damage than IBIC. The effect of ion induced damage with channeling techniques of CCM and CSTIM is discussed further in Chapter 5.

1.5.2. Effects of a Low Ion Dose on Electronic Properties

For a dose of less than 10^4 MeV light ions/μm^2 (10^{12} ions/cm^2), the crystal lattice is essentially intact. Only ion induced changes in the electrical properties of the crystal as measured with IBIC are detectable, whereas structural changes as measured with ion channeling are not. However, a similar dose of heavy ions may cause a significant disruption of the crystal lattice, as was shown in Figure 1.17.

The electronic distinction between simple, ion induced Frenkel defects distributed randomly throughout the lattice and a localized concentration of defects forming a cluster is shown in Figure 1.19. Here the defects introduce two localized states into the energy band gap of a semiconductor above the Fermi level, E_F. If the defects are generated homogeneously through the material as in Figure 1.19a, their levels do not significantly warp the energy band gap, but they may shift the Fermi level uniformly in the material. A defect cluster can be thought of as a large group of closely spaced simple defects, as shown in Figure 1.19b. A cluster can introduce such a large, localized concentration of

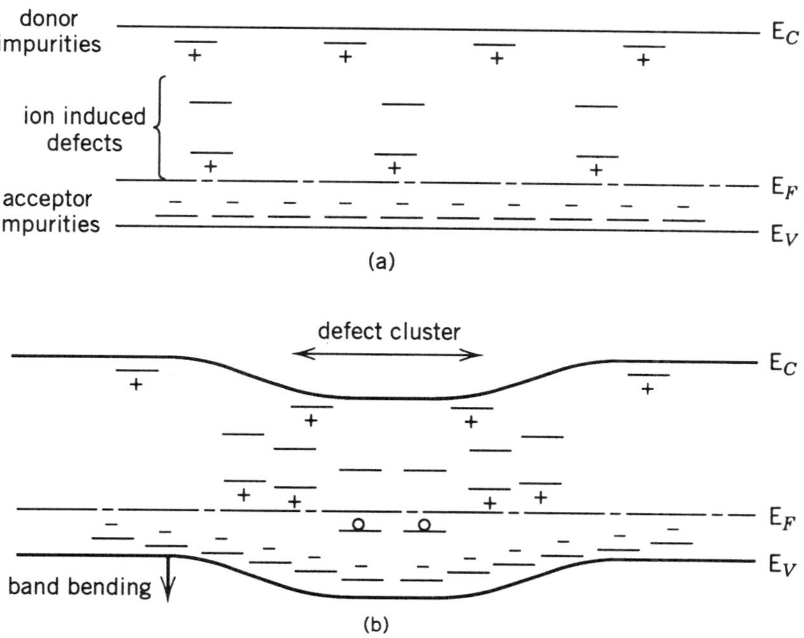

Figure 1.19. Schematic showing the difference between (a) Frenkel defects and (b) defect clusters introducing two defect energy levels between the conduction band edge (E_C) and the valence band edge (E_V) in p-type silicon. The signs indicate the defect charge state. Modified form from Ref. 33 with permission (© 1980 John Wiley and Sons Inc.)

energy levels that the edges of the band in their vicinity may be warped as a result of the charge attracted to the defect levels, until the Fermi level crosses the lower level of the defect. The effects of trapping and recombination of charge carriers at defects are further described in Chapter 6, which treats the damage introduced by MeV ^{1}H ions and ^{4}He ions as simple Frenkel defects.

Ion induced defects in silicon decrease the carrier lifetime and increase the resistivity, and a brief discussion of the effect of ion irradiation of semiconductor devices is now given, based on Ref. 33. If the semiconductor resistivity is increased then the reverse breakdown voltage of the irradiated device is consequently increased. In *pn* junctions, ion induced damage increases the minority carrier concentration and consequently the reverse leakage current goes up. In bipolar transistors, ion induced defects decrease the transistor gain because they cause increased recombination, which lowers the probability of electron transit between the base and the drain. An increase in resistivity in the collector region caused by defects can raise the collector-base breakdown voltage and increase the saturation resistance. Charge trapped in the surface insulating and passivation layers and the presence of surface states produce enhanced carrier recombination, which further decreases the transistor gain. In a junction field effect transistor, ion induced damage increases the channel resistance by increasing the resistivity of the silicon substrate. However, the static characteristics do not depend on the carrier lifetime, so ion irradiation does not have as great an effect as it does on bipolar transistors.

1.5.3. Effects of a High Ion Dose on Structural Properties

For a dose greater than 10^8 MeV light ions/μm^2 (10^{16} ions/cm^2), the ion induced defect density is so high that the crystal lattice starts to lose its well-ordered long-range physical structure, and damage related effects can be detected with ion channeling [56].

When making measurements of ion induced damage with ion channeling, it is important to distinguish between two entirely different mechanisms that lead to dechanneling. As is well known, ion implantation causes swelling of the sample. At any boundary between implanted and unimplanted regions of the sample, the crystal planes are tilted as a result of the implantation-induced swelling. As the swelling increases with increasing dose, the tilt becomes sufficient to dechannel the incident ion beam. For many materials, the width of the tilted region around the edges of the irradiated area is approximately 10 μm. This is of little consequence for ion channeling analysis with unfocused beams, because the amount of tilted material is usually a small proportion of the total beam area. However, for microprobe analysis, where the analyzed region may be as small as 100×100 μm^2 this can be a significant effect.

A large number of studies have been carried out on the effects of the intense microprobe irradiation on thick, single crystals with apparently conflicting results [57]. These conflicting results can be resolved [58,59] by recognizing the contribution made by swelling, as distinct from actual lattice damage owing to displaced nuclei. The contribution from swelling-induced dechanneling becomes more significant as the irradiated area becomes smaller. The use of an unscanned microprobe a few microns in diameter measures the dechanneling primarily from the swelling of the crystal, rather than from the damage to the crystal itself.

An extreme example of microprobe-induced swelling in diamond is shown in Figure 1.20 [60]. A beam of 3 MeV ^{1}H ions was first scanned over a 120 $\times$

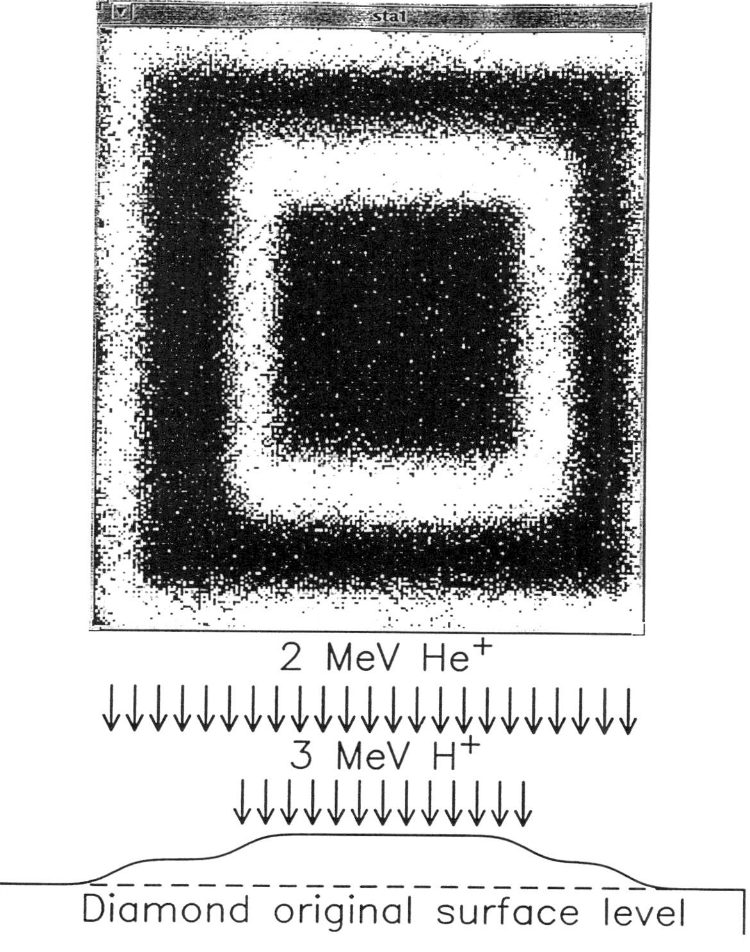

Figure 1.20. CCM image of a $\langle 011 \rangle$ diamond showing dechanneling by the swollen edges of an initial 120 $\times$ 120 μm^2 area due to irradiation with 3 MeV ^{1}H ions. This region was imaged using 2 MeV ^{4}He ions over a 230 $\times$ 230 μm^2 area, which has also caused the crystal to swell. White regions correspond to high yield from dechanneling, dark for low yield from good crystal. Reprinted from Ref. 60 with kind permission from Elsevier Science B.V., Amsterdam, The Netherlands.

120-μm^2 area to produce a swollen region. This region was then imaged with a 2 MeV ^{4}He beam focused to a spot size of 2 μm. The dechanneling from the tilted edges of the initial ^{1}H irradiation are clearly seen, as is dechanneling from the similarly tilted edges of the larger area irradiated with ^{4}He ions. The diamond crystal has otherwise not been damaged, at least as shown by measurements of the $\chi_{\min}$ from the regions of the crystal within the scanned area, but away from the tilted edges of the swollen region. This is illustrated in Figures 1.21 and 1.22, which respectively show, as a function of 2 MeV ^{4}He ion and 1.4 MeV ^{1}H ion dose, the contribution to the $\chi_{\min}$ from the swollen edges of the irradiated area and the contribution from lattice damage itself. The contribution from the swelling does not become significant until the crystal planes are tilted by an amount comparable to the channeling critical angle. It is possible to measure a $\chi_{\min}$ in the tilted crystal by focusing a beam on the tilted region and tilting the entire crystal to allow channeling once more. This measurement shows that the tilted material has a $\chi_{\min}$ characteristic of good crystal [58].

The critical dose at which swelling-induced dechanneling becomes significant has been denoted D_s, which depends weakly on the experimental parameters and is typically 1 to 4×10^{17} ions/cm^2. Swelling is thus an almost inevitable consequence of microprobe imaging if CCM images with good statistics are to

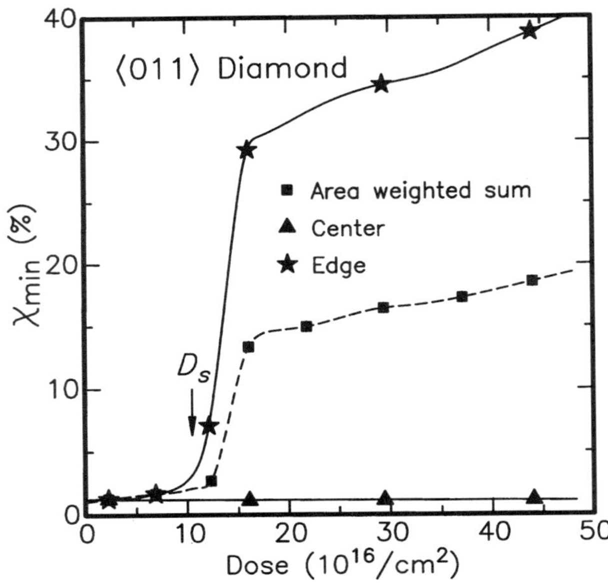

Figure 1.21. Increase in $\chi_{\min}$ with dose in a 60×60 μm^2 area of $\langle 011 \rangle$ diamond irradiated with 2 MeV ^{4}He ions. The $\chi_{\min}$ was measured at the scan center (triangles) and swollen edge (stars). The drastic increase in the $\chi_{\min}$ at the critical dose D_s is due to tilting of the crystal planes by more than the channeling critical angle. The lines are to guide the eye.

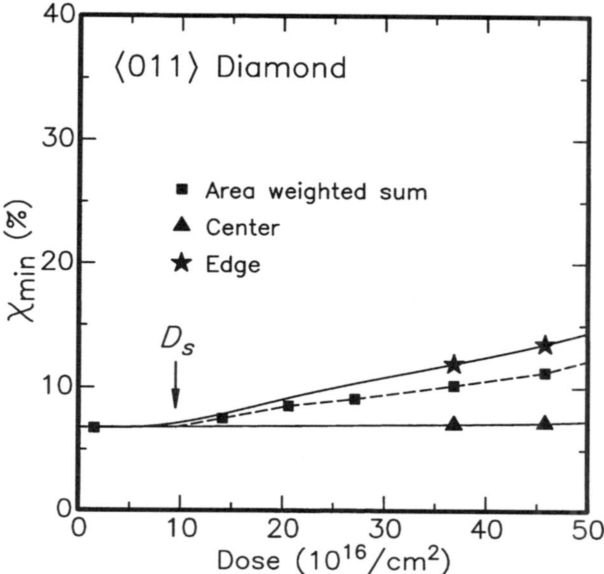

Figure 1.22. Increase in χ_{min} with dose in a $60 \times 60 \ \mu m^2$ area of $\langle 011 \rangle$ diamond irradiated with 1.4 MeV ^{1}H ions. The χ_{min} was measured at the scan center (triangles) and swollen edge (stars). The dechanneling effect is much less severe than in Figure 1.21.

be obtained. The width of the swollen region seen in a CCM image is typically approximately 10 μm, so clearly any scan over a region of interest should be at least this much larger.

Apart from the effects of swelling, the χ_{min} for both silicon and diamond crystals is reasonably insensitive to lattice damage, measured to doses of 5×10^{17} ions/cm^2. However, other semiconductors such as gallium arsenide and mercury–cadmium–telluride suffer severe lattice damage from point defects created by the ion beam. At the same dose, the damage is generally less severe at lower beam fluxes because of room temperature annealing [61,62]. Finally, it is also possible for the irradiation to anneal pre-existing damage. For an investigation of this effect in radiation-damaged diamond, see Ref. 58.

1.6. COMPARISON BETWEEN MeV IONS AND keV ELECTRONS

keV electrons interact mainly by causing electronic excitation of the atomic electrons in a manner similar to the energy loss of MeV ions, which was described in Section 1.1. According to the scattering theory described in Section 1.2, the maximum energy transferred from a particle of mass M_1 with energy

E_1 to a stationary particle with mass M_2 occurs when the incident particle is scattered through $180°$, as given by Eq. (1.6). In this case, the energy transferred to the particle M_2 is $E_2 = [1 - K(\theta_S)]E_1$, which is equal to

$$E_2 = \frac{4M_1 M_2 E_1}{(M_2 + M_1)^2} \tag{1.21}$$

A MeV ^{1}H ion can transfer a maximum fraction of approximately $E_1/465$ to an atomic electron. The small fraction of energy transferred results in the ion being deflected only through a small angle. In comparison, if a keV electron collides with an atomic electron, the electron might transfer all its energy in a single collision, resulting in the electron being scattered through a large angle.

The predominance of large scattering angles results in an electron beam that is finely focused at the sample surface becoming much larger with increasing depth. This limits the spatial resolution attainable in thick layers in electron microscopy techniques such as electron probe microanalysis (EPMA), electron backscattering and electron beam induced current (EBIC). Conversely, although the much greater mass of MeV light ions makes them much more difficult to focus than electrons, they suffer from less lateral scattering in the material and the focused MeV ion beam stays well collimated for a much longer distance in the material.

The number of keV electrons that penetrate a given thickness of material gradually falls off with increasing depth, so there is no well-defined electron range [63]. The Gruen range, R_e, is defined as the depth to which an electron would penetrate if it suffered no large angle scattering events and can be expressed by

$$R_e = \frac{0.043}{\rho} E^{1.75} \tag{1.22}$$

where R_e is in micrometers, ρ is the mass density (g/cm^3) and the electron energy E is in keV. This is sufficiently accurate for electron energies between 20 and 200 keV, but it has been further refined [64]. The electron range as a function of incident energy in amorphous silicon, based on Eq. (1.22), was shown in Figure 1.5. For an energy above 100 keV, an electron travels more than a hundred times farther than the ions of the same energy, so it obviously loses energy at a much lower rate per micron. From Eq. (1.2), the ratio of the rate of electronic energy loss by an electron to that of a ^{1}H ion (proton) is approximately

$$\frac{(dE/dz)_e}{(dE/dz)_p} = \frac{v_p^2 \ln v_e^2}{v_e^2 \ln v_p^2} \tag{1.23}$$

so, for the same energy, the rate of energy loss of the electron is about six hundred times less than a ^{1}H ion. Comparing the rates of energy loss of a 38 keV electron and a 3 MeV ^{1}H ion, as shown in Figure 1.4, the rate of electron energy loss is a seventh that of the ^{1}H ion.

The volume over which the electrons lose energy can be approximated by a sphere with a diameter of the Gruen range R_e. For 38 keV electrons shown in Figure 1.5, where $R_e \sim 12$ μm, the maximum lateral extent over which the electrons are stopped is approximately ± 6 μm. This is approximately ten times larger than the maximum lateral extent for 3 MeV ^{4}He ions which have the same range. This point is discussed further in Chapter 6 where the charge generation volumes of keV electrons for EBIC and MeV ions for IBIC microscopy are compared.

The amounts of damage caused by MeV light ions and keV electrons are also of interest for comparing the merits of the different types of microscopy. The amount of energy that an electron can transfer to an atomic nucleus in a pure Coulomb interaction can be determined from Eq. (1.21) with $M_1 = m_e$. The maximum energy transfer to the atomic nucleus is less than 2 eV for a 10 keV electron in silicon. Since the total amount of energy required to displace an atomic nucleus in silicon is approximately 20 eV, the displacement threshold electron energy is approximately 140 keV. An example of a measurement to determine the threshold electron displacement energy is given in Ref. 65. Even electrons with an energy greater than this only lose, on average, a small fraction of that energy in the production of displaced atomic nuclei. For electron energies up to approximately 5 MeV, isolated Frenkel defects are mainly produced, and in silicon these mainly consist of vacancy and vacancy-related isolated defects that are randomly distributed throughout the irradiated volume of the crystal. The typical operating energy range of a scanning electron microscope of 1 to 40 keV is therefore well below the displacement energy; so little material modification occurs, apart from beam heating effects.

REFERENCES

1. H. Geiger and E. Marsden, *Phil. Mag.* **25:**606 (1913).
2. E. Rutherford, *Phil. Mag.* **21:**669 (1911).
3. N. Bohr, *Phil. Mag.* **25:**10 (1913).
4. N. Bohr, *Phil. Mag.* **30:**581 (1915).
5. J.F. Ziegler and W.K. Chu, *Atom. Data Nucl. Data Tables* **13:**463 (1974).
6. H.H. Anderson and J.F. Ziegler, *Hydrogen Stopping Powers and Ranges in All Elements.* Pergamon, New York (1977).
7. J.F. Ziegler, J.P. Biersack, and U. Littmark, *The Stopping and Range of Ions in Solids.* Pergamon, New York (1985).
8. J.B. Marion and F.C. Young, *Nuclear Reaction Analysis Graphs and Tables.* North–Holland, Amsterdam (1968).

9. L.C. Northcliffe and R.F. Schilling, *Nucl. Data Tables* **A7**:223 (1970).

10. E.C. Montenegro, S.A. Cruz, and C. Varga–Aburto, *Phys. Lett.* **92A**:195 (1982).

11. J.F. Janni, *Atom. Data Nucl. Data Tables* **27**:147 (1982).

12. U. Fano, *Annu. Rev. Nucl. Sci.* **13**:1 (1963).

13. J. Lindhard, *Proc. Roy. Soc. Lond.* **A311**:11 (1969).

14. H.E. Schiott, *Interaction of Energetic Charged Particles with Solids*, Brookhaven National Laboratory, Rept. BNL–50336 (1973).

15. W.K. Chu, J.W. Mayer, and M.A. Nicolet, *Backscattering Spectrometry*. Academic Press, New York (1978).

16. J.W. Mayer and E. Rimini, eds., *Ion Beam Handbook for Materials Analysis*, Academic Press, New York (1977).

17. L.C. Feldman and J.W. Mayer, *Fundamentals of Surface and Thin Film Analysis*, North–Holland, Amsterdam (1986).

18. M.T. Robinson and O.S. Oen, *Proc. Conf. Le Bombardement Ionique*, ed. J.J. Trillat, CNRS, Paris (1962).

19. G. Dearnaley, *IEEE Trans. Nuc. Sci.* **11**(3):249 (1964).

20. D.V. Morgan, ed., *Channeling Theory, Observations and Applications*. Wiley, New York (1974).

21. L.C. Feldman, J.W. Mayer, and S. T. Picraux, *Materials Analysis by Ion Channeling*, Academic Press, New York (1982).

22. J.P. Biersack and L.G. Haggmark, *Nucl. Instr. Meth.* **174**:257 (1980).

23. M. Posselt and J.P. Biersack, *Nucl. Instr. Meth.* **B64**:706 (1992).

24. J.H. Barrett, *Phys. Rev.* **B3**:1527 (1971).

25. J.H. Barrett, *Nucl. Instr. Meth.* **B44**:367 (1990).

26. A. Dygo, P.J.M. Smulders, and D.O. Boerma, *Nucl. Instr. Meth.* **B64**:701 (1992).

27. N. Bohr, *Phys. Rev.* **58**:654 (1940).

28. N Bohr, *Phys. Rev.* **59**:270 (1941).

29. L.C. Northcliffe, *Phys. Rev.* **120**:1744 (1960).

30. J. Lindhard, M. Scharff, and H.E. Schiott, *Mat. Fis. Medd. Dan. Vid. Selsk* **33**(14): (1963).

31. J. Lindhard, V. Nielson, M. Scharff, and P.V. Thomsen, *Mat. Fis. Medd. Dan. Vid. Selsk* **33**(10): (1963).

32. O.B. Firsov, *Zh. Eksperim. Teor. Fiz.* **36**:1517 (1959).

33. V.A.J. van Lint, T.M. Flanagan, R.E. Leadon, J.A. Naber, and V.C. Rogers, *Mechanisms of Radiation Effects in Electronic Materials*. Wiley, New York (1980).

34. H.A. Bethe, *Ann. Phys.* **5**:325 (1930).

35. F. Bloch, *Ann. Phys.* **16**:285 (1933).

36. W.K. Chu and D. Powers, *Phys. Lett.* **40A**:23 (1972).

37. N. Bohr, *Mat. Fis. Medd. Dan. Vid. Selsk* **18**(8): (1948).

38. G. Molière, *Z. Naturforsch* **2a**:133 (1947).

39. L.H. Thomas, *Proc. Symp. Phil. Soc.* **23**:542 (1927).

40. J. Lindhard and M. Scharff, *Mat. Fis. Medd. Dan. Vid. Selsk* **27**(15): (1953).

41. E. Bonderup and P. Hvelplund, *Phys. Rev.* **A4**:562 (1971).

42. H.M. Rosenberg, *The Solid State*. Oxford University Press, New York (1988).

43. C. Hammond, *Introduction to Crystallography*. Oxford University Press, New York (1990).

44. M.T. Robinson and O.S. Oen, *Appl. Phys. Lett.* **2**(2):30 (1963).

45. J. Stark, *Phys. Z.* **13:**973 (1912).

46. D. Gemmell, *Rev. Mod. Phys.* **46**(1):129 (1974).

47. B.R. Appleton, C. Erginsoy, and W.M. Gibson, *Phys. Rev.* **161**(2):330 (1967).

48. J.A. Davies, J. Denhartog, and J.L. Whitton, *Phys. Rev.* **165**(2):345 (1968).

49. L.C. Feldman and B.R. Appleton, *Phys. Rev.* **B8**(3):935 (1973).

50. G. Foti, F. Grasso, R. Quattrocchi, and E. Rimini, *Phys. Rev.* **B3**(7):2169 (1971).

51. C. Erginsoy, *Phys. Rev. Letts.* **15**(8):360 (1965).

52. W.M. Gibson, C. Erginsoy, and H.E. Wegener, *Phys. Rev. Lett.* **15**(8):357 (1965).

53. J.W. Corbett, *Electron Radiation Damage in Semiconductors and Metals*. Academic Press, New York (1966).

54. S.M. Sze, *Physics of Semiconductor Devices*, Wiley–Interscience, New York (1981).

55. M.W. Bench, I.M. Robertson, and M.A. Kirk, *Nucl. Instr. Meth.* **B59/60:**372 (1991).

56. E. Rimini, J. Haskell, and J.W. Mayer, *Appl. Phys. Lett.* **20:**237 (1972).

57. D.N. Jamieson, R.A. Brown, C.G. Ryan, and J.S. Williams, *Nucl. Instr. Meth.* **B54:**213 (1991).

58. S.P. Dooley and D.N. Jamieson, *Nucl. Instr. Meth.* **B66:**369 (1992).

59. S.P. Dooley, D.N. Jamieson, and S. Prawer, *Nucl. Instr. Meth.* **B77:**484 (1993).

60. D.N. Jamieson, M.B.H. Breese, and A. Saint, *Nucl. Instr. Meth.* **B85:**676 (1994).

61. R.A. Brown, J.C. McCallum, and J.S. Williams, *Nucl. Instr. Meth.* **B54:**197 (1991).

62. S.P. Russo, P.N. Johnston, R.G. Elliman, S.P. Dooley, D.N. Jamieson, and G.N. Pain, *Nucl. Instr. Meth.* **B64:**251 (1992).

63. J.J. Marshall and A.G. Ward, *Can. J. Res.* **A15:**39 (1937).

64. T.E. Everhart and P.H. Hoff, *J. Appl. Phys.* **42:**5837 (1971).

65. J.J. Laferski and P. Rappaport, *Phys. Rev.* **111:**432 (1958).

2

PRINCIPLES OF THE NUCLEAR MICROPROBE

2.1. INTRODUCTION

It is the purpose of this chapter to discuss how the various ion–solid interactions, introduced in Chapter 1, are used in a nuclear microprobe to make images. It introduces the idea of nuclear microscopy and compares it with the long established microscopy techniques involving visible light and electrons. This chapter begins with a basic discussion of the general scattering processes involved in ion–solid interactions in the context of imaging. Each specific process is discussed again in more detail in Chapters 4 to 6 for the relevant analytical technique.

After the general discussion of imaging, this chapter continues with a discussion of the essential features of a nuclear microprobe system. Excellent descriptions of these features have appeared in previous books, such as the review of nuclear microprobe applications by Watt and Grime [1]. Comprehensive descriptions of all of the technology for the acceleration of ion beams, as well as the associated particle and radiation detectors are also available—see for example the books by England [2] or Livingston and Blewett [3]. It is not the purpose of this chapter to cover this material in detail once again. Instead, it provides a brief overview, with an emphasis on those features that are particularly important for materials analysis with a nuclear microprobe.

2.1.1. Scattering

Scattering is the basis of microscopy with light, keV electrons and MeV ions. In each case, incident particles scatter from a sample and images can be formed from the scattered particles. In some cases, images can also be formed from new particles induced from, or created within, the sample. The general scattering process can be written as

$$\text{Sample}\,(p, p')\,\text{Sample}'$$

where p stands for the incident particle and p' is the scattered particle, which may not be the same as p. Also the scattering process may not leave the sample unchanged after scattering.

The number of scattered particles can then be written

$$Y(\theta_S) = Q_i N_s \int_\Omega \sigma(\theta_S)\, d\Omega \tag{2.1}$$

where $Y(\theta_S)$ is the yield of particles at a particular scattering angle θ_S, Ω is the solid angle of the detector used to detect the scattered particles, Q_i is the number of incident particles on the sample, N_s is the areal density of the scattering centers, and $\sigma(\theta_S)$, with units of area, is the scattering cross-section. The actual number of particles detected depends on the efficiency of the detector.

Microscopy techniques that involve keV electrons or light take advantage of the very large scattering cross-sections when these particles interact with matter. This means a high-quality image can be obtained in real time, making it possible to view the sample directly. This is done in the conventional optical microscope and the transmission electron microscope (TEM). In both of these microscopes, the scattered particles are focused into an image after the scattering process has occurred. The scattering process is

$$\text{Sample}\,(h\nu, h\nu)\,\text{Sample} \qquad\qquad \text{Light}$$

or

$$\text{Sample}\,(e, e)\,\text{Sample} \qquad\qquad \text{Electrons}$$

The essential components of an optical microscope are identified in Figure 2.1. Analogous components can also be identified in a TEM, as shown in Figure 2.2.

The focusing strategy is different in both a scanning electron microscope (SEM) [4] and a nuclear microprobe. The incident charged particles are focused before interacting with the sample. The reason for this is that the scattered particles, which are not necessarily the same as the beam particles, cannot usually

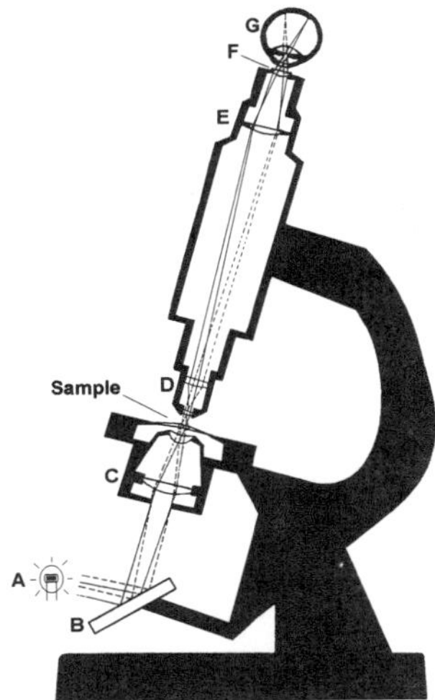

Figure 2.1. The essential components of an optical microscope (analogous components in a TEM are in parenthesis): A, light source (electron source); B, mirror; C, condenser lens assembly (magnetic solenoid condenser lens); D, objective lens (magnetic solenoid objective lens); E, projector lens (magnetic solenoid projector lens); F, eyepiece lens (second magnetic solenoid projector lens); G, final image on observer's retina (scintillation screen).

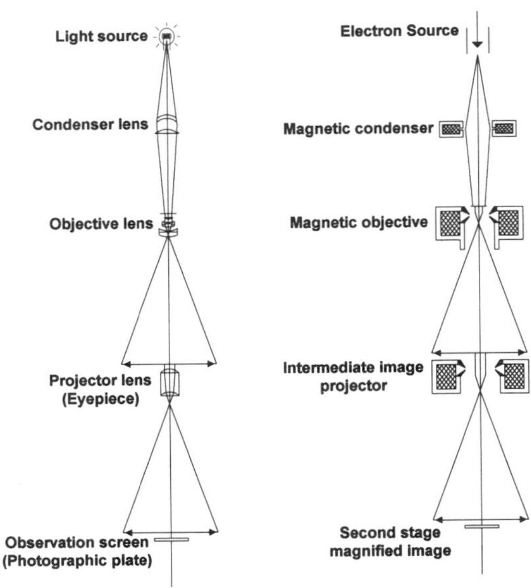

Figure 2.2. The analogous components in an optical microscope and a TEM. The major difference is that the optical microscope uses glass lenses, but the TEM uses magnetic lenses.

be focused into useful images. Despite the different focusing strategy, the major components of a nuclear microprobe are also analogous to those of the optical microscope. The analogous components are identified on a diagram of a nuclear microprobe, as shown in Figure 2.3. There is also an obvious difference in scale. The individual components are discussed in detail in Section 2.2.

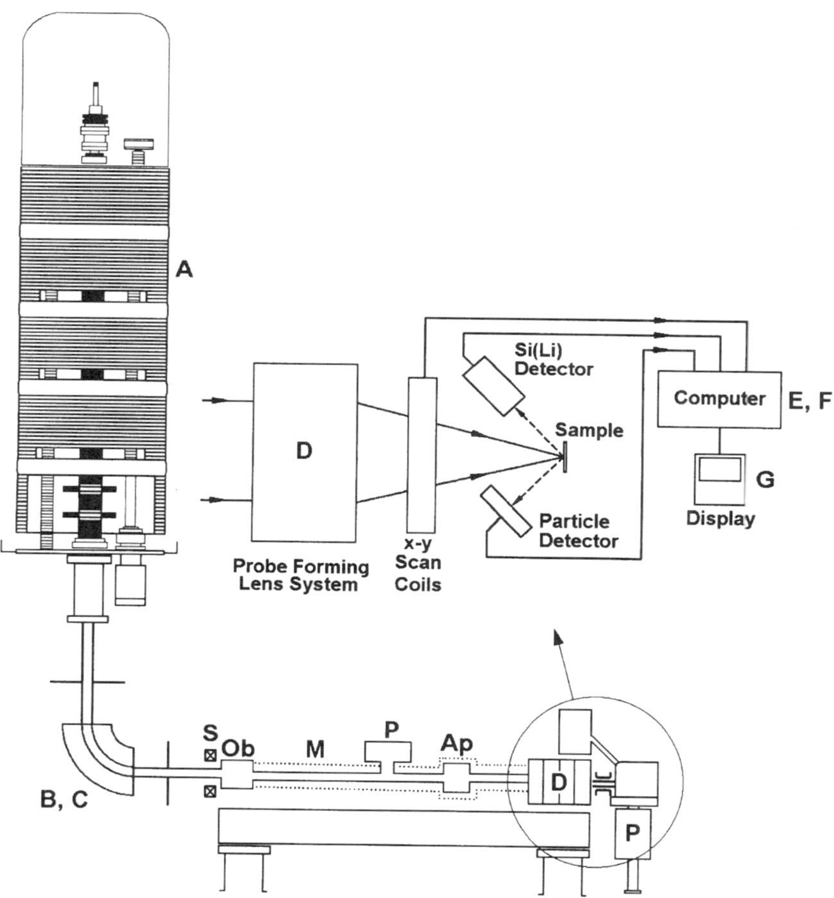

Figure 2.3. Schematic diagram of a nuclear microprobe system. The lettering scheme identifies the analogous components with those of a optical microscope depicted in Figure 2.1: A, ion source and accelerator; B, analyzing magnet; C, condenser lens (part of the double focusing analyzing magnet in this example); D, probe-forming lens system; E, scanning and data acquisition system; F, computer; G, data display on computer monitor. Also shown are S, beam steerer; Ob, object collimator; Ap, aperture collimator; P, vacuum pumps; M, magnetic shielding. The length of the system from the analyzing magnet to the sample is ~9 m.

In a nuclear microprobe, as with a SEM, the charged particle beam is focused into a small spot, called a probe. To make an image, the probe is scanned over the region of interest on the sample. Appropriate detectors measure scattered particles, induced radiation, or other signals produced from the sample. The scattered particles of interest are not limited to just the scattered incident particles as was the case with the TEM. Scattering processes that are commonly employed to form images with MeV ions, denoted X, are:

Sample (X, X) Sample	Back or forward scattering
Sample $(X, \text{X-ray})$ Sample	PIXE
Sample $(X, X'\gamma)$ Sample$'$	NRA
Sample (X, eh) Sample	IBIC
Sample $(X, h\nu)$ Sample	IBIL

The signals from all detectors are digitized by a data acquisition system and recorded in an energy spectrum. Features of interest in the sample usually produce a characteristic signal in the energy spectrum. In a SEM and a nuclear microprobe, the simplest way to form an image is to map the intensity of those characteristic energy signals as a function of the position of the probe on the sample. The characteristic signal is usually identified by a window in the energy spectrum defined for the relevant range in energy.

2.1.2. Historical Background

The history associated with the original development of the first nuclear microprobe at Harwell is described fully in Ref. 1 (Chapter 1) and is briefly reviewed here. Work at Harwell in the late 1960s using nuclear reaction analysis and PIXE was frustrated by the low current density attainable using unfocused MeV ion beams for analysis of small regions, with a consequently unacceptably long measurement period required. At about the same time, there was considerable interest in the development of high-voltage electron microscopes operating up to several MeV. The use of magnetic and electrostatic quadrupole lenses for focusing high-energy electron beams was under careful consideration [see for example 5].

At Harwell these two topics of research lead to the suggestion by Ferguson that it might also be possible to focus high-energy ion beams with quadrupole lenses to give a usefully large beam current density in small areas for PIXE and nuclear reaction analysis.

The initial concern from the available published literature was that spherical aberration of the quadrupole lenses would severely limit the focused beam spot size to several tens of microns. Because of this, the first microprobe used a Russian quadruplet lens configuration, which was chosen to give minimum effects of spherical aberration. This configuration was so named because the focusing properties were derived by Yavor, Dymnikov, and their colleagues in Leningrad (now St. Petersburg) in the mid-1960s [6].

The first microprobe at Harwell, built by Cookson in the early 1970s [7], achieved a beam spot size of less than 4 μm, which far exceeded expectations. Since that time, the technology and applications of nuclear microprobes have advanced considerably. A conference series devoted to nuclear microprobe technology and applications was established in 1987. References to the proceedings for these conferences can be found in Appendix 1.

By the mid 1990s, more than 40 nuclear microprobes were in use worldwide. Many of these were constructed specifically for applications in the field of materials science.

2.1.3. Nuclear Microscopy

In nuclear microscopy, the images formed from the wide variety of ion scattering processes display features of the sample that cannot readily be imaged by other techniques. The most outstanding example of this is the ability of MeV ions to penetrate deep beneath the surface of the sample, with low scattering, to produce signals from hidden features. Furthermore, high-energy ion scattering often involves a relatively simple interaction with the sample. As discussed in Chapter 1, this can allow the scattering cross-section to be calculated analytically. This is particularly true for the scattering of MeV ions themselves where the Rutherford scattering cross-section can often be used to measure the sample stoichiometry. Knowledge of the stopping power of the sample also allows depth profiles of elements to be extracted from the energy spectrum.

The scattering processes exploited with nuclear microscopy to form images may be divided into high-beam current and low-beam current techniques. This division is made on the basis of the size of the scattering cross-section and also on the detection method. The detectors for scattered particles or induced X-rays are usually located a few centimeters away from the sample and collect a fraction of the available particles. The fraction detected depends on the solid angle of the detector as well as its efficiency. In other cases, the sample itself is the detector. For example, electrical contacts on a semiconductor sample provide a signal related to the number of electron–hole pairs, *eh*, produced by each incident ion. This is usually a very efficient process.

High-current techniques are those such as backscattering spectrometry, PIXE, NRA, ERDA, CCM, and IBIL, where a beam current of 100 pA or more is required to obtain good images in a reasonable time (~1 hr). Low-current techniques are those such as IBIC, STIM and CSTIM, where useful images can be obtained in a reasonable time with just a few fA of beam. As discussed in detail in Chapters 4, 5, and 6, use of a low beam current is possible because almost every ion in the incident beam produces a signal that is detected. For this reason, the low-beam-current techniques are often also called single-particle techniques.

Although the images provide useful information about the spatial variations in the sample, it is the energy spectrum that usually contains the most important quantitative information. Therefore, the data acquisition system should allow for the extraction of energy spectra corresponding to particular regions of interest in the sample. These regions may cover only part of the imaged area. In fact, one of the most common applications of the nuclear microprobe is to obtain data that allows the energy spectrum from neighboring regions of small samples to be compared.

2.1.4. Data

To get the most information about the sample from the nuclear microprobe, a sophisticated data acquisition and analysis system is required. The data acquired consists of spatially resolved images and energy spectra. Typically, more than one detector will be in use at the same time. The incoming data consists of energy signals from each detector that have been tagged with the corresponding (x, y) coordinate of the probe on the sample. These data can be displayed in a variety of modes, illustrated schematically in Figure 2.4.

During the data collection on-line period, it is usual for the operator to be able to see the total energy spectrum from each detector as the spectra accumulate. A total energy spectrum is simply the histogram of energy signals from a particular detector without regard to the probe position on the sample. The total energy spectrum can only be used as an approximate guide to the structure of the sample, since it is the superposition of all the different energy spectra from the different subregions of the sample within the scanned area.

It is also usual for one or more images that represent intensity maps of the signal in predefined windows in the energy spectra to be displayed. These images are usually provided for preliminary evaluation of the sample. A more sophisticated approach to on-line imaging is the dynamic analysis technique for PIXE, discussed in Section 4.1.

Images generated from the single energy windows usually do not fully exploit the richness of the data set. They may show features that are difficult to interpret because of interfering energy signals or for other reasons. Therefore it is usual for considerable additional data processing to be done when data collection is completed (i.e., off-line).

To facilitate off-line processing, the data set can be sorted into a data array, for each detector in use, with coordinates (x, y, E). This is shown in Figure 2.4. At each coordinate of the data array is the corresponding number of counts, or yield, Y. Within a particular scanned area, the sample is likely to display many different features. The area occupied by these features is designated the region of interest. To take full advantage of the quantitative nature of the data set, it is necessary to obtain the energy spectrum from each region of interest. A convenient method is to allow shapes to be drawn on an image by use of a pointing device, such as a mouse. The analysis system then provides the energy spectrum from that region of the sample alone. This process is

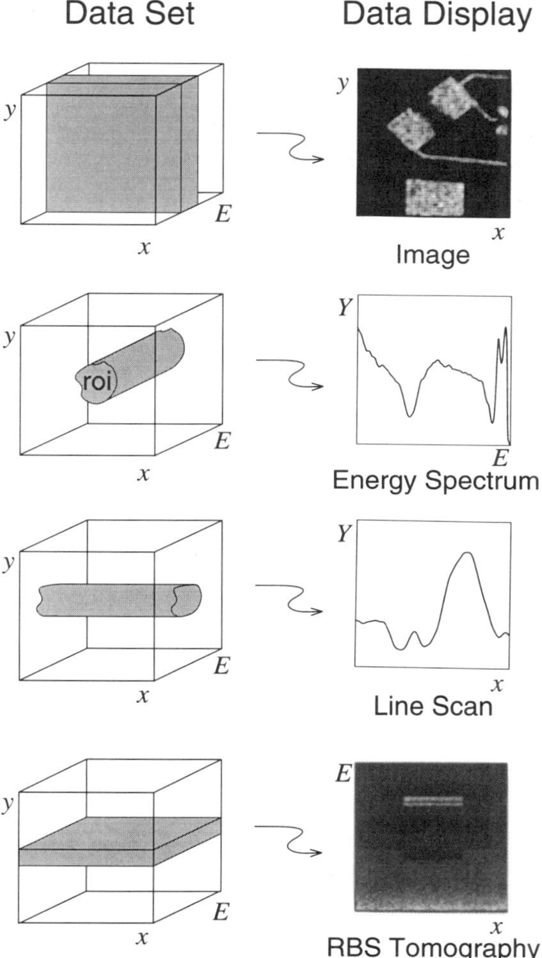

Figure 2.4. The data provided by a nuclear microprobe data acquisition system may be displayed in a variety of modes as illustrated here.

illustrated in Figure 2.4. The shapes, representing different regions of interest, are drawn on a selected image from a window in the total energy spectrum. The resulting energy spectra from the regions of interest are then produced from the stored data array. These energy spectra can then be quantitatively analyzed to provide useful information about the regions of interest in the sample.

It is also possible to obtain the energy spectrum from a region of interest by allowing the probe to dwell there for as long as necessary. In this case, a relatively simple data acquisition and analysis system will suffice. However, this method has the disadvantage that the intense radiation from the stationary probe might damage the sample before useful data can be obtained.

In some samples, it is of interest to know the yield of a particular signal as a function of position across the sample. In this case a narrow rectangle is usually drawn on an image, and the corresponding yield as a function of position is extracted from the data array. This is designated as a line scan in Figure 2.4.

Other data display options are possible. For example, if the detector in use is for backscattered particles, then a tomographic image [8,9] can be obtained by setting a window in the x or y spectrum and producing a map of the yield as a function of y or x and E, as shown in Figure 2.4. Tomographic imaging is discussed further in Section 4.8. Also, if the data set is sorted in such a way that preserves the time-order of the incoming events, then it is also possible to produce graphs of the change in the sample structure with increasing beam dose. This is particularly valuable for the study of beam damage, such as for IBIC, discussed in Chapters 6 and 7.

2.2. NUCLEAR MICROPROBE COMPONENTS

The important components of a nuclear microprobe are the accelerator, the nuclear microprobe beam-line, the probe-forming lens system, the sample chamber and detectors, the scanning system, and the data acquisition and analysis system. These components were identified in Figure 2.3. A single-ended Van de Graaff accelerator was shown, although other types of accelerators may be employed, including tandem accelerators or, less often, cyclotrons [10]. In an electrostatic accelerator, the low energy (keV) beam from an ion source is accelerated by the accelerating column. The ion source and accelerator take the place of the light source in an optical microscope (identified by an A in Figures 2.1 and 2.3). When the ion beam leaves the accelerator, it passes into a bending or analyzer magnet. On some systems, a condenser lens is used to transport the beam from the exit of the accelerator into the bending magnet. This takes the place of the mirror and condenser lens in an optical microscope (B and C respectively, in Figures 2.1 and 2.3). The bending magnet serves several purposes. It can be used to switch the beam between several different beamlines that share the same accelerator. It is also used to fix the energy of the beam from an electrostatic accelerator. This is accomplished with a set of slits on the downstream side of the bending magnet. The slits are oriented so that, at a constant magnetic field, an energy increase will deflect the beam onto one slit, and a decrease onto the other slit. A suitable feedback mechanism maintains the terminal potential to keep the beam current balanced on the two slits. After the bending magnet, the beam enters the microprobe beam line itself.

At the entrance of the beam line is a collimator, which may be a set of slits or a set of diaphragms of various sizes. This is known as the object collimator (designated Ob in Figure 2.3). A demagnified image of the object collimator is eventually focused onto the sample for analysis by the probe-forming lens system.

The divergence of the beam entering the probe-forming lens system is typically limited by an aperture collimator, which again may be a set of slits or a set of diaphragms of various sizes (designated Ap in Figure 2.3). The aperture collimator is usually located just upstream of the probe-forming lens system itself (D in Figure 2.3).

The distance from the probe-forming lens system to the sample position is the image distance, and the distance from the object collimator to the probe-forming lens system is the object distance. The probe-forming lens system is usually located as close as possible to the sample, so that the object distance is long and the image distance is short. This will give a large demagnification factor to ensure that the smallest possible image of the object diaphragm is focused into the probe on the sample. The sample is normally located close to the image plane of the probe-forming lens system. A detailed discussion of the ion optics of the probe-forming lens system can be found in Chapter 3.

The sample is normally located inside a chamber that is equipped with an array of different types of detectors, as well as stages for manipulation of the sample position and orientation. The various signals from the different detectors are recorded by the data acquisition system, which also monitors the position of the probe on the sample by a suitable interface to the scanning system (E, F, and G in Figure 2.3). Each of the key components are now discussed in greater detail.

2.2.1. Accelerators

The most desirable features of an accelerator for nuclear microprobe operation is that it be easy to operate and require minimal attention while it is running. It should provide a stable beam current and provide a beam with an energy spread less than 100 eV per MeV of beam energy. The beam should also be as bright as possible, as is discussed in the following section.

Many nuclear microprobe systems have been attached to accelerators that were originally dedicated to low-energy nuclear physics. As the frontier of nuclear physics moved to higher energies, many of these laboratories are now dedicated full time to nuclear microprobe analysis. An example of this is the Melbourne system [11,12]. Other laboratories are equipped with accelerators specifically purchased either for nuclear microprobe applications, or for ion beam analysis in general, such as the present systems in Oxford [13], Lund [14], and Sydney [15,16].

Two types of electrostatic accelerators are commonly used. Single-ended accelerators operate with the ion source inside the terminal. In this case, the terminal potential is positive and the ion source produces positive ions that are then accelerated away from the terminal. Tandem, or double-ended, accelerators operate with a negative ion source at close to ground potential. The negative ions are then drawn in toward the positive terminal potential, where they are stripped to positive ions and accelerated, for a second time, away from the terminal. This has the advantage that the accelerated ions can pick up two or more times the terminal potential, depending on the charge state of the stripped ion. A further advantage is that the ion source is not inside the terminal, which simplifies maintenance. Consequently a tandem accelerator can be more compact and operate at a lower terminal potential compared with a single–ended machine that produces ions at the same energy. For example, a tandem accelerator with a maximum terminal potential of 1 MV, such as the National Electrostatics Corporation (NEC) model 3SDH, can produce 3 MeV $^4\mathrm{He}^{2+}$ from an ion source that produces $^4\mathrm{He}^-$. As another example, the NEC 5SDH-02 machine, with a terminal potential of 1.75 MV, can produce 3.5 MeV $^1\mathrm{H}^+$ ions from an off–axis duoplasmatron ion source that produces $^1\mathrm{H}^-$ ions. A disadvantage of a tandem accelerator is that the range of light negative ions that are readily available is more restricted than is the range of positive ions.

Tandem accelerators also suffer from the disadvantage that the stripper system, which is used to change the negative ions to positive ions, invariably degrades the brightness of the final accelerated ion beam. The energy spread can also be increased because of energy straggling. This can then potentially degrade the final probe resolution because of the chromatic aberrations of the probe-forming lens system. Both gas and carbon foil stripper systems have been used. Use of gas stripping for a nuclear microprobe requires the optimum gas pressure in the stripper canal to be determined by experiment, since it is a function of many of the other parameters of the system. The optimum pressure for N_2 gas stripping, measured for the Oxford microprobe system, was $\sim 10^{-5}$ mbar which represented a trade-off between stripping efficiency (which increases with pressure) and brightness (which decreases with pressure) [17]. For foil strippers, measurements on the Lund microprobe system showed that ultrathin carbon foils (thickness 0.5 $\mu\mathrm{g/cm}^2$) gave superior performance compared with gas stripping [18]. The thickness of these thin foils is equivalent to only ten carbon layers which makes them very difficult to handle because of their fragility.

The need to use a stripper system has not prevented the Oxford nuclear microprobe, equipped with a NEC 5SDH-02 tandem accelerator, from achieving the best probe resolution to date with a current of 100 pA of MeV $^1\mathrm{H}$ ions focused into 330 nm [19]. It can, therefore, be concluded that the degradation imposed by the stripper system is not the most significant limitation to achieving high-resolution probes.

Because of the need to operate the column at megavolt potentials, most electrostatic accelerators are installed inside a pressure vessel filled with an insulating gas. For most stable operation, this gas should be cooled, by pumping through an external circuit, and also filtered, scrubbed, and dried. Filtering is particularly important to keep the environment of the accelerator itself free from loose debris that may cause unpredictable discharges or other unstable operation. Scrubbing removes potentially damaging corona discharge breakdown products from the insulating gas. Commonly, SF_6 is employed as the insulating gas, because of its exceptional resistance to breakdown, however CO_2, N_2, and SF_6 mixtures are also used.

2.2.2. Ion Sources

The crucial parameter associated with the ion source and accelerator for the operation of a nuclear microprobe is the beam brightness. This is simply a measure of the number of ions that pass through a given area with a given maximum divergence at a given energy. For a nuclear microprobe, it is convenient to define brightness B as

$$B = \frac{i}{(A_o)(A_a/D^2)E_o} \tag{2.2}$$

where i is the beam current at energy E_o that will pass through an object collimator of area A_o and an aperture collimator of area A_a, located a distance D from the object collimator. When defined in this way, B is often called the normalized or reduced brightness, because of the beam energy on the denominator. A convenient unit for the brightness is $pA/(\mu m^2\ mrad^2\ MeV)$. If the ion beam consists of $^1H_2^+$, it is convenient to multiply B by two and introduce the unit particle–picoamp, since each molecule results in the scattering of two essentially independent $^1H^+$ ions from the sample. In general, the brightness is a strong function of A_o, A_a, D, and E_o as discussed later in this section.

In conventional electrostatic accelerators, two types of ion sources are most often used for nuclear microprobes. One type is the radio-frequency (RF) ion source and the other is the duoplasmatron source. In a RF ion source, gas atoms in a bottle are ionized into a plasma by a RF field. The bottle of the Melbourne source is shown in Figure 2.5. Also visible in the figure are the copper rings that couple the plasma in the bottle to the RF power supply and the magnetic solenoid used to shape the plasma. Visible beneath the ion-source bottle are the electrodes of an einzel lens used to inject the ion beam into the accelerator. For light ions, the gas is typically $^1H_2$ or ^4He, or other isotopes, and mixtures such as N_2/CO or CH_3/H_2 can be used to produce heavier ions. A small positive bias voltage across the ion source bottle propels the positive ions out through a canal. The structure of the canal is the key to successful operation of such an ion source with a nuclear microprobe. This is because the brightness of the beam from the accelerator, and hence the resolution of the focused probe, depends on the geometry of the canal. During operation of the ion source, the canal becomes sputtered by the ions, causing the canal walls to erode. This usually has the effect of increasing the amount of current provided by the ion source, but decreasing the brightness of the beam. The net result is that the amount of beam current available in the sample chamber, focused to a probe of a given diameter, may actually decrease. Although considerable research has been done on the design of long-lasting canals for high brightness beams, more work still needs to be done.

Figure 2.5. The radio frequency ion source on the Melbourne 5U Pelletron accelerator.

When used in a single–ended machine, a RF ion source is usually configured to produce approximately 10 μA of beam current. This ensures that the source will operate for a relatively long time without requiring maintenance. The longevity of the source is a particularly important issue when the source is used in a single–ended machine, where maintenance requires opening the pressure vessel to gain access to the terminal. However trouble-free operation for more than 1800 hr with $^1H^+$ ions can be achieved in practice.

When used with a double-ended accelerator, the RF ion source requires a charge-exchange canal to convert the positive ions into negative ions prior to injection into the accelerator. The charge-exchange canal is typically filled with the vapor of an alkaline metal, such as lithium or rubidium. Negative ions are produced with an efficiency of approximately 2%. In this case, the ion source is usually configured to produce more than 100 μA of positive beam for injection into the charge-exchange canal. The ion source requires careful attention to operational procedures to prevent the metal vapor from contaminating the rest of the system, particularly the electrodes of the einzel lenses also found in the ion source. Experience has shown that maintenance of these sources can be a hazardous operation.

In a duoplasmatron ion source, a hot filament is located in a chamber filled with gas, usually H_2. A low-voltage arc is struck between the filament and a plate to produce a plasma that is confined by an axial solenoidal magnetic field. The plate contains a small aperture to extract an ion beam from the plasma. The aperture can be positioned on the axis to extract positive ions from the core of the plasma or off the axis to extract negative ions from the surrounding envelope. The filament can be a loop of plain molybdenum or tungsten wire, or it can be given a coating of special materials that enhance the brightness of the source. A duoplasmatron source can be used to produce $^4He^+$ ions, but, in this case, the filament deteriorates more quickly than when producing ^1H ions.

An alternative ion source, used less commonly for nuclear microprobes, is the Penning ionization gauge (PIG) source. Like the RF and duoplasmatron sources, the PIG source also involves the extraction of ions from a plasma. In this case, the plasma is excited by electrons confined in a strong magnetic field. An extraction electrode draws out positive ions from an aperture. From measurements made on the Faure nuclear microprobe system [20], it has been shown that the PIG source can produce a beam of $^4He^+$ ions, brighter by about a factor of four compared with a duoplasmatron source.

2.2.2.1. Brightness Measurements

Measurements of the brightness of ion sources reveal that the brightness is generally not independent of A_o and A_a. It is therefore essential to quote measured values for the brightness with the corresponding values for E_o, A_o, A_a, and D. As an example, measured brightness values for the Melbourne and Oxford nuclear microprobe systems are shown in Figure 2.6 for 3 MeV $^1H^+$ ions. This shows the very dramatic peaking in the brightness that occurs in both cases as the size of the aperture collimator is reduced. Notice that the brightness values shown in the figure are from measurements made in the microprobe beam line, not directly from the ion source itself. Other measurements have shown that the brightness measured in these two places is about the same [21].

These brightness measurements suggest therefore that the brightness peaking effect is a general property of the highly collimated MeV ion beam used in a nuclear microprobe system. This is fortunate, since it allows nuclear microprobes to reach a good probe resolution despite the intrinsic aberration of the lenses needed for MeV ion beams. Simplistic theoretical analysis of the size of the focused probe, based on the theoretical properties of the probe-forming lens system, often do not look promising because they neglect the brightness peaking effect. The best method of theoretical calculation of the probe size from the measured brightness distribution is still a matter of active research.

For analysis with ^1H ions, single-ended accelerators can produce molecu-

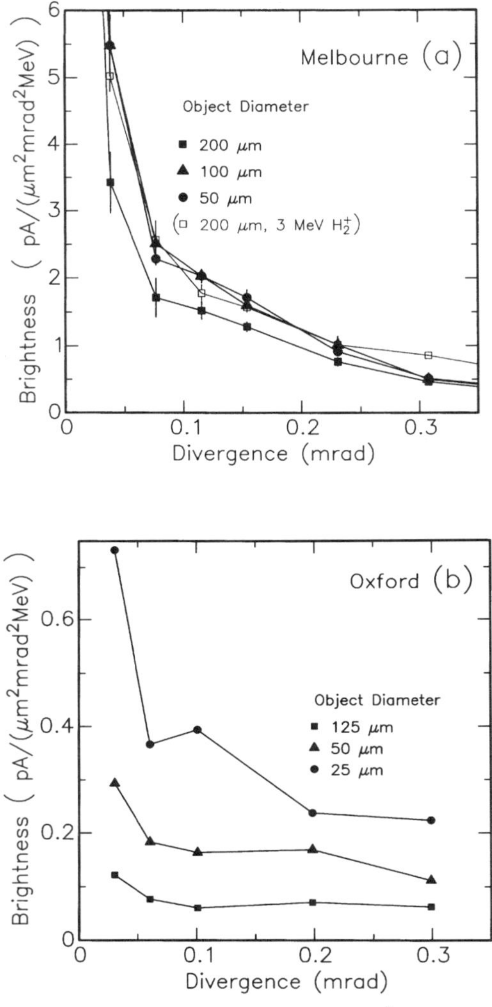

Figure 2.6. (a) Measured beam brightness for 3 MeV $^1H^+$ ions from the Melbourne Pelletron accelerator (radio frequency ion source, single-ended accelerator). The brightness is clearly a strong function of the divergence and hence also the size of the aperture collimator. (b) Measured beam brightness for 3 MeV $^1H^+$ ions from the Oxford system (duoplasmatron ion source, double-ended accelerator).

lar hydrogen ions, $^1H_2^+$. These offer the possibility of brighter beams. Measurements on the Faure system [20] with a duoplasmatron ion source showed that the molecular ion beam was a factor 24 times brighter compared with the $^1H^+$ beam. This is obviously a significant advantage for a nuclear microprobe because for the same beam current compared with ^1H ions it allows the use of smaller collimators and hence the reduction in the probe size.

2.2.2.2. Alternative Ion Sources Some new types of ion sources have been used to produce ions for nuclear microprobe systems. These are often specially optimized to produce particular ions. For example, several liquid metal ion sources have been developed for heavy–ion backscattering measurements (for an example, see Figure 4.9), or for sample modification by focused-beam irradiation. The metals are selected for their low melting points and include lithium [22], lithium–beryllium [23], or gallium [24]. These ion sources do not, as yet, appear to be in routine operation. They do, however, hold great promise, since brightnesses have been reported that are superior to RF ion source under some conditions [25].

Another ion source, presently under development, is the field-ionization source. This consists of a very sharp emitter tip immersed in hydrogen or helium gas. When the tip is biased with a high voltage, a positive ion beam is emitted from the tip. Since the tip radius is very small, typically 0.1 μm or less, the brightness can be many orders of magnitude larger than a RF ion source. Unfortunately, many practical obstacles remain to be overcome, including the short life of the tip (hours) and the low beam current (up to 20 nA) [26]. Use of a field ionization source would also require optimization of the ion optics of the ion source itself [27–29].

2.2.3. Bending Magnet and Condenser Lens

Once the beam has been accelerated, it must be transported into the microprobe beam line. This is usually accomplished by at least one bending magnet. If the distance between the exit of the accelerator and the bending magnet is large, it is usual to include a quadrupole lens system that fills the role of a condenser lens in a conventional optical microscope. This is typically a quadrupole doublet or triplet, and it is located between the exit of the accelerator and the entrance of the bending magnet. It is not essential for these lenses to be of high precision.

If a quadrupole lens condenser system is employed, it is essential that an associated beam steerer also be used. This steerer is denoted "S" in Figure 2.3. The steerer is necessary because the quadrupole lenses will invariably steer the beam off-axis, owing to possible misalignments between the axis of the condenser lens system and axis of the beam. It is very difficult to get this alignment exactly right. Furthermore, the beam axis may change as the ion source ages. Therefore, the steerer is essential to steer the beam into the tightly collimated microprobe beam line.

The bending magnet can be a simple dipole field, or a more sophisticated double-focusing 90° spectrometer magnet. The momentum dispersion of the bending magnet can reduce the momentum spread of the beam entering the beam line. This is advantageous, since it reduces the chromatic aberration introduced by the probe-forming lens system. However, from empirical measurements of the probe resolution at a variety of laboratories, the diameter of the final focused probe is not very sensitive to the dispersion of the bending magnet. In the case of a double-focusing magnet, the object collimator of the nuclear microprobe beam line itself is usually located close to the image plane. This will ensure that the maximum beam is transmitted into the microprobe beam line. This means that the double focusing magnet also fills the role of the condenser lens system. It is also beneficial to have a weak steerer located just upstream of the object collimator to correct for small misalignments between the beam direction and the axis of the microprobe itself. Indeed, on the Melbourne system, the angle between the beam axis and the microprobe beam line appears to change from day to day, as well as a function of the warm-up time of the accelerator. The weak steerer easily corrects this misalignment.

2.2.4. Collimators

2.2.4.1. The Object Collimator The object collimator can consist of either slits or diaphragms. If slits are used, it is advantageous to have four independent jaws to allow for accurate alignment with the axis of the system, but one independent, precision vertical and horizontal jaw is satisfactory in practice. Alternatively, the object collimator can consist of diaphragms made from a durable metal alloy, containing platinum or iridium. In either case, a range of openings are usually possible, from 300 μm, down to fractions of a micron when low-beam current techniques are employed.

The use of slits or diaphragms is usually determined by the nature of the probe-forming lens system. An orthomorphic probe-forming lens system, such as in Melbourne, has the same demagnification in both the vertical and horizontal planes and so is better suited for circular diaphragms. The Oxford system is a high-excitation triplet with differing demagnification factors in the two planes. This system uses slits to collimate the beam since rectangularly shaped object collimators can take better advantage of the differing demagnification factors of the triplet.

To avoid excessive heating of the object collimators, a monitor Faraday cup is typically located just upstream of the object collimator. This serves to reduce the beam current from a few microamps down to a few nanoamps. This is to minimize damage to the delicate object collimators themselves. The monitor cup will have a small hole to admit the beam onto the object collimator. This is typically around 300 μm in size. The monitor cup should be equipped with cooling to dissipate the considerable power deposited by the ion beam.

Overheating of slits should be avoided, since thermal expansion can cut off the beam. This is a particularly frustrating problem when small slit openings are employed for low-current techniques. Overheating can also cause deterioration of the smooth surface of the collimators that can result in increased random scattering [30]. This can cause problems when high-resolution, low-current techniques are employed, because the scattered ions contribute to a random background signal that reduces contrast in the images as well as degrades resolution of the probe.

Another inevitable problem is the deterioration of the object collimators from ion implantation of the beam itself, which causes the collimator material to swell. This is a severe problem for small diaphragms where the swelling may cause the opening to become obstructed. The only cure for this problem is regular replacement of the collimators, perhaps as often as once a year for daily operation with ^{1}H ions. This problem is more severe for heavy ions.

2.2.4.2. The Aperture Collimator The main purpose of the aperture collimator is to limit the divergence of the beam entering the probe-forming lens system. This is because the ion trajectories with the largest divergence generally suffer the worst aberration when focused into the probe. The aperture collimator is usually located a few meters downstream of the object collimator and may be a set of slits or simple diaphragms. The size of the aperture collimator is dictated by the brightness of the beam from the accelerator, the size of the aberrations of the probe-forming lens system, the size of the scattering cross-section being used to analyze the sample, and the size of the desired probe. Given the typical beam brightness available to a nuclear microprobe system, a range of aperture sizes from 4 mm down to tens of microns are typically available. The smallest aperture collimators are used when low-beam-current techniques are employed.

In all quadrupole probe-forming lens systems, which consist of lenses that converge in either the vertical or horizontal planes, the spherical aberration has a greater effect on ion trajectories that travel in the diagonal planes. Consequently, if spherical aberration is the limiting aberration, square or rectangular aperture collimators should be oriented so that the slits are at 45° to the horizontal and vertical planes to more fully collimate the most divergent ion trajectories. A circular aperture collimator is close to ideal in orthomorphic systems.

In some systems devoted to applications of low beam current techniques, an additional collimator is used. This is known as the antiscatter collimator and is typically installed immediately before the entrance of the probe-forming lens system. Its purpose is to remove ions scattered from the earlier collimators that would otherwise not be correctly focused, since even a single misplaced ion can cause problems with low beam current techniques.

2.2.5. Sample Chamber

The design of the sample chamber is a complicated art for a nuclear microprobe system. This is because samples with a very wide range of characteristics must be accommodated, as must several detectors for the radiation and particles produced from the sample. The geometry of the detection system may range from the detection of transmitted ions through thin samples, to the detection of photons at backward angles from thick, opaque samples.

A plan of the sample chamber configuration used in Oxford, which is similar to configurations used elsewhere, is shown in Figure 2.7. This shows the locations of the most common detectors used to measure the signals for PIXE, backscattering spectrometry, NRA, and STIM. Two optical microscopes are used to provide information during focusing of the beam and to locate sample features: a front-viewing zoom microscope and a rear-viewing microscope with four objectives mounted on a rotating turret. This turret can be removed when it is not required. The rear-viewing microscope is useful for transparent samples. Figure 2.7 can be compared with the photograph of the chamber shown in Figure 2.8.

Because the chamber design is intended to be reconfigured for many different measurements, in practice only those detectors relevant to a particular measurement are in the chamber, allowing them to be located at their optimum positions. Also, the sample stage is mounted on the top flange of the chamber. This allows the entire top flange to be simply substituted when alternative sample stages are required. Experience has shown that this simple reconfiguration process is an efficient operating method.

Each component of the sample chamber is now discussed in turn.

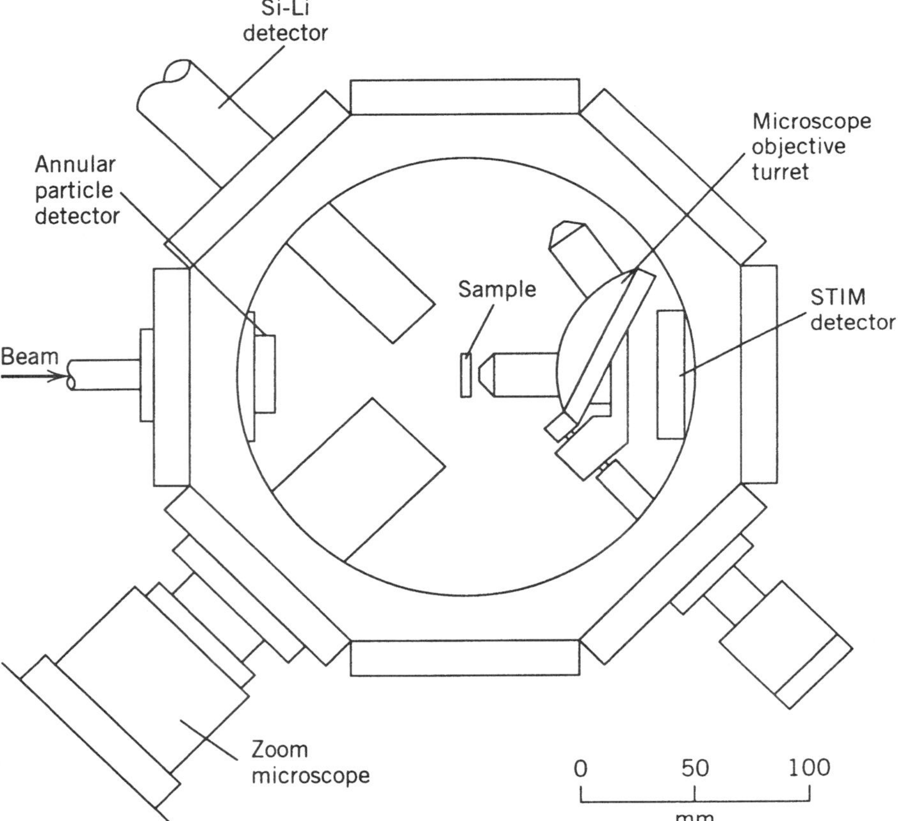

Figure 2.7. The configuration of the Oxford sample chamber, similar to many chambers in use elsewhere. Modified from Ref. 13, with kind permission of Elsevier Science B.V., Amsterdam, The Netherlands.

Figure 2.8. A plan-view photograph of the Oxford sample chamber with the lid removed. The major components are identified on Figure 2.7. A variety of sample stages can be mounted on the lid. For example, an eucentric goniometer for ion channeling measurements is shown in Figure 5.12.

2.2.5.1. The Sample Stage   This should provide motion transverse to the beam direction over a range of a centimeter or so in two directions, to an accuracy of 5 μm or less, with a long-term stability of 1 μm or less. For very high resolution work, typically done with low beam currents, the sample stage needs to be stable to less than 50 nm. A stability of 1 μm is relatively easy to achieve; however, 50 nm is much more difficult. For IBIC measurements, the sample stage should be equipped with connecting leads that pass out of the vacuum system for the electrical signals measured from the sample. A goniometer, suitable for ion channeling experiments on single crystal samples, is discussed in Chapter 5 and shown in Figure 5.12.

2.2.5.2. Provision for Fast Sample Changing   Sample changing can be done by simply mounting more than one sample on the sample stage. Alternatively, or as well, the chamber can be equipped with efficient vacuum pumps to allow fast pump down if changing samples requires venting the chamber. Some chamber designs incorporate an air lock, which avoids venting the chamber at all.

2.2.5.3. Detectors for Ions or Photons   Commonly used detectors include a Si(Li) detector for X-rays, a semiconductor charged-particle detector for particles, and a channeltron for ion induced electrons. More elaborate and sophisticated systems are discussed in Section 2.2.6.

2.2.5.4. Optical Microscope   The optical microscope serves two purposes. The first is to observe the ion beam scintillations to focus the probe-forming lens system. The second is to locate various regions of interest on the sample. A wide variety of optical microscope systems are in use [see the review in Ref. 31]. For thin, transparent samples, it is possible to use a microscope with an internal objective at 0° and external eyepieces as shown in Figures 2.7 and 2.8. For thick, opaque samples, which form the bulk of the samples discussed in this book, it is better to employ a front-viewing microscope.

The most desirable location for a front-viewing microscope is at a scattering angle of 180°. This can be achieved with a suitable objective drilled with a hole to admit the beam. The microscope would then occupy the best position for a backscattered ion detector with good mass resolution. Objectives on retractable arms are available but tend to suffer from mechanical complexity. A good compromise for the front–viewing microscope is to use a re–entrant port at a scattering angle of 135°. This allows a close-viewing geometry, with all microscope components located outside the vacuum system. The 45°-viewing geometry is not a problem in most flat samples. It is necessary to keep in mind that such a system will allow the entire range of the beam in a transparent, scintillating, sample to be seen. Therefore, a correctly focused probe will appear as a line with a length proportional to the range of the ions in the scintillator and a width equal to the diameter of the probe. The 45°-viewing geometry can be less satisfactory for transparent samples containing buried inclusions. This is because scintillations produced by the ion beam on the surface of the sample will appear at a different position in the field of view compared with the inclusion. The simplicity of the arrangement usually outweighs these two disadvantages.

The microscope should have a field of view no wider than 1 mm at maximum magnification. This gives sufficient magnification for both focusing the ion beam into the probe as well as locating regions of interest on the sample for analysis. It is also very advantageous to equip the microscope with a sensitive charge-coupled device (CCD) camera. This can often provide an image on a video monitor that has superior contrast compared with that seen by the eye directly in the eyepiece of the microscope.

2.2.5.5. Charge Integration System This is essential for reproducible quantitative elemental analysis. It can be accomplished by electrical isolation of the sample stage from the chamber with a connection to a current digitizer via an insulated feedthrough. To suppress ion induced electrons, the connection to the sample stage can be made through a bias battery of approximately +400 V. The positive terminal of the battery is connected to the sample stage and the negative terminal of the battery is connected to the current digitizer. The optimum bias voltage needs to be determined by experiment. Alternative schemes for ion induced electron suppression consist of negatively biased rings placed approximately 5 mm upstream of the sample through which the beam passes. However, this ring can obstruct the detectors at backward angles.

To reduce problems caused by the charging of insulating samples, such as arcing to ground and large bremsstrahlung background in the PIXE spectrum, a hot filament can be used to flood the sample with electrons. This neutralizes charge buildup on the sample. However, after a few minutes of this, metal from the filament is deposited on the sample, as will readily be seen by the PIXE spectrum. For the same reason, coating the sample with a conducting gold or carbon layer can cause problems with contamination. An alternative scheme to minimize these problems is to reduce the size of the sample and mount it on a conducting backing. It may also be beneficial to place a grounded copper grid on the sample surface. The grid should be positioned so that the regions of interest are visible in the grid holes.

Other schemes include the use of rotating vanes that chop the beam to periodically sample the current. These suffer from being mechanically complex, as well as potentially degrading the image quality if the chopper frequency beats with the scan frequency.

2.2.6. Detectors

The sample chamber is typically equipped with an array of detectors. Unlike a conventional ion beam analysis system, the detectors typically have a relatively large solid angle. In the case of detectors for scattered particles, this is often at the expense of energy resolution because the kinematic energy spread can be larger than the intrinsic resolution of the detector and associated electronics.

Large detectors are essential for backscattering spectrometry, PIXE, NRA, and ERDA because of the relatively low beam currents, typically less than 100 pA, available in high resolution probes, as well as the need to collect sufficient statistics from each point in the region of interest to produce a reasonable image.

The X-ray detector is typically a Si(Li) detector with an energy resolution of 160 eV or better. Good resolution is essential not only to resolve closely spaced X-ray peaks but also to improve the signal-to-noise ratio for trace elements. The Si(Li) detector is usually located at a backward angle of 135° and can have a have a detector crystal as large as 80 mm^2, although it appears that most commercial detector manufacturers in the mid 1990s are experiencing problems reliably fabricating such large crystals. Si(Li) detectors are discussed further in Section 4.1.

Scattered and transmitted beam particles are typically detected with semiconductor charged-particle detectors. Owing to their relatively low cost compared with a Si(Li) detector, these may be employed in arrays to maximize the solid angle. Detectors up to 400 mm^2 in close geometry can be used, resulting in a solid angle of more than 1000 msr. For generating images, degraded energy resolution caused by a large solid angle is not necessarily a problem because the image usually derives from a window in the energy spectrum that may be many times wider than the detector resolution. It is often convenient to routinely use two detectors, one with a large solid angle (>100 msr) for imaging and the other with a small solid angle (<30 msr) for superior energy resolution spectra.

Particle detectors can be mounted at forward and backward scattering angles. Owing to the possibility of severe damage to the forward detector if it is exposed to a high dose of beam, as is readily done with a few seconds exposure to a beam intended for PIXE or backscattering analysis, this detector is often an unencapsulated photodiode. These have the advantage of being inexpensive and have an energy resolution comparable to a surface barrier detector. Their small surface area is not usually a problem because they can be used in a close geometry. Provision is also usually made for collimation of the forward detector for CSTIM measurements. This is discussed further in Chapter 5.

An annular semiconductor detector at a steep backward angle is commonly used to measure the energy of ions that can be backscattered or generated as charged particle reaction products with NRA. In this position, the kinematic energy spread for scattering from a given nucleus over the diameter of the detector is at a minimum, which improves the energy resolution for large area detectors. Also, the variation in the kinematic energy spread for scattering from different mass nuclei is at a maximum, which improves the mass resolution of the energy signals. A full discussion of the detector geometry to optimize a variety of signals is given in Bird and Williams [32].

A semiconductor detector is also used for ERDA in a location similar to that of the STIM detector, but mounted at a shallow angle of typically 20° with respect to the beam axis. Another common detector found in the sample chamber is for ion induced electrons. These can consist of channel electron multipliers or a scintillator with a photomultiplier. Further descriptions are given in Section 4.6.

More exotic detector systems include those in Melbourne and Lund, which are used to detect the visible light from the sample for IBIL. This is accomplished in several ways. A simple method used in Melbourne is to fit a monochromator to the front-viewing microscope with a phototube configured in single-photon-counting mode to measure the light intensity. This is a convenient and relatively sensitive system, except for the extremes of the visible spectrum, which are strongly attenuated by the optics of the microscope. Ideally, the monochromator should be fitted with a focal plane CCD array to detect the light intensity. Alternatively, a single photomultiplier tube can be used, but this suffers from the disadvantage that the monochromator must be scanned to produce a light wavelength spectrum. Panchromatic systems simply measure the total light intensity, independent of the wavelength. This can be done with a photomultiplier tube looking into the chamber through a simple light pipe. An appropriate photomultiplier tube for IBIL work is the Hamamatsu type R943-02, which has a broad, flat spectral response between 160 and 930 nm, as well as the advantage of a low dark current and a high sensitivity.

2.2.7. Probe-Forming Lens System

A full discussion of this most crucial part of the nuclear microprobe appears in Chapter 3. However, it is briefly introduced here.

2.2.7.1. *Quadrupole Lenses*

2.2.7.1. Quadrupole Lenses The probe-forming lens system typically consists of two or more precision quadrupole lenses. The manufacture of suitable lenses requires precision machining techniques, otherwise the lens suffers from mechanical defects that result in undesirable field components. Mechanical tolerances should be less than 25 μm [33]. Further discussion of this important issue can be found in Chapter 3.

Many different precision machining techniques have been applied to the construction of nuclear microprobe lenses. One of the best utilizes a computer controlled wire-cutting spark-erosion milling machine. This technique was pioneered by the Oxford microprobe group [34] and was used to fabricate the Oxford Microbeams integral lens. The main characteristic of these lenses is that the magnetic yoke is a single piece of iron, without joints. Similar lenses have been constructed in Melbourne, for use in a quadruplet probe-forming lens system with lenses of two different lengths. A photograph of two of the assembled lenses is shown in Figure 2.9. The computer control has allowed a complicated pole profile to be cut that consists of a combination of straight lines, circles, and hyperbolas. The Heidelberg group also employed the wire cutting technique to fabricate the individual subcomponents of the yoke that was assembled from several parts [21].

The design of an optimum probe-forming lens system is complicated, particularly when taking into consideration the performance of the accelerator that provides the beam. Nevertheless, several general conclusions are possible. For the brightness distribution typical of present accelerators, the probe-forming

Figure 2.9. Two completed magnetic quadrupole lenses of different lengths for use in a Russian antisymmetric quadruplet probe-forming lens system.

lens system should be optimized to produce the strongest demagnification possible. This will ensure the highest resolution probe and is usually achieved with a system of three or four quadrupole lenses. The Oxford triplet system is a good example of a strongly demagnifying system in which a demagnification as high as 80 can be achieved.

Other design considerations are also important. For example, a high demagnification can be associated with a steep convergence angle on the sample. This can cause problems with measurements involving ion channeling with a large beam current. This issue is discussed in Section 5.3.4.

2.2.7.2. Quadrupole Lens Alignment Procedure

2.2.7.2. Quadrupole Lens Alignment Procedure   Probe-forming lens systems will only achieve optimum performance if careful attention is paid to accurate alignment of each lens in the system. In this respect a simple doublet system may offer advantages. Outlined here is the procedure for alignment of the probe-forming lens system.

Step 1. The first step is to ensure that the beam line itself is correctly aligned. In principle, this can be accomplished optically by a reference laser beam or a theodolite. However, this can lead to problems, because the ion beam may not follow the same path, owing to weak residual magnetic fields. These may remain even after the beam line is adequately shielded from external a.c. and d.c. magnetic fields.

Also, the laser suffers from severe diffraction from the collimators in the system, which makes alignment difficult. A superior procedure is to use the ion beam itself. The beam line may be constructed, piece by piece, starting from the object collimator. As each additional component is added to extend the line, the precise alignment should be checked by ensuring that the ion beam is centered on the end of the component. This may be easily done by temporarily blanking the component with a quartz window, then bringing the ion beam onto the center of the window. The final goal of this process should be to bring the ion beam into the center of the sample stage in the sample chamber. This point should coincide with the center of the field of view of the optical microscope used to locate the region of interest on the sample.

When this has been accomplished, the axis of the ion beam is defined by a line from the center of the object collimators to the center of the field of view of the microscope. The next step is to bring the quadrupole lenses onto that line.

Step 2. Each quadrupole lens should be checked for excessive parasitic multipole fields. Methods for doing this are discussed in Chapter 3. Then the individual quadrupole lenses in the probe-forming lens system should be aligned with the beam. This procedure requires careful attention to ensure accurate alignment.

First, an aperture collimator should be selected that is sufficiently small to be uniformly illuminated by the beam from the accelerator. This is important to ensure reproducible results. The size of the object collimator is not critical. The unfocused beam should be observed in the sample chamber with the optical microscope to verify uniform illumination. An appropriate scintillator screen, such as the type discussed in Chapter 3, can make this observation easier. The center of the illuminated region should be noted, perhaps with reference to a graticule in the eyepiece of the microscope. It should be on the center of the field of view of the microscope if the aperture collimator is correctly aligned with the beam line.

Next, each individual quadrupole lens should be brought to a vertical line focus. The horizontal position of the lens should then be adjusted to bring the line focus onto the center of the field of view. This process should then be repeated for a horizontal line focus and a corresponding vertical movement. This should bring the axes of each lens onto the beam line.

Step 3. With each lens now aligned with the beam line, it is possible to adjust each lens to ensure accurate rotational alignment. This is done by selecting one lens as a reference lens and rotating alignment of the remaining lenses with it. This is done by temporarily forming a doublet with each of the remaining lenses with the reference lens, of opposite polarity. The doublet should be adjusted to focus the ion beam to the smallest spot possible. Then, with the reference lens fixed, the other lens should be rotationally adjusted until the focused spot is again as small as possible.

When this is done for all lenses in the system, the probe-forming system should be reconfigured, as appropriate. Then, one final rotational alignment should be made with any one of the lenses in the system to again minimize the spot size. The purpose of this is to correct the effect of any residual rotational misalignment.

2.2.7.3. Lens Focusing   The most convenient way to focus the probe-forming lens system is by observation of the beam on a quartz screen with the optical microscope. This allows focusing of the probe to an accuracy of approximately 1 μm. During this procedure, it is also possible to crudely diagnose lens–system problems. Rotational misalignment shows up as a tendency for the probe to form a diagonal line when the lens field strengths are adjusted through the optimum values. More serious are three- or four-lobed beam spots that result from parasitic multipole field components, as discussed in detail in Chapter 3. These can arise from shorted turns in the lenses, stray charge on insulators inside the vacuum system near the beam path, unshielded inhomogeneous d.c. fields, or other factors.

To focus accurately below 1 μm, assuming the lens system is adequate, it is convenient to scan a test structure containing features at the scale of 1 μm [1]. An appropriate structure is a 2,000 mesh copper grid. Small adjustments to the lens fields can then be made with reference to the resulting images to find the optimum focus.

2.2.8. Vacuum, Magnetic Shielding, and Vibration

The vacuum system on a nuclear microprobe should be as clean as possible, especially if trace-element analysis is contemplated. In Melbourne, this is achieved with all-metal seals and ion pumps. A good vacuum is essential to limit the effect of gas scattering, which can degrade high spatial resolution measurements, particularly those that employ low-beam-current techniques, such as STIM, CSTIM, and IBIC. Measurements in Melbourne showed that with a vacuum of 10^{-7} Torr, the proportion of $^1H_2$ molecular ions broken up over a flight path of 8.6 m was 0.3% [36]. This measurement was not able to distinguish between molecules broken up by slit scattering or by gas scattering, but still gives a useful indication of the size of the halo from the scattering processes. For high-resolution applications, a beam halo more intense than this may not be tolerable. A good, clean vacuum also helps prevent the buildup of carbon on the sample surface under the irradiated region. This can degrade the quality of channeling measurements, as well as produce a carbon peak in backscattering and NRA spectra.

For high-resolution measurements with submicron probes, sample vibration can be a problem. Ion pumps have no moving parts and can produce a clean vacuum, which are advantages. They have the disadvantage that the strong in-homogeneous d.c. magnetic field can introduce aberrations into the focused probe. However, this can be significantly reduced by appropriate shielding, as discussed in Chapter 3.

Magnetic shielding of the entire beam-line is also essential to reduce undesirable effects from stray a.c. fields. These arise from all mains powered equipment in the laboratory including the artificial lighting. It is particularly important to shield the line from fields generated by earth loops. In Oxford, it was found that a defective diffusion pump heater coil could produce an earth leakage current in a power cable running parallel to the beam line at a distance of 2 m, which was sufficiently strong to degrade the resolution of the probe by more than 1 μm. Measurements made on the Singapore nuclear microprobe system showed that the region close to the entrance of the probe-forming lens system was most sensitive to stray fields causing a degradation of the probe resolution [35].

To minimize vibration, the entire beam line can be constructed on a heavy iron girder, which is isolated from the floor by antivibration mountings. Efficient vibration isolation of all vacuum pumps is also essential.

2.2.9. Nuclear Microprobe Systems

The parameters of three nuclear microprobe systems appear in Table 2.1. Each system takes a different approach to the probe-forming lens system. The Mel-

TABLE 2.1. Nuclear Microprobe Probe-Forming Lens Systems

		Melbourne [11, 12]	Oxford [13]	Sydney [15, 16]
Accelerator		NEC 5U[a] Pelletron	NEC 5SDH-02[a] Tandem	3-MV Tandetron General Ionex[b]
Object distance[c] (mm)		8260	6980	4360
Image distance[c] (mm)		290	150	310
Quadrupole lens type		Magnetic	Magnetic	Electrostatic
Lens system configuration		Quadruplet	Triplet	Quadruplet
Lens lengths (mm)	1	30	100	75
	2	60	100	75
	3	60	100	75
	4	30		75
Lens bore radius (mm)		6	7.5	3.2
Turns per pole		14	12	n.a.
Working distance[d] (mm)		150	170	150
Demagnifications[e]	$1/M_x$	27	83	14
	$1/M_y$	27	24	14
Focal length (mm)	f_x	290	−77	195
	f_y	290	270	195

[a] NEC, National Electrostatics Corporation.
[b] Now High Voltage Engineering Europa B.V.
[c] Measured from the center of the probe-forming lens system.
[d] Measured from the exit of the last lens.
[e] The optical properties of probe-forming lens system are discussed in Chapter 3.

bourne system is a Russian antisymmetric quadruplet of magnetic quadrupole lenses (Figure 2.10), the Sydney system is similar but employs electrostatic lenses (Figure 2.11), and the Oxford system is a high excitation triplet of magnetic quadrupole lenses (Figure 2.8).

Each system has been designed with a different main goal. In the Melbourne system, the central design goal was to achieve a relatively shallow beam convergence angle in the probe for CCM imaging and to also maintain a strong demagnification in both vertical and horizontal planes. In Oxford, a high excitation triplet configuration is used that has very large demagnifications in both planes. The exceptionally large demagnification of the Oxford system in one plane is as a result of the beam cross-over in that plane. This takes full advantage of the brightness peaking effect, discussed in Section 2.2.2, to achieve very high-resolution probes. In Sydney, an electrostatic system is used which makes the fields in the probe-forming lens system independent of the mass of the ions

Figure 2.10. A photograph of the probe-forming lens system and the sample chamber of the new Melbourne nuclear microprobe. The chamber is fitted with an eucentric goniometer and stepper motors for computer control. The lenses are mounted on a moveable trolley and are shown in the long image distance position where they are used to focus probes of MeV $^{14}N^+$ ions. The normal position, for MeV ^{1}H or ^{4}He ions, is close to the sample chamber.

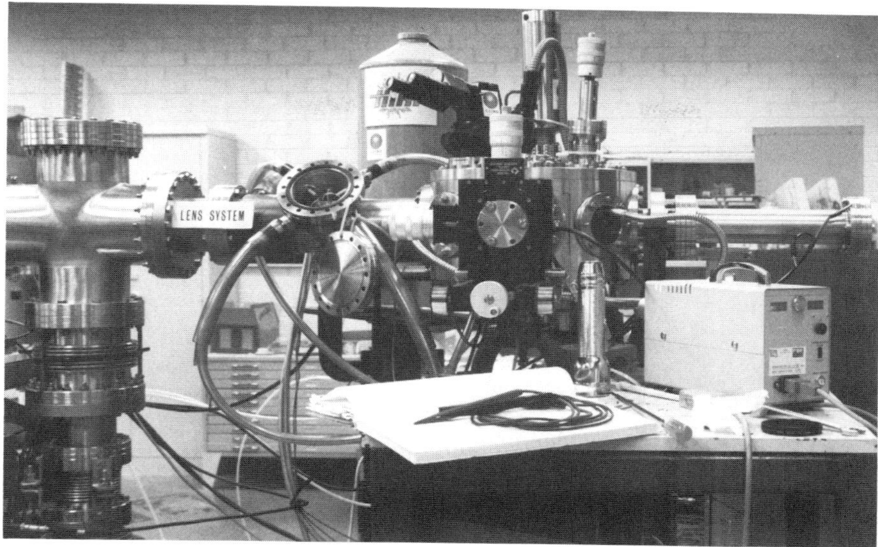

Figure 2.11. The end-station of the Sydney electrostatic microprobe system. The flange covering the manifold in which the lenses are mounted is shown removed.

(see also Chapter 3). This is required for their measurements with heavy ion beams and also leads to practical advantages when switching on and focusing the lenses, because they are free from hysteresis.

2.3. DATA ACQUISITION, SCANNING AND CONTROL SYSTEMS

Remarkably, there is no standard nuclear microprobe data acquisition system. Each laboratory has developed its own system, based on different hardware and software. Also, rapid technological change in computer hardware has resulted in a high rate of change in data acquisition systems. A comprehensive review describes the many systems developed [37].

Despite the lack of standardization, several general properties are usually incorporated. Figure 2.12 shows the data acquisition task divided between two computers: one acquires the data, and the other controls the system hardware and displays the data to the experimenter. In some systems, the functions of these two computers are combined in a single machine.

2.3.1. Scanning System

Scanning the focused probe over the region of interest on the sample is usually done with a magnetic scanning system. This can be more versatile than an electrostatic system where very large potentials can be required to scan large

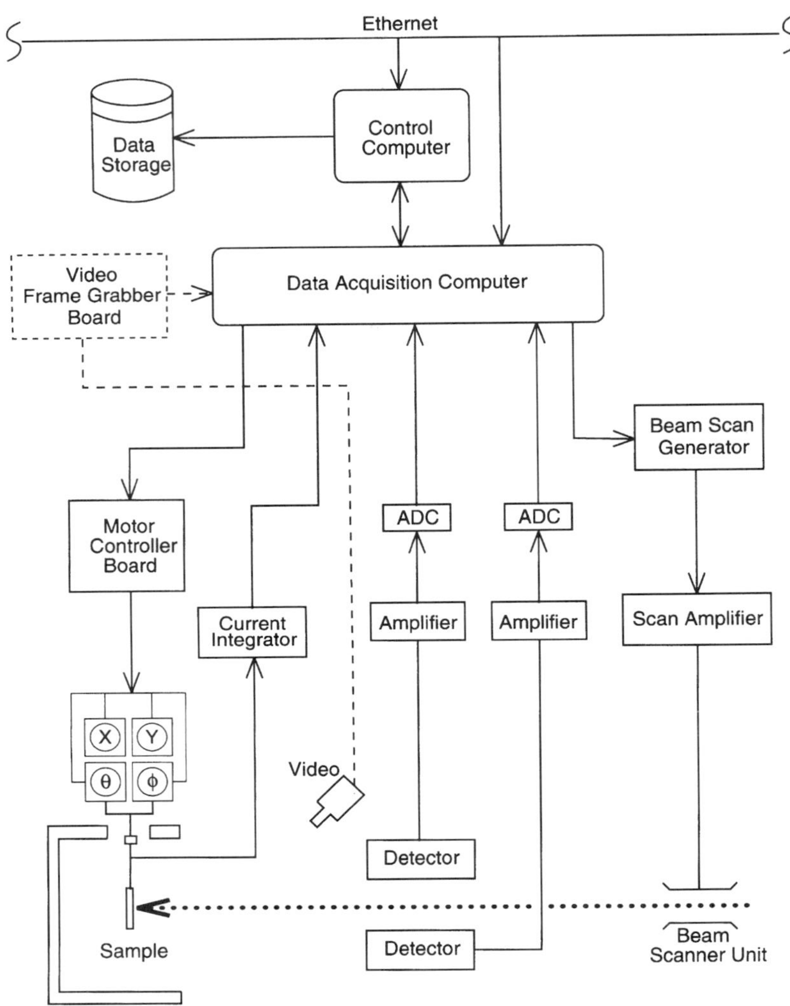

Figure 2.12. The essential features of a nuclear microprobe data acquisition and control system. Not all such systems incorporate all the features shown here. Only two detectors are shown here, but most systems have provision for simultaneous data acquisition from more than two detectors.

areas (approximately 1×1 mm^2 or more). The magnetic scan coils are usually mounted outside the vacuum system and can be furnished with several tappings to produce scan sizes on several scales. A precaution to observe is that the scan frequency be kept below 15 Hz. Otherwise eddy currents in the metal vacuum pipe will introduce a phase delay in the position of the probe compared with the current in the scan coils. This has a seriously degrading effect on the spatial resolution of the images. It is possible to use faster scan frequencies if the scan coils are mounted on a short section of electrically insulating vacuum tubing; however, precautions must be taken to prevent the tube from becoming charged. Such stray charge will also degrade the spatial resolution of the probe. A photograph of the air-core saddle coils used in Melbourne for scans of approximately $400 \times 400 \ \mu\text{m}^2$ scans is shown in Figure 2.13.

2.3.2. Data Acquisition

The data acquisition system must be able to collect the signal from one or more detectors and, preferably, store this signal along with the corresponding spatial coordinate, or pixel, of the probe on the sample. The need to store the correct spatial coordinate from any randomly triggered detector usually causes the most problems in the construction of a suitable data acquisition system because of the inevitable delay between the production of an event from the sample, and the arrival of a digitized signal from the detector. There are two general solutions to this problem.

1. A scan system generates a digital raster pattern, with a dwell on each pixel gated on the current integration. This ensures that each pixel within the scan area receives the same dose of beam. All detected signals can therefore be stored with the appropriate (x, y) coordinate of the pixel. A

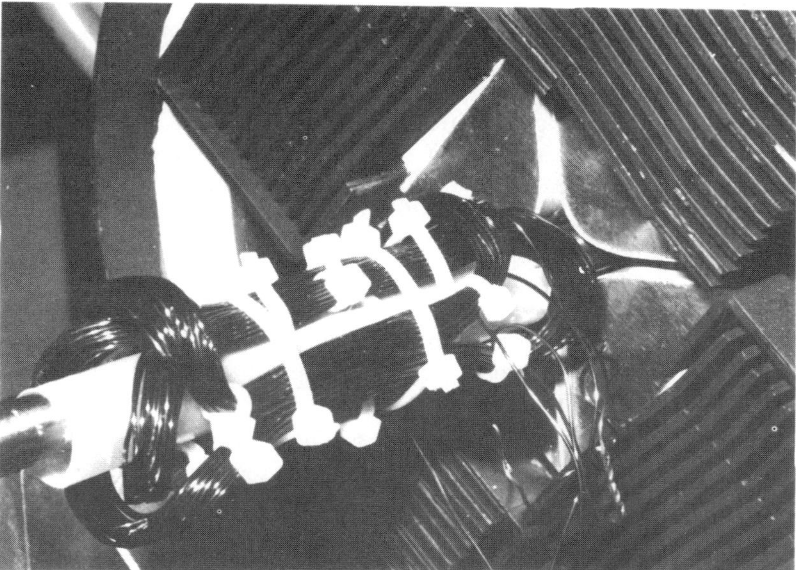

Figure 2.13. The air core scan coils used to provide scans of up to ~$400 \times 400 \ \mu\text{m}^2$ on the Melbourne system. The peak coil current is about 1 amp, and the vacuum tube diameter is 10 mm.

disadvantage of this system is that parts of the sample can be missed if the step size of the scan is large compared with the probe resolution.

2. A random scan pattern ensures that all portions of the region of interest on the sample are irradiated. However, the delay between the production of the event in the sample and the digitization of the signal means that the true spatial coordinate of the event is not the same as the scan coordinate at the time the event is ready to be processed by the computer. This problem can be overcome with appropriate buffering of the scan system.

A further requirement of the data acquisition system is the need to display the incoming images in real-time, allowing the experimenter to monitor the data and make adjustments to the sample position if necessary. The heavy computational and data transfer burden of the display task usually requires a separate computer from the one involved in the actual data collection. Systems that integrate both collection and display functions in a single computer invariably suffer from considerable deadtime.

Several schemes exist for the processing and storage of data coming from the detectors. The simplest acquires and stores the data in map mode. In this case, "windows" in the energy spectra from the detectors are predefined. Only the intensity of the signals from these windows are stored as images; signals from outside the windows are only included in the total energy spectra. This mode has the advantage that the measured data is relatively compact but the disadvantage that unknown features of the sample can be overlooked. It also has the disadvantage that spectra associated with subregions within the scan area cannot be retrieved from the measured data set. However, coarse energy spectra can be extracted by forming a histogram from the intensity of a given region in images for a range of different energy windows.

A more versatile data acquisition scheme acquires and stores every event detected from the sample, along with its spatial coordinate. This is known as an event-by-event mode or occasionally list mode. This mode can result in an essentially open-ended size for the data file, which can impose a considerable burden on data storage space, because data sets can readily be 50 MB or more in size. The concept of storing all available signals from all detectors on a nuclear microprobe system was originally named total quantitative scanning analysis (TQSA) by the Melbourne group, who developed and applied the concept for a nuclear microprobe [38, 39].

The advantage of the event-by-event scheme is that all signals from the detectors are recorded, allowing unexpected signals from the sample to be imaged at the conclusion of data taking. This is important because subtle features of the sample may not be observable until the measurement is concluded. The event-by-event scheme also allows more flexibility in the extraction of energy spectra from regions of interest of the sample. An example of the computer software for the event-by-event scheme is given in Figure 2.14. The sample in this figure was a microelectronic device, and total x, y, and energy

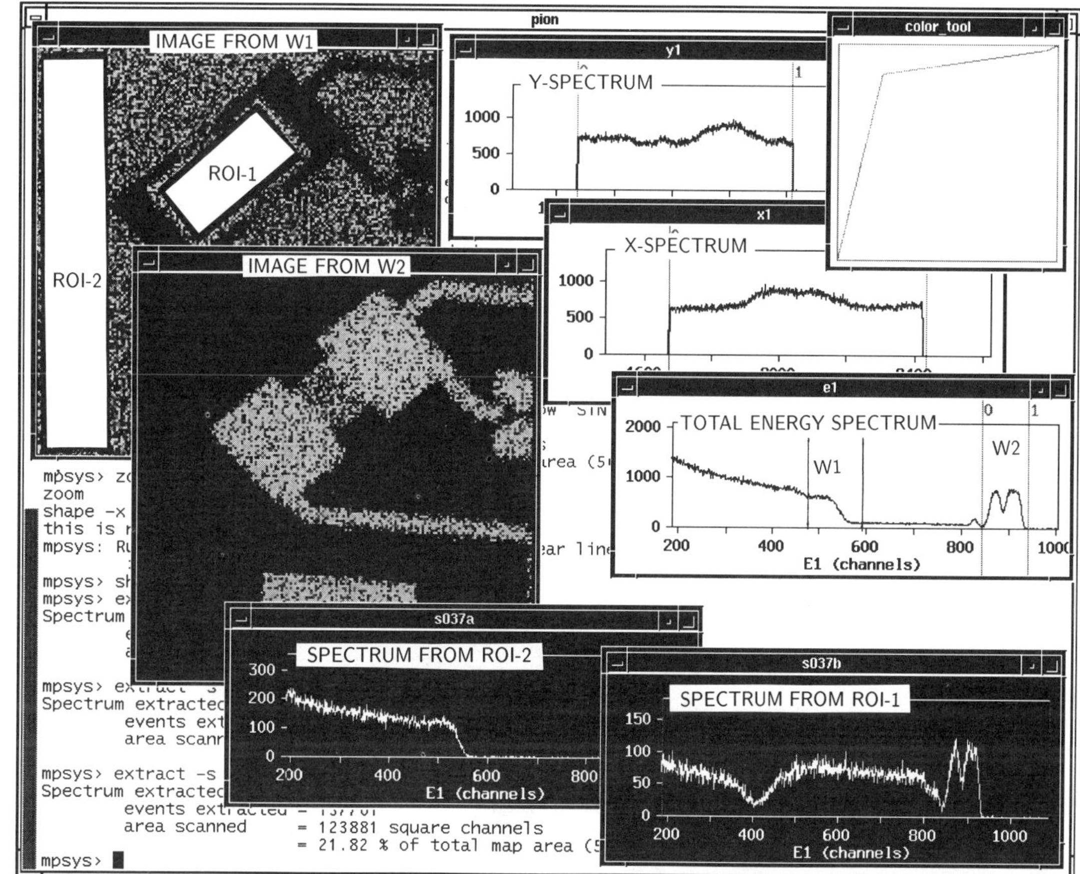

Figure 2.14. The data acquisition and analysis system on the Melbourne system showing shapes defining regions of interest on selected images and the corresponding energy spectra.

E spectra are shown. In this case, the energy spectra are for backscattered particles. Two windows have been set in the total energy spectrum "W1" and "W2." Images from the integrated yield from these windows are shown, corresponding to the gold surface layers and a deeper layer. A pointing device has been used to define two regions of interest in one of the images, "ROI-1" and "ROI-2." The corresponding energy spectra from those two regions alone have also been extracted from the data set and displayed. The color tool, seen top right, is used to change the relationship between the intensity in the images and the grey scale of the display. This is very convenient for the emphasis of subtle features in the images. Considerable additional processing can also be performed.

2.3.3. System Control

It is also desirable for the data acquisition system to control the nuclear microprobe itself to give the greatest degree of flexibility for measurements. For example, with the channeling technique, accurate control over the sample orientation is important. This is usually accomplished by computer control of stepper motors connected to the sample stage goniometer. This function can be integrated with the data acquisition system to form a virtual instrument control panel. An example of this is shown in Figure 2.15, which displays the control panel of a system to control a channeling goniometer. It shows (clockwise from top left) a data window for a detector signal, beam current and data rate gauges (x, y, θ, ϕ), motor position controls, and a video image of the sample. The position of the beam on the sample can easily be changed by pointing to the desired location in the video image. More complicated sequences of sample stage motion and data taking processes can be made with stored lists of instructions. Some systems have been developed that also control the functioning of the entire microprobe beam line and associated accelerator [40].

2.3.4. Data Processing

Spectra extracted from a data set that correspond to regions of interest can be processed in many ways. One of the most important is to normalize the spectra to the area of the region of interest to which they correspond. In addition, the full range of conventional analysis techniques normally applied to spectra from broad beam analysis can be used. Details of these, in the context of the measurement techniques, appear in Chapter 4.

A complete discussion of image processing may be found in standard texts [such as Ref. 41]. Some of these have been applied to nuclear microprobe images [42]. Image processing is often motivated by the need to improve the contrast in images that typically consist of low statistics. The low statistics are a result of the typically low cross-sections involved in the scattering process (such as backscattering and PIXE). In other cases, such as STIM and IBIL images, background noise can degrade the image unless appropriate image processing

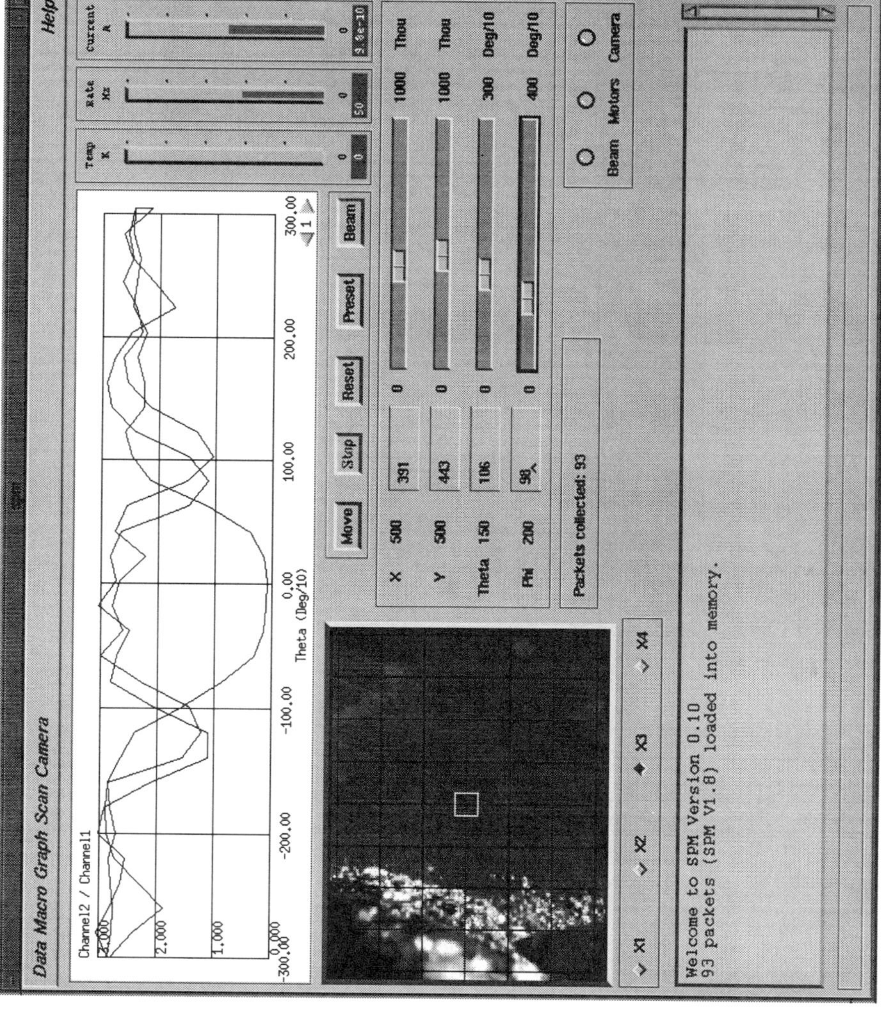

Figure 2.15. The virtual instrument control panel for a four-axis sample-stage goniometer in the Melbourne nuclear microprobe system.

is used, which can often be as simple as a background subtraction (see Section 5.4.6).

One of the most beneficial processes is to simply smooth the image by replacing each pixel by the average of the surrounding pixels. This has the disadvantage that the spatial resolution of the image is degraded. More sophisticated smoothing is possible including convolution with Gaussians, cylinders, sombrero functions, etc. Multiple images can also be combined. This can be done for several purposes. For example, for thin samples, the image of the intensity of a characteristic X-ray line can be divided by the image of the carbon signal from the backscattering spectrum. This can have the effect of normalizing the X-ray signal to the local sample mass, thus correcting for random variations in the sample. Again, this topic is further discussed in Chapter 4.

2.4. A DAY IN THE LIFE OF A NUCLEAR MICROPROBE

This section describes the typical processes involved in an analysis of the Oxford nuclear microprobe. As an example, a CSTIM or CCM experiment is described that can routinely be performed in a single day. As is common with other systems, the Oxford system is used on a daily basis by researchers from a very wide variety of disciplines, not just those interested in materials applications. The system has been specifically designed to maximize the time available for analysis and minimize the time required to set up the system for a specific measurement.

From a cold start, production of a focused ion beam in the sample chamber is a routine process and can be achieved in approximately $\frac{1}{2}$ hr by an experienced operator, even allowing for focusing of the quadrupole lenses. This typically allows 10 hr of analysis time for use in a single day. The sample chamber can be reconfigured relatively easily to replace a standard translation stage with a goniometer for channeling analysis if required. The chamber is also designed so that additional experimental equipment, such as an arm to open a detector shutter or a screen to allow channeling patterns to be viewed, can be added or removed via attachments to the chamber side ports. On a typical day, it is possible to have reconfigured the sample chamber, loaded the sample, focused the beam, and aligned a particular part of a sample for channeling analysis, within 1 hr. Initial sample alignment can be achieved optically, and then by observation of a data display, located close to the sample chamber, of the total signal production from one of the detectors. Exact alignment is usually achieved from windowed images produced by the data acquisition system. Typically, the beam current can be stable for several hours, and the microprobe left to take data unattended during long runs. Small adjustments to the ion source and beam steerers are usually necessary during the course of a day to maintain the current, however. The microprobe is shut down at the end of each day, a process taking $\frac{1}{4}$ hr.

In addition to the ease of set-up for the experimental hardware, microprobe operation is greatly enhanced by having dedicated data acquisition software. At

Oxford [43], the PC-based data acquisition system has been designed for ease of use by researchers outside the dedicated microprobe group. Procedures are used that allow large amounts of data (perhaps several hundred runs in a single week) to be stored for off-line processing. For CCM and CSTIM, with a data collection time of 20 min/run, it is possible to generate data for up to, perhaps, 25 regions of interest from several different samples in a day. For map-mode data, with 25 windows required per run, this can generate over 30 MB of information per run. For data taken in event-by-event mode, a count rate of 2 kHz produces 2 MB of data every 5 min and the run may continue for 1 hr. As the data accumulate, and the images become clearer, new, interesting, and unexpected features of the sample become visible.

At this point, the data analysis and interpretation can begin.

REFERENCES

1. F. Watt and G.W. Grime, eds., *Principles and Applications of High Energy Ion Microbeams*. Adam Hilger, Bristol (1987).

2. J.B.A. England, *Techniques in Nuclear Structure Physics*, Parts 1 & 2. Macmillan, New York (1974).

3. M.S. Livingston and J.P. Blewett, *Particle Accelerators*. McGraw Hill, New York (1962).

4. J.I. Goldstein and H. Yakowitz, eds., *Practical Scanning Electron Microscopy*. Plenum Press, New York (1977).

5. *Proceedings of the AMU-5 Workshop on High Voltage Electron Microscopy*, held at Argonne National Laboratory, 5-7275 (1966).

6. A.D. Dymnikov and S.Y. Yavor, *Sov. Phys.-Tech. Phys.* **8:**639 (1964).

7. J.A. Cookson, A.T.G. Ferguson, and F. Pilling, *J. Radioanal. Chem.* **12:**39 (1972).

8. A. Kinomura, M. Takai, T. Matsuo, M. Satou, S. Namba, and A. Chayahara, *Jpn. J. Appl. Phys.* **28:**L1286 (1989).

9. M. Takai, *Scan. Microsc.* **6:**147 (1992).

10. J.A. van der Heide, R.J.L.J. de Regt, W.A.M. Gudden, P. Magendans, H.L. Hagedoorn, P.H.A. Mutsaers, A.V.G. Mangnus, A.J.R. Aendenroomer, L.C. de Folter and M.J.A. de Voigt, *Nucl. Instr. Meth.* **B64:**336 (1992).

11. G.J.F. Legge, C.D. McKenzie, and A.P. Mazzolini, *J. Microsc.* **117:**185 (1979).

12. D.N. Jamieson, M.B.H. Breese, and A. Saint, *Nucl. Instr. Meth.* **B85:**676 (1994).

13. G.W. Grime, M. Dawson, M. Marsh, I.C. McArthur, and F. Watt, *Nucl. Instr. Meth.* **B54:**52 (1991).

14. K.G. Malmqvist, G. Hyltén, M. Hult, K. Håkansson, J.M. Knox, N.P.-O. Larsson, C. Nilsson, J. Pallon, R. Schofield, E. Swietlicki, U.A.S. Tapper, and C. Yang, *Nucl. Instr. Meth.* **B77:**3 (1993).

15. S.H. Sie and C.G. Ryan, *Nucl. Instr. Meth.* **B15:**664 (1986).

16. S.H. Sie, C.G. Ryan, D.R. Cousens, and G.F. Souter, *Nucl. Instr. Meth.* **B45:**543 (1990).

17. G.W. Grime, F. Watt, and D.N. Jamieson, *Nucl. Instr. Meth.* **B45:**508 (1990).

18. U.A.S. Tapper, R. Hellborg, and K.G. Malmqvist, *Nucl. Instr. Meth.* **B34:**407 (1988).

19. G.W. Grime and F. Watt, *Nucl. Instr. Meth.* **B75:**495 (1993).

20. U.A.S. Tapper, W.R. McMurray, G.F. Ackerman, C. Churns, G. De Villiers, D. Fourie, P.J. Groenewald, J. Kritzinger, C.A. Pineda, J. Pilcher, H. Schmitt, K. Springhorn, and T. Swart, *Nucl. Instr. Meth.* **B77:**17 (1993).

21. K. Traxel, P. Arndt, J. Bohsung, K.-U. Braun-Dullaeus, M. Maetz, D. Reimold, H. Schiebler, and A. Wallianos, *Nucl. Instr. Meth.* **B104:**19 (1995).

22. M.B.H. Breese, J.A. Cookson, J.M. Cole, and A.E. Ledbury, *Nucl. Instr. Meth.* **B54:**12 (1991).

23. M. Takai, R. Mimura, H. Sawaragi, and R. Aihara, *Nucl. Instr. Meth.* **B77:**25 (1993).

24. P.M. Read, J.A. Cookson, and G.D. Alton, *Nucl. Instr. Meth.* **B24/25:**627 (1987).

25. P.M. Read, G.D. Alton, and J.T Maskrey, *Nucl. Instr. Meth.* **B30:**293 (1988).

26. G.L. Allan, G.J.F. Legge, and J. Zhu, *Nucl. Instr. Meth.* **B34:**122 (1988).

27. R.A. Colman and G.J.F. Legge, *Optik* **95:**99 (1993).

28. R.A. Colman and G.J.F. Legge, *Nucl. Instr. Meth.* **B73:**561 (1993).

29. R.A. Colman, G.L. Allan, and G.J.F. Legge, *Rev. Sci. Intr.* **63:**5653 (1992).

30. B.E. Fischer, *Nucl. Instr. Meth.* **B30:**284 (1988).

31. C.G. Ryan and W.L. Griffin, *Nucl. Instr. Meth.* **B77:**381 (1993).

32. J.R. Bird and J.S. Williams, eds., *Ion Beams for Materials Analysis.* Academic Press, Orlando (1989).

33. F.W. Martin, *Nucl. Instr. Meth.* **B54:**17 (1991).

34. D.N. Jamieson, G.W. Grime, and F. Watt, *Nucl. Instr. Meth.* **B40/41:**669 (1989).

35. F. Watt, T.F. Choo, K.K. Lee, T. Osipowicz, I. Orlic, and S.M. Tang, *Nucl. Instr. Meth.* **B104:**101 (1995).

36. G.S. Bench, PhD thesis, University of Melbourne, (1991).

37. N.E.G. Lövestam, *Nucl. Instr. Meth.* **B77:**71 (1993).

38. G.J.F. Legge and I. Hammond, *J. Microsc.* **117:**209 (1979).

39. P.M. O'Brien, G. Moloney, A. O'Connor, and G.J.F. Legge, *Nucl. Instr. Meth.* **B77:**52 (1993).

40. M.L. Roberts, T.L. Moore, R.S. Hornady, and J.C. Davis, *Nucl. Instr. Meth.* **B54:**1 (1991).

41. J.C. Russ, *The Image Processing Handbook.* CRC Press, Boca Raton (1992).

42. N.E.G. Lövestam and E. Swietlicki, *Scan. Microsc.* **6**(3):607 (1992).

43. G.W. Grime and M. Dawson, *Nucl. Instrum. Meth.* **B89:**223 (1994).

3

MICROPROBE ION OPTICS

3.1. PROBE-FORMING LENS SYSTEMS AND QUADRUPOLE LENSES

For quantitative analysis with a nuclear microprobe system, the probe must not only be bright but also tightly focused. This can only be achieved if the probe-forming lens system is of the highest possible quality, which means the minimum of parasitic aberrations. Experience has shown that parasitic aberration is by far the most significant limitation to focusing high-resolution probes.

To see why this is such a problem it is necessary to take a close look at the ion optics of a nuclear microprobe. A major practical difficulty lies with the lenses required to focus the MeV ions into a small probe. In an electron microscope, magnetic solenoid lenses are typically used with the electron beam passing along the axis of the solenoid. Solenoids are circularly symmetric and relatively easy to make to high precision. However, in the center of a solenoid, the strongest component of the magnetic field is parallel to the axis and hence has little effect on the beam. This is because the magnetic field lines, $\mathbf{B}$, are nearly parallel to the velocity vector, $\mathbf{v}$, of the beam and the magnetic force, F, is therefore small:

$$\mathbf{F} = Ze\mathbf{v} \times \mathbf{B}. \tag{3.1}$$

where Ze is the charge on a beam particle and $\times$ designates a vector cross-product. Most focusing of the beam is done by the fringe field at each end of the solenoid.

Protons are approximately 1,860 times heavier than electrons, but with the same magnitude of charge. Because the magnetic field scales as the square root of the mass, the magnetic fields required must be stronger by a factor of 43. The situation is even worse when probes of heavier ions are contemplated. For this reason, normal solenoid lenses are generally too weak for focusing MeV ions. However, superconducting solenoid lenses have been employed with limited success [1–3], although there has been a recent resurgence of interest with a new system to be constructed in Germany [4]. Detailed analysis [5] of solenoid systems suggest that a factor-of-two improvement in probe resolution may be possible compared with quadrupole-based systems, provided parasitic aberrations are ignored.

Instead, most probe-forming lenses for MeV ions use quadrupole lenses. Electrostatic, magnetic, or combined field lenses for achromatic operation may be employed, and typical pole configurations are shown in Figure 3.1. Magnetic lenses have the advantage that the lenses may be mounted outside the vacuum system, which is convenient for alignment with the beam axis. In quadrupole lenses, most of the magnetic field lines are at right angles to the ion trajectories, hence the full field is utilized to focus the beam.

An insight into the differing focusing properties of magnetic and electrostatic lenses may be gained by considering the ratio of the pole tip fields, B or V, required to focus heavier ions of different charge states compared with ^{1}H ions (subscript p):

$$\frac{B}{B_p} = \frac{1}{Z} \sqrt{\frac{ME}{M_p E_p}} \qquad \text{magnetic} \qquad\qquad (3.2a)$$

$$\frac{V}{V_p} = \frac{1}{Z} \frac{E}{E_p} \qquad\qquad \text{electrostatic} \qquad\qquad (3.2b)$$

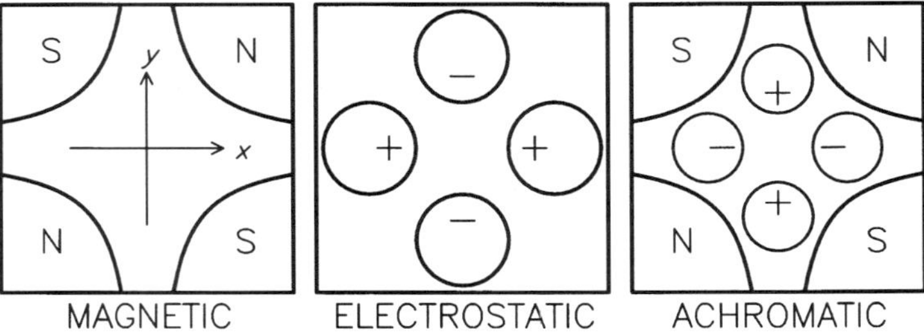

Figure 3.1. The coordinate system and pole configurations for magnetic, electrostatic and achromatic quadrupole lenses. The axes form a left-handed system, so that the z axis points into the page. The configurations shown are for positive lenses that converge in the xoz plane; the beam travels in the direction of the z axis (i.e. into the page). For the achromatic lens, the electrostatic field, considered alone, would form a negative focus; the superimposed magnetic field, of twice the bending strength of the electrostatic field, results in a net positive focus.

where Z is the charge state of the ion, which has energy E and mass M, compared with the proton of energy E_p and mass M_p. For electrostatic lenses, the focusing field is independent of the mass of the ions. This is an obvious advantage when heavy-ion probes are required. Equations (3.2) also reveal that electrostatic lenses are more sensitive to chromatic aberration than magnetic lenses, which can be a problem in some situations.

It is desirable to focus the MeV ions into a 1 μm diameter probe, or smaller. However, it is very difficult to manufacture the required fourfold symmetric quadrupole shape to a sufficient degree of precision. If the magnetic field departs from ideal fourfold symmetry, perhaps due to mechanical defects, aberrations prevent the achievement of a 1 μm probe. As discussed in Chapter 2, exotic machining techniques have been employed to produce the required fourfold symmetric pole profiles to an accuracy of 2 μm.

A single magnetic quadrupole lens acts on the ion beam in quite a peculiar way: the lens squeezes the beam in the one plane and expands it in the perpendicular plane. Consequently, a system of two or more quadrupole lenses, with alternating polarities, is necessary to focus the beam to a high-resolution probe. Furthermore, quadrupole lenses suffer from relatively large spherical aberration. This is the aberration that causes rays (or ion trajectories), which enter the lens with a large divergence to be bent too much compared with paraxial rays. Aberration of a system of quadrupole lenses can be significant, but it is possible to minimize its effect by several strategies, such as limiting how far off-axis the ions can go inside the lenses or correction of aberrations with extra multipole lenses.

With a carefully crafted lens system, it is possible to focus the ion beam into a 1 μm diameter probe with sufficient beam current for analysis, typically 100 pA. This is done by using the probe-forming lens system with a long object distance and a short image distance to produce a demagnified image (the probe) of the object collimator.

It is desirable to employ a lens system with the largest possible demagnifiying power. However, small probes cannot be achieved if the lens system suffers from large aberrations. Extensive work on the optimization of the configuration of the probe-forming lens system itself has already been done [6–8]. A wide variety of systems are presently in use, including doublets, triplets, and quadruplets.

The work concerned with the study of the ion optics of nuclear microprobe lens systems has drawn heavily on the very large body of literature on ion optics. The classic texts on electron and ion optics go into the theory of ion optics in great detail. Indeed, it is an unfortunate consequence of the wealth of detail that the excellent work can, at times, be inacessible to workers whose primary motivation is to apply the nuclear microprobe to microanalysis.

Some of the most important and thorough previous works on focused beams utilizing quadrupole lenses are the classic texts by Hawkes [9,10]. Comprehensive treatments of all aspects of the optics of charged particle beams have been published by Steffen [11], Grivet [12], Banford [13], Septier [14,15], El Kareh and El Kareh [16], Szilagyi [17], and Wollnik [18]. This list is not exclusive, but the material in this chapter draws freely on these much more general works.

It is the purpose of the rest of this chapter to discuss the theory and practice of ion optics with the aim of investigating the effect that various aberrations have on the probe resolution. It is hoped that this will allow nuclear microprobe practitioners to diagnose the performance of their probe-forming system to eliminate faulty lenses that may be limiting resolution. Most of the theory presented here has been incorporated into a computer code, PRAM: propagate rays and aberrations with matrices.

3.2. ION OPTICS

This section describes the general mathematical formalism used throughout this chapter for the analysis of nuclear microprobe ion optics. Section 3.2.1 introduces the aberration coefficients that will be the subject of the rest of this chapter. Section 3.2.2 discusses the computation of the aberration coefficients of multipole lenses in general terms. The later sections of this chapter provide equations for the aberration coefficients and discuss experimental measurement methods.

3.2.1. Aberration Coefficients

A nuclear microprobe system may be considered to be a sequence of optical elements, stacked along the z axis (i.e., the beam axis), each of which has an effect on the beam. The elements may be drift spaces, quadrupole lenses, sextupole lenses, octupole lenses, and so on. The axis of the optical system is defined to be the z axis, $(x, y) = (0, 0)$. A ray entering a microprobe system may be defined by a vector:

$$\Lambda_o = \begin{bmatrix} x_o \\ \theta_o \\ y_o \\ \phi_o \end{bmatrix} \tag{3.3a}$$

where the subscript o denotes the origin of the system, $z = 0$, usually taken to be the position of the object collimator. x_o and y_o are the coordinates of the ray at this point, and θ_o and ϕ_o are the divergences of the ray projected onto the xoz and yoz planes respectively. x, y, and z form a left-handed coordinate system, and ions move along the z axis in the direction of positive z.

The coordinates of the ray in the image plane, $z = z_i$, may be written

$$\Lambda_i = \begin{bmatrix} x_i \\ \theta_i \\ y_i \\ \phi_i \end{bmatrix} \tag{3.3b}$$

It is possible to define a series expansion for Λ_i in terms of the components of Λ_o:

$$x_i = \sum_{\rho=0}^{\infty} \sum_{\sigma=0}^{\infty} \sum_{\tau=0}^{\infty} \sum_{\nu=0}^{\infty} (x/x^\rho \theta^\sigma y^\tau \phi^\nu) x_o^\rho \theta_o^\sigma y_o^\tau \phi_o^\nu \tag{3.4a}$$

$$\theta_i = \sum_{\rho=0}^{\infty} \sum_{\sigma=0}^{\infty} \sum_{\tau=0}^{\infty} \sum_{\nu=0}^{\infty} (\theta/x^\rho \theta^\sigma y^\tau \phi^\nu) x_o^\rho \theta_o^\sigma y_o^\tau \phi_o^\nu \tag{3.4b}$$

$$y_i = \sum_{\rho=0}^{\infty} \sum_{\sigma=0}^{\infty} \sum_{\tau=0}^{\infty} \sum_{\nu=0}^{\infty} (y/x^\rho \theta^\sigma y^\tau \phi^\nu) x_o^\rho \theta_o^\sigma y_o^\tau \phi_o^\nu \tag{3.4c}$$

$$\phi_i = \sum_{\rho=0}^{\infty} \sum_{\sigma=0}^{\infty} \sum_{\tau=0}^{\infty} \sum_{\nu=0}^{\infty} (\phi/x^\rho \theta^\sigma y^\tau \phi^\nu) x_o^\rho \theta_o^\sigma y_o^\tau \phi_o^\nu \tag{3.4d}$$

where the notation is similar to that of Brown [19]:

$$(a/x^\rho \theta^\sigma y^\tau \phi^\nu) \equiv \frac{1}{\rho! \, \sigma! \, \tau! \, \nu!} \left. \frac{\partial^{\rho+\sigma+\tau+\nu} a_i}{\partial x^\rho \partial \theta^\sigma \partial y^\tau \partial \phi^\nu} \right|_{x=x_o, \, \theta=\theta_o, \, y=y_o, \, \phi=\phi_o} \tag{3.5}$$

where $a = x$, θ, y, or ϕ. The symbols $(a/x^\rho \theta^\sigma y^\tau \phi^\nu)$ are known as the aberration coefficients.

Many of the aberration coefficients are equal to zero. Some coefficients are nonzero only when parasitic aberration is present in the lens system. Parasitic aberrations are the result of lens misalignments, power supply ripple, momentum spread of the beam and many other undesirable effects. They will be discussed further in Sections 3.4 and 3.5.

3.2.2. Multipole Field Lenses

Each nonzero aberration coefficient $(a/x^\rho \theta^\sigma y^\tau \phi^\nu)$ in Eqs. (3.4) arises from the action of multipole field components in the lenses [e.g., see 20]. These may be computed from the field profile within the lens. Discussed here are magnetic multipoles; however, the theory is similar for electrostatic multipoles. A magnetic multipole lens which has $2n$ poles has a field profile:

$$B(r) = B_o \left(\frac{r}{r_o} \right)^{n-1} \tag{3.6}$$

where $B(r)$ is the magnetic field as a function of r, the distance from the lens axis along a radius that passes through the center of a pole; B_o is the magnetic field on the center of the pole tip; and r_o is the bore radius of the lens (i.e., the distance from the lens axis to the center of a pole tip).

For a $2n$-pole magnetic lens, the magnetic scalar potential may be obtained by use of the lens symmetry conditions and the source-independent Laplace equation $\nabla^2\phi = 0$ [e.g., see 21]. It may be written

$$\phi_n(r, \alpha) = \frac{-k_{2n}r^n}{n} \sin(n\alpha), \tag{3.7}$$

where $n = 1, 2, 3, 4, \ldots$, correspond to dipoles, quadrupoles, sextupoles, octupoles, and so on, k_{2n} is a constant, and (r, α, z) form a right-handed cylindrical coordinate system. With ϕ_n defined in this way, the first pole encountered along a radius as α increases from zero (anticlockwise), looking downstream, will be a north pole (see Figure 3.1). Equation (3.7) applies to lenses with hyperbolic pole tips, but is a good approximation for cylindrical pole tips as are commonly used in practice.

By calculation of the magnetic field in a multipole lens from Eq. (3.7) and by the use of Ampère's law, it is possible to show

$$B_o = n\mu_o NI/r_o \tag{3.8}$$

where μ_o is the permeability of the vacuum, N is the number of turns per pole, and I is the current in each turn. This equation assumes that the reluctance of the magnetic circuit in the iron parts of the lens is negligible compared with the reluctance of the bore, and that there are no "parasitic" reluctances due to poorly fitted components of the lens. It also assumes there is no saturation of the iron parts of the lens. These effects combine to reduce B_o by less than 5% compared with Eq. (3.8) in a well-designed quadrupole lens.

A $2n$-pole lens introduces terms into the expansions of the image coordinates of order $n - 1$ and higher; for example, quadrupole lenses introduce mainly first-order terms. If two or more quadrupoles are in the system, they can be adjusted to make the first-order terms (x/θ) and (y/ϕ) zero, thus making x_i and y_i independent of θ_o and ϕ_o to first order. Such a system is focused. However, the quadrupoles also introduce aberrations of third and higher order.

In probe-forming lens systems, it is usually only necessary to examine aberration coefficients leading to aberrations of x_i and y_i, because the image size alone is important. However, terms higher than first order in the expansions of θ_i and ϕ_i are generally small compared with the first-order terms. This is not the case with the higher-than-first-order expansions of x_i and y_i. The first-order theory is discussed in Sections 3.3 and 3.4, and higher-order terms in Section 3.5.

3.3. FIRST-ORDER THEORY

In this section, the first-order theory of the optics of a nuclear microprobe system is considered. Section 3.3.1 defines transfer matrices for optical elements of lens systems, and Section 3.3.2 defines the cardinal elements (the focal length and principal plane positions), the magnifications, and demagnifications, and obtains these quantities from the first-order transfer matrix of the lens system.

3.3.1. Transfer Matrices

The first-order transformation of a ray vector through an optical element may be written in the general form

$$\Lambda_2 = M_E \Lambda_1 \tag{3.9}$$

where M_E is the 4×4 transfer matrix of the element. From Eqs. (3.4), the components of M_E can be identified as the first-order coefficients:

$$M_E = \begin{bmatrix} (x/x) & (x/\theta) & (x/y) & (x/\phi) \\ (\theta/x) & (\theta/\theta) & (\theta/y) & (\theta/\phi) \\ (y/x) & (y/\theta) & (y/y) & (y/\phi) \\ (\phi/x) & (\phi/\theta) & (\phi/y) & (\phi/\phi) \end{bmatrix} \tag{3.10}$$

A consequence of Liouville's theorem is that the determinant of this matrix is always equal to unity, which is another way of saying the phase space volume is conserved [e.g., see 11, 13, for a proof of this theorem].

3.3.1.1. Drift Spaces The simplest optical element, a drift space, has a transfer matrix given by

$$M_D = \begin{bmatrix} 1 & D & 0 & 0 \\ 0 & 1 & 0 & 0 \\ 0 & 0 & 1 & D \\ 0 & 0 & 0 & 1 \end{bmatrix} \tag{3.11}$$

where D is the length of the drift space. The matrix (3.11) is a first-order expansion in θ and ϕ around the direction of the beam axis, because, for example, $D\theta$ is the first-order term in a power series expansion of $D\tan\theta$. Equation (3.11) is a good first approximation for systems in which the rays have small divergences, as in a nuclear microprobe system. It changes x and y by $D\theta$ and $D\phi$, respectively, but leaves θ and ϕ unchanged.

3.3.1.2. Quadrupole Lenses   The transfer matrix of a quadrupole may be found from the equations of motion of a particle in a quadrupole field, taken to first order. For a quadrupole of bore radius r_Q and physical pole length l in the z direction, which converges in the xoz plane, the transfer matrix is

$$M_{Q+} = \begin{bmatrix} \cos(\beta L_Q) & (1/\beta)\sin(\beta L_Q) & 0 & 0 \\ -\beta\sin(\beta L_Q) & \cos(\beta L_Q) & 0 & 0 \\ 0 & 0 & \cosh(\beta L_Q) & (1/\beta)\sinh(\beta L_Q) \\ 0 & 0 & \beta\sinh(\beta L_Q) & \cosh(\beta L_Q) \end{bmatrix} \tag{3.12a}$$

and for a quadrupole that diverges in the xoz plane

$$M_{Q-} = \begin{bmatrix} \cosh(\beta L_Q) & (1/\beta)\sinh(\beta L_Q) & 0 & 0 \\ \beta\sinh(\beta L_Q) & \cosh(\beta L_Q) & 0 & 0 \\ 0 & 0 & \cos(\beta L_Q) & (1/\beta)\sin(\beta L_Q) \\ 0 & 0 & -\beta\sin(\beta L_Q) & \cos(\beta L_Q) \end{bmatrix} \tag{3.12b}$$

where

$$L_Q = l + 1.1 r_Q \tag{3.13}$$

L_Q is the quadrupole lens effective length given by Grivet and Septier [22]. This takes into account the extra focusing power in the fringe fields, which extend beyond the actual physical pole. β is the strength of the quadrupole, given by

$$\beta = \sqrt{\frac{|B_Q|}{r_Q}\frac{Ze}{p}} \tag{3.14}$$

where B_Q is the field on the center of a pole tip, r_Q is the bore radius, and p is the relativistic beam momentum, given by

$$p = \sqrt{2MZe\phi_a + (Ze\phi_a/c)^2}, \tag{3.15}$$

where $Ze\phi_a$ is the ion energy, M is the ion mass, and c is the speed of light. The use of the relativistic momentum for a system with 3 MeV ^{1}H ions results in a 0.1% increase of the quadrupole fields compared with the nonrelativistic fields, because 3 MeV ^{1}H ions travel at approximately 7% of the speed of light.

A quadrupole is termed positive if it converges in the xoz plane and negative if it diverges in the xoz plane. This information is often conveniently encoded as the sign of the parameter β. Of course, in using this convention, the sign must be removed from β (i.e., its modulus taken) before using it in the transfer matrices (3.12).

β has units of m^{-1}, so it is possible to define a dimensionless quadrupole strength parameter βL_Q. This parameter may be used as a guide to the optical thickness of a quadrupole lens: $\beta L_Q \gtrsim 1$ indicates a strong lens.

The above transfer matrices apply to electrostatic quadrupoles if β is defined by

$$\beta = \frac{1}{r_Q} \sqrt{\frac{|V_Q|}{\phi_a}} \tag{3.16}$$

where V_Q is the lens pole tip voltage.

For achromatic quadrupole lenses, in which there are superimposed magnetic and electrostatic fields in the achromatic configuration, then the same transfer matrices apply if β is replaced by

$$\beta = \sqrt{|\beta_M^2 - \beta_E^2|} \tag{3.17}$$

where β_M^2 is the magnetic contribution (3.14) and β_E^2 is the electrostatic contribution (3.16). The sign assigned to β, and hence the polarity of the lens, is the same sign as $\beta_M^2 - \beta_E^2$.

The transfer matrix of sextupole lenses and higher-order multipole lenses is the same as a drift space since they have no first-order effect.

The transfer matrix of the entire focusing system may be built up by application of the matrix of each element in turn. For example, for a system of four quadrupoles, the transfer matrix is given by

$$\Lambda_i = M_{\text{system}}\Lambda_o = M_{Di}M_{Q4}M_{D43}M_{Q3}M_{D32}M_{Q2}M_{D21}M_{Q1}M_{Do}\Lambda_o \tag{3.18}$$

where M_{Do} is the transfer matrix of the drift space between the object and the first quadrupole; M_{Q1}, M_{Q2}, M_{Q3}, M_{Q4} are the transfer matrices of each quadrupole; M_{D43}, M_{D32}, and M_{D21} are the transfer matrices of the drift spaces between the quadrupoles; and M_{Di} is the transfer matrix of the drift space from the last quadrupole to the image. The complete transfer matrix of the system, M_{system}, describes all the first-order properties of the system.

3.3.2. The Cardinal Elements

The cardinal elements (focal length and principal plane positions), the magnifications, and the demagnifications may be obtained from the first-order transfer matrix (3.18) of the complete lens system. The cardinal elements are defined for a lens system in which there is no coupling between the xoz and yoz planes; that is, $(x/y) = (x/\phi) = (\theta/y) = (\theta/\phi) = (y/x) = (y/\theta) = (\phi/x) = (\phi/\theta) = 0$. This condition applies in systems consisting of quadrupoles and drift spaces in which the quadrupoles are correctly aligned.

3.3.2.1. Focal Lengths   The focal lengths are given by

$$f_x = \frac{-1}{(\theta/x)} \tag{3.19a}$$

$$f_y = \frac{-1}{(\phi/y)} \tag{3.19b}$$

where the focal lengths are measured (downstream if positive) from the position of the exit principal planes.

3.3.2.2. Principal Planes   The principal plane positions, measured from the start of the system in the case of the entrance principal planes, and from the end of the system in the case of the exit principal planes, are given by:

$$\text{Entrance} \quad P_{x1} = \frac{(\theta/\theta) - 1}{(\theta/x)} \tag{3.20a}$$

$$P_{y1} = \frac{(\phi/\phi) - 1}{(\phi/y)} \tag{3.20b}$$

$$\text{Exit} \quad P_{x2} = \frac{(x/x) - 1}{(\theta/x)} \tag{3.20c}$$

$$P_{y2} = \frac{(y/y) - 1}{(\phi/y)} \tag{3.20d}$$

where the distances are measured (toward the center of the lens system if positive) from the entrance or exit of the system. This means that P_{x2} and P_{y2} are defined differently compared with the usual definitions of the principal planes in visible light optics.

The positions of these planes are shown in Figure 3.2 for two contrasting systems: a low-excitation system and a high-excitation system. In the high-excitation system, the beam crosses over the axis in one plane, which puts the exit principal plane very close to the image plane. This leads to a large magnitude demagnification in that direction.

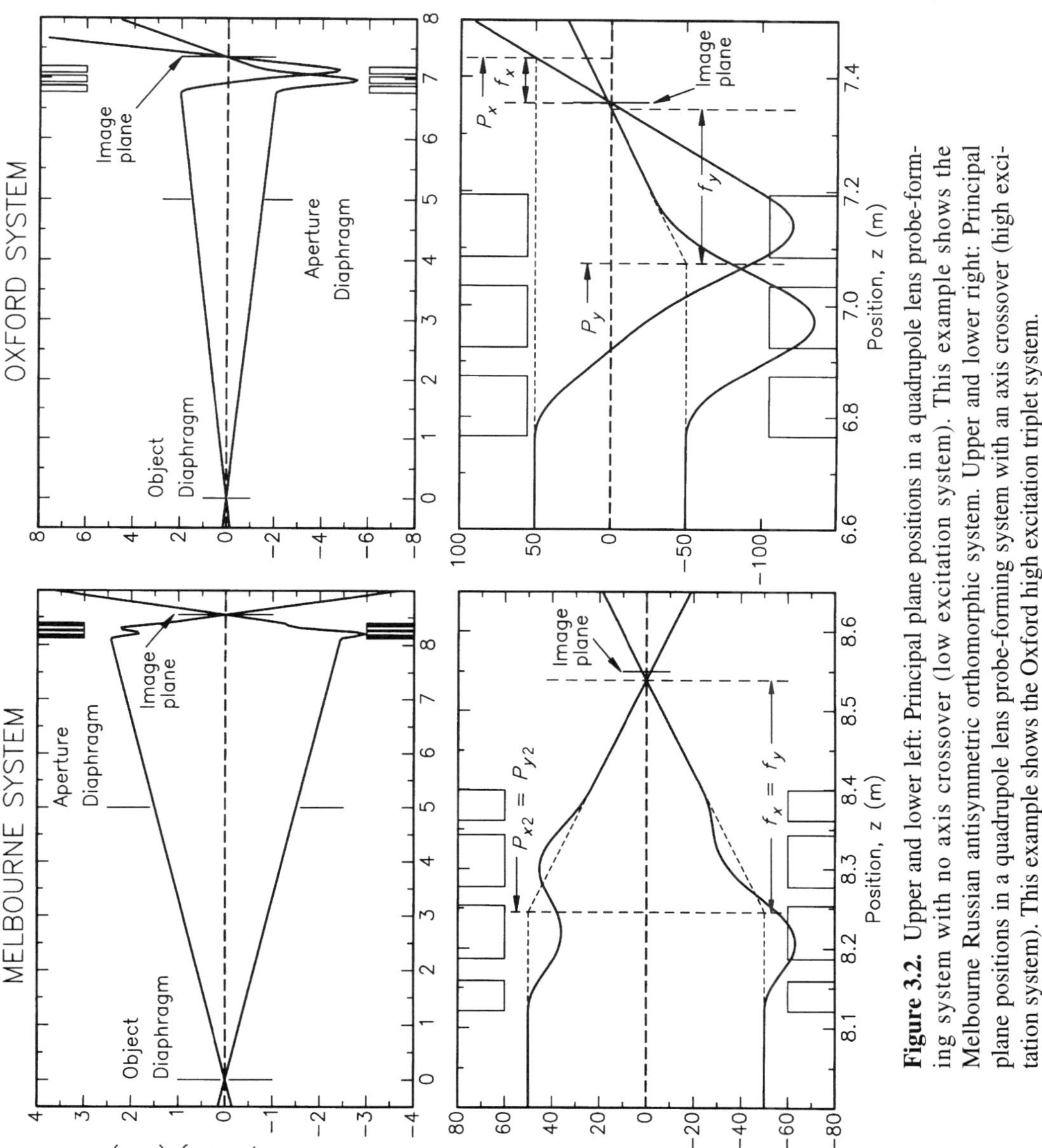

Figure 3.2. Upper and lower left: Principal plane positions in a quadrupole lens probe-forming system with no axis crossover (low excitation system). This example shows the Melbourne Russian antisymmetric orthomorphic system. Upper and lower right: Principal plane positions in a quadrupole lens probe-forming system with an axis crossover (high excitation system). This example shows the Oxford high excitation triplet system.

From the above definitions, it can be seen that the focal lengths of a drift space and elements of sextupole and higher-order lenses are infinite. The principal plane positions in these elements all lie at the center of the element.

3.3.2.3. Magnification and Demagnification

The magnification and demagnification of a lens system, expressed in terms of the elements of the transfer matrix, are defined by

$$\text{Magnification} \quad M_x = (\theta/\theta)^{-1} \tag{3.21a}$$

$$M_y = (\phi/\phi)^{-1} \tag{3.21b}$$

$$\text{Demagnification} \quad D_x = (\theta/\theta) \tag{3.21c}$$

$$D_y = (\phi/\phi) \tag{3.21d}$$

provided the system is stigmatic with $(x/\theta) = (y/\phi) = 0$. An orthomorphic system has $D_x = D_y$. The probe in a nuclear microprobe system is the demagnified image of the object collimator. The sign of the magnification is not important, since it does not matter if the image is inverted or erect. Throughout the rest of this chapter, the word *image* refers to the probe itself.

It is desirable that the demagnifications be as large in absolute value as possible so that a greatly demagnified image of the object collimator is focused on the target, although it is important that the aberrations are not too large.

3.3.2.4. Astigmatism Coefficients

The distances from the point at which the transfer matrix is calculated to the x and y image planes are

$$C_{10} = \frac{(x/\theta)}{(\theta/\theta)} \tag{3.22a}$$

$$C_{01} = \frac{(y/\phi)}{(\phi/\phi)} \tag{3.22b}$$

The system is stigmatic if $C_{10} = C_{01}$, and the transfer matrix has been calculated at the common image plane of a stigmatic system if $C_{10} = C_{01} = 0$. In the latter case the system is said to be in focus. When this is so, then

$$(\theta/\theta) = (x/x)^{-1} \tag{3.23a}$$

$$(\phi/\phi) = (y/y)^{-1} \tag{3.23b}$$

as a consequence of Liouville's theorem. The quantities (x/θ) and (y/ϕ) are termed the astigmatism coefficients. The quantities (x/ϕ) and (y/θ), which are only nonzero if the quadrupoles are rotationally misaligned, are called the skew astigmatism coefficients.

In computer programs that make use of these equations, the quadrupole field strengths are typically adjusted by a numerical optimization routine to minimize the reduced astigmatism coefficient, $\sqrt{(x/\theta)^2 + (y/\phi)^2}$. When this has been reduced to below 10^{-8} m/rad the system is in focus.

Note that, in a system that contains only a single quadrupole lens, it is not possible to make both (x/θ) and (y/ϕ) zero simultaneously. This is because a single quadrupole lens can only produce a line focus.

For systems of three or more quadrupole lenses, there is more than one combination of field strengths that can stigmatize the system. The solution with the higher field strengths usually produces a system with one or more axis crossovers. Such systems can have large magnitude demagnifications, at the expense of large aberrations.

3.4. LOW-ORDER ABERRATIONS

In this section, we consider some new aberration coefficients that are a measure of the sensitivity of the image size of a probe-forming lens system to the following parasitic effects: the energy spread of the beam (chromatic aberration), the alignment of each quadrupole within the lens system; and the power supply ripple or the error in the adjustment of each lens to its optimum value for a focus (excitation aberration). Chromatic aberration is not normally considered to be a parasitic aberration of the system, but it is very convenient to treat it as such, because its effect on the beam is similar to the other parasitic effects listed here.

The misalignment aberration considered here is from lens rotation misalignment (rotation aberration). This is the most important lens misalignment; error in the x and y position of a quadrupole leads mainly to a displacement of the image rather than an increase in size. Misalignments of the actual pole tips within the lens generally introduce higher-order aberrations, and are discussed in Section 3.5.1.

The dominant effect of chromatic, excitation, and rotation aberrations may be calculated easily with first-order matrix theory, because the aberrations all manifest themselves most strongly as increases in the astigmatism coefficients (x/θ), (x/ϕ), (y/θ), and (y/ϕ) of the first-order transfer matrix. In addition, beam energy fluctuations and lens power supply instabilities have the same effect as time-dependent parasitic quadrupole field components in the system.

The discussion begins, in Section 3.4.1, by considering each of these parasitic effects in the theoretical formalism outlined in the previous sections; connection with experiment is made in Section 3.4.2.

3.4.1. Low-order Aberrations: Theory

With chromatic and excitation aberrations, the dominant effect is the increase in magnitude of the astigmatism coefficients (x/θ) and (y/ϕ), whereas rotation aberrations mainly cause an increase in magnitude of the skew astigmatism coefficients (x/ϕ) and (y/θ). The relevant astigmatism coefficient increases at a linear rate as a function of the magnitude of the respective aberration parameter (i.e., at a rate proportional to the beam energy error δ, the quadrupole excitation error E_j, or the rotation error ρ_j, where j is a number that counts each quadrupole lens in the system). For a system that otherwise would be in focus, it is possible to define parasitic aberration coefficients, in addition to the first-order (non-aberration) term:

$$
\begin{array}{lll}
 & x_i = & y_i = \\
\text{First order} & (x/x)x_o & (y/y)y_o \\
\text{Chromatic} & +(x/\theta\delta)\theta_o\delta & +(y/\phi\delta)\phi_o\delta \\
\text{Excitation} & +\sum_{j=1}^{k}(x/\theta E_j)\theta_o E_j & +\sum_{j=1}^{k}(y/\phi E_j)\phi_o E_j \\
\text{Rotation} & +\sum_{j=1}^{k}(x/\phi\rho_j)\phi_o\rho_j & +\sum_{j=1}^{k}(y/\theta\rho_j)\theta_o\rho_j
\end{array}
\tag{3.24}
$$

where k is the number of lenses in the system. For systems in which two or more lenses are powered by a single power supply, it is possible to define combined excitational aberration coefficients, in place of the terms in E_j, as follows:

$$
\sum_{j=1}^{p}(x/\theta G_j)\theta_o G_j \quad \text{in the } xoz \text{ plane} \tag{3.25a}
$$

and

$$
\sum_{j=1}^{p}(y/\phi G_j)\phi_o G_j \quad \text{in the } yoz \text{ plane} \tag{3.25b}
$$

where p is the number of power supplies and G_j is the half peak-to-peak value of the ripple in power supply j, or the error in the adjustment of power supply j to its optimum value.

The above expressions for the parasitic aberrations neglect any coefficients that depend on the x or y coordinate of the ray vector. These contribute significantly less aberration to the image than θ_o- and ϕ_o-dependent coefficients because, in a microprobe system, the object collimator (which defines x_o and y_o) is usually considerably smaller than the aperture collimator (which defines θ_o and ϕ_o). Because the magnitudes of the parasitic aberration coefficients give the sensitivity of the resolution of the microprobe to the various image aberrations, it is desirable that the coefficients be as small in absolute value as possible.

The parasitic aberration coefficients may be calculated most easily by introduction of a known perturbation to a system that had previously been focused. The aberration coefficient may then be calculated from the rate of change of the relevant astigmatism coefficient. Application of a chromatic or excitation perturbation is straightforward. A rotation perturbation may be applied to a single quadrupole by transformation of the quadrupole transfer matrix with a pre- and post-rotation matrix R_j:

$$M'_{Qj} = R_j^{-1} M_{Qj} R_j \qquad (3.26)$$

where R_j is of the form

$$R_j = \begin{bmatrix} -\cos\rho_j & 0 & \sin\rho_j & 0 \\ 0 & \cos\rho_j & 0 & \sin\rho_j \\ -\sin\rho_j & 0 & \cos\rho_j & 0 \\ 0 & -\sin\rho_j & 0 & \cos\rho_j \end{bmatrix}. \qquad (3.27)$$

One problem with this method is that unless the perturbations δ, E_j, G_j, and ρ_j are chosen carefully, rounding errors in the computation lead to considerable errors in the resulting coefficients.

When making the comparison of the maximum probe size predicted by all contributions of Eqs. (3.24) and (3.25), it is important to realize that the actual intensity distribution of the focused probe must be taken into consideration when assessing the effect of each parasitic aberration on the size of the probe. At first glance, it may seem possible to calculate the best probe size, taking into consideration any parasitic aberration, by simply substituting the maximum beam divergence for θ_o and ϕ_o, as well as the greatest possible power supply error or energy fluctuation for G_j and δ. However, in a real system, there may not actually be an ion that simultaneously has both the maximum divergence and the maximum δ. Futhermore, experimental measurements of the beam brightness reveal that the beam intensity is not uniform as a function of divergence or object position (see Figure 2.6). Consequently, simplistic calculations of the minimum probe size achievable by a given system on the basis of theoretical aberration coefficients can be misleading. A simple method of assessing the effects of different aberrations on the probe size is still a matter of active research.

3.4.2. Low-order Aberrations: Experiment

The parasitic aberration coefficients can be measured in much the same way as they were calculated, that is, by application of a perturbation to a focused system and measurement of the resulting increase in size of the image. This method was applied by Cookson [23] to measure the rotation aberration of the Harwell quadruplet and by Grime et al. [24] to measure the parasitic aberration coefficients of the Oxford triplet. The image diameter can be measured by a calibrated graticule in a microscope eyepiece. It is found that the image size increase is linear with the size of the relevant perturbation. Care must be taken that the divergence-limiting aperture collimator is fully illuminated by the beam.

Some example measurements are presented here. These were done on an early form of the Melbourne system, which had a working distance of 350 mm and a demagnification of 16. The perturbations introduced to measure each aberration coefficient were (a) changes in the individual quadrupole excitations to introduce E_j perturbations; (b) changes in the currents of the power supplies connected to each pair of quadrupoles to introduce G_j perturbations; (c) changes in the accelerator energy to introduce δ perturbations; and (d) rotations of individual quadrupoles about the z axis to introduce ρ_j perturbations. Graphs of the image radius along the x and y axes as a function of the relevant perturbation are shown in Figure 3.3. The parasitic aberration coefficients were extracted from straight lines fitted to these data, and are presented in Table 3.1.

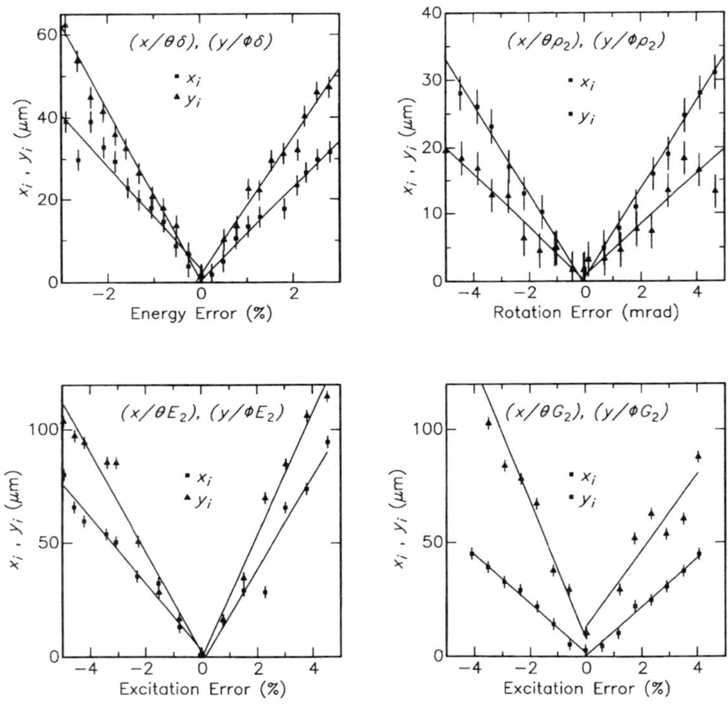

Figure 3.3. Measurement of the coefficients that give the sensitivity of the probe size to the parasitic aberrations of the system. Shown here are the parasitic aberrations that depend on the first power of the divergence. These are for an early version of the Melbourne quadruplet system with magnetic quadrupole lenses (total system length 8.6 m; lenses 1 and 4 of length L_Q = 32 mm; lenses 2 and 3 of length L_Q = 57 mm; bore radius r_o = 6.35 mm; drift length between lens poles 34 mm; image distance from last pole 253 mm). The coefficients measured here are as follows clockwise, from top left: chromatic; rotation; coupled power supply error on lenses #2 and #3; power supply error on lens #2 alone.

TABLE 3.1. Magnification and Low-Order Parasitic Aberration Coefficients of the Melbourne System

Aberration Coefficient	Theoretical Value	Magnitude Measured	Aberration Coefficient	Theoretical Value	Magnitude Measured
(x/x)	$-1/16$	$1/15$	(y/y)	$-1/16$	$1/12$
$(x/\theta\delta)$	154	158	$(y/\phi\delta)$	170	184
$(x/\phi\rho_1)$	-26	21	$(y/\theta\rho_1)$	-26	22
$(x/\phi\rho_2)$	63	66	$(y/\theta\rho_2)$	63	52
$(x/\phi\rho_3)$	-51	58	$(y/\theta\rho_3)$	-51	40
$(x/\phi\rho_4)$	14	13	$(y/\theta\rho_4)$	14	9
$(x/\theta E_1)$	-123	79	$(y/\phi E_1)$	133	90
$(x/\theta E_2)$	225	180	$(y/\phi E_2)$	-436	350
$(x/\theta E_3)$	-331	310	$(y/\phi E_3)$	201	170
$(x/\theta E_4)$	76	47	$(y/\phi E_4)$	-69	50
$(x/\theta G_1)$	-47	35	$(y/\phi G_1)$	63	74
$(x/\theta G_2)$	-108	110	$(y/\phi G_2)$	-236	240

The measured values are uncertain by approximately 10% as a result of the measurement method. Units: x and y in μm; θ and ϕ in mrad; δ in %; ρ_i in mrad; E_j and G_j in %. The Melbourne system employed a Russian antisymmetric quadruplet of magnetic quadrupole lenses (total system length 8.6 m; lenses 1 and 4 length $l = 32$ mm; lenses 2 and 3 length $l = 57$ mm; bore radius $r_Q = 6.35$ mm; drift length between lens poles 34 mm; image distance from last pole 253 mm).

Also shown in the table are calculated coefficients obtained from the first-order theory described in the previous section, using the computer program PRAM. The calculated results from PRAM, as well as other ion optics programs such as TRANSPORT [25] and OXRAY [6] (more recently superseded by TRAX) agree within 2%. This is extremely good considering the differences between the actual models used in the four programs, although the good agreement is probably not surprising, given that the four coefficients can all be calculated with first-order theory. The experimental coefficients agree with the theoretical coefficients to within 20%, on average. This probably reflects the accuracy of the spot size measurement method.

As a guide to the constraints imposed on the size of the various parasitic effects, the coefficients for the Melbourne system with a shorter working distance than before (demagnification of 20) were used to calculate the size of the parasitic effect required to increase the image diameter by 1 μm. This was done for divergence half angles $\theta_o = \phi_o = 0.1$ mrad, which is typical of the divergence routinely used. The results, given in Table 3.2, show that the chromatic aberration would contribute the most aberration to the image for these divergence angles, assuming that the beam phase space is uniformly filled, which is not necessarily a correct assumption. Despite this, with reasonable assumptions, all parasitic effects contribute less than 1 μm to the image diameter (see also the notes attached to Table 3.2). As the aperture increases in size, the parasitic aberrations increase linearly; however, parasitic aberrations from lens defects, which depend on the square of the divergence, rapidly become dominant.

TABLE 3.2. The Calculated Size of the Parasitic Effect Required for a 1-μm-Diameter Contribution to the Image Size[a]

Description of Parasitic Quality	Symbol Used	Value for 1-μm Contribution	Value in Melb. System
Object diameter	$2t_o$	20.3 μm[b]	Variable
Chromatic momentum spread	δ	±0.028%	±0.005%[c]
Power supply #1 stability	G_1	±0.068%	±0.002%[d]
Power supply #2 stability	G_2	±0.020%	±0.002%[d]
Rotation of quadrupole #1	ρ_1	±0.21 mrad	Less[e]
Rotation of quadrupole #2	ρ_2	±0.089 mrad	Less[e]
Rotation of quadrupole #3	ρ_3	±0.12 mrad	Less[e]
Rotation of quadrupole #4	ρ_4	±0.46 mrad	Less[e]

[a]The calculation was done for the Melbourne system with a working distance of 240 mm, and an aperture of diameter 1 mm. With this aperture, the divergence half angle is $\theta_o = \phi_o = 0.1$ mrad.
[b]With an object diameter of 20 μm, and an aperture diameter of 1 mm, the beam current is 100 pA, the minimum required for PIXE at a reasonable count rate on thin biological targets.
[c]The energy stability of the Melbourne Pelletron was measured by Davies [58], yielding $\delta = \pm 0.005\%$ at 1 MeV.
[d]The power supplies of the Melbourne quadrupole lenses can be assumed to have an overall long-term stability of 15 ppm [59]. In this case, $G_2 = 0.002\%$.
[e]The yoke diameter of a quadrupole is 305 mm. Therefore, a rotation of $\rho = 0.089$ mrad corresponds to a displacement of a point on the edge of the yoke by 14 μm. This corresponds to approximately a quarter of a turn of the barrel of the micrometer fitted to the yoke of the quadrupole. Therefore, there is sufficient rotation sensitivity to align the quadrupoles to better than 0.089 mrad.

3.5. HIGHER ORDER ABERRATIONS

Nuclear microprobes that utilize quadrupole lenses in their probe-forming lens systems suffer from two types of higher-order divergence-dependent aberration. One type is aberration intrinsic to perfectly constructed and aligned lenses, such as spherical aberration. The other is parasitic aberration introduced by mechanical imperfections in the lens, as well as lens misalignments. For example, the Melbourne lenses, when used with a beam of large divergence, suffer from parasitic aberration that is comparable in magnitude to the intrinsic aberration [26]. If small probe sizes are to be achieved, it is essential to be able to identify and eliminate aberration in the probe-forming lens system.

The divergence-dependent aberrations of a stigmatic quadrupole lens system may be written:

$$x_i = (x/x)x_0 + (x/\theta^3)\theta_o^3 \quad + (x/\theta\phi^2)\theta_o\phi_o^2 + \Delta x_i \tag{3.28a}$$

$$y_i = (y/y)y_0 + (y/\theta^2\phi)\theta_o^2\phi_o + (y/\phi^3)\phi_o^3 \quad + \Delta y_i \tag{3.28b}$$

where (x/θ^3), $(x/\theta\phi^2)M_y = (y/\theta^2\phi)M_x$, and (y/ϕ^3) are the intrinsic spherical aberration coefficients, and Δx_i and Δy_i are the additional ray displacements in the ray positions, owing to parasitic multipole field components from lens defects, as well as any actual multipole lenses present in the system. Simple analytical expressions exist for the spherical aberration coefficients derived for the thin-lens model, $\beta L_Q \leq 0.9$ [27] and discussed below. Alternatively, the coefficients may be calculated to greater accuracy with third-order computer programs [24,28].

A defect in a quadrupole lens results in the superposition of parasitic multipole field components on the main quadrupole field. The parasitic components introduced by defects in the lenses can be obtained theoretically [6,29,30]; however, it is important to measure the defects directly. The parasitic multipole field components of a quadrupole lens may be determined directly from accurate measurements of the field as a function of polar angle in the bore of the defective quadrupole [31].

Alternatively, the components may be determined from the effect they have on the focused image of the defective quadrupole [26]; this measurement method is discussed further in Section 3.6. The latter measurement method showed 0.01% to 0.3% parasitic sextupole field components and 0.02% to 0.5% parasitic octupole field components in the fields of the Melbourne quadrupoles. This should be regarded as the maximum tolerated in high-quality lenses. These components degrade the resolution of the probe when beams of large divergence are used. Additional multipole lenses can, in principle, be used to correct these aberrations at the cost of great additional system complexity. In a practical system, minimization of the parasitic aberrations by more careful lens construction techniques is more desirable.

Aberration coefficients of order n in the divergence are introduced into a magnetic lens system by multipole fields equivalent to the superposition of two types of magnetic $2(n+1)$ pole lenses. One type, a normal magnetic multipole, is a $2(n+1)$ pole lens oriented with poles symmetrically placed at angles of $\pm j\pi/(2(n+1))$ ($j = 1, 3, \ldots, 2n+1$) to the x axis. The other type, a skew magnetic multipole, is the same as a normal multipole but is rotated by $\pi/(2(n+1))$ about the z axis (the beam direction). Electrostatic multipoles behave similarly. A more sophisticated approach introduces the concept of a multipole phase angle, α_n. In this case the combined effect of a normal multipole and a skew multipole can be considered as the effect of a single multipole rotated by the appropriate phase angle around the z axis. For example, a single magnetic $2n$-pole lens has a phase angle $\alpha_n = 0°$ when there is a north pole at $(360/(4n))°$ above and a south pole at $(360/(4n))°$ below the positive x axis [32]. In general, parasitic multipole field components can have an arbitrary phase angle.

Each multipole introduces several nth-order aberration coefficients, not all of which are necessarily independent. The number of additional separate $2(n+1)$ pole lenses required to correct parasitic and intrinsic aberration of order n is the same as the number of independent aberration coefficients. It is usually not necessary to consider parasitic skew duodecapoles or parasitic multipoles above fifth order, as these are small compared with the lower order components. Also, no fourth-order mulitpoles need be considered, since lens defects that introduce these components also introduce second-order components which are likely to have a large effect on the system.

The ratio between the values of each of the many aberration coefficients introduced by a single parasitic multipole field component depends on its physical location in the system. Therefore, it is possible to correct the parasitic aberration of order n in a single quadrupole with a normal multipole and a skew multipole each with $2(n+1)$ poles, provided that they are superimposed on top of the quadrupole itself. An alternative scheme is to use a single multipole with a variable phase angle. Correction is more challenging if remote multipoles are used.

What follows now is a more complete discussion of the aberrations of a quadrupole probe-forming lens system above first order, again treating the theoretical (Section 3.5.1) and experimental (Section 3.5.2) aspects in turn.

3.5.1. Higher Order Aberrations: Theory

The first- and high-order contributions to the displacement of a ray in the image plane of a lens system which suffers from spherical aberration are as follows:

	$\Delta x_i =$	$\Delta y_i =$
Astigmatism from a focusing error	$(x/\theta)\theta_o$	$(y/\phi)\phi_o$
Rotation misalignment (skew quadrupole)	$+ (x/\phi)\phi_o$	$+ (y/\theta)\theta_o$
Parasitic sextupole	$+ (x/\theta^2)\theta_o^2$ $+ (x/\phi^2)\phi_o^2$	$+ (y/\phi\theta)\phi_o\theta_o$
Parasitic skew sextupole	$+ (x/\theta\phi)\theta_o\phi_o$	$+ (y/\phi^2)\phi_o^2$ $+ (y/\theta^2)\theta_o^2$
Parasitic octupole and intrinsic spherical aberration	$+ (x/\theta^3)\theta_o^3$ $+ (x/\theta\phi^2)\theta_o\phi_o^2$	$+ (y/\phi^3)\phi_o^3$ $+ (y/\phi\theta^2)\phi_o\theta_o^2$
Parasitic skew octupole	$+ (x/\theta^2\phi)\theta_o^2\phi_o$ $+ (x/\phi^3)\phi_o^3$	$+ (y/\phi^2\theta)\phi_o^2\theta_o$ $+ (y/\theta^3)\theta_o^3$
Parasitic duodecapole	$+ (x/\theta^5)\theta_o^5$ $+ (x/\theta^3\phi^2)\theta_o^3\phi_o^2$ $+ (x/\theta\phi^4)\theta_o\phi_o^4$	$+ (y/\phi^5)\phi_o^5$ $+ (y/\phi^3\theta^2)\phi_o^3\theta_o^2$ $+ (y/\phi\theta^4)\phi_o\theta_o^4$

$$(3.29)$$

θ_o and ϕ_o are the initial divergences of the ray projected onto the xoz and yoz planes (z is the beam direction); the ray is assumed to originate from a point on the axis.

For convenience, the terms that depend on θ alone in the xoz plane and ϕ alone in the yoz plane are refered to as the principal terms. All others, that depend on θ and ϕ together, are referred to as the cross-terms.

In stigmatic quadrupole lens systems the cross-terms are related:

$$(x/\theta\phi^2)M_y = (y/\theta^2\phi)M_x \qquad (3.30)$$

Fifth-order divergence-dependent coefficients cannot be easily computed from analytical formulae; instead, numerical calculations must be performed. As a check of the accuracy of such calculations, the cross-terms to fifth order should satisfy the relationships of Foster [33]:

$$(x/\theta^3\phi^2) = 2(y/\theta^4\phi)$$
$$(y/\theta^2\phi^3) = 2(x/\theta\phi^4)$$

It has been found that coefficients from OXRAY satisfy these relationships.

The fifth-order aberrations are likely to be present in most practical quadrupoles because they arise from the use of spherical pole tips, rather than the ideal hyperbolic pole tips. Parzen [34] showed that it was possible to choose the ratio of the lens bore radius to the pole tip radius to make the fifth-order component of the field vanish over a large percentage of the lens bore. Parzen's work showed that for the elimination of the fifth-order aberration:

$$R = 1.094\, r_Q$$

where R is the pole tip radius and r_Q is the bore radius.

Fifth-order coefficients can also be introduced by the inevitable fringe fields from quadrupole lenses. Experimental measurements of the fringe fields from high-quality magnetic quadrupole lenses reveal the presence of duodecapole field components. It may be possible to eliminate these field components by careful shaping of the edges of the pole pieces.

3.5.1.1. Second Order: Sextupole Field Components

It is most convenient to calculate the second-order aberration coefficients from a second-order transfer matrix, which is defined by Brown [19], except here terms that depend on the energy spread of the beam are neglected. Therefore,

$$\begin{pmatrix} \Lambda_i \\ \Xi_i \end{pmatrix} = \begin{pmatrix} M_Q & S \\ 0 & M_Q^2 \end{pmatrix} \begin{pmatrix} \Lambda_o \\ \Xi_o \end{pmatrix}$$

where

$$\Lambda_a \equiv (x_a \quad \theta_a \quad y_a \quad \phi_a)^T$$

(where $a = i$ or o) was defined in Eqs. (3.3);

$$\Xi_a \equiv (\; x_a^2 \quad x_a\theta_a \quad x_a y_a \quad x_a\phi_a \quad \theta_a^2 \quad \theta_a y_a \quad \theta_a\phi_a \quad y_a^2 \quad y_a\phi_a \quad \phi_a^2 \;)^T$$

M_Q contains the elements of the first-order transfer matrix; M_Q^2 is a 10×10 matrix, formed from the elements of M_Q as follows:

$$M_{Qij}^2 = \sum_{k=1,4} \sum_{l=1,4} [M_{Qik}M_{Qjl} + (1 - \delta_{kl})M_{Qjk}M_{Qil}]$$

where δ_{kl} is the Kroneker delta function that has zero value if $k \neq l$ and one if $k = l$ and

$$S \equiv \begin{pmatrix} S_{111} & S_{112} & S_{113} & S_{114} & S_{122} & S_{123} & S_{124} & S_{133} & S_{134} & S_{144} \\ S_{211} & S_{212} & S_{213} & S_{214} & S_{222} & S_{223} & S_{224} & S_{233} & S_{234} & S_{244} \\ S_{311} & S_{312} & S_{313} & S_{314} & S_{322} & S_{323} & S_{324} & S_{333} & S_{334} & S_{344} \\ S_{411} & S_{412} & S_{413} & S_{414} & S_{422} & S_{423} & S_{424} & S_{433} & S_{434} & S_{444} \end{pmatrix}$$

are the sextupole field component matrix elements, the nonzero terms of which are

$$\begin{aligned}
S_{111} &= -\tfrac{1}{2}k_s^2 L_S^2 & S_{211} &= -k_s^2 L_S & S_{313} &= k_s^2 L_S^2 & S_{413} &= 2k_s^2 L_S \\
S_{112} &= -\tfrac{1}{3}k_s^2 L_S^3 & S_{212} &= k_s^2 L_S^2 & S_{314} &= \tfrac{1}{3}k_s^2 L_S^3 & S_{414} &= k_s^2 L_S^2 \\
S_{122} &= -\tfrac{1}{12}k_s^2 L_S^4 & S_{222} &= -\tfrac{1}{3}k_s^2 L_S^3 & S_{323} &= \tfrac{1}{3}k_s^2 L_S^3 & S_{423} &= k_s^2 L_S^2 \\
S_{133} &= \tfrac{1}{2}k_s^2 L_S^2 & S_{233} &= k_s^2 L_S & S_{324} &= \tfrac{1}{6}k_s^2 L_S^4 & S_{424} &= \tfrac{2}{3}k_s^2 L_S^3 \\
S_{134} &= \tfrac{1}{3}k_s^2 L_S^3 & S_{234} &= k_s^2 L_S^2 \\
S_{144} &= \tfrac{1}{12}k_s^2 L_S^4 & S_{244} &= \tfrac{1}{3}k_s^2 L_S^3
\end{aligned}$$

where

$$k_s^2 = \frac{B_S}{r_S^2}\frac{Ze}{p}$$

B_S is the sextupole field strength at the lens bore radius (i.e., on the pole tip), r_S is the bore radius of the sextupole (the same as the bore radius of the quadrupole lens, r_Q, for superimposed parasitic fields), Ze is the ion charge, p is the relativistic beam momentum, and L_S is the effective length of the sextupole (the same as the quadrupole lens, L_Q, for superimposed parasitic fields).

Several special cases can be recognized:

For a sextupole lens the M_Q matrix is that of a drift space.

For a perfect quadrupole lens the S_{ijk} elements are all zero

For a quadrupole lens containing parasitic sextupole field components, the S_{ijk} elements depend on the strength of the parasitic sextupole components. In this case $L_S = L_Q$ and $r_S = r_Q$.

3.5.1.2. Third Order: Spherical Aberration

In the thin lens model, the spherical aberration coefficients of a thin quadrupole lens have been derived by Dymnikov et al. [27]:

$$C_p = \frac{2 - 2n + 3n^2}{6} \frac{\psi_Q}{L_Q} \left[\frac{a_x^2}{f_x} \right]^2 \tag{3.31a}$$

$$C_s = -\frac{2 + 2n - n^2}{2} \frac{\psi_Q}{L_Q} \left[\frac{a_x^2 a_y^2}{f_x f_y} \right] \tag{3.31b}$$

$$D_p = \frac{2 - 2n + 3n^2}{6} \frac{\psi_Q}{L_Q} \left[\frac{a_y^2}{f_y} \right]^2 \tag{3.31c}$$

$$D_s = C_s \tag{3.31d}$$

where:

$$C_p(\theta/\theta) = (x/\theta^3)$$

$$C_s(\theta/\theta) = (x/\theta\phi^2)$$

$$D_p(\phi/\phi) = (y/\phi^3)$$

$$D_s(\phi/\phi) = (y/\theta^2\phi)$$

and here

$$n = \frac{\beta_E^2}{\beta_M^2 - \beta_E^2}$$

($n = 0$ for magnetic lenses, $n = -1$ for electrostatic lenses, and $n = 1$ for achromatic lenses, which are superimposed magnetic and electrostatic lenses); a_x and a_y are the lens object distances in the xoz and yoz planes; f_x and f_y are the lens focal lengths in the xoz and yoz planes; and L_Q is the lens effective length defined in Eq. (3.13). ψ_Q is a normalization-independent correction factor that allows for the shape of the quadrupole fringe fields and is given by

$$\psi_Q = \frac{\displaystyle\int_{-\infty}^{+\infty} f^2(z)\,dz}{\displaystyle\int_{-\infty}^{+\infty} f(z)\,dz},$$

where $f(z)$ is the measured field profile of the lens. The quantity, ψ_Q, measured for some Melbourne quadrupole lenses, was $\psi_Q = 0.93 \pm 0.01$ for quadrupoles with $L_Q = 60$ mm, and $\psi_Q = 0.86 \pm 0.02$ for quadrupoles with $L_Q = 40$ mm. For both lenses, $r_o = 6.35$ mm. However, for simplicity, ψ_Q can be approximated by unity for all calculations.

3.5.1.3. Third Order: Octupole Field Components   If the field profile of an octupole lens is modeled as rectangular in the z direction, then the spherical aberration coefficients are those derived by Yavor et al. [35]:

$$C_p = -\frac{\psi_O \gamma}{5}[(L_O + a_x)^5 - a_x^5], \tag{3.32a}$$

$$\begin{aligned}
C_s = 3\psi_O \gamma \Big[& a_x^2 a_y^2 L_O + a_x a_y(a_x + a_y)L_O^2 \\
& + \frac{1}{3}(a_x^2 + 4a_x a_y + a_y^2)L_O^3 \\
& + \frac{1}{2}(a_x + a_y)L_O^4 + \frac{1}{5}L_O^5 \Big].
\end{aligned} \tag{3.32b}$$

D_p and D_s may be obtained from the expressions for C_p and C_s by replacing a_x by a_y and a_y by a_x, where a_x and a_y are the object distances of the octupole (the object may be real or virtual). L_O is the octupole effective length, which measurements on octupole lenses in Melbourne was found to be

$$L_O = l + 0.53r_O$$

where l is the actual octupole length and r_O is the octupole bore radius. γ is the octupole strength parameter, given by

$$\gamma = \frac{B_O Z e}{r_O^3 p}$$

where B_O is the field on the center of an octupole field tip, p is the relativistic beam momentum, and r_O is the octupole bore radius. ψ_O is defined in the same way as ψ_Q for a quadrupole lens. Measurements in Melbourne for a magnetic octupole lens with $L_O = 13$ mm gave $\psi_O = 0.86 \pm 0.02$, but this may be approximated by unity for simplicity.

The spherical aberration coefficients of a system of k lenses may be obtained by summation of the spherical aberration coefficients of the individual lenses in the system by use of the spherical aberration coefficient summation formulae given by Yavor et al. [36]:

$$C_{p(k)} = \sum_{j=1}^{k} \frac{{}_jC_p}{M_{x(j-1)}^4} \tag{3.33a}$$

$$C_{s(k)} = \sum_{j=1}^{k} \frac{{}_jC_s}{M_{x(j-1)}^2 M_{y(j-1)}^2} \tag{3.33b}$$

$$D_{p(k)} = \sum_{j=1}^{k} \frac{{}_jD_p}{M_{y(j-1)}^4} \tag{3.33c}$$

$$D_{s(k)} = C_{s(k)} \tag{3.33d}$$

where ${}_jC_p$, ${}_jC_s$, ${}_jD_p$, and ${}_jD_s$ are the spherical aberration coefficients of the jth lens, and $M_{x(j-1)}$ and $M_{y(j-1)}$ are the combined magnifications of the first $j-1$ lenses of the system, given by

$$M_{x(j-1)} = \prod_{i=1}^{j-1} {}_iM_x$$

$$M_{y(j-1)} = \prod_{i=1}^{j-1} {}_iM_y$$

where ${}_iM_x$ and ${}_iM_y$ are the magnifications of the ith lens.

The equations were incorporated into a beam optics computer program: PRAM. This program was first tested by application to several systems for which theoretical or experimentally measured spherical aberration coefficients have been published. For a series of magnetic quadrupole multiplets [24,29,30], PRAM calculates spherical aberration coefficients that are on average a factor of 1.72 greater than those measured. For a range of electrostatic and magnetic quadruplets [9,23,35], PRAM calculates spherical aberration coefficients that are on the average a factor of 0.88 less than those calculated by the authors using analytical techniques. The agreement between the PRAM calculations and the published coefficients is extremely good considering the simplicity of the model and that some of the published systems required quite large quadrupole excitations, for which the thin-lens model was not expected to be appropriate.

3.5.1.4. Fifth Order: Duodecopole Field Components

No analytic expressions exist for the three fifth-order coefficients of a duodecapole, but the numerical raytracing program OXRAY, or TRAX, may be used to calculate the effect of a duodecapole superimposed on a quadrupole lens.

3.5.2. Higher-order Aberrations: Experiment

There are two basic approaches toward the measurement of the higher order properties of a quadrupole or multipole, lens. The first approach is to make an actual measurement of the multipole field components by careful measurement of the magnetic field. Because this does not require the ion beam, it is termed the *off-line method*. The second is to make use of the ion beam itself to measure the multipole field components and is hence called the *on-line method*.

3.5.2.1. Off-line Method: Field Mapping   A field-mapping device can make measurements of the magnetic field at many points within the lens bore, using a Hall probe that is small compared with the size of the lens bore. The strength of the multipole field components may thus be determined. This method also allows any variations in the strength of the multipole components as a function of z to be determined [37]. This is useful, because multipole components often arise only in the fringe fields of the lens. A variant of this approach is to measure the average multipole components using a rotating coil [38,39]. This has the advantage of simplicity, but does not accurately provide information about the z dependence of the field distribution unless the coil is short compared with the lens length.

Once the multipole field components in each lens of the system have been obtained, the aberration coefficients can be calculated using the equations in Section 3.5.1, and hence the effect of these components on the focused probe can be determined.

3.5.2.2. On-line Method: Use of the Ion Beam   A different approach, which is more general, is to employ the actual ion beam itself to measure the lens aberrations. There are several ways of doing this.

The most straightforward method, and least satisfactory, is to make direct observations of the image spot focused by the quadrupole lens system. Often, the image shapes characteristic of contamination from large sextupole fields (three-lobed image) or large octupole fields (four-lobed image) can be identified. Some theoretical image shapes are shown in Figures 3.4 and 3.5. However, this observation method is usually unsatisfactory because it is difficult to interpret the aberrations present in the lens system from the appearance of the image spot, since more than one field component may be present.

A superior method is the scanned-ray pencil method. A pencil of rays from the object collimator is scanned over the entrance of the lens or lenses under test [40]. At each position, the pencil will have a particular divergence θ_o and ϕ_o. The resulting ray position in the image plane can then be measured to determine x_i and y_i, and hence the aberration coefficients can be deduced. This method can be time-consuming and does not provide a complete and instantaneous picture of all lens aberration.

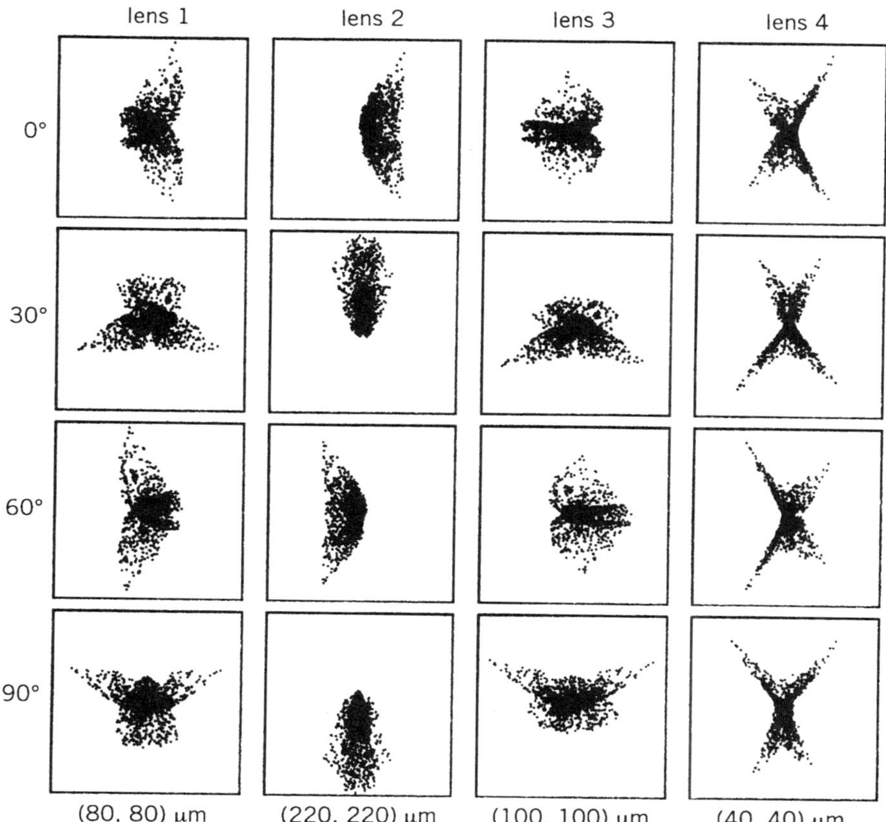

Figure 3.4. Image plots for a quadruplet lens configuration, showing the effect of a 1% sextupole field in each lens in turn, at a phase angle of $\alpha_3 = 0°$, 30°, 60°, and 90°. The box lengths in the x and y directions are shown at the base of each column. Note that spherical aberration dominates the image in column 4.

The grid shadow method, discussed in the next section, is more satisfactory. This is essentially a generalization of the scanned-ray pencil method. This method involves placement of a fine grid in or near the Gaussian image plane and observation of the shadow pattern on a screen located downstream. The structure of the shadow pattern is very sensitive to the aberrations present in the system. The patterns produced can readily be interpreted with reference to theoretical patterns to identify the major parasitic field components in the system. Quantitative analysis is also possible, as described below.

Before discussion of the measurement method, some results can be summarized. As yet, with the probable exception of the Oxford Integral Lenses [41], parasitic aberration is the main limitation to achieving higher resolution probes. On most systems, to achieve a 1 μm probe with 100 pA of beam, the

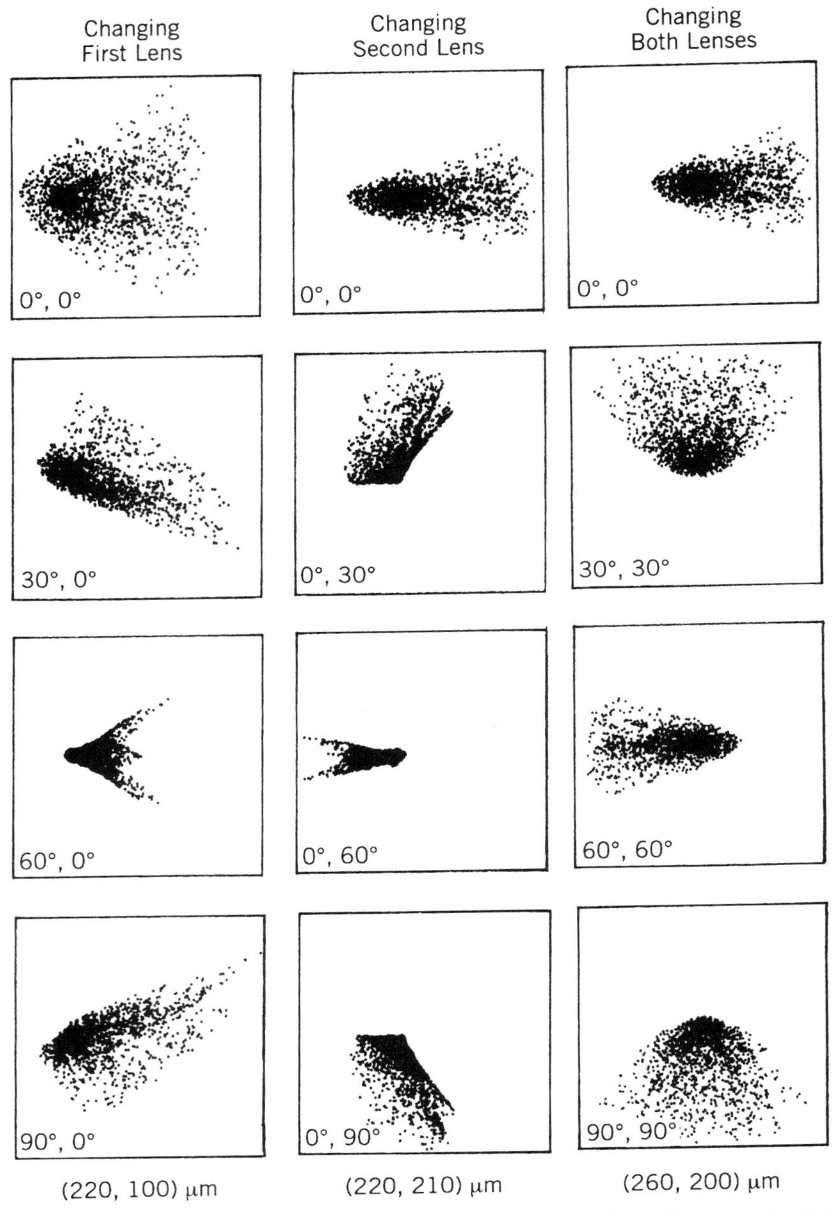

Figure 3.5. Image plots for a doublet lens configuration with a 1% sextupole field in each lens. The sextupole phase angle of each lens is shown in the bottom-left corner of each image, with the first lens encountered by the beam shown first. The box lengths in the x and y direction are shown at the base of each column. The top three images are the same case, plotted to three different scales.

amount of parasitic sextupole and octupole field contamination must not be greater than approximately 0.3%, although the precise level that can be tolerated is a function of ion source brightness and lens system configuration. By contrast, experience has shown that lens systems with approximately 10% sextupole contamination are unable to achieve probes smaller than 20 μm [42].

3.5.2.3. Effect of Parasitic Sextupole Fields on Resolution   The effect on the size of the focused probe of a doublet, triplet, and quadruplet lens configurations that include parasitic sextupole fields with different phase angles has been considered in detail elsewhere [32] and will be only briefly considered here. As the sextupole field ($n = 3$) component is rotated through increasing phase angle, α_3, the second-order aberration coefficients in Eqs. (3.29) all vary sinusoidally with a period of 120°, representing one complete rotation of the sextupole field.

Figure 3.4 shows image plots from a Russian quadruplet lens configuration of the Harwell microprobe [43], with a sextupole field strength of 1% on each individual lens in turn, at phase angles of $\alpha_3 = 0°$, 30°, 60°, and 90°. These image plots were generated by transforming several thousand object coordinates to their image coordinates under the effect of the relevant sextupole aberrations. Similar plots for triplet and doublet configurations are given in Ref. 32, together with tables for the variation in the resultant image size for the different sextupole phase angles.

Figure 3.5 shows the effect on the image shape of a 1% sextupole field in both lenses of a doublet configuration of the same Harwell microprobe at different sextupole phase angles. Whereas the primary effect of a single sextupole field present in a lens configuration is to rotate the aberrated image, this is not necessarily the case when more than one sextupole field is present in the configuration. Results of similar image plots for triplet and quadruplet configurations have been tabulated [32] and show that the size of the aberrated probe is very sensitive to the phase angle of the sextupole fields.

3.6. GRID SHADOW METHOD

The grid shadow method has had a long history in the study of aberrations of electron–optical quadrupole lenses [44]. Discussed here is a variation of the method employed by Deltrap [29], Hardy [30], and Okyama and Kawakatsu [45].

The experimental method is very simple. A fine grid is placed in the converging image plane of the lens system under test; this is shown schematically in Figure 3.6. For nuclear microprobe systems with MeV ions, a 2000-mesh grid (period 12.7 μm) is satisfactory. A smaller period grid will lead to greater sensitivity. The grid need not be of stopping thickness for the ion beam. A 2-μm thick copper grid gives good contrast in the shadows with 3 MeV ^{1}H ions, even though the ion range in copper is 50 μm. The shadows can be cast on a glass screen, although brighter images can be obtained with screens made from ionoluminescent chemicals such as $ZnSiO_4$:Mn.

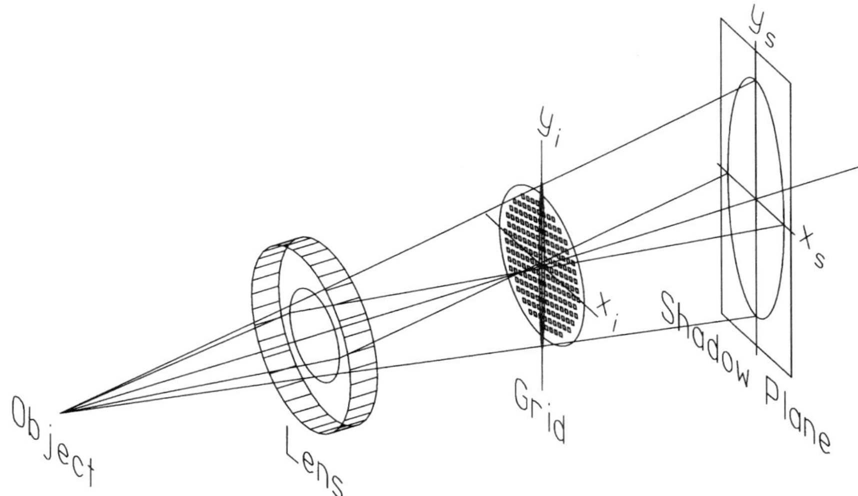

Figure 3.6. The grid shadow method for a single quadrupole lens. The rays diverge from the object collimator (assumed to be small), and are focused to a line in the converging image plane (x_i, y_i). The grid, located in the converging image plane, is rotated about the z axis by a few degrees. The line is spread because of spherical aberration, which causes the rays in the xoz plane to cross over the axis upstream of the grid. Downstream of the grid, the rays diverge to the shadow plane (x_s, y_s), where the grid shadow pattern is observed. In the theoretical and experimental shadow patterns of single quadrupole lens systems discussed in this chapter, only the central section of the shadow pattern is analyzed.

The grid shadow pattern is characteristic of the aberrations of the lens system. Because the beam itself is used to diagnose the lens system under test, the method can be applied to magnetic, electrostatic, or achromatic lens systems. The field-mapping methods can typically only be applied to magnetic systems.

In Section 3.6.1, we consider the diagnosis of lens aberrations in systems that consist of a single quadrupole lens. This is the essential first step in the diagnosis of an entire system, so that faulty lenses can be identified and eliminated. Section 3.6.2, considers the diagnosis of aberrations of the probe-forming system as a whole. Other applications of the grid shadow method are discussed in Section 3.6.3.

3.6.1. Single Quadrupole Systems

Presented here is the grid shadow method for the measurement of beam-divergence-dependent parasitic and intrinsic aberrations of single quadrupole lens systems. The method may also be used to measure the effect of actual multipole lenses in the system.

A single quadrupole lens produces a real image in one plane and a virtual image in the perpendicular plane. Only aberration of the real image of a point object are considered. In the real image plane, the beam is always focused to a line. Aberration in the perpendicular plane may be studied by simply reversing the current through the quadrupole.

The discussion that follows describes how the shadow pattern may be interpreted and used to make quantitative measurements.

The contributions to the aberration of a ray in the real image plane of a single quadrupole lens that converges in the xoz plane, are, in general, given by Eqs. (3.29); but, for a single positive quadrupole lens, Eq. (3.29b) may be written

$$\Delta y_i = (y/\phi)\phi + \text{insignificant higher-order terms} \tag{3.34}$$

The higher-order terms are insignificant compared with the very large contribution from (y/ϕ), since the lens diverges in the yoz plane.

3.6.1.1. Effects of Pole Misalignments

The type of pole misalignment that gives rise to each order aberration coefficient and the corresponding parasitic multipole component has been discussed in detail [46, and references therein], but may be briefly summarized as follows: (a) A single radial or angular pole displacement introduces sextupole and octupole components. (b) Radial or angular displacements of adjacent poles introduce sextupole components. (c) Radial or angular displacements of opposite poles introduce octupole components. (d) Other lens defects, such as missing pole windings, iron inhomogeneities, or flux leakage from joints, will result in virtual pole displacements with the same effect as actual pole displacements. Although correction of these defects by deliberate perturbations of the lens structure is possible [47], the procedure is difficult and not recommended. Here, the emphasis is on the diagnosis of problems, and it is assumed they can be fixed by replacement of the offending lens with a more perfect lens.

3.6.1.2. Transformation from the Image Plane to the Shadow Plane

For simplicity, the following regions of the line image of a positive lens that are close to the x axis are considered. In this region, rays have a very small divergence ϕ_o in the yoz plane. Therefore, the aberration cross-terms that depend on ϕ_o will not be considered further.

The transformation from the image plane $(x_i, \theta_i, y_i, \phi_i)$ to the shadow plane $(x_s, \theta_s, y_s, \phi_s)$ is the drift space transformation:

$$x_s = x_i + D\theta_i \tag{3.34a}$$

$$\theta_s = \theta_i \tag{3.34b}$$

$$y_s = y_i + D\phi_i \tag{3.34c}$$

$$\phi_s = \phi_i \tag{3.34d}$$

where D is the distance from the real image plane to the shadow plane (see also

Figure 3.6). To third order, in a lens system which suffers only from spherical aberration or contains an octupole field component, these equations enable θ_i and ϕ_i to be eliminated from the terms of Eqs. (3.28) and (3.34), neglecting object size, to yield

$$x_s = \frac{D}{M_x}\left(\frac{x_i}{(x/\theta^3)}\right)^{1/3} + x_i \tag{3.35a}$$

$$y_s = \frac{[(y/\phi) + D/M_y]y_i}{(y/\phi)} \tag{3.35b}$$

All cross-terms have been neglected, because only the central region of the shadow plane has been considered.

3.6.1.3. Application of the Theoretical Results

If the magnitude of aberration introduced by the coefficient (x/θ^3), x_i, is small compared with the displacement of the ray in the shadow plane, x_s, then x_i can be neglected compared with other terms in Eq. (3.35a). This has been done by some workers, who studied single quadrupole systems with a large value of D. However, in some systems, D cannot be made arbitrarily large. Also, spherical aberration, which remains even if all parasitic aberration is eliminated, may be compensated by actual octupole lenses introduced in the system. These octupole lenses may introduce large (x/θ^3) terms. Therefore, to retain generality, x_i is retained in Eq. (3.35a).

Some workers have placed grids in the real plane with the grid bars aligned with the x and y axes [30], or rotated about the z axis [45]. Figure 3.7 shows theoretical grid shadow patterns calculated with Eqs. (3.35) for these two cases. With the grid aligned with the x and y axes, the period of the vertical grid bar shadows changes at a rate determined by the value of (x/θ^3). With the grid rotated about the z axis, the grid bar curves take on the shape of cubic curves. The latter case offers a superior method for identification for the aberration in

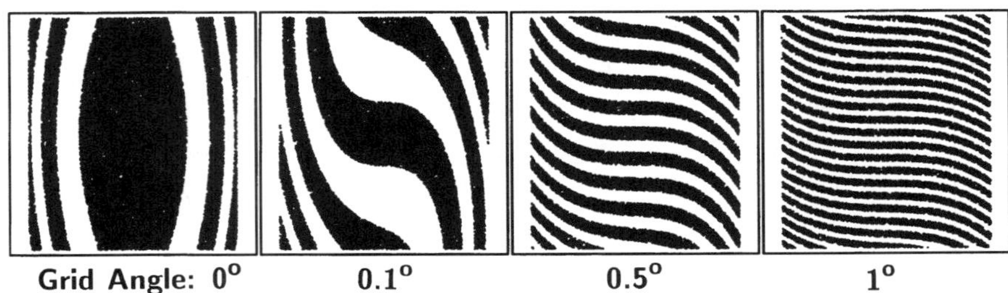

Figure 3.7. Theoretical grid shadow patterns showing the effect of the grid rotation angle on the shadow pattern for a single quadrupole lens. The parameters used were as follows: object drift space 8.27 m; lens effective length $L_Q = 0.0666$ m; image drift space 0.21 m; $D = 0.2$ m; grid period $P = 12.7$ μm; shadow plane area 10×10 mm^2. The lens was assumed to suffer from only intrinsic spherical aberration.

the system, because the cubic curve is easier to observe than subtle variations in the period of the vertical grid bar shadows. It can also be seen from Figure 3.7 that the sensitivity of the method decreases as the grid is rotated further from alignment with the axis. The optimum angle for a particular system can easily be determined by experiment.

The shape of the curved grid bar shadows may be obtained from the image-plane-to-shadow transform shown by Eqs. (3.35), together with the two possible equations for lines through the center of the grid bars in the image plane

$$y_i = x_i \tan \alpha \qquad\qquad (3.36a)$$

$$y_i = -x_i \cotan \alpha \qquad\qquad (3.36b)$$

where α is the rotation angle of the grid relative to the x axis.

These equations neglect the y_i intercept of the grid bar, because the shape of the grid bar shadows is insensitive to this quantity near the x axis, where ϕ_i is small. Equation (3.36b) is the most useful, because grid bars that form shallow angles to the y axis have the most curvature in the shadow plane and are therefore most sensitive to the value of (x/θ^3). In this case, the equation for the shape of a curved grid bar shadow may be written

$$x_s = -\frac{D}{M_x} \left(\frac{y_s}{(x/\theta^3)\eta \cotan \alpha} \right)^{1/3} - \frac{y_s}{\eta \cotan \alpha}, \qquad (3.37)$$

where

$$\eta = \frac{(y/\phi) + D/M_y}{(y/\phi)}$$

Equation (3.37) again shows that the grid bar shadows are shaped like cubic curves. If the period of the grid bar shadows along the y_s axis is denoted by P_s, then from Equation (3.35b),

$$\eta = \frac{P_s \cos \alpha}{P} \qquad\qquad (3.38)$$

where P is the actual period of the grid.

3.6.1.4. *Measurement of the Aberration Coefficients* Equation (3.38) allows the quantity η to be obtained from experimentally measured quantities. Equations (3.37) and (3.38) show how the aberration coefficient (x/θ^3) may be measured: First, η may be calculated from a measurement of the period of the wide grid bar shadows along the y_s axis using Eq. (3.38). Then the grid bar shadow closest to the axis of the quadrupole can be digitized, and Eq.

(3.37) fitted to the resultant data points with (x/θ^3) as a free parameter. It is not necessary to choose the precisely central grid bar shadow for this, because the shapes of all the grid bar shadows in the central region of the pattern are approximately the same for small parasitic fields. This method of measurement of (x/θ^3) has the advantage that all parameters required to determine (x/θ^3) are obtained from the grid shadow pattern.

If (x/θ^3) is small and the grid is rotated away from alignment with the x and y axes by a large amount, then the grid bar shadows form almost straight horizontal lines and the cubic curve shape Eq. (3.37) is difficult to observe. As shown by Figure 3.7, when the grid is rotated to closer alignment with the x and y axes, the cubic curve shape becomes more pronounced as the ends of the grid bar shadows become further displaced from the original almost-straight lines. Eventually, the ends of the shadows are as far from the x axis as they are from the y axis. At this grid angle, the curvature is very easy to observe, provided that the grid bars are thin and cast thin shadows. This grid angle is given by

$$\alpha = \arctan \left| \frac{D}{\eta\,(x/\theta^3)\,\theta_{\max}^3\,M_x^3} \right| \tag{3.39}$$

The width of the grid bars will be thin enough to allow the grid pattern to be easily observable, provided that the period of the grid is such that

$$P < 4(x/\theta^3)\,\theta_{\max}^3\,M_x^3\,\cos\alpha \tag{3.40}$$

where $\theta_{\max}$ is the greatest possible divergence of the incident beam in the xoz plane. Equations (3.39) and (3.40) allow the grid rotation angle and period to be chosen for an experiment to measure (x/θ^3).

3.6.1.5. Aberrations Other than Spherical　　So far, only the effect of (x/θ^3) on the grid shadow pattern has been discussed. If aberrations other than spherical aberration are present, then the calculation of the theoretical grid shadow pattern by analytical methods is very laborious. An alternative approach is to use a numerical method. Equations (3.28) may be used to "raytrace" a large number of rays from the object collimator to the image plane using a Monte Carlo method. Appropriate ray vectors can be randomly generated, subject to the constraint that each ray passes through the aperture collimator of the system. Grime and Watt [6] showed that this method of building up an image gives results in good agreement with actual numerical raytracing of the individual rays through the system and is considerably less time consuming. Once an image consisting of a few thousand rays has been built up as described, the rays that would intercept a grid bar can be removed and the remainder transformed to the shadow plane with a simple drift space transformation. A plot of these rays forms the grid shadow pattern.

Figure 3.8 shows theoretical grid shadow patterns calculated in this way for a positive quadrupole lens. The figure shows just the central region of the entire shadow pattern, and only one aberration coefficient is nonzero, so that the characteristic pattern can be seen. The central region of the shadow will just display the effect of the principal coefficient from each multipole or for small parasitic contamination. This is the coefficient of the highest order in divergence angle

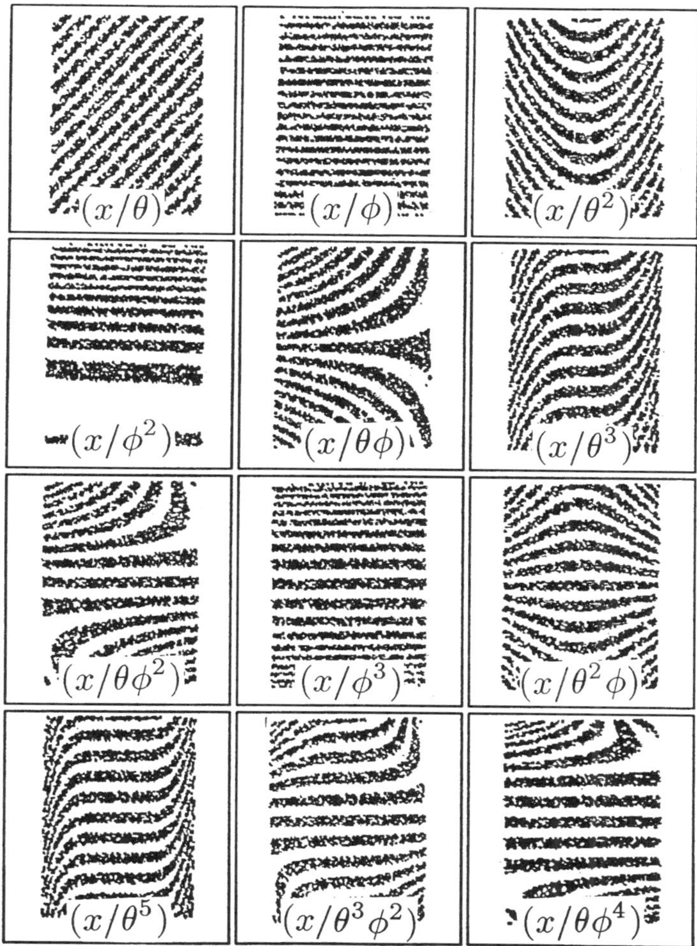

Figure 3.8. Theoretical grid shadow patterns for a single quadrupole lens showing the effect of a variety of aberration coefficients. In a real quadrupole lens, not all of these are necessarily nonzero. Each coefficient was assumed to contribute 50 μm of aberration in the absence of all other possible aberrations. Comparison of experimental shadow patterns allows the dominant aberration coefficient to be identified, and hence the effect, parasitic or otherwise, that will limit the resolution of the lens. The parameters were similar to those in Figure 3.7, except that the grid angle was fixed at $\alpha = 2°$.

θ. However, contamination by very complicated, or strong parasitic field distributions can produce patterns where the other coefficients dominate the pattern, even in the central region. A further complication is that the shape of the shadow pattern for a reversal of the polarity of the lens depends on the phase angle of the parasitic field component. The relationship between the positive and negative excitation shadow patterns has been discussed elsewhere [47].

Experimental grid shadow patterns often resemble a combination of the theoretical patterns, because more than one parasitic multipole component may be present in the quadrupole field. The theoretical patterns can be used as guides for the identification of the most significant parasitic aberration in the experimental grid shadow patterns (this will be discussed further below).

The whole shape of the theoretical grid shadow patterns for each of the parasitic aberration coefficients may be determined. The method is similiar to that used above for the case where only (x/θ^3) was assumed to be nonzero. In this case, a simplification is possible because the parasitic aberration x_i is usually small compared with the width of the shadow pattern x_s. Therefore, it is possible to neglect the term in x_i on the right-hand side of Eq. (3.34a). Then the combined effect of the principal parasitic aberration coefficients causes the central grid bar shadows to take the shape of a polynomial:

$$y_s = -\left(\eta \cotan \alpha \sum_{\sigma=1}^{5} \frac{(x/\theta^\sigma)M_x^\sigma}{D^\sigma} \right)(x_s^\sigma) \tag{3.41}$$

and it is assumed that $(x/\theta^4) = 0$. Equation (3.41) follows as a generalization of Eq. (3.37). Experimental measurement of the aberration coefficients from the grid shadow pattern is possible, by simply fitting Eq. (3.41) with the coefficients as free parameters. The other parameters in the equation, η, M_x, and α depend only on the geometry of the system.

The parasitic cross-terms cause the central regions of the grid bar shadows away from the x_s axis to become inclined, with gradients given by

$$y_s = -\frac{\eta\, Y_s^\tau M_x^\sigma M_y^\tau (x/\theta^\sigma \phi^\tau)\cotan \alpha}{D^\sigma[(y/\phi)M_y + D]} x_s^\sigma + Y_s \tag{3.42}$$

where Y_s is the intercept with the y_s axis of the grid bar shadow. The inclination is seen to increase as Y_s increases. For regions distant from the y_s axis, where x_s is large, Eq. (3.42) does not apply. Often, only those axial regions of the grid bar shadow patterns of negligible inclination are investigated, and the parasitic cross-terms do not then need to be considered further.

3.6.1.6. Single Quadrupole Lens with Single Multipole Lens

As an experimental test of Eq. (3.41), a series of shadow patterns can be produced with a lens system that consists of a single quadrupole lens with a single multipole lens. Representative experimental images obtained with a multipole lens configured as a sextupole, octupole, and quadrupole–duodecapole are shown in Figure 3.9. Also shown are a shadow of a grid with all lenses turned off, the effect of a "parasitic" quadrupole (that is, a quadrupole defocus), and a quadrupole alone.

The quadrupole was initially focused onto the grid from observation of the grid shadow pattern using the method of Okayama and Kawakatsu [45]. This was done with the multipole switched off and degaussed so that residual fields did not influence the grid shadow pattern. When the optimum focus was obtained, the grid bar shadows apppeared as a series of parallel horizontal lines, which indicated that the grid shadow pattern was insensitive to the small intrinsic and parasitic aberrations of the quadrupole. This was as expected, because a relatively coarse grid rotated at a relatively steep angle was used for the measurement.

The grid shadow patterns in Figure 3.9 clearly show the effect of the principal aberration coefficient (x/θ^n) of each of the multipoles. The first three shadow patterns (Figures 3.9a, b, and c), show that as the strength of the quadrupole is increased to achieve a focus on the grid, the shadows of the individual grid bars rotate until they are all parallel with the x axis. The three following shadow patterns (Figures 3.9d, e, and f), show that the effect of a sextupole, octupole, or duodecapole lens is to cause the previously straight lines in the shadow pattern to become curved, as expected. Although the differences between the curvature for the octupole and duodecapole are subtle, a cubic curve gives a superior fit to the octupole pattern, and a fifth order curve gives a superior fit to the duodecapole pattern.

An extension of the method is to use the observation of the shadow pattern to tune the system to achieve correction of the principal parasitic aberration coefficients. The method could also be extended to study the cross-coefficients by observation of the off-axis regions of the shadow pattern.

3.6.1.7. Single Quadrupole Lens with Single Octupole Lens Taking the quadrupole–octupole system from the previous example and increasing the strength of the octupole considerably allows a further test of Eqs. (3.37) and (3.42). Experimental grid shadow patterns, taken as a function of the octupole field strength, are shown in Figure 3.10.

The increasing curvature of the ends of the grid bar shadows, as the octupole strength was increased, indicated that (x/θ^3) increased as expected. Figure 3.10e shows the effect of a misalignment between the axis of the quadrupole and the axis of the octupole. The central region of the grid bar shadows becomes inclined to the axis, owing to the effect of $(x/\theta\phi^2)$. This is in good qualitative agreement with Eq. (3.42).

The result of fitting Eq. (3.37) to a digitization of the central grid bar shadows of Figure 3.10 allows (x/θ^3) to be measured as a function of the octupole lens pole-tip field strength. The results are in good agreement with Eqs. (3.32),

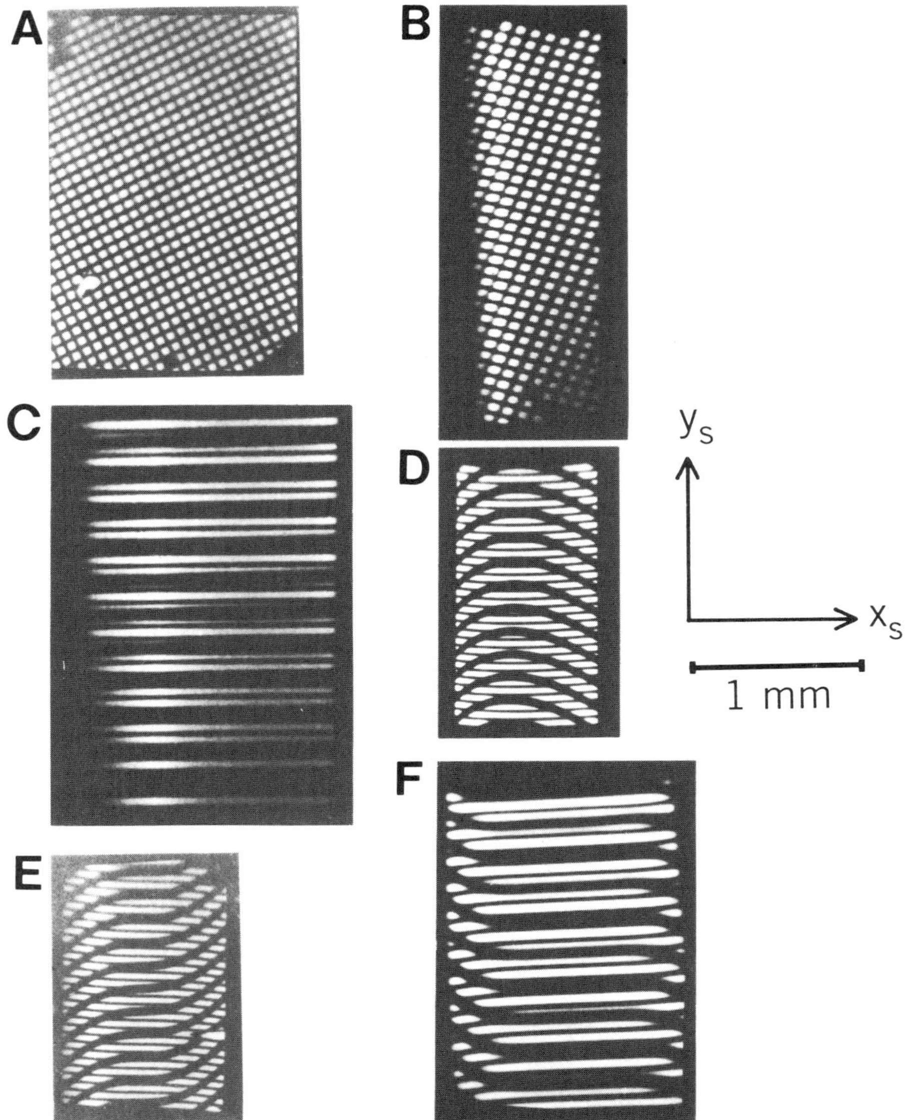

Figure 3.9. Experimental shadow patterns of single quadrupole–single multipole lens systems. These show that the dominant effect of each multipole is to introduce an aberration coefficient of the form (x/θ^n). (a) Shadow of grid with all lenses off. (b) Quadrupole lens on, with focus downstream of grid with dominant coefficient (x/θ). (c) Quadrupole lens in focus on grid, with no aberration visible. (d) Quadrupole with sextupole lens: (x/θ^2). (e) Quadrupole with octupole lens: (x/θ^3). (f) Octupole lens configured as a quadrupole lens, to generate large parasitic duodecapole field components: (x/θ^5). The grid period was 63.5 μm, rotated by $20°$; other parameters were similar to those of Figure 3.7. Reprinted from Ref. 60 with kind permission from Elsevier Science B.V., Amsterdam, The Netherlands.

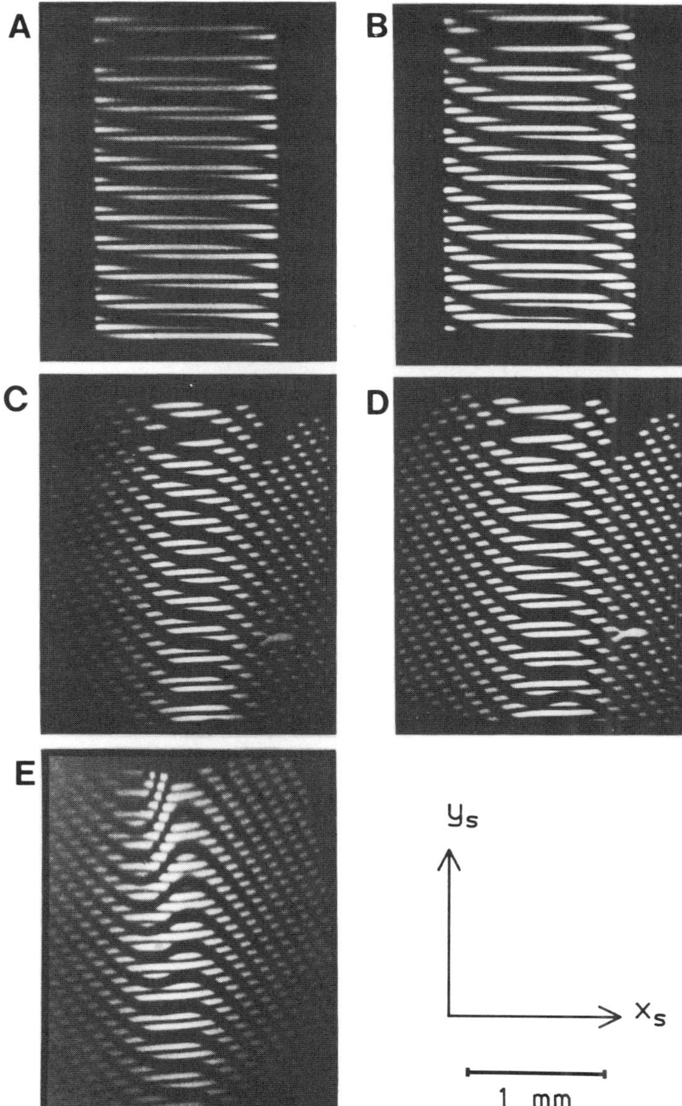

Figure 3.10. Experimental shadow patterns of a single quadrupole–single octupole lens system, showing the effect of increasing the octupole strength. The pattern shape is dominated by the effect of (x/θ^3); but when the octupole field is strong, effects from $(x/\theta\phi^2)$ can also be seen. The parameters were similar to those of Figure 3.8. Reprinted from Ref. 26 with kind permission from Elsevier Science B.V., Amsterdam, The Netherlands.

as calculated by PRAM. They also agree well with the numerical raytracing program OXRAY [6]. It can be concluded that the single octupole lens used for this measurement performed as expected.

3.6.1.8. Single Quadrupole Lenses   The grid shadow patterns of quadrupole singlet systems are more complicated than patterns of the quadrupole–octupole system, because of the small size of the intrinsic and parasitic aberrations of a single quadrupole. This requires the grid shadow method to be employed in its most sensitive configuration. Typically, this requires the use of a 2,000-mesh grid, rotated about the z axis by $1°$ or less. The optimum angle is readily achieved simply by observation of the shadow pattern, and slowly rotating the quadrupole lens itself toward perfect alignment with the grid. When the angle is sufficiently small, the curvature of the shadow pattern will be observed. Of course, it is also possible to rotate the grid itself, but in magnetic quadrupole lens systems, the lenses are usually equipped with fine lens rotation mechanisms for alignment purposes.

It is often found that the exact form of a singlet grid shadow pattern is not reproducible. This is due to two factors. The first is that the parasitic aberration is probably partially introduced by pole-tip field inhomogeneities that are influenced by hysteresis. The second is that random residual fields could be produced by other inactive lenses on the beam line.

It is always preferable to examine each singlet with other lenses removed from the beam line because of the influence of residual fields from the other lenses. Unfortunately, the opportunity to do this in practice is usually not available. The true singlet result can instead be taken to be the average of a series of independent measurements. The effects from the random residual fields will then be reduced.

Some singlet shadow patterns for the Melbourne system are shown in Figure 3.11. These patterns, when compared with the theoretical patterns of Figure 3.8, reveal parasitic sextupole, skew sextupole, and octupole field components.

An interesting set of shadow patterns from the Oxford system [41] is shown in Figure 3.12. The top pattern is from one of the lenses of the Oxford system in use between 1980 and 1988. It shows a curvature, characteristic of parasitic sextupole field contamination, arising from a mechanical defect. The middle pattern is characteristic of the other two lenses in the same system, which shows only high-order duodecapole field components. When the defective lens was replaced with a good spare lens, the best resolution of the probe-forming lens system improved from 1 μm to 0.5 μm at the same probe current. This was an exceptionally good result for an experiment that took only a couple of hours work. The lower pattern in Figure 3.12 is from a new Oxford "Integral" lens, which likewise shows only high-order duodecapole field contamination. Duodecapole components inevitably arise in the fringe field regions of such lenses with square edges. They are not usually a problem in practical probe-forming lens systems, because their effects can be readily minimized by collimation of the beam.

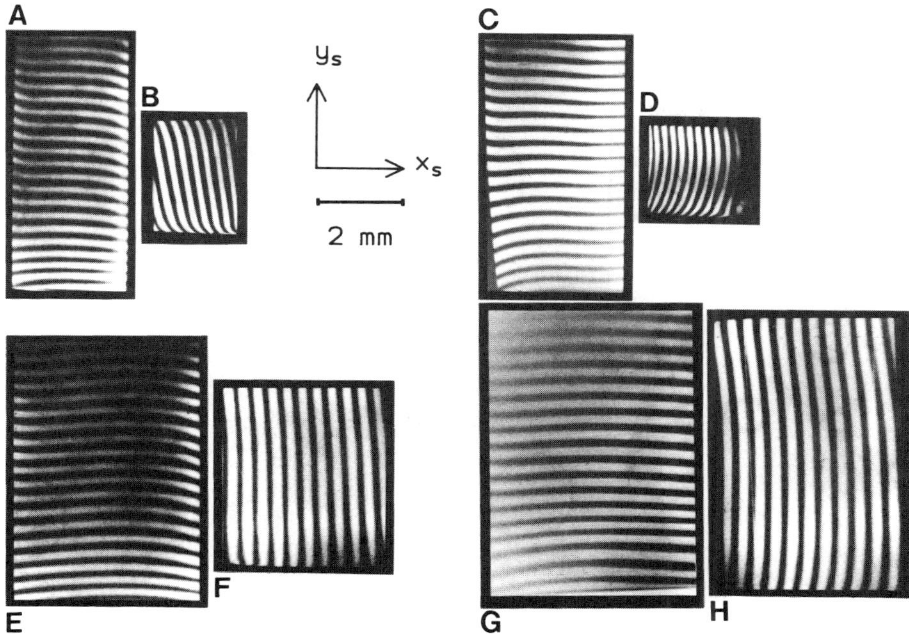

Figure 3.11. Experimental shadow patterns of single quadrupole lens systems from an early version of the Melbourne nuclear microprobe showing the parasitic aberration typical of high-quality quadrupole lenses. The dominant aberration coefficient is either (x/θ^2), $(x/\theta\phi)$, or (x/θ^3). The lens system was originally configured as a Russian anti-symmetric quadruplet (total system length 8.6 m; lenses 1 and 4 length l = 32 mm; lenses 2 and 3 length l = 57 mm; bore radius r_0 = 6.35 mm; drift length between lens poles 34 mm; image distance from last pole 253 mm; grid period 12.7 μm; grid angle approximately 2°). Reprinted from Ref. 61 with kind permission from Elsevier Science B.V., Amsterdam, The Netherlands.

The utility of the grid shadow method to electrostatic and achromatic lenses is shown in Figure 3.13. These patterns were obtained from one of the achromatic quadrupole lenses in the Lund triplet system [48,49]. When the achromatic lens is operated with the magnetic (electrostatic) component alone, the grid shadow patterns reveal considerable parasitic octupole (skew sextupole) field components. These possibly arise from mechanical defects but could also be due to charge build-up on the insulators of the electrostatic lens from beam halo. Despite this, the achromatic lens is remarkably insensitive to the beam energy, as the shadow patterns show.

3.6.1.9. *Reduced Aberration Coefficients*

After quantitative analysis of these patterns, it is convenient to convert the measured aberration coefficient into an effective parasitic multipole field component. This can be accomplished

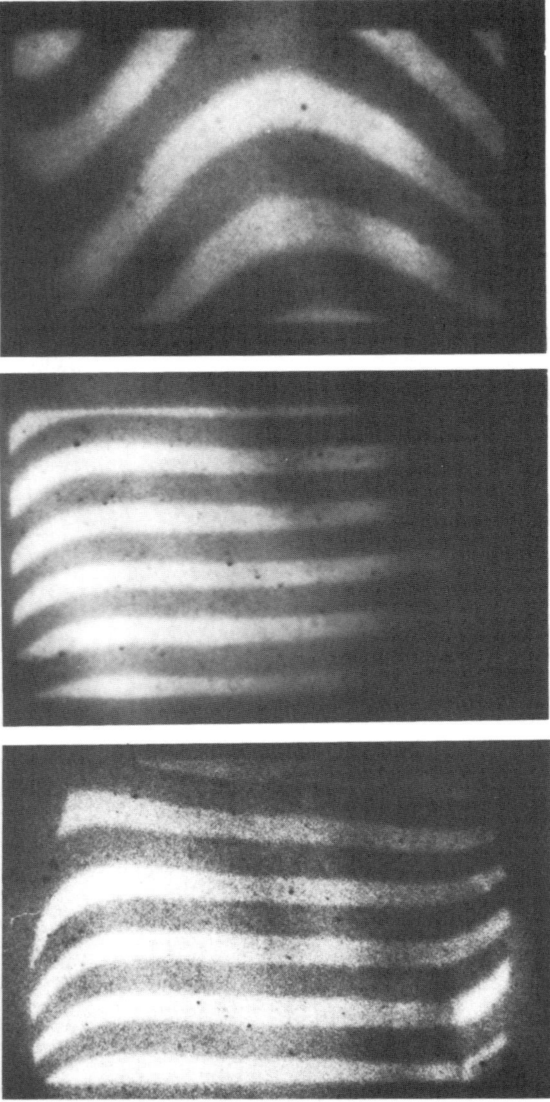

Figure 3.12. Experimental shadow patterns of single quadrupole lenses of the Oxford high excitation triplet. Top: A defective quadrupole lens showing large parasitic sextupole field contamination. Middle: An otherwise identical lens showing just duodecapole field components that probably arises from the fringe field regions. Bottom: A new integral lens, again showing duodecapole field components alone. Reprinted from Ref. 41 with kind permission from Elsevier Science B.V., Amsterdam, The Netherlands.

by use of calculated reduced-aberration coefficients that relate the width of a quadrupole line focus to the strength of any parasitic multipole fields.

For the principal parasitic terms, Eq. (3.29a) may be rewritten:

$$\Delta x_i =$$

Parasitic sextupole	$(x/\theta^2 B_S)\theta_o^2 B_S + (x/\phi^2 B_S)\phi_o^2 B_S$	
Parasitic skew sextupole	$+(x/\theta\phi B_{Sk})\theta_o\phi_o B_{Sk}$	(3.43)
Parasitic octupole	$+(x/\theta^3 B_O)\theta_o^3 B_O$	
Parasitic duodecapole	$+(x/\theta^5 B_D)\theta_o^5 B_D,$	

where $(x/\theta^2 B_S)$, $(x/\theta\phi B_{Sk})$, $(x/\theta^3 B_O)$, and $(x/\theta^5 B_D)$ are the reduced parasitic multipole aberration coefficients, and B_S, B_{Sk}, B_O, and B_D are the strengths of the parasitic multipole components in the quadrupole fields.

These reduced coefficients are determined by the geometric parameters of the singlet systems, using the theory discussed in Sections 3.2 and 3.5. They can be used to calculate the field strength of the parasitic sextupole and octupole components of the singlet system from the measured values of the parasitic aberration coefficients:

$$B_S = \frac{(x/\theta^2)}{(x/\theta^2 B_S)}$$

$$B_O = \frac{(x/\theta^3) - (x/\theta^3)_{\text{int}}}{(x/\theta^3 B_O)}$$

The percentage parasitic field component may then be calculated from the theoretical pole tip fields of the quadrupoles calculated from simple first-order matrix theory.

Representative values for the reduced aberration coefficients for the Melbourne system appear in Table 3.3. Using these coefficients and the measured values of (x/θ^n), it was shown that the quadrupole lenses suffer from parasitic sextupole

field contaminants of 0.05% to 0.3% and parasitic octupole field contaminants of 0.02% to 0.4% of the main quadrupole field [26]. These values should be regarded as the upper limit acceptable in a high-quality probe-forming lens system that is capable of a 1 μm probe, with sufficient beam current for analytical applications (~100 pA).

3.6.2. Multiple Quadrupole Systems

As was the case for the single quadrupole lens systems, the grid shadow method is applied to study the aberrations of the probe-forming lens system itself by observation of the shadow pattern of a grid placed at or near the Gaussian image plane. If the method is to be useful, then several grid apertures must be

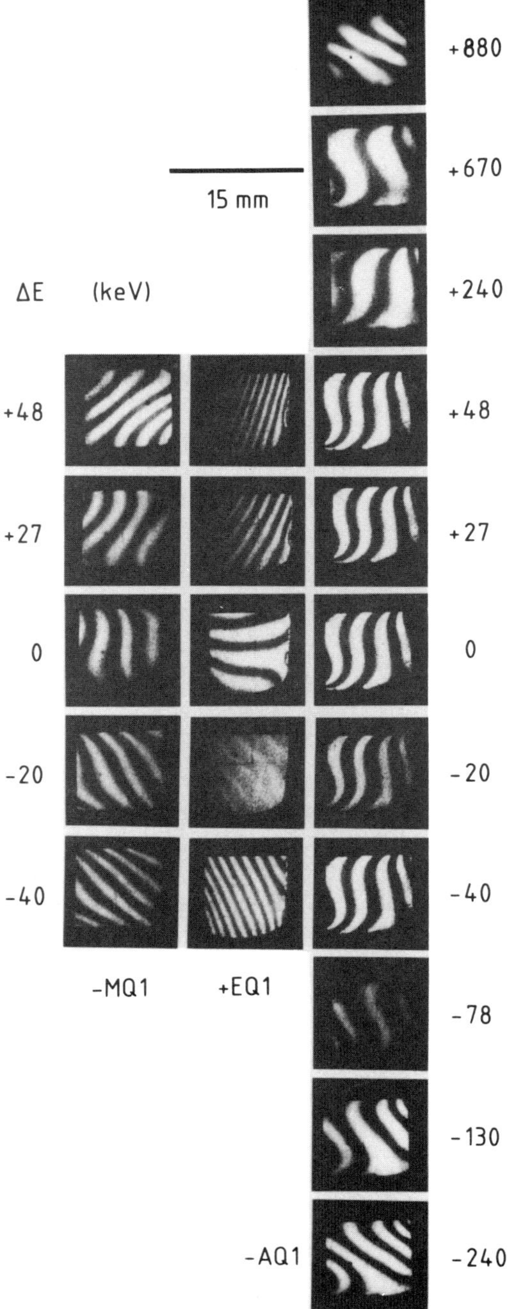

15 mm
ΔE   (keV)
+880
+670
+240
+48
+27
0
-20
-40
-78
-130
-240
+48
+27
0
-20
-40
-MQ1
+EQ1
-AQ1

TABLE 3.3. Reduced Parasitic Aberration Coefficients of the Four Quadrupole Lens Singlet Systems of the Melbourne System

Coefficient	Quadrupole			
	1	2	3	4
$(x/\theta^2 B_S)$	-109	-112	-114	-114
$(x/\theta\phi B_{Sk})$	219	224	228	228
$(x/\theta^3 B_O)$	132	130	131	135

Units: x in m; θ in rad; B_S, B_{Sk} and B_O in % of B_Q.

shadowed by the aberrated image. The number of grid apertures shadowed can always be increased by moving the grid away from the Gaussian image plane at the cost of reduced sensitivity of the shadow pattern to the image aberrations. It is useful to have at least four grid apertures shadowed, in which case the grid period should be given by

$$P \le \tfrac{1}{4}[(x/\theta^3)(x/\theta\phi^2)(y/\theta^2\phi)(y/\phi^3)]^{1/4}\theta^3_{\max} \tag{3.44}$$

where P is the period of the grid, $[(x/\theta^3)(x/\theta\phi^2)(y/\theta^2\phi)(y/\phi^3)]^{1/4}$ is the geometric mean spherical aberration coefficient, and $\theta_{\max}$ is the maximum initial divergence of the beam. Equation (3.44) assumes that there is no parasitic aberration. In most practical systems, a 2,000-mesh grid is satisfactory for the grid shadow method.

The width of the penumbral shadow of a grid bar must also be kept narrower than the width of the shadow of a grid aperture, by ensuring the object collimator size satisfies

$$t_{ox} \le \left| \frac{P}{2M_x} \right| \tag{3.45}$$

Figure 3.13. Experimental shadow patterns of a Lund achromatic quadrupole lens, consisting of superimposed electrostatic and magnetic fields. Patterns are shown, as a function of beam energy departure from optimum, for the magnetic component, the electrostatic component, and the entire lens in the achromatic configuration. The shape of the shadow pattern is dominated by parasitic aberration, but nevertheless the focus of the lens system in the achromatic configuration is remarkably insensitive to the energy of the ion beam. Reprinted from Ref. 49 with kind permission from Elsevier Science B.V., Amsterdam, The Netherlands.

where t_{ox} is the object radius or half-width and M_x is the magnification in the xoz plane, and similarly in the yoz plane.

To provide an idea of the grid shadow patterns characteristic of typical aberrations in a probe-forming lens system, theoretical shadow patterns calculated by the Monte Carlo method described earlier are shown in Figures 3.14 and 3.15.

Shadow patterns shown in Figure 3.14 are for the following cases: an ideal system, a system with spherical aberration, a system with spherical and chro-

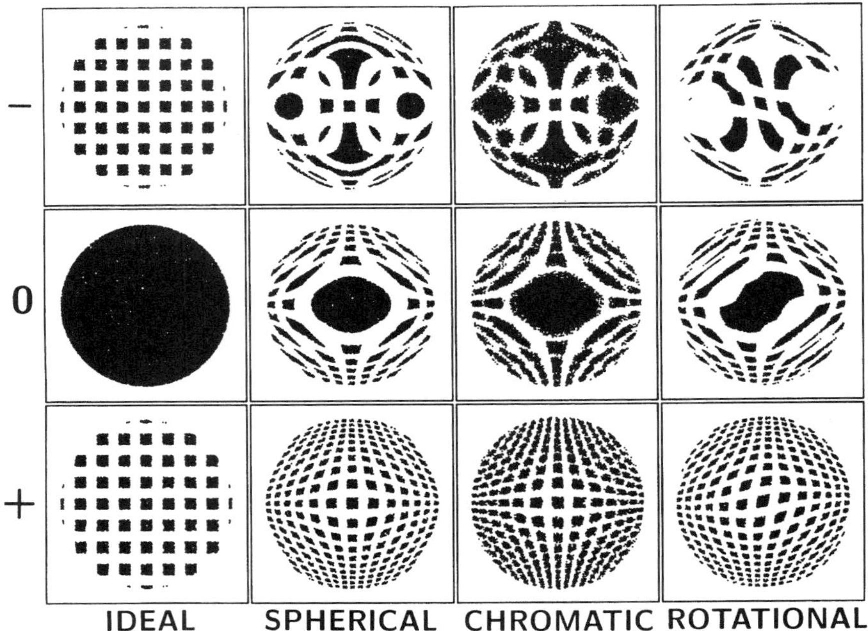

Figure 3.14. Theoretical grid shadow patterns of a Russian antisymmetric quadrupole lens quadruplet showing a variety of effects. The top row (−) shows the effect of the grid being displaced upstream of the Gaussian image plane; the middle row (0) of the grid being in the Gaussian image plane; and the bottom row (+) of the grid being displaced downstream from the image plane. The first column shows the patterns for a perfect lens system that suffers from no aberration. The second column shows a system that suffers only from the intrinsic spherical aberration. The third column shows the effect of up to 0.1% chromatic momentum error, randomly distributed in the incident rays; notice the blurring of the pattern toward the edges. The fourth column shows the effect of a rotational misalignment of 0.3 mrad (a rotational displacement of only 30 μm on the edge of a 200 mm diameter yoke!) in the second quadrupole lens of the quadruplet. The other parameters were as follows: object drift space 8.12 m; lenses 1 and 4 effective length $L_Q = 38$ mm; lenses 2 and 3 effective length $L_Q = 67$ mm; drift between lenses 25 mm; image drift 146 mm; $D = 200$ mm; grid period 12.7 μm.

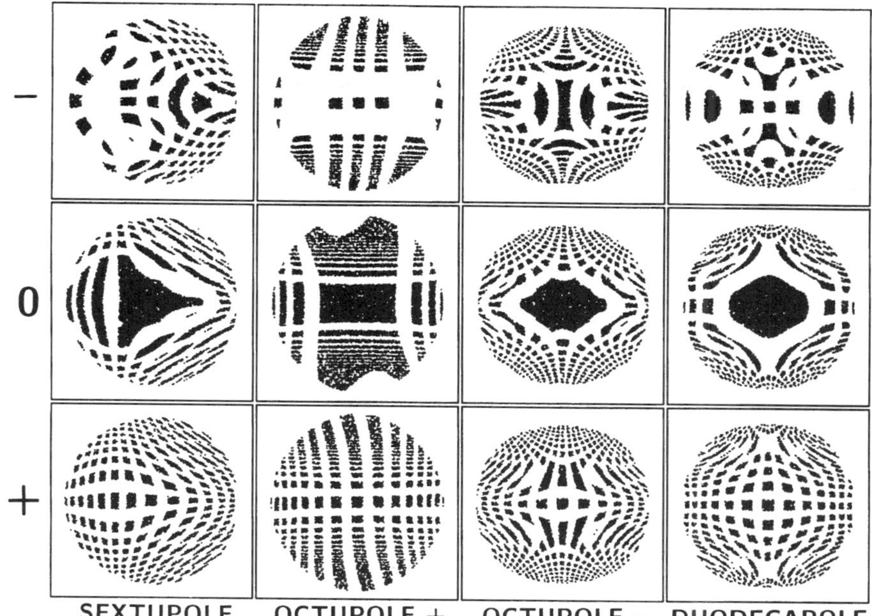

Figure 3.15. Theoretical grid shadow patterns of a Russian antisymmetric quadrupole quadruplet, showing the effect of parasitic sextupole, octupole, and duodecapole field components in lens number 2. The rows show the effects of changing the grid position, as with Figure 3.14. The first column is for a parasitic sextupole field of strength 0.25%. The second and third columns are for parasitic octupole fields of strengths +0.5% and −0.5%, respectively. The effect of a parasitic octupole is more complicated than for the other multipoles, because it can partially correct, or reinforce, the effect of the intrinsic spherical aberration. The fourth column is for a parasitic duodecapole field of strength 10%. Other parameters are the same as for Figure 3.14.

matic aberration, and a system with spherical aberration and a single lens rotational misalignment. For a grid located in the Gaussian image plane of an ideal system suffering from no aberration whatsoever, all rays can pass through a grid aperture, and, hence, the shadow pattern is a featureless disc. A small displacement of the grid, upstream or downstream, produces straight lines in the shadow pattern; spherical aberration, calculated with the equations in Section 3.5, and intrinsic to a real system, produces the characteristic curvature.

The chromatic aberration produces an effect similar to that of a large object collimator that causes a wide penumbra. However, the effect becomes more pronounced toward the edges of the pattern as it becomes increasingly blurred.

The effects of parasitic multipoles are shown in Figure 3.15. A parasitic sextupole produces the characteristic asymmetrical pattern. The effect of a parasitic octupole field is more complicated, because it can partically correct, or enhance, the effect of the intrinsic spherical aberration of the system.

In the theoretical patterns, it can be seen that the pattern for the grid located upstream of the Gaussian image plane is much more complicated than the corresponding pattern for the grid located downstream of the image plane. This is particularly true if the grid is located between the Gaussian image plane and the circle of least confusion. These positions should be avoided in an experimental situation.

Experimental patterns for the Melbourne system are shown in Figure 3.16. The combined effect of the intrinsic spherical aberration of the system, together with the parasitic aberration of each lens, is responsible for the appearance of the patterns. The patterns show that the spherical aberration, together with the effects from the parasitic octupole field components, dominate. The effects from the parasitic sextupole field components are relatively small.

A similar set of patterns is shown in Figure 3.17 for the Sydney electrostatic quadruplet. Once again, it can be seen that parasitic octupole field components, along with spherical aberration, dominate the patterns.

3.6.2.1. *Extracting the Spherical Aberration Coefficients*   If spherical aberration is the dominant aberration in the system, or there are parasitic octupole components, then it is possible to apply the method of Deltrap [29] and Hardy [30] to extract spherical aberration coefficients from the grid shadow pattern. This method is now briefly outlined.

If the shadow plane is located some way downstream from the Gaussian image plane, then it is possible to simplify the analysis of the shadow pattern. This is because it is possible to neglect the magnitude of the aberration in the Gaussian image plane compared with the displacement of a ray in the shadow plane. This is only possible if

$$D \gg \left| \frac{(x/\theta^3)}{(\theta/\theta)} \theta_o^2 \right| + \left| \frac{(x/\theta)}{(\theta/\theta)} \right| \tag{3.46}$$

where D is the distance from the grid to the shadow plane, $(x/\theta)/(\theta/\theta)$ is the distance from the grid to the Gaussian image plane, (x/θ^3) is the principal spherical aberration coefficient, and (θ/θ) is the demagnification. In this case it is possible to write

$$\frac{x_i}{x_s} = \left| \frac{(x/\theta^3)}{D^3(\theta/\theta^3)} x_s^2 + \frac{(x/\theta)}{D(\theta/\theta)} + \frac{(x/\theta\phi^2)}{D^3(\theta/\theta)(\phi/\phi)^2} y_s^2 \right| \tag{3.47a}$$

$$\frac{y_i}{y_s} = \left| \frac{(y/\phi^3)}{D^3(\phi/\phi)^3} y_s^2 + \frac{(y/\phi)}{D(\phi/\phi)} + \frac{(y/\theta^2\phi)}{D^3(\theta/\theta)^2(\phi/\phi)} x_s^2 \right| \tag{3.47b}$$

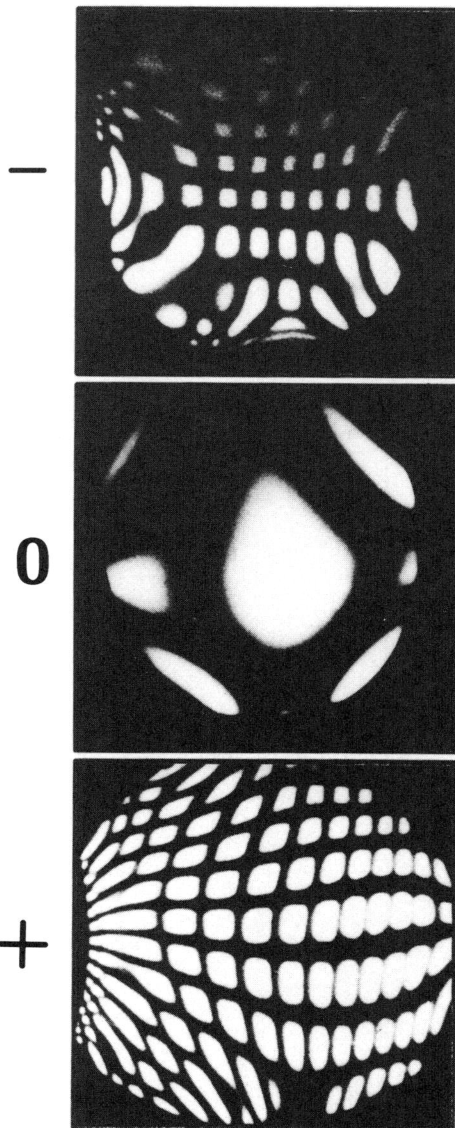

Figure 3.16. Experimental shadow patterns of an early version of the Melbourne Russian antisymmetric quadruplet, that show the combined effect of spherical aberration and parasitic octupole or duodecapole field components. Parameters are similar to those of Figure 3.14. Adapted from Ref. 61 with kind permission from Elsevier Science B.V., Amsterdam, The Metherlands.

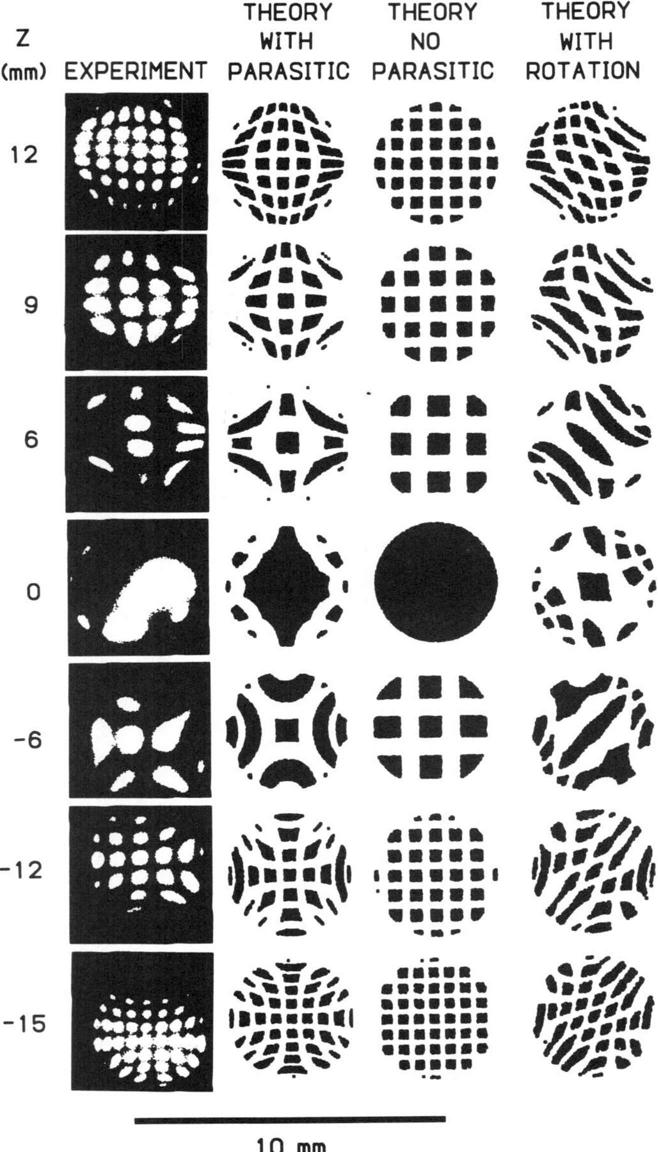

Figure 3.17. Experimental shadow patterns of the Sydney electrostatic Russian anti-symmetric lens system showing the effect of residual parasitic octupole field components. Reprinted from Ref. 62 with kind permission from Elsevier Science B.V., Amsterdam, The Netherlands.

If the grid shadow pattern is regular and not too complicated, it is possible to identify a point on the grid that transforms to a particular point on the shadow pattern. In fact, a line of constant y_s could be drawn on the grid shadow pattern, and all the x_s points at which it crosses the center of the grid bar shadows identified with the corresponding x_i points obtained from the actual coordinates on the grid bars themselves. Then, a graph of $Y = x_i/x_s$ as a function of $X = x_s^2$ has gradient $(x/\theta^3)/[D^3(\theta/\theta)^3]$, with Y intercept

$$Y'_x = \frac{(x/\theta)}{D(\theta/\theta)} + \frac{(x/\theta\phi^2)}{D^3(\theta/\theta)(\phi/\phi)^2}\, y_s^2.$$

If lines of constant y_s are drawn for several different y_s values, then the resulting Y intercepts, Y'_x, plotted as a function of $X'_x = y_s^2$, have Y' intercept $(x/\theta)/(D(\theta/\theta))$, with gradient $(x/\theta\phi^2)/(D^3(\theta/\theta)(\phi/\phi)^2)$. Hence (x/θ^3), $(x/\theta\phi^2)$, and (x/θ) can be measured. The process can be repeated for lines of constant x_s, to measure (y/ϕ^3), $(y/\theta^2\phi)$, and (y/ϕ).

A problem with this method is that it is important to identify the optical axis accurately; otherwise, a term in x_s^{-1} appears in Eq. (3.47a), and a term in y_s^{-1} appears in Eq. (3.47b). These terms make the plots of $Y = x_i/x_s$ (or y_i/y_s) as a function of $X = x_s^2$ (or y_s^2) nonlinear, and, therefore, introduce inaccuracies in the measurement of the spherical aberration coefficients.

A further effect that could make the graphs of Y as a function of X nonlinear is parasitic aberration. So it is essential that the individual lenses are tested first, and defective lenses replaced, before using the grid shadow method on the whole system.

An application of the grid shadow method is illustrated in Figures 3.18 and 3.19. The experimental patterns were obtained from a version of the Melbourne system in which a single octupole lens was used to correct the spherical aberration cross-terms, along with the superimposed cross-terms contributed by parasitic octupole field components. The measured coefficients, listed in Table 3.4, show that the cross-terms were successfully corrected.

3.6.3. Other Applications of the Grid Shadow Method

Recent measurements on both the Melbourne and Shanghai systems, which use identical lens systems, show that the parasitic sextupole field contamination is less than 0.3% of the quadrupole pole-tip field [50]. This level of sextupole field contamination was also seen, using the grid shadow method, in the lenses of the magnetic quadrupole doublet at Eindhoven University [51]. The method has also been used as an aid to precise focusing of the probe, and in identifying the residual parasitic sextupole components and lens rotational misalignments in a magnetic quadrupole doublet [52].

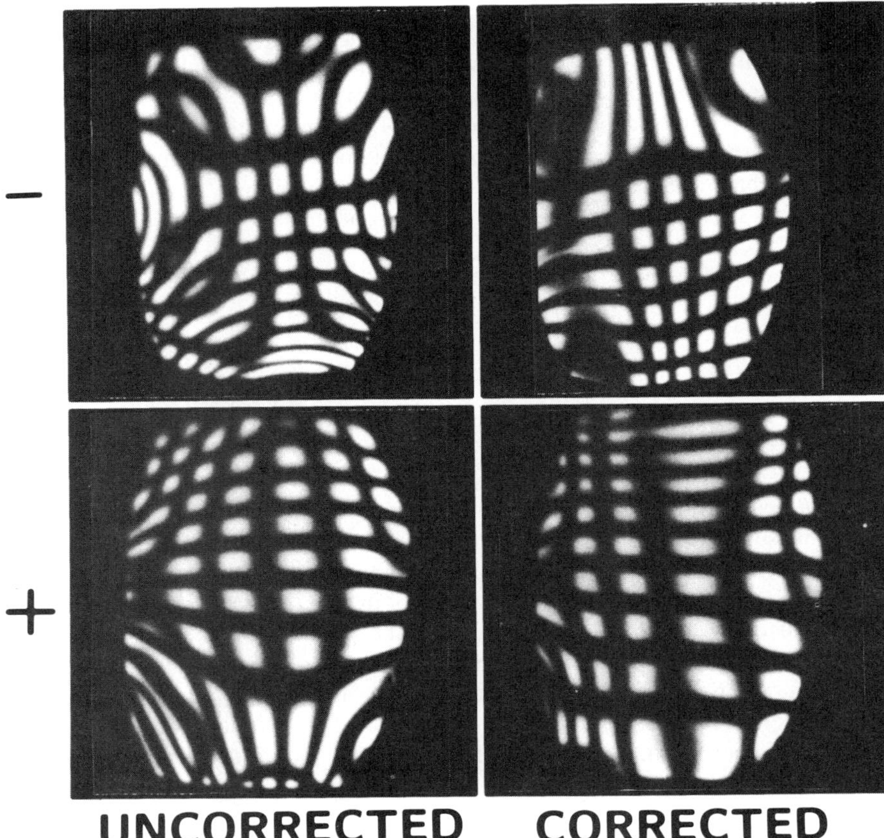

Figure 3.18. Experimental shadow patterns of a version of the Melbourne system, with parameters similar to those of Figure 3.14, except that an octupole lens of length $l = 10$ mm was inserted between the center two quadrupole lenses. Shadow patterns with the grid upstream $(-)$ and downstream $(+)$ of the Gaussian image plane. Left column: All octupole lenses off. Right column: The octupole was used to correct the spherical aberration cross-terms. Reprinted from Ref. 61 with kind permission from Elsevier Science B.V., Amsterdam, The Netherlands.

Extraneous, inhomogeneous, magnetic, and electrostatic fields can permeate a microprobe beamline and degrade resolution. These can often be identified from a grid shadow pattern, because such inhomogeneous fields do not usually produce the patterns typical of the parasitic multipole components in a quadrupole lens. For example, this method has been used to show that an ion pump can significantly perturb the beam in a UHV microprobe system, requiring the introduction of magnetic shielding [53]. The severely distorted grid shadow patterns, shown in Figure 3.20, were a certain indication of this problem.

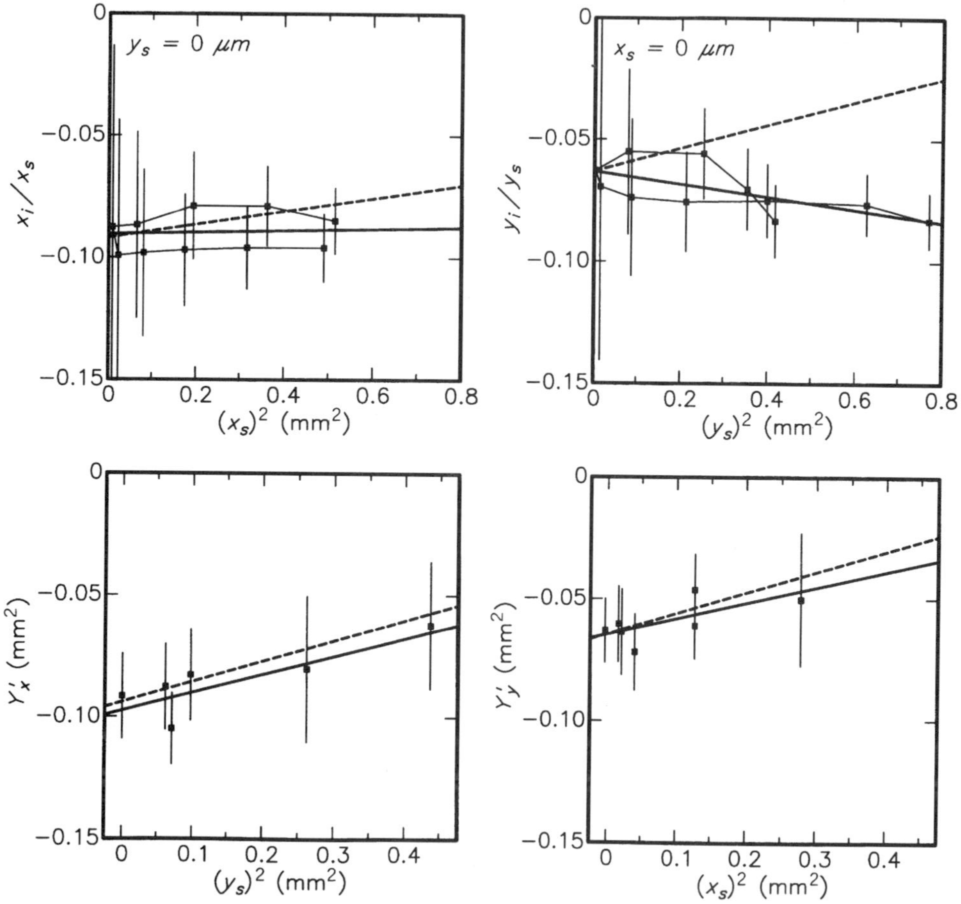

Figure 3.19. Measurement of the third-order aberration coefficients from the shadow patterns in Figure 3.18, using the method described in the text. The dashed line shows the result for the intrinsic spherical aberration alone. Adapted from Ref. 61 with kind permission from Elsevier Science B.V., Amsterdam, The Netherlands.

TABLE 3.4. Experimental and Theoretical Spherical Aberration Coefficients the Uncorrected (All Octupoles Off) and Corrected (a Single Octupole Used to Correct the Spherical Aberration Cross-Terms) System[a]

System	(x/θ^3)	$(x/\theta\phi^2)$	$(y/\theta^2\phi)$	(y/ϕ^3)
Octupoles off				
Experiment	-71 ± 45	-300 ± 52	-210 ± 130	36 ± 19
Theory	-104	-322	-322	-199
Octupole 2 on				
Experiment	-210	24	-70	-204
Theory	-180	0	0	-345

[a]Parasitic aberration prevented accurate measurement of the spherical aberration coefficients. The experimental errors reflect the reproducibility of the measurement. The errors on the experimental coefficients for the system with octupole 2 on can be assumed to be the same in magnitude as those for the octupole off all units μm/mrad^3.

SOURCE. Reproduced from Ref. 61 with the kind permission of Elsevier Science B.V., Amsterdam, The Netherlands.

In the Sydney system, an ungrounded aperture located upstream of the lens system itself was found to become charged, due to beam halo. The resulting electric field generated a parasitic skew quadrupole field that was readily seen from the shadow patterns, as shown in Figure 3.21. Deliberate biasing of the offending aperture identified the problem. The problem was cured by grounding the aperture.

3.7. FURTHER CONSIDERATIONS

Although the probe-forming lens system must be of the best possible quality, minimization of mechanical vibration and stray magnetic fields is essential for high-resolution probes. It has been reported that the limitations to decreasing the probe size in the Oxford system are mechanical vibrations and stray fields, which limit the effective probe resolution to no smaller than 0.2 μm [54]. In Melbourne, this limit has been reduced to less than 100 nm [55] for some applications, because of the use of ion pumps containing no moving parts, as well as magnetic shielding.

The present state of the art for a 100 pA beam, the presupposed minimum necessary for PIXE or RBS, is 0.33 μm [56]. This was achieved in Oxford after careful optimization of the accelerator and the microprobe beam line itself. Elimination of the stray a.c. fields and vibration would probably allow this figure to be significantly improved. For a discussion of further possible future paths to improvement, which mainly involve improvements to the brightness of the ion source, see Legge et al. [57].

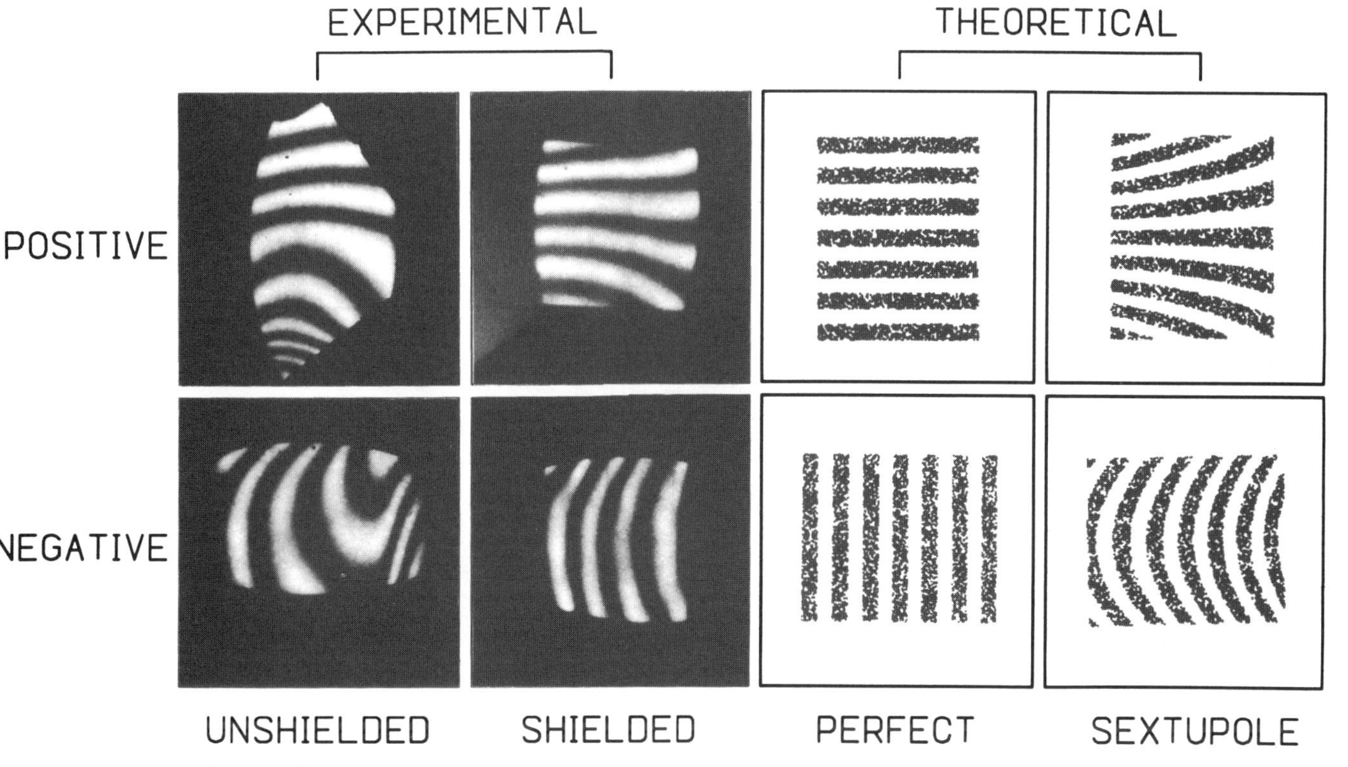

Figure 3.20. Experimental shadow patterns of a single-quadrupole system, in which a large stray magnetic field from an unshielded ion pump was present. When a magnetic shield was installed, the shadow patterns showed the sextupole and skew-sextupole field contamination from the quadrupole lens alone. Reprinted from Ref. 53 with kind permission from Elsevier Science B.V., Amsterdam, The Netherlands.

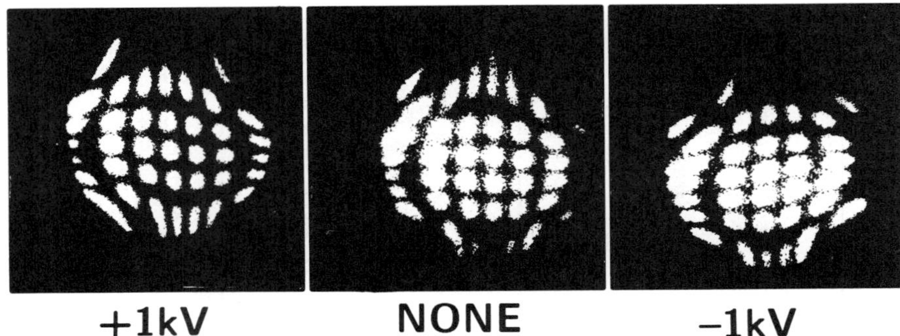

Figure 3.21. Experimental shadow patterns from the Sydney quadruplet, showing the effect of stray electric charge on an isolated collimator located upstream of the lens system. The charge distribution has produced an effect equivalent to a skew quadrupole field.

REFERENCES

1. C.J. Maggiore, *Scan Electr Microsc.* **1**:439 (1980).

2. C.J. Maggiore, *Nucl. Instr. Meth.* **191**:199 (1981).

3. H. Koyama-Ito and L. Grodzins, *Nucl. Instr. Meth* **174**:331 (1980).

4. J. Meijer, A. Stephan, H.H. Bukow, C. Rolfs, and F. Bruhn. *Nucl. Instr. Meth.* **B104**:77 (1995).

5. M.R. Dix, "A study of the ion-optics of a superconducting solenoid lens." Honours thesis, School of Physics, The University of Melbourne, (1983).

6. G.W. Grime and F. Watt, *Beam Optics of Quadrupole Probe-forming Systems.* Adam Hilger, Bristol, 1984.

7. A.D. Dymnikov, D.N. Jamieson, and G.J.F. Legge, *Nucl. Instr. Meth.* **B104**:64 (1995).

8. V.A. Brazhnik, A.D. Dymnikov, D.N. Jamieson, S.A. Lebed, G.J.F. Legge, A.G. Ponomarev, and V.E. Storizhko. *Nucl. Instr. Meth.* **B104**:92 (1995).

9. P.W. Hawkes, *Quadrupole Optics* (Springer Tracts in Modern Optics, vol. **42**). Springer, Berlin, 1966.

10. P.W. Hawkes, *Quadrupoles in Electron Lens Design.* Academic Press, New York, 1970.

11. K.G. Steffen, *High Energy Beam Optics* (Monographs and Texts in Physics and Astronomy, vol. **17**) Interscience, New York, 1965.

12. P. Grivet, *Electron Optics.* Pergamon Press, New York, 1965.

13. A.P. Banford, *The Transport of Charged Particle Beams.* E. & F. N. Spon, London, 1966.

14. A. Septier, ed., *Focusing of Charged Particles*, Vols. I and II. Academic Press, London, 1967.

15. A. Septier, ed., *Applied Charged Particle Optics* (Advances in Electronics and Electron Physics, Supplement 13A). Academic Press, London, 1980.

16. A.B. El Kareh and J.C.J. El Kareh, *Electron Beams, Lenses and Optics*, Vol. II. Academic Press, New York, 1970.

17. M. Szilagyi, *Electron and Ion Optics*. Plenum Press, New York, 1988.

18. H. Wollnik, *Optics of Charged Particles*. Academic Press, London, 1987.

19. K.L. Brown, A First- and Second-Order Matrix Theory for the Design of Beam Transport Systems and Charged Particle Spectrometers, SLAC 75 (Stanford Linear Accelerator Center, 1967).

20. H. Wollnik, *Nucl. Instr. Meth.* **103**:479 (1972).

21. H. Nakabushi and T. Matsuo, *Nucl. Instr. Meth.* **198**:207 (1982).

22. P. Grivet and A. Septier, *Nucl. Instr. Meth.* **6**:126, 243 (1960).

23. J.A. Cookson, *Nucl. Instr. Meth.* **165**:477 (1979).

24. G.W. Grime, F. Watt, G.D. Blower, J. Takacs, and D.N. Jamieson, *Nucl. Instr. Meth.* **197**:97 (1982).

25. K.L. Brown, D.C. Carey, C. Iselin, and F. Rothacker, *TRANSPORT, a computer program for designing charged particle beam transport systems*, CERN 80-04, Geneva (1980).

26. D.N. Jamieson and G.J.F. Legge, *Nucl. Instr. Meth. B* **29**:544 (1987).

27. A.D. Dymnikov, T.Y. Fishkova, and S.Y. Yavor, *Nucl. Instr. Meth.* **37**:268 (1965).

28. D. Heck, Kernforschungszentrum, Karlsruhe Annual Report KFK2379, pp. 108, 130 (1976).

29. J.H.M. Deltrap, PhD. Thesis, Cambridge University (1964).

30. D.F. Hardy, PhD. Thesis, Cambridge University (1967).

31. G.R. Moloney, Honours thesis, School of Physics, The University of Melbourne (1986).

32. M.B.H. Breese and J.A. Cookson, *Nucl. Instr. Meth. B* **61**:343 (1991).

33. A. Foster, PhD. Thesis, University of London (1968); referred to in Ref. 10, p. 66.

34. G. Parzen, Magnetic fields for transporting charged particle beams, BNL 50536, ISA 76–13 (1976).

35. S.Y. Yavor, T.Y. Fishkova, E.V. Shpak, and L.A. Baranova, *Nucl. Instr. Meth.* **76**:181 (1969).

36. S.Y. Yavor, L.P. Ovsyannikova, and L.A. Baranova, *Nucl. Instr. Meth.* **99**:103 (1972).

37. G.R. Moloney, D.N. Jamieson, and G.J.F. Legge, *Nucl. Instr. Meth. B* **54**:24 (1991).

38. J. Cobb and R. Cole, Proceedings of the International Symposium on Magnetic Technology, Stanford, CA (1965), p. 431.

39. M.B.H. Breese, D.N. Jamieson, and J.A. Cookson, *Nucl. Instr. Meth. B* **54**:28 (1991).

40. F.W. Martin, *Nucl. Instr. Meth. B* **54**:17 (1991).

41. D.N. Jamieson, G.W. Grime, and F. Watt, *Nucl. Instr. Meth. B* **40/41**:669 (1989).

42. M.B.H. Breese and D.N. Jamieson, *Nucl. Instr. Meth.* **B104**:81 (1995).

43. P.M. Read, J.A. Cookson, and G.D. Alton, *Nucl. Instr. Meth. B* **24/25**:627 (1987).

44. A.D. Dymnikov, T.Y. Fishkova, and S.Y. Yavor, *Radio Eng. Elect. Phys. USA* **9**:1515 (1964).

45. S. Okayama and H. Kawakatsu, *J. Phys. E* **16**:166 (1983).

46. M.B.H. Breese, D.N. Jamieson, and J.A. Cookson, *Nucl. Instr. Meth. B* **47**:443 (1990).

47. M.B.H. Breese and D.N. Jamieson, *Nucl. Instr. Meth. B* **83**:394 (1993).

48. U.A.S. Tapper and B.R. Nielsen, *Nucl. Instr. Meth. B* **44**:219 (1989).

49. D.N. Jamieson and U.A.S. Tapper, *Nucl. Instr. Meth. B* **44**:227 (1989).

50. D.N. Jamieson, J. Zhu, P. Mao, and R. Lu, *Nucl. Instr. Meth.* **B104**:86 (1995).

51. J.A. van der Heide, R.J.L.J. de Regt, W.A.M. Gudden, P. Magendans, H.L. Hagedoorn, P.H.A. Mutsaers, A.V.G. Mangnus, A.J.R. Aendenroomer, L.C. de Folter, and M.J.A. de Voigt, *Nucl. Instr. Meth. B* **64**:336 (1992).

52. E. Swietlicki, N.E.G. Lövestam, and U. Wätjen, *Nucl. Instr. Meth. B* **61**:230 (1991).

53. U.A.S. Tapper, D.N. Jamieson, E. Swietlicki, N.E.G. Lövestam, and T. Hansson, *Nucl. Instr. Meth. B* **62**:155 (1991).

54. M.B.H. Breese, P.J.C. King, G.W. Grime, and F. Watt, *Inst. Phys. Conf. Ser.* **117**:101 (1991).

55. G.S. Bench and G.J.F. Legge, *Nucl. Instr. Meth. B* **40/41**:655 (1989).

56. G.W. Grime and F. Watt, *Nucl. Instr. Meth. B* **75**:495 (1993).

57. G.J.F. Legge, J.S. Laird, L.M. Mason, A. Saint, M. Cholewa, and D. N. Jamieson, *Nucl. Instr. Meth. B* **77**:153 (1993).

58. P. Davies, Honours Thesis, School of Physics, The University of Melbourne (1975).

59. G.J.F. Legge, D.N. Jamieson, P.M.J. O'Brien, and A.P. Mazzolini, *Nucl. Instr. Meth.* **197**:85 (1982).

60. D.N. Jamieson and G.J.F. Legge, *Nucl. Instr. Meth. B* **30**:235 (1988).

61. D.N. Jamieson and G.J.F. Legge, *Nucl. Instr. Meth. B* **34**:411 (1988).

62. D.N. Jamieson, C.G. Ryan, and S.H. Sie, *Nucl. Instr. Meth. B* **54**:33 (1991).

4

ANALYTICAL TECHNIQUES

This chapter first describes particle induced X–ray emission (PIXE), backscattering spectrometry, nuclear reaction analysis (NRA), and elastic recoil detection analysis (ERDA) for the detection and quantification of the elemental composition of materials. These are compared with other techniques available for elemental analysis. Several books cover the theory and practice of techniques with unfocused ion beams in great detail [1–9], and other good reference sources are the conference proceedings listed in Appendix 1. Focused ion beam applications of these methods using the nuclear microprobe are also described in many of the same conference proceedings, as well as Refs. 8–11. Aspects particularly relevant to using these methods for elemental analysis with a nuclear microprobe from a variety of research fields are described in this chapter.

The measured signals from these four methods rely on interactions between an ion and the inner-shell electrons or atomic nuclei of the sample to generate the measured signals. These processes occur infrequently, so these methods consequently need a large beam current for analysis, and an important criterion is the probability per incident ion, or cross-section, for producing a particular analytical signal, as introduced in Chapters 1 and 2. The larger the signal cross-section, the less beam current is needed to make a measurement in a given time. This consideration is important for nuclear microprobe analysis because the spatial resolution of the focused beam spot limits the amount of beam current (described in Chapter 2). The cross-sections for PIXE are generally the highest of these four techniques and 100 pA of beam current is usually adequate for analysis. Typically, NRA and ERDA have the lowest cross-sections and so require several nanoamps of beam current.

Other nuclear microprobe techniques, which do not necessarily give elemental information, are then described. These methods are ion induced electron imaging, scanning transmission ion microscopy (STIM), ion microtomography (IMT), and ion beam induced luminescence (IBIL). These methods rely on using a focused beam, because they are most useful in giving spatially resolved information. Every incident ion can produce a measurable signal with these methods; therefore, a much smaller beam current of 1 fA to 10 pA is usually adequate for analysis, enabling a reduction in beam spot size to approximately 100 nm on the sample surface. Ion beam induced charge (IBIC) microscopy is discussed separately in Chapter 6. Ion channeling, a process that can be used in conjunction with the above analytical methods for the analysis of crystalline material, is described in Chapter 5.

4.1. PARTICLE INDUCED X–RAY EMISSION

Particle induced X–ray emission analysis detects X–rays generated in the sample by MeV ions. This is the most commonly used nuclear microprobe technique and has been widely applied to trace element analysis in the biomedical and geological fields [11]. Accounts of the theory and applications of PIXE using focused and unfocused MeV ion beams are given in Refs. 1, 8, 9, 11–16, and materials applications in Ref. 11.

A vacancy is created in the inner electron shells of an atom if energy greater than the electron-binding energy, which is typically several keV, is supplied. There are many ways of exciting X–rays with different forms of incident energy, such as bombardment with X–rays to give X–ray fluorescence [17]. The most common analytical method, however, uses a keV electron beam in an electron microprobe [18,19]. MeV light ions and keV electrons have a high cross-section for ejecting K, L, or M shell electrons because their velocity approaches the inner shell electron velocity [12,20]. An inner shell vacancy exists for about 10^{-17} s before being filled by an electron transition from an outer shell with subsequent emission of either an X–ray and/or an Auger electron (the energy of the electron can give information on the chemical composition of the sample using an electron microprobe [21]). The energy (and hence wavelength) of the emitted X–ray is unique to the originating element, so the measured X–ray energy (or wavelength) spectrum allows the elements present in the sample to be identified. With PIXE, the measured X–ray yield is nearly independent of the chemical state or bonding within the sample and the X–ray production cross-sections are well known; therefore, trace element concentrations of less than 1 ppm can be detected and quantified.

The generation of various X–ray lines caused by de-excitation of electrons falling from higher shells is shown schematically in Figure 4.1. For a vacancy created in the K shell, a K_α X–ray is emitted if an L shell electron fills the vacancy, and a more energetic K_β X–ray is emitted if an M or N shell electron fills the vacancy. There is a higher probability of emitting a K_α X–ray than a K_β X–ray. Similarly L_α, L_β, and L_γ X–rays are caused by an L shell

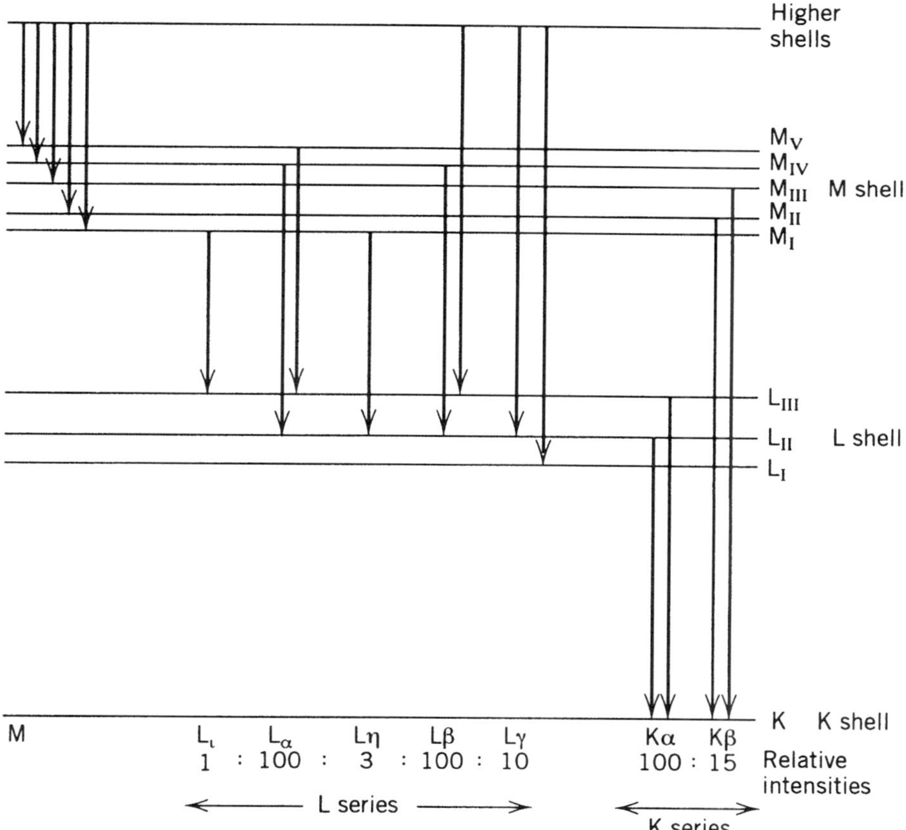

Figure 4.1. Schematic of the generation of various X–rays lines from different electron shells. The approximate relative intensities of the X–ray lines are also shown. Reprinted from Ref. 11 with permission of IOP Publishing Ltd.

vacancy being filled by an electron transition from a higher shell. Figure 4.2 shows the energies of the main X–ray line groups as a function of atomic number; an approximate estimate of the emitted X–ray energy (in keV) can be obtained using the relationship:

$$E_{K\alpha} \sim \frac{Z^2}{100} \quad E_{L\alpha} \sim \frac{Z^2}{750} \quad E_{M\alpha} \sim \frac{Z^2}{3000} \tag{4.1}$$

Quantitative determination of trace element concentrations using PIXE relies on an accurate knowledge of the electron-shell ionization cross-sections. Figure 4.3 shows the variation of the cross-sections for K and L shell ionization as a function of atomic number for 1.5 MeV and 3 MeV ^{1}H ions [12]. A number of

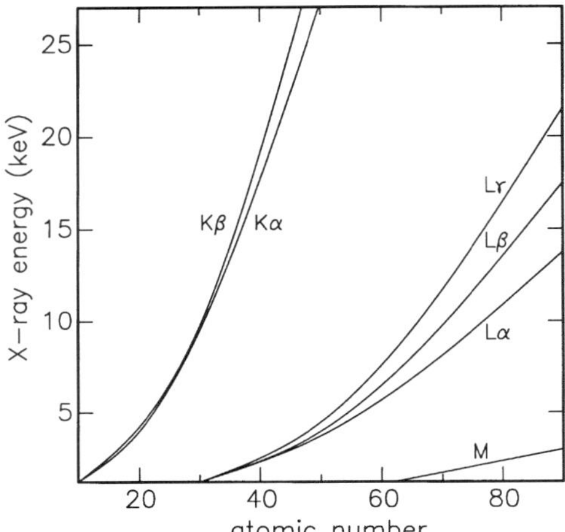

Figure 4.2. Energy of the main X–ray lines as a function of sample atomic number.

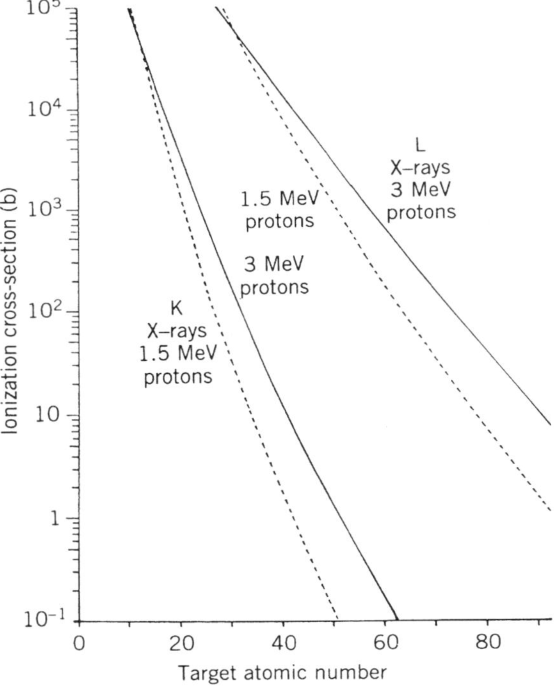

Figure 4.3. Calculated cross-sections for K and L shell ionizations as a function of sample atomic number for 1.5 MeV ^{1}H ions (dashed lines) and 3 MeV ^{1}H ions (solid lines). Reprinted from Ref. 11 with permission of IOP Publishing Ltd.

different approaches have been used to calculate the ionization cross-sections [22], and accurate values are given in Ref. 23. Table 4.1 shows the ionization cross-sections for the K shell of copper for MeV ^{1}H ions, MeV ^{4}He ions, and keV electrons. The ionization cross-section for electrons has a maximum at approximately 30 keV, whereas the ionization cross-sections for ^{1}H and ^{4}He ions peak at higher energies. The large ionization cross-section for ^{1}H ions compared with heavier ions of the same energy has resulted in the former being the most commonly used ion. Thus, PIXE is frequently used to refer specifically to proton induced X–ray emission.

The fraction of K or L shell vacancies that give rise to an X–ray, and not an Auger electron, is called the fluorescence yield ω_K or ω_L. The fraction of these K or L X–rays that give the α, β, or γ line is given by the branching ratio b. Numerical values for X–ray energies, fluorescence yields, and branching ratios are tabulated in Refs. 1, 8, 14, 24, and 25. When a thin homogeneous sample of areal density N_s, containing a fraction p by weight of an element A, is bombarded with MeV ions the total X–ray yield, Y_T, per ion for a particular line is, following from Eq. (2.1):

$$Y_T = \left(\frac{N_A p N_s}{A} \right) \sigma b \epsilon \omega \tag{4.2}$$

where $\sigma, b, \omega, \epsilon$, and N_A are the ionization cross-section, branching ratio, fluorescent yield, and detection efficiency for the particular X–ray line. N_A is Avogadro's number.

Determination of the elemental concentrations from measured yields is usually carried out with computer fitting programs such as PIXAN (26), GUPIX (27), or GEOPIXE (28). The result is elemental concentrations that can be determined in thin samples with a typical accuracy of 10%, or 5% if thin-sample standards of known concentrations are used. Quantitative analysis of thick samples is more difficult [14,29–31] because the energy lost by the beam in the sample and the attenuation of X–rays as they leave the sample must be taken into account.

TABLE 4.1. Ionization Cross-Sections in Barns for the K Shell of Copper

Electrons		^{1}H Ions		^{4}He Ions	
Energy (keV)	σ	Energy (MeV)	σ	Energy (MeV)	σ
10	87	1	17	4	53
20	384	2	99	6	171
30	477	3	220	8	355
40	445	4	345	10	583

SOURCE. Reprinted, by permission, from S.A.E. Johannson and J.L. Campbell, *PIXE: A Novel Technique for Elemental Analysis.* Copyright © 1988 by John Wiley & Sons, Inc.

PIXE analysis generally uses a Si(Li) detector to measure the energy of the emitted X–rays. An important feature of such a detector is that it is simultaneously sensitive to a range of different X–ray energies with a typical energy resolution of better than 160 eV. This resolution is insufficient to measure the different components of the α, β, and γ lines (shown in Figure 4.1) from which information on the chemical nature of the sample can be deduced [32]. The Si(Li) detector is usually separated from the microprobe chamber vacuum by a thin (less than 25 μm) beryllium window. Because X–rays emitted from the sample must pass through this window, the detection efficiency for X–rays below 2 keV is poor, as they suffer strong attenuation. This low–energy X–ray attenuation and limited energy resolution of the Si(Li) detector limits the practical uses of PIXE to the detection of elements with higher atomic number than sodium. The use of thicker absorber foils to stop ions backscattering from thick samples into the detector crystal increases the minimum detectable atomic number still further.

The Si(Li) detector efficiency also starts to decrease for X–ray energies greater than 20 keV due to incomplete X–ray absorption within the thin silicon crystal in the detector. Thus, PIXE analysis is rarely used to measure X–ray energies greater than 35 keV because of the incomplete absorption and the decreasing X–ray production cross-section for higher energy X–rays.

The dominant source of background in the X–ray energy spectrum measured using a Si(Li) detector in an electron microprobe is the bremsstrahlung radiation emitted by the incident keV electrons as they undergo many scattering events with the atomic electrons. This limits the elemental sensitivity of electron probe microanalysis (EPMA) to approximately 500 ppm using a Si(Li) detector. A MeV ^{1}H ion beam generates many orders of magnitude less brehmsstrahlung radiation [33] than a keV electron beam, giving PIXE an analytical sensitivity of better than 0.1 ppm for many elements in a low atomic number matrix. An example of the intrinsically lower background using PIXE compared with EPMA is illustrated in Figure 4.4, which shows X–ray energy spectra of a pollen tube measured using both methods [34]. The high background under the electron induced X–ray spectrum resulted in several trace elements being undetected, although they are clearly visible in the PIXE spectrum. A similar demonstration [35] used X–rays measured from the mineral xenotine.

The sensitivity of electron microprobes can be improved by more than a factor of ten using a crystal spectrometer, which measures the wavelength of the X–rays, rather than a Si(Li) detector which measures the X–ray energy. However, this necessitates sequential measurement of the X–ray peaks and use of a large electron beam current owing to the low spectrometer efficiency. The relatively large X–ray production cross-sections for PIXE, combined with X–ray detection using a Si(Li) detector, allows a beam current of approximately 100 pA to be used for analysis in the nuclear microprobe, enabling a spatial resolution of 400 nm to be achieved [36]. Applications in this case could include imaging of the distribution of a heavy element from the M shell X–rays, or of light elements from their K shell X–rays. However, for rapid

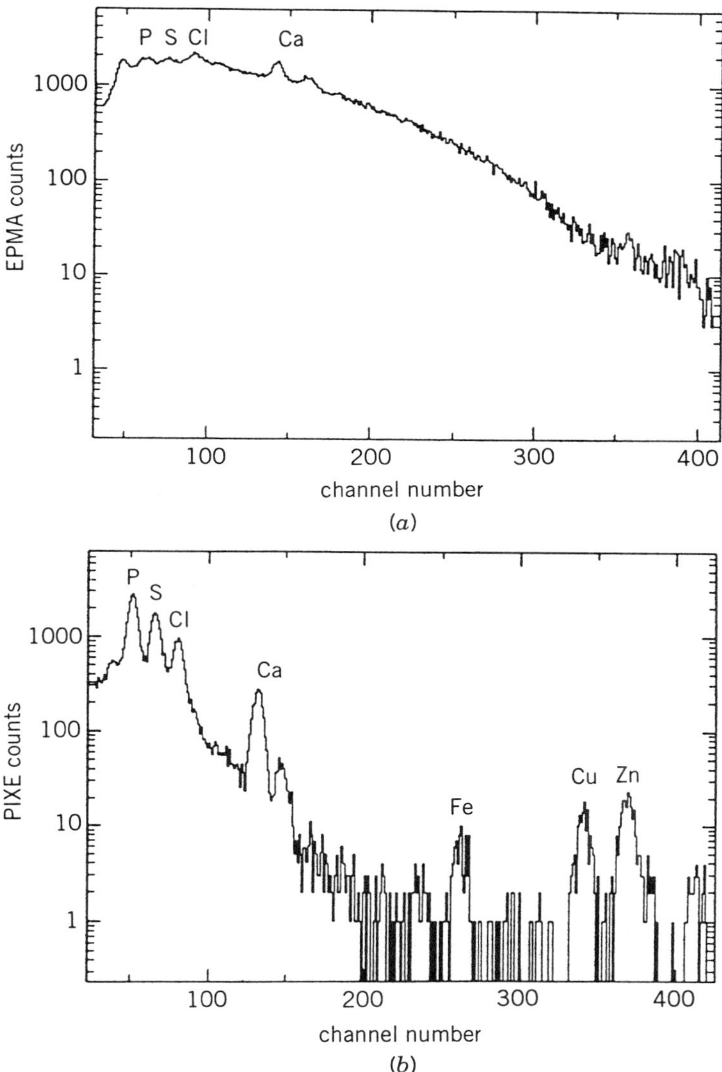

Figure 4.4. Comparison of X–ray spectra of a pollen tube measured using (a) EPMA and (b) PIXE. Reprinted from Ref. 34 with permission of K. Traxel.

(i.e., ~10 min), precision measurements of trace element concentration (i.e., ~0.1 ppm), in general, a beam current of 10 nA is desirable. In many systems, this can be focused into a probe of less than 10 μm.

A further advantage of PIXE, in some situations, is the large analytical depth of up to 50 μm from the long range of MeV ^{1}H ions. The high penetrating power of MeV ^{1}H ions enables another interesting variation of PIXE analysis, whereby the beam is transmitted from the vacuum chamber into air through a thin foil [37]. This external beam analysis is very useful for studies of large or delicate samples, such as paintings or manuscripts [38,39] which cannot be placed inside the microprobe chamber, although positional and spatial resolution in such systems is relatively poor.

Much of the development work of PIXE with the nuclear microprobe was carried out for biomedical and geological applications; Figures 4.5 and 4.6 show

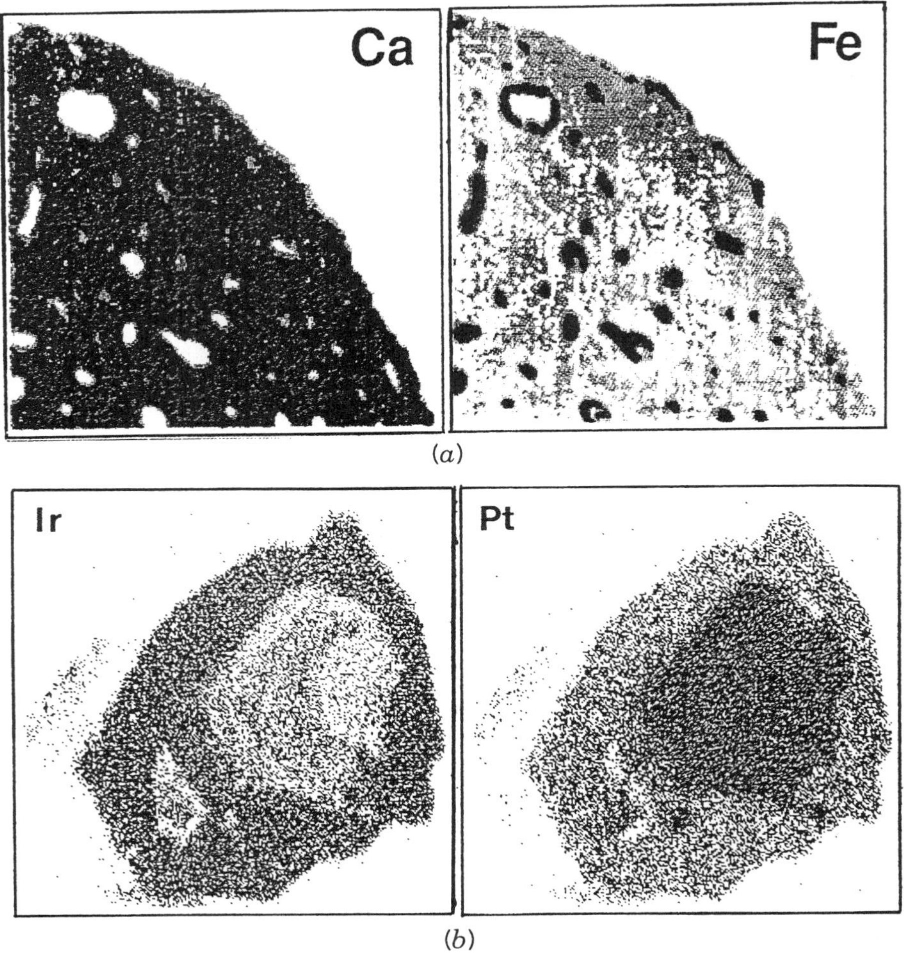

Figure 4.5. (a) 2.5 × 2.5 mm^2 PIXE images showing the distributions of calcium and iron within a sample of bone [40]. (b) 0.5 × 0.5 mm^2 PIXE images showing the distributions of iridium and platinum in a mineral [41]. Figures reprinted with kind permission from Elsevier Science B.V., Amsterdam, The Netherlands.

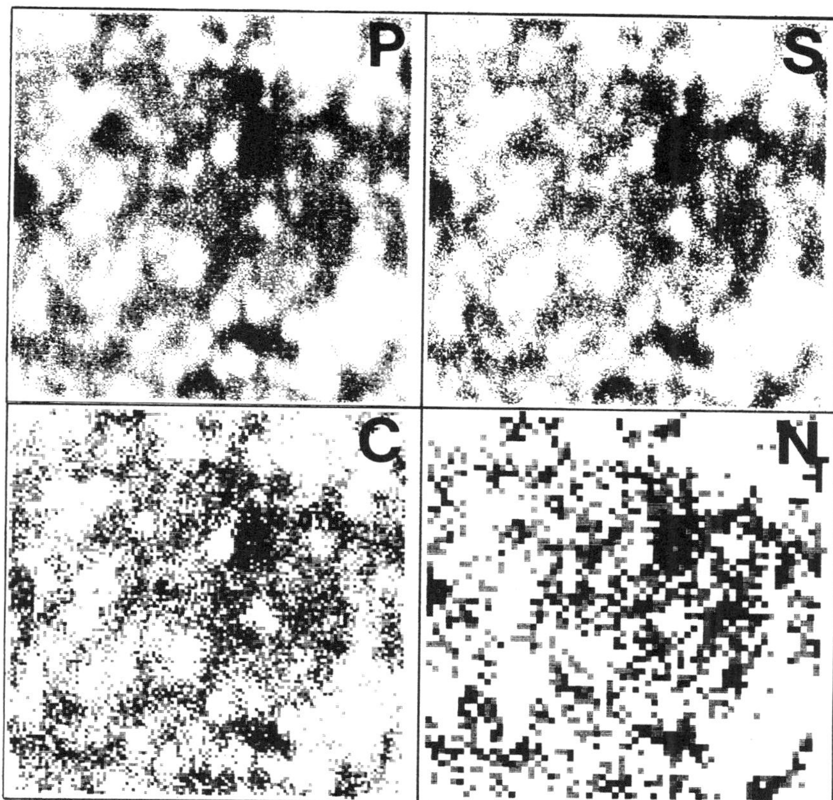

Figure 4.6. $100 \times 100 \ \mu\text{m}^2$ images of unstained brain tissue containing a senile plaque showing the distributions of phosphorous and sulfur measured with PIXE, and carbon and nitrogen measured with backscattering spectrometry. Reprinted from Ref. 43 with permission from J.P. Landsberg.

various examples of PIXE images from these research fields. Figure 4.5a shows images of the calcium and iron distributions from bones excavated from the Mary Rose shipwreck [40], as part of a study into the effects of bone contamination. The iron lining of the calcium bone cavities is clearly shown here, and movement of a range of metals through the bone structure was investigated. Figure 4.5b shows images of the platinum and iridium distributions from a mineral sample [41], which were part of an investigation into the trace element composition of heterogeneous grains. This same study compared the capabilities of PIXE and EPMA to image the element distribution in small grains, and showed that PIXE had significant advantages for low elemental concentrations. Figure 4.6 shows the distribution of phosphorous and sulfur in a thin section of brain tissue, which was part of a study into the possible role of aluminium in Alzheimers disease [42,43]. Also shown are the carbon and nitrogen distributions within this same area, as measured with backscattering spectrometry. A STIM image of this same area is shown later, in Figure 4.27e.

4.1.1. Dynamic Particle Induced X–Ray Emission Analysis

In most applications of PIXE using a nuclear microprobe, images of elemental distributions are typically produced by simply windowing a characteristic line in the X–ray energy spectrum and producing an image of the intensity of all counts within the window. For signals of major elements on a weak background this procedure produces accurate images of the elemental distribution, although some background subtraction is often required. However, this procedure can lead to significantly erroneous images owing to several effects that include the overlap of X–ray lines contributed by other elements, overlap of signals contributed by pulse pile-up from intense lower energy X–ray lines, variations in the intensity of the continuum bremsstrahlung background, overlap with incomplete charge collection tails on higher energy X–ray lines, and overlap with escape peaks from higher energy X–ray lines. Furthermore, a single window in the energy spectrum does not necessarily take advantage of the total signal from a particular element, which may contribute more than one X–ray line at several different energies.

In the case of simple peak overlap, off-line image processing can be used to remove the interference. Pallan and Knox [44] produced true calcium concentration images from a window in the energy spectrum surrounding the calcium K_α line by subtracting a scaled potassium-shell K_α image, produced from another window, which was uncontaminated and hence could correct for the overlap of the potassium-shell K_β with the calcium K_α line. Another example of a similar procedure is given in Ref. 45. Other overlap problems, such as manganese/chromium, cobalt/iron, and selenium/lead, cannot be treated so easily [45].

An additional problem with the use of a simple windowing technique is that the image must be converted from an image of X–ray peak intensity into, ideally, an image of element concentration. All these problems can be overcome by dynamic analysis [46,47], which is based on event-by-event acquisition. This procedure treats each count in the in-coming energy spectrum as the index of a vector from a precalculated matrix. Each coefficient in the vector gives the concentration per count of each element, or background, in the sample. These coefficients are then added to each accumulating image of the elemental concentration distributions. In some cases, the coefficients are negative to subtract overlap and background. The images are then free from interferences from overlapping X–ray lines and are background subtracted at each pixel.

The essence of the method is as follows: A PIXE energy spectrum, **S**, with counts y_i in channel i, can be fitted, by the method of least squares, with a model function f. This comprises the sum of the line-shape functions for each element in the sample, scaled by the signal peak area a_k of each element k, together with pile-up and background contributions. Then, f is related to $\mathbf{S}$ by the matrix equation

$$\boldsymbol{\alpha}\mathbf{a} = \boldsymbol{\beta}\mathbf{S} \qquad (4.3)$$

where $\boldsymbol{\alpha}$ and $\boldsymbol{\beta}$ are matrices of the partial derivatives of f with respect to the peak area a_k:

$$\alpha_{jk} = \Sigma_i w_i^{-1} \beta_{ji}\beta_{ki} \qquad (4.4)$$

$$\beta_{ji} = w_i (\partial f_i/\partial a_j) \qquad (4.5)$$

$\mathbf{a}$ is the vector of elemental signal peak areas and w_i is a weighting coefficient. The precise form of the weighting coefficient, which is intended to be a measure of the accuracy of the measurement of each element, is not critical. Satisfactory results can be obtained with unit weights ($w_i = 1$) [48] or, better, the statistical weights ($w_i = f_i^{-1}$) [49]. It is important that the weights $w_i = y_i^{-1}$ are not used, because this will force f to pass preferentially through regions of the spectrum where the statistics are poor, leading to errors fitting the peaks where the statistics are good [48,49].

The peak area a_k of element k, is then related to the elemental concentration C_k (for thick samples) or areal density (for thin samples) by

$$a_k = Q\Omega \epsilon_k T_k Y_k C_k \qquad (4.6)$$

where Q is the integrated dose, Ω is the detector solid angle, ϵ_k is the detector efficiency, T_k is the attenuation of the filter on the X–ray detector for the characteristic X–ray line, and Y_k is the yield from the element. Hence the concentration vector, $\mathbf{C}$, of the single element's concentrations C_k can be found from the matrix equation

$$\mathbf{C} = Q^{-1}\boldsymbol{\Gamma}\mathbf{S} \qquad (4.7)$$

where

$$\Gamma_{ki} = (\Omega \epsilon_k T_k Y_k)^{-1}\Sigma_j \alpha_{kj}^{-1}\beta_{ji}$$

Notice that the Γ matrix depends weakly on f, the function fitted to the original spectrum, through the weights w. However, in practice, Γ may be determined once from a representative sample of a particular group of samples [46]. Γ may then be applied to the analysis of each of the other samples in the group. An example of the Γ matrix is shown below.

The generalization of the above analysis to the production of true elemental images from samples with overlapping X–ray lines is then straightforward. The concentration vector becomes a function of (x, y), $C_k(x, y)$. Each count arriving in the energy spectrum $\mathbf{S}$ can be used to increment the intensity of elemental image M_k at pixel (x, y), determined by the present beam coordinate on the sample, by an amount

$$\delta M_k(x, y) = Q(x, y)^{-1}\Gamma_{ki} \tag{4.8}$$

where i is the channel number of the count. The concentration distribution, $C_k(x, y)$, is then related to this image by

$$C_k(x, y) = Q(x, y)^{-1}M_k(x, y) \tag{4.9}$$

where $Q(x, y)$ is the accumulated beam dose at that pixel.

The method has been tested by analyzing a test sample of manganese–chromium–cobalt–iron–copper, and each element could be separately imaged despite the overlapping X–ray lines [46], as shown in Figure 4.7. It has also been tested on a sample of pyrite (FeS_2) containing a variety of trace elements, including arsenic, gold, zinc, tellurium, and silver. In this case, an image of the true gold distribution could be obtained free from interference between the gold L_α and zinc K_β lines, and the gold L_β and silver K_β lines [46], as shown in Figure 4.8. Some representative rows of a Γ matrix for a different geological sample are also shown in Figure 4.8. Note the regions of positive and negative coefficients that eliminate interferences. Further examples of this procedure are shown in Ref. 47.

Figure 4.7. Top: Raw X–ray intensity images from windows placed in the energy spectrum from a 3 MeV ^{1}H ion irradiated test sample of pure metal foils. Notice the contribution of chromium to the manganese image, manganese to the iron image, and iron to the cobalt image. Bottom: The same data set treated with the dynamic analysis method. The effect of the interferences have been significantly eliminated. Reprinted with the permission of C.G. Ryan.

Total
Mn
Cr
Co
Fe
Cu
Cr
Mn
Fe
Co
Cu
Relative Conc.
1.0
0.8
0.6
0.4
0.2
0.0

RAW

Total
Mn
Cr
Co
Fe
Cu
Cr
Mn
Fe
Co
Cu
Relative Conc.
1.0
0.8
0.6
0.4
0.2
0.0

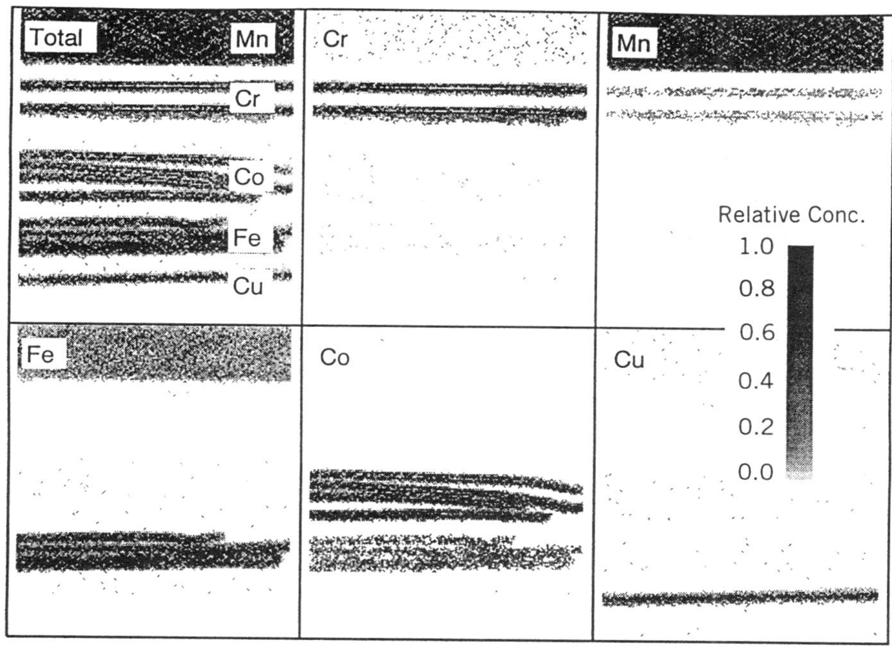

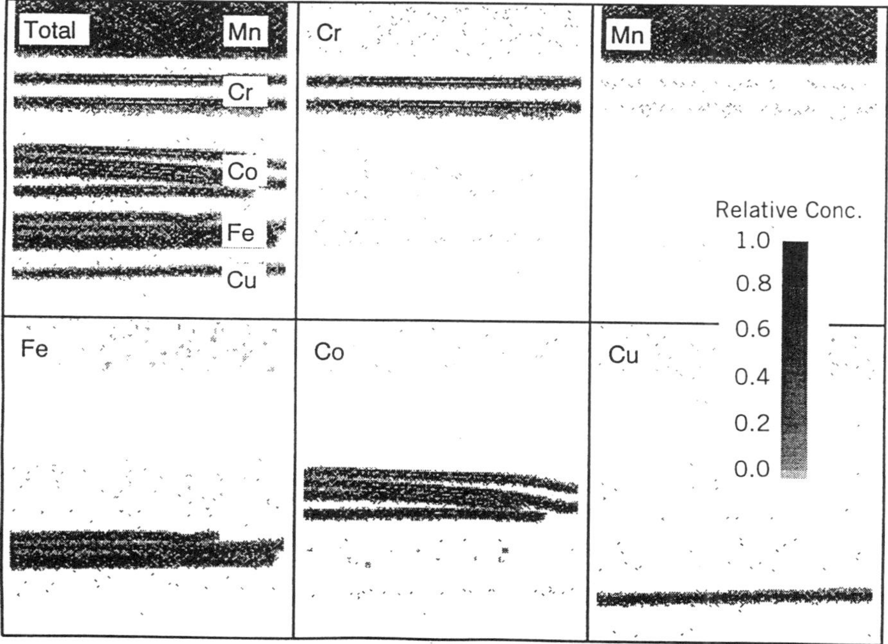

DYNAMIC

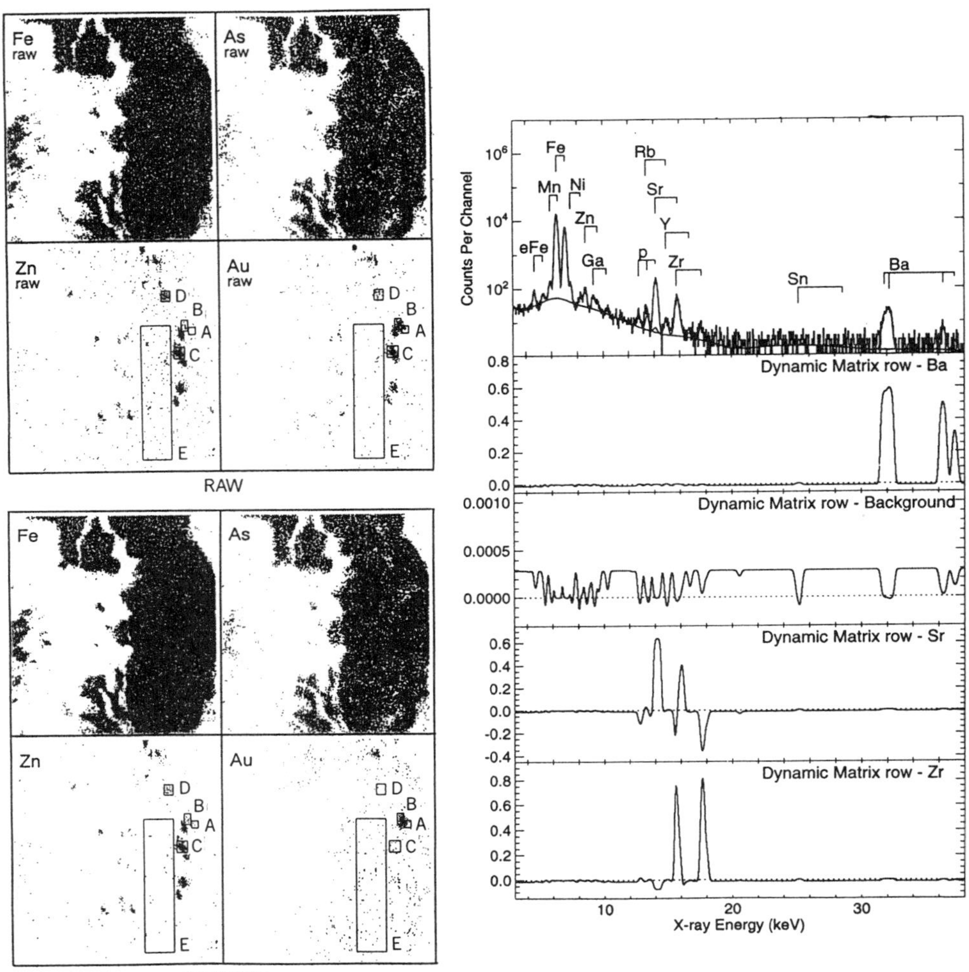

Figure 4.8. Top left: Raw X–ray intensity images from windows placed in the energy spectrum from a 3 MeV ^{1}H ion irradiated pyrite (FeS$_2$) sample. Notice how the gold and zinc images are very similar because of X–ray line interferences. Bottom left: The same data set treated with the dynamic analysis method. The gold image is now free from interference from zinc and a native gold inclusion can be seen, together with a lower concentration distribution of gold in the pyrite itself. Right: Rows of the Γ matrix used to produce the true images of barium, strontium, and zirconium in a different ore sample. Notice the negative parts of the strontium spectrum used to eliminate zirconium interference. Reprinted with the permission of C.G. Ryan.

4.2. BACKSCATTERING SPECTROMETRY

With backscattering spectrometry, the number and energy of elastically backscattered ions, typically 2 MeV ^{4}He ions, are measured to determine the sample stoichiometry and elemental depth distributions. This widely used technique has found many applications in studies of interfaces and diffusion profiles of thin films [4]. Many aspects of the basic theory and applications are comprehensively described in Ref. 2.

An analogous technique in the scanning electron microscope is the generation of electron backscattering images, where the backscattered electron intensity depends both on the atomic number and the topography of the sample. In some cases, electron backscattering can give even stronger topographical contrast than secondary electron imaging. However, quantitative elemental analysis is difficult using electron backscattering, because neither the energy of the detected electrons nor the backscattering yield changes much between different elements present.

In comparison, the energy of a backscattered ion depends strongly on the sample nucleus involved in the collision; so, elemental identification with ion backscattering spectrometry is based on measuring the energy of the backscattered ions. The sample stoichiometry is determined by measuring the number of backscattered ions from each element. Quantitative analysis is possible because the scattering cross-sections of the ions with the sample nuclei and scattering kinematics are both well known, as described in Section 1.2. Backscattering spectrometry is confined to detection of elements that are heavier than the incident ions, but with incident ^{4}He ions, this limitation only rules out measuring hydrogen and helium. In a forward scattering geometry, this limitation can be overcome for thin samples.

The energy lost by the ion during its path in and out of the sample enables the depth variations of the structure to be measured. The ion loses an energy of ΔE_{in} along its path into the sample and ΔE_{out}, after being backscattered from a depth z, along its path out of the sample, where

$$\Delta E_{\text{in}} = \int_o^z \left(\frac{dE}{dz} \right) dz \quad \text{and} \quad \Delta E_{\text{out}} = \int_{z/(\cos\,\theta_S)}^o \left(\frac{dE}{dz} \right) dz \qquad (4.10)$$

where θ_S is the angle through which the incident ion is scattered. The measured backscattered ion energy E_m from a given element at a depth z is thus

$$E_m = K(\theta_S)(E_o - \Delta E_{\text{in}}) - \Delta E_{\text{out}} \qquad (4.11)$$

where $K(\theta_S)$ is the kinematic factor given in Eq. (1.6). The measured backscattering ion energy spectrum can be converted to a depth spectrum for each element present, using the rate of ion energy loss in the sample and the different kinematic factors for the different elements present. The elemental composition, stoichiometry, and depth profiles are usually determined using computer simulations of the measured backscattering spectrum, such as RUMP [50].

The mass resolution between the different elements present in the sample is greatest at $\theta_S = 180°$, so the semiconductor detector used for backscattering spectrometry is commonly placed at angles close to this, as shown in Figure 2.7. In general, the mass resolution for light elements is much better than the mass resolution for heavy elements, so it is easier to differentiate between different light elements present than between different heavy elements present. Mass resolution with backscattering spectrometry can be increased using ions heavier than MeV ^{4}He ions [51], but this introduces some limitations.

Although the mass resolution is best at scattering angles close to 180°, scattering angles closer to 90° are commonly used to maximize the depth resolution of a measurement. This is because the path-length out of the sample is greater, so ΔE_{out} in Eq. (4.11) is more easily distinguished. Scattering angles close to 90° can only be used for a limited class of samples, which consist of a thin metal or amorphous layer on a light-element substrate, and which are optically flat over the area analyzed.

The depth and mass resolution attainable with backscattering spectrometry depends on the accuracy with which the energy of the ions recoiling from a given depth can be measured, and this can be degraded in several ways. The energy straggle of the ion beam moving through the sample (described in Section 1.3) is one limitation. Energy straggle increases the energy spread of the ions with path-length in the sample, so the energy resolution is consequently best for ions scattering from close to the sample surface. There is also a kinematic energy spread owing to the range of scattering angles accepted by the semiconductor detector. This can be minimized by locating the detector close to 180° and only accepting a narrow angular range of recoiling ions. This, however, limits the detector solid angle and leads to a low counting rate, necessitating a longer measurement time or a larger beam current. The energy resolution of the semiconductor detector and its electronics is typically 15 keV for a backscattered MeV light ion. With 2 MeV ^{4}He ions scattered through a steep backward angle into a detector with a narrow angular acceptance, a depth resolution of 20 nm is attainable. With a shallow scattering angle, a depth resolution of ~10 nm is possible under highly restrictive conditions. The respective merits of using shallow-angle and large-angle backscattering geometries are summarized in Table 4.2.

An example of the use of heavier ions at a shallow angle to maximize the depth resolution is given in Figure 4.9. This shows an early example of a backscattering spectrum for 2 MeV ^{7}Li ions on a multilayer structure comprising 10 tantalum/silicon bilayers in which the tantalum and silicon layer thicknesses are 7 nm and 11 nm, respectively. The shallow scattering angle and the use of ^{7}Li ions resulted in a very restricted depth penetration, and only the first three tantalum layers were detected. The measured depth resolution in this geometry using 2 MeV ^{7}Li ions was 8.5 nm for silicon, compared with 10 nm for 2 MeV ^{4}He ions in an identical geometry.

TABLE 4.2. Comparison of Different Angles for Backscattering Spectrometry

	$\theta_S \sim 90°$	$\theta_S \sim 180°$
Mass resolution	Poor	Good
Depth resolution	Good	Poor
Scattering cross-section	Large	Small
Kinematic spread	Large	Small
Energy Straggle	Large	Small

Using a nuclear microprobe, backscattering spectra can be measured from precisely located regions as small as 1 μm across, such as metallization contact holes on microcircuit devices. Figure 4.10 shows measured and simulated backscattering spectra obtained from a platinum silicide contact pad, using 2 MeV ^{4}He ions. The dotted line is a simulation based on the intended contact pad structure shown in the inset; incomplete silicide formation together with some etching of the platinum layer were indicated by comparison of the measured spectrum with the simulated spectrum. Further examples of such spectra are shown in Chapter 7.

Examples of backscattering images used for biomedical research are shown

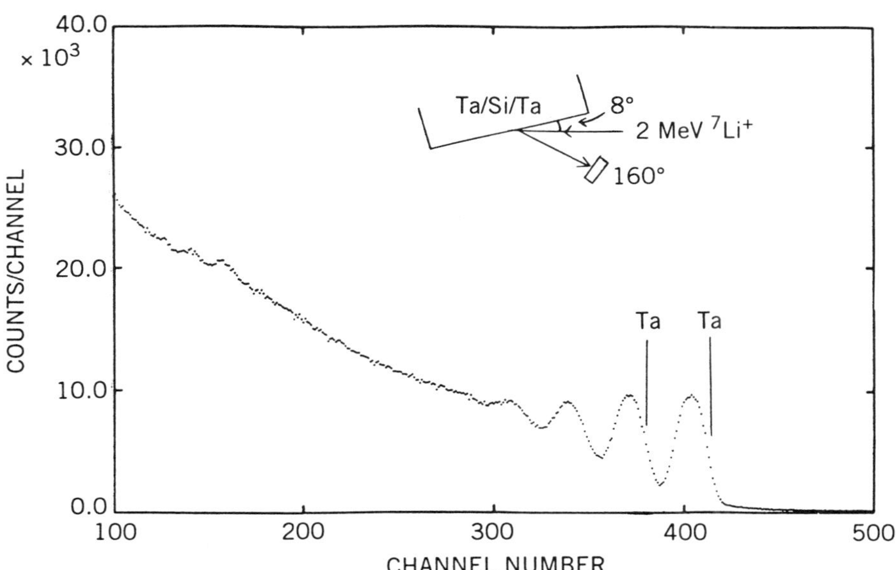

Figure 4.9. Shallow angle backscattering spectrum for 2 MeV ^{7}Li ions at an angle of 8° to the tantalum/silicon multilayer sample surface. Reprinted from Ref. 52 with kind permission from Elsevier Science B.V., Amsterdam, The Netherlands.

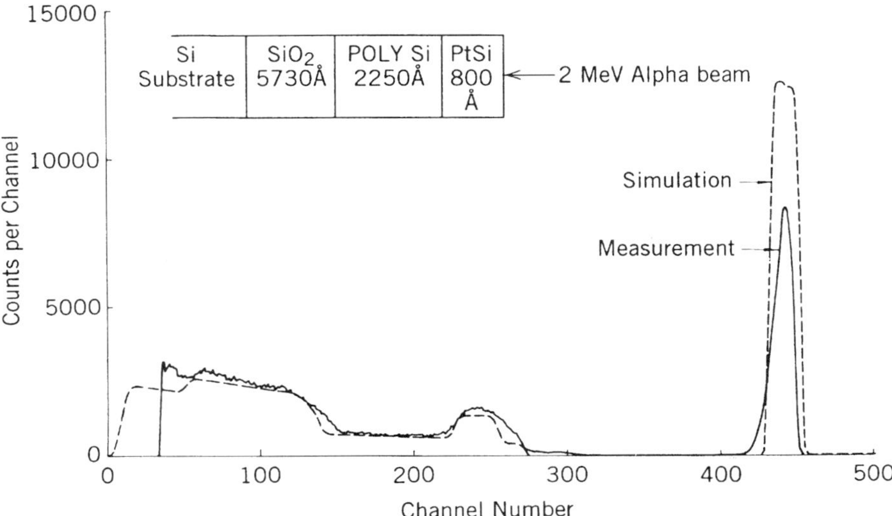

Figure 4.10. Measured and simulated backscattering spectra using 2 MeV ^{4}He ions from a platinum silicide contact pad. The dotted line is a simulation based on the intended structure shown in the inset. Reprinted from Ref. 53 with kind permission from Elsevier Science B.V., Amsterdam, The Netherlands.

in Figure 4.6, where the distributions of carbon and nitrogen in unstained brain tissue containing a senile plaque core [42,43] complement the phosphorous and sulfur PIXE images from this same area. Because the tissue section was only approximately 5 μm thick, there was no energy overlap between the ions backscattered from carbon and from nitrogen, enabling the elemental distributions to be individually imaged. Another example of a backscattering image, which was generated using 2 MeV ^{4}He ions from a pyrite crystal [54], is shown in Figure 4.11. This was part of a study into the use of ion channeling to measure the lattice location of elements such as nickel and gold in minerals. These backscattering images allow regions of the crystal that are free of inclusions and defects to be identified, allowing the substitutionality of nickel and gold to be measured in regions of good crystal.

Another example of depth-resolved backscattering images is shown in Figure 4.12. The sample was a graphite limiter tile from the inner wall of a fusion reactor, coated with a 21 μm thick titanium–carbon layer [55]. A region, analyzed using 6 MeV ^{1}H ions, is displayed as a 64 × 64 pixel image showing the distributions of titanium and carbon from depths of 3, 10, and 17 μm within the coating. A crack across the tile coating is seen running in the middle of the images. Other examples of backscattering images are shown in Chapters 7 and 9.

Because 2 MeV ^{4}He ions generally do not have enough energy to generate resonant scattering effects or nuclear reactions, they are ideal for backscattering spectrometry. However, MeV ^{1}H ions can generate both resonant scattering

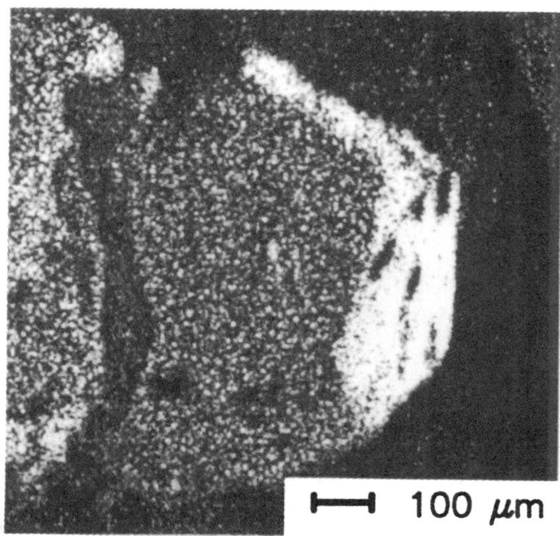

Figure 4.11. Backscattering image of a pyrite crystal. Reprinted from Ref. 54 with kind permission from Elsevier Science B.V., Amsterdam, The Netherlands.

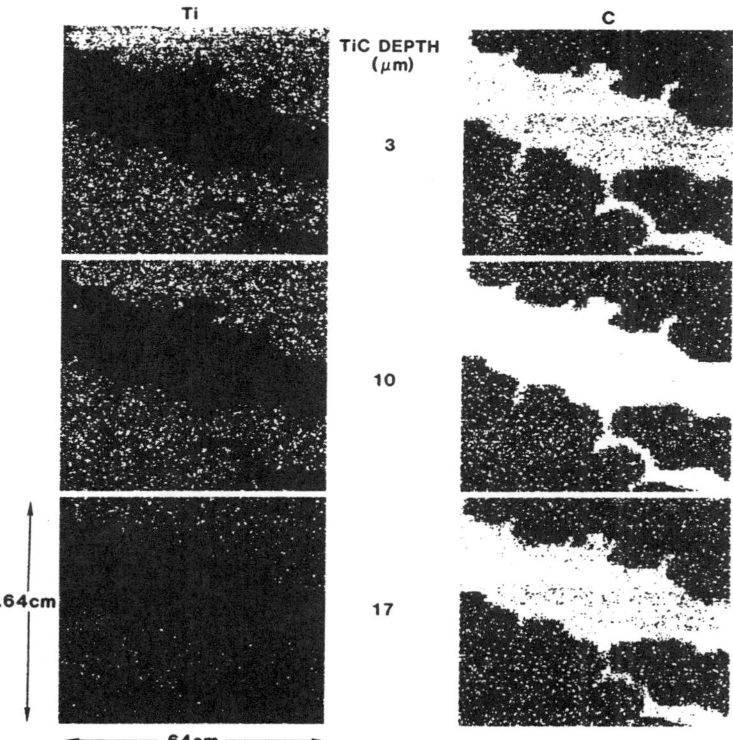

Figure 4.12. Depth-resolved backscattering images at depths of 3, 10, and 17 μm showing the concentration of titanium and carbon from a TiC coated limiter tile from the inner wall of a fusion reactor. Reprinted from Ref. 55 (© 1983 IEEE).

effects and nuclear reactions, so backscattering spectrometry using these ions is more complex, as the resonances complicate interpretation of the measured spectrum. Figure 4.13 is an example of this behavior observed in backscattering spectra from a multiphase superconducting YBaCuO crystal. The spectrum in Figure 4.13a shows a large hump corresponding to resonant scattering from oxygen at about 3.5 MeV, leading to a much larger number of ions scattered from oxygen at this energy; with 2 MeV ^{4}He ions, the Rutherford scattering cross-section gives only a very small oxygen signal in Figure 4.13b.

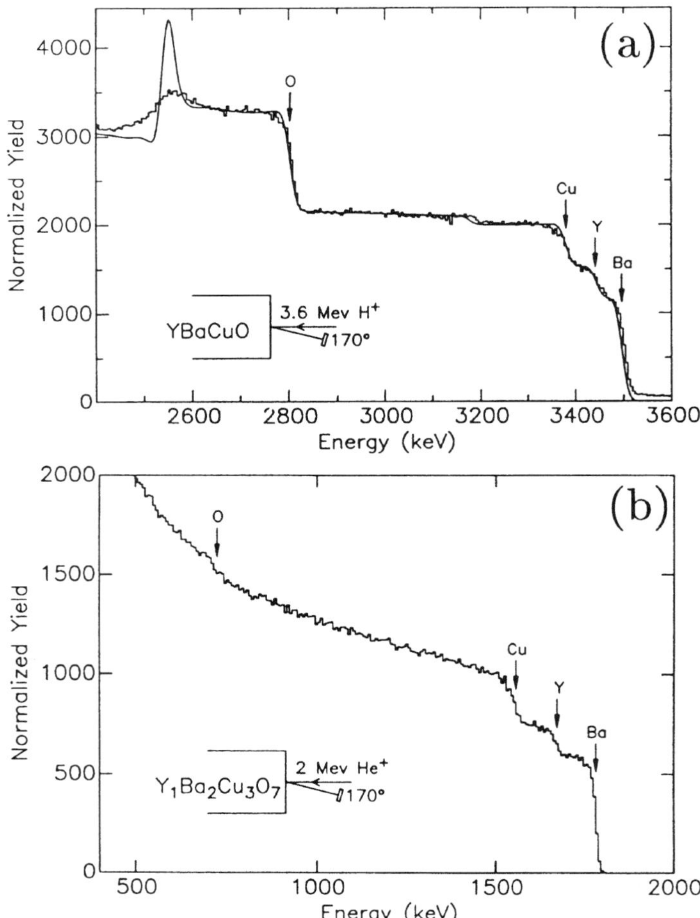

Figure 4.13. Measured backscattering spectra from a multiphase YBaCuO crystal using (a) 3.6 MeV ^{1}H ions and (b) 2 MeV ^{4}He ions scattered through 170°. The ^{1}H ion spectrum shows the effect of strong resonance and has been fitted using the measured oxygen cross-sections shown in Figure 4.15a. Reprinted from Ref. 56 with kind permission from Elsevier Science B.V., Amsterdam, The Netherlands.

4.3. NUCLEAR REACTION ANALYSIS

An ion with an energy of several MeV can overcome the Coulomb barrier of the atomic nucleus and approach to within a distance comparable with the nuclear radius, as described in Section 1.2. The energy at which this can be expected to occur, based on the assumption of Coulomb repulsion between two positive charges, is shown in Figure 4.14 for ^{1}H and ^{4}He ions. The probability of the ion interacting with the atomic nucleus is greatest for light ions incident on light elements present in the sample. When this occurs, there can be deviations from Rutherford scattering cross-sections, as shown in Figure 4.13a for MeV ^{1}H ions on oxygen [56]. This resonant scattering results in the ion being scattered with an energy given by Eq. (1.7), but with a different scattering cross-section from that given by Eq. (1.6). The sample nucleus involved in the collision remains unaltered. Another possibility is that there may be a nuclear reaction, which causes a structural change to the sample nucleus itself, and reaction products such as high-energy ^{1}H ions, ^{4}He ions, neutrons [57], or γ-rays [58,59] can be emitted and detected.

Figure 4.14 gives a guide to the ion energy above which a particular nuclear interaction can have resonant behavior. When the condition is reached, the interaction of ion and sample nucleus is very sensitive to the beam energy, and there are sharp variations in the magnitude and angular directions of scattering as the beam energy is varied. Figure 4.15 shows the measured scattering cross-sections for oxygen and silicon for scattering angles of 110°, 150°, and 170°,

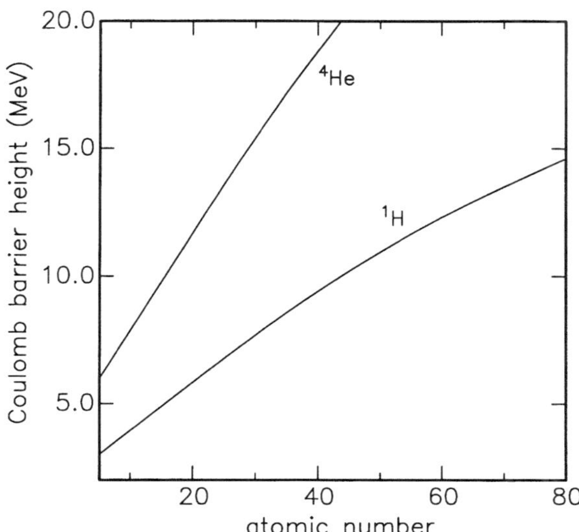

Figure 4.14. Coulomb barrier height as a function of sample atomic number for ^{1}H and ^{4}He ions.

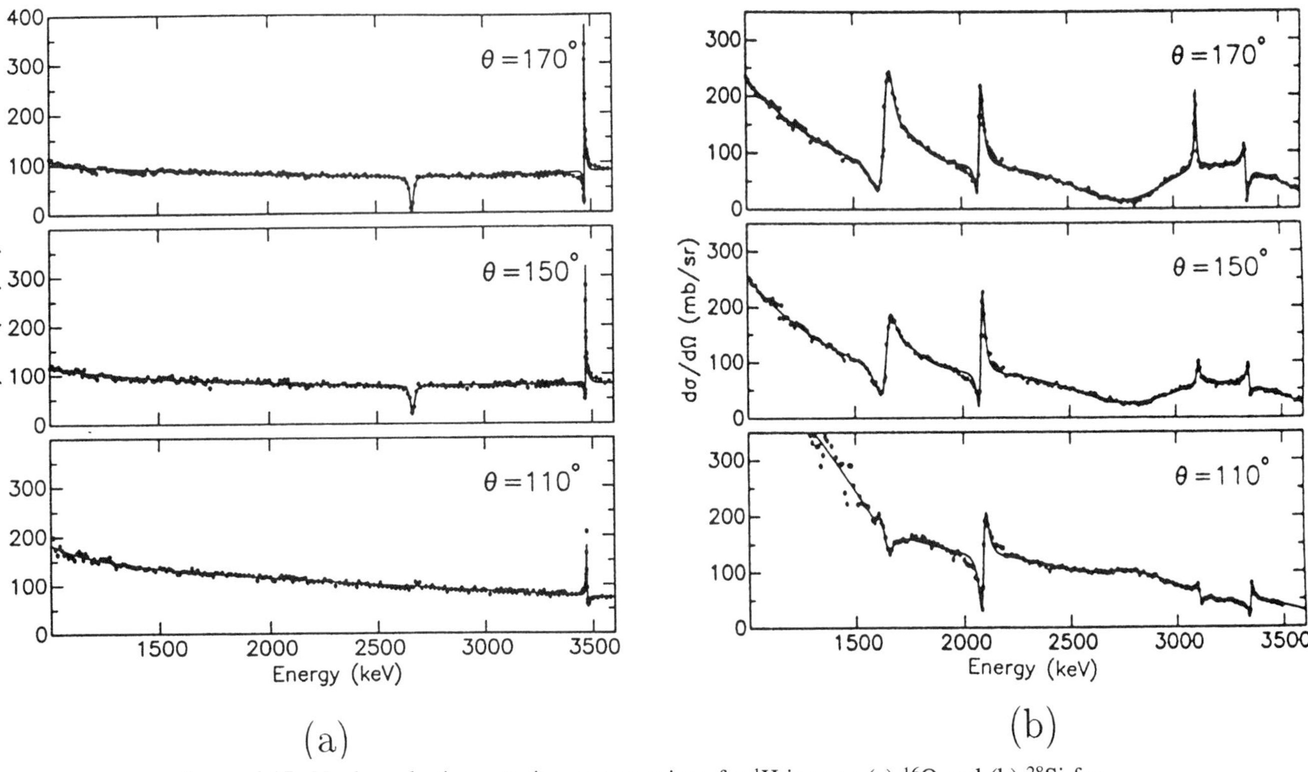

Figure 4.15. Nuclear elastic scattering cross-sections for ^{1}H ions on (a) ^{16}O and (b) ^{28}Si for detector angles of 110°, 150°, and 170°. Reprinted from Ref. 56 with kind permission from Elsevier Science B.V., Amsterdam, The Netherlands.

as a function of incident ^{1}H ion energy [55]. The measured scattering cross-sections shown in Figure 4.15a were used to fit the measured backscattered spectrum in Figure 4.13a, demonstrating that resonance scattering effects can be incorporated into backscattering simulations; this is important for nuclear microprobe applications, which regularly use MeV ^{1}H ions for analysis. Figure 4.14b shows that for silicon, there are scattering peaks at several resonance energies that change in intensity with different scattering angles.

In some cases, the increase in the resonant scattering cross-section for a particular light element is enough to make generation of spatially resolved images with a nuclear microprobe more practical. For example, Figure 4.16a shows an image of the oxygen distribution in a SiO_2–BN composite material using an oxygen resonant scattering peak [60]. The large resonant cross-section of 500 mb/sr enabled the oxygen distribution to be imaged in a reasonable time period, whereas this was not possible at lower incident ion energy because of the much lower non-resonant scattering cross-section. The corresponding silicon distribution in this area is also shown, which enabled the stoichiometry of the grains to be measured. Carbon and nitrogen distributions in SiC–Si_3N_4 composites have also been similarly imaged using 3 MeV ^{1}H ions [61].

4.3.1. Charged Particle Detection

A typical NRA measurement uses a beam of several nanoamps of 1 to 10 MeV light ions, such as ^{1}H (proton, p), ^{2}H (deuteron, d) or ^{3}He ions, to generate MeV protons or α-particles as charged-particle reaction products from light elements present in a matrix of heavy elements that produce no nuclear reactions.

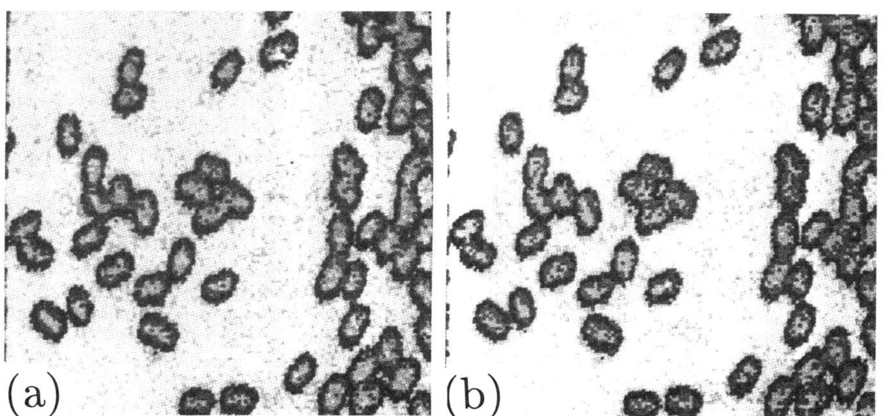

Figure 4.16. $170 \times 170 \ \mu m^2$ area of a SiO_2–BN composite material, showing (a) the oxygen distribution using a resonant scattering peak at 3.05 MeV with ^{4}He ions and (b) the silicon distribution. Reprinted from Ref. 60 with kind permission from Elsevier Science B.V., Amsterdam, The Netherlands.

This process can represented, for example, by

$$^{12}\text{C}(d,\ p)^{13}\text{C} \quad Q \text{ value} = 2.722 \text{ MeV} \tag{4.12}$$

Here a deuteron interacts with a ^{12}C nucleus in the sample, and a high-energy proton is emitted as the measured reaction product. The ^{12}C nucleus is changed into ^{13}C, and the total mass difference between the initial and final particles involved is given by the Q value of the reaction. Table 4.3 shows some commonly used reactions in which the measured reaction product is either a high-energy proton or an α-particle. A wide range of light elements present can be measured, provided that the correct particle and energy are chosen. The reactions with larger Q values provide extra energy to be shared by the product particles. References 1, 3, and 62 list many nuclear reactions and cross-sections commonly used, and many papers on NRA and the underlying processes can be found in the conference proceedings listed in Appendix 1.

The energies of the charged particle reaction products are usually measured with a large-area semiconductor detector, such as the annular detector shown in Figure 2.7. An absorber layer is often placed in front of the detector to stop backscattered ions from being measured, as this would degrade the analytical sensitivity. If the only ions entering the detector are from the particular nuclear reaction being excited, there is very little background in the measured spectrum, enabling NRA to achieve sensitivities of 0.01 ppm in favorable cases. The cross-

TABLE 4.3. Nuclear Reactions Commonly Used for Charged Particle Detection

Reaction	Q-Value (MeV)	Incident Energy (MeV)	Emitted Energy (MeV)	Lab Cross-Section σ (mb/sr)
$^2\text{H}(d,p)^3\text{H}$	4.032	1.0	2.3	5.2
$^2\text{H}(^3\text{He},p)^4\text{He}$	18.352	0.75	13.0	61
$^3\text{He}(d,p)^4\text{He}$	18.352	0.45	13.6	64
$^6\text{Li}(d,\alpha)^4\text{He}$	22.374	0.70	9.7	6
$^7\text{Li}(p,\alpha)^4\text{He}$	17.347	1.5	7.7	1.5
$^9\text{Be}(d,\alpha)^7\text{Li}$	7.153	0.6	4.1	~1
$^{11}\text{B}(p,\alpha)^8\text{Be}$	8.586	0.65	5.57	0.12
	5.65	0.65	3.7	90
$^{12}\text{C}(d,p)^{13}\text{C}$	2.722	1.2	3.1	35
$^{13}\text{C}(d,p)^{14}\text{C}$	5.951	0.64	5.8	0.4
$^{14}\text{N}(d,\alpha)^{12}\text{C}$	13.574	1.5	9.9	0.6
	9.146	1.2	6.7	1.3
$^{15}\text{N}(p,\alpha)^{12}\text{C}$	4.964	0.8	3.9	~15
$^{16}\text{O}(d,p)^{17}\text{O}$	1.917	0.9	2.4	0.74
	1.05	0.9	1.6	4.5
$^{18}\text{O}(p,\alpha)^{15}\text{N}$	3.98	0.73	3.4	15
$^{19}\text{F}(p,\alpha)^{16}\text{O}$	1.25	1.25	6.9	0.5
$^{31}\text{P}(p,\alpha)^{28}\text{Si}$	1.917	1.51	2.7	16

SOURCE. Reprinted from Ref. 1 with permission from Academic Press, New York.

sections for a specific reaction is usually low, necessitating the use of a large ion beam current for analysis. The desire to increase the beam current density for NRA with MeV ion beams was one of the main reasons for the development of the nuclear microprobe in the late 1960s at Harwell [63]. Most recent nuclear microprobe applications of NRA usually involve measurement of light element concentrations at specific points on the sample surface [64–66], rather than producing images, which would require much better counting statistics. Chapter 8 of Ref. 11 and Refs. 52 and 67 list many applications of NRA using a nuclear microprobe for the measurement of the main isotopes of most elements up to an atomic number of 13. Much of the work is devoted to the measurement of various light-element distributions in iron- and nickel-based alloys, and nitrogen and carbon measurements in various matrices using (d, p) reactions.

Nuclear reaction analysis has been widely used for nuclear microprobe applications in metallurgy [52,68] because of its ability to pick out light elements, such as carbon, in heavier metal matrices. Typical NRA energy spectra obtained from the ^{12}C(d, p) reaction are shown in Figure 4.17 [69]. Spectrum a is from quartz with a thin carbon layer on top, which gives a sharp peak from the ^{12}C. There are various lower energy peaks from oxygen and silicon nuclear reactions. Spectrum b is from a sample with a uniform carbon content: the spectrum has a peak approximately 0.5 MeV wide with a shape determined by the

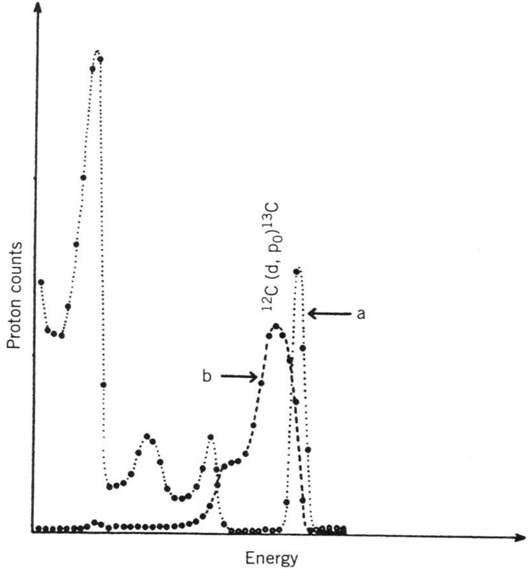

Figure 4.17. Measured ^{1}H ion energy spectra from two materials bombarded with 1.3 MeV deuterons. (a) Quartz with a thin-surface carbon film, (b) standard steel. Reprinted from Ref. 69 with kind permission from Elsevier Science B.V., Amsterdam, The Netherlands.

varying cross-section for deuterons reacting with ^{12}C at different depths in the sample. Light elements present can be depth profiled by taking into account the energy losses of the charged particle reaction products during their path out of the sample, in a similar manner to depth profiling using backscattering spectrometry. Deconvolution of the measured energy spectrum to determine the element depth distribution to accuracies of 100 nm over depths of a few microns is a well-established technique [70].

An application of NRA with a nuclear microprobe in which spatially resolved images were generated used the ^{12}C(d, p) reaction is [71]. This has a reasonably large cross-section of 35 mb/sr and was used to generate depth-resolved images of carbon fibers embedded in a metal matrix, as shown in Figure 4.18. Another

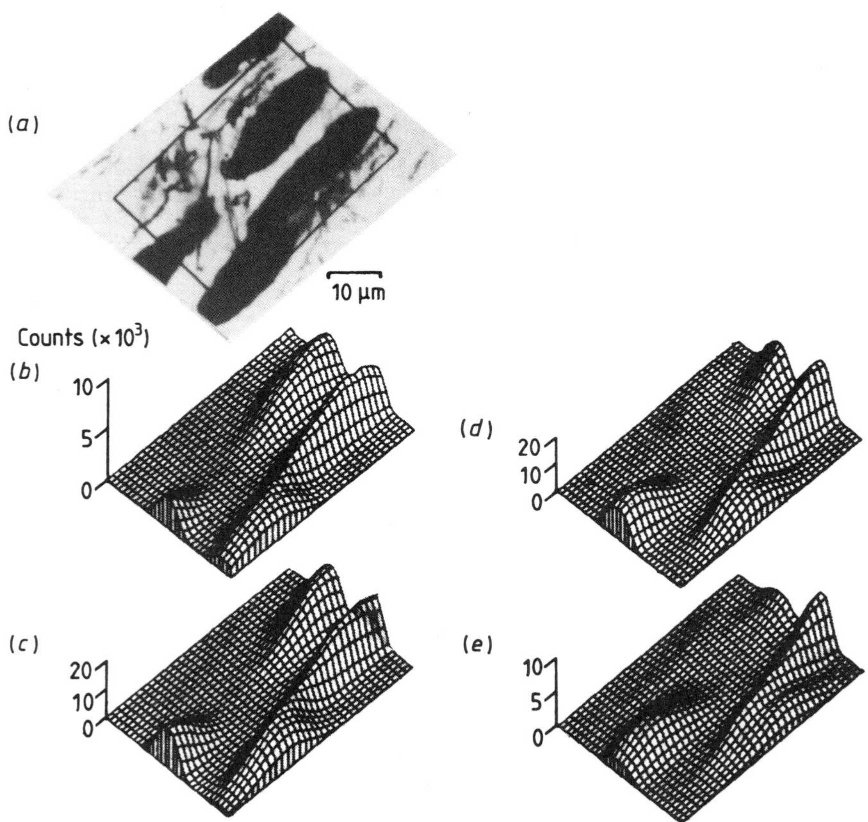

Figure 4.18. (a) Optical image showing the surface of the sample with carbon fibers embedded in a metal matrix. The rest of the figure shows the strength of the carbon signal from measured ^{1}H ions corresponding to carbon at depths of (b) 0 to 0.8 μm, (c) 0.8 to 2.1 μm, (d) 2.1 to 3.2 μm, (e) 3.2 to 4.4 μm. Reprinted from Ref. 71 with permission of D. Heck.

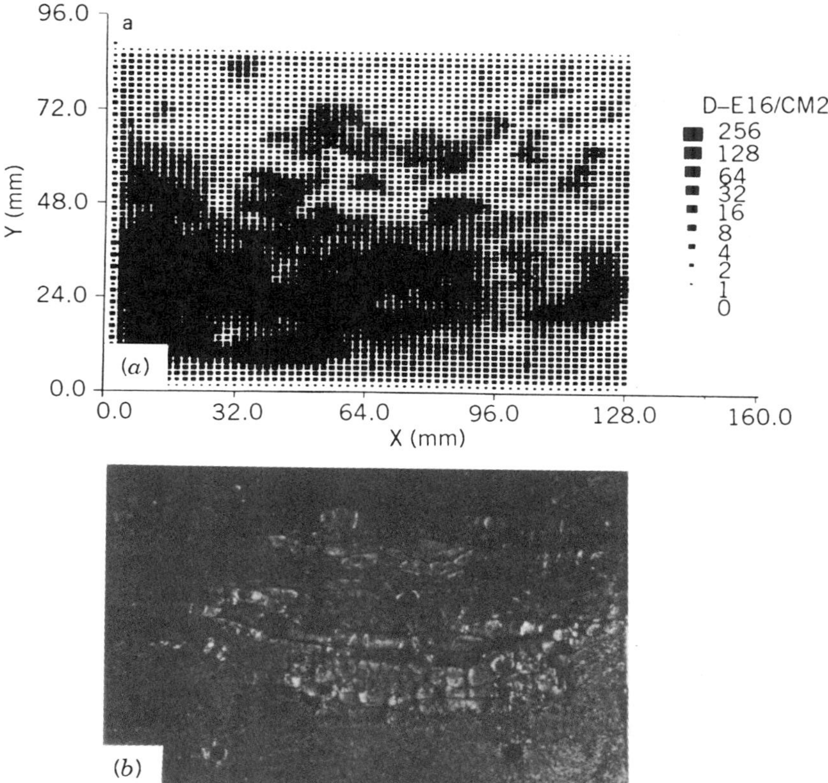

Figure 4.19. (a) NRA image showing the deuterium concentration deposited on the plasma facing surface of a limiter tile after removal from a fusion reactor. (b) Optical image showing the correspondence between the surface morphology and the deuterium distribution. Reprinted from Ref. 72 with kind permission from Elsevier Science B.V., Amsterdam, The Netherlands.

example, given in Figure 4.19a, shows the measured quantitative distribution of deuterium across the surface of a wall tile from a Tokamak fusion reactor [72]. The deuterium distribution was imaged using 750 keV ^{3}He ions using the d(^{3}He,p) reaction. Figure 4.19b shows an optical image of this same area, of which the upper two thirds had been severely damaged by high-energy electrons produced within the plasma; this analysis clearly showed the most damaged region contained the most deuterium.

4.4. ELASTIC RECOIL DETECTION ANALYSIS

A heavy MeV ion can eject lighter atomic nuclei from the sample by elastic scattering. With ERDA [73], the energy of forward-recoiling light nuclei, displaced from the sample by heavier incident ions, is measured using a glancing-angle geometry, enabling quantitative elemental analysis and depth profiles with a resolution of approximately 5 nm. The elastically scattered incident ions are stopped in an absorber layer in front of the detector, whereas the lighter recoiling sample nuclei are transmitted through the absorber layer because of their lower rate of energy loss. The energy of the displaced sample nuclei is reduced by the energy they lose in traveling up to the sample surface through a shallow angle. This enables ERDA to give depth-resolved elemental profiles, in a similar manner as in backscattering spectrometry. Here the kinematic factor, $K_E(\theta_S)$, is defined as the fraction of the ion energy that is transferred to the sample nucleus. For an ion of mass M_1, the kinematic factor of the recoiling nucleus of mass M_2 at an angle θ_S is

$$K_E(\theta_S) = \frac{4M_1 M_2}{(1 + M_1/M_2)^2} \cos^2 \theta_S \qquad (4.13)$$

At a shallow scattering angle, the analytical depth is small because of the reduced ion penetration depth; however, the depth resolution for analysis close to the surface is good. Typical ion beams used for ERDA are MeV ^{4}He ions for the measurement and depth profiling of hydrogen, or high energy (~30 MeV) ^{35}Cl ions for the measurement of a range of lighter elements.

The energy resolution attainable using ERDA is limited by the large energy straggling of the heavy ions entering the sample, the kinematic spread of the displaced light ions scattered through different angles, and the energy straggle of the displaced nuclei leaving the sample and passing through the absorber layer. The poor energy resolution of semiconductor detectors for the measurement of ions heavier than helium has resulted in many ERDA measurements using time-of-flight detectors [74] or magnetic spectrographs [75]. The relatively low cross-sections for forward scattering of sample nuclei necessitates using a high beam current of tens of nanoamps. This, together with the glancing angle geometry required for ERDA, makes good spatial resolution difficult to achieve with a nuclear microprobe. Examples of ERDA with a nuclear microprobe include the measurement of hydrogen distributions across a sectioned catalyst pellet [76], imaging of the distribution of hydrogen in silicon [54], a study of hydrogen mobility [77], and hydrogen measurements in melt inclusions [78].

4.5. COMPARISON WITH OTHER TECHNIQUES FOR ELEMENTAL ANALYSIS

There are many other techniques available for elemental analysis, each with its own strengths and drawbacks. The relative merits of using nuclear microprobe methods can only be properly judged by comparing them with these other techniques. Detailed reviews and comparisons of microprobe techniques may be found elsewhere [4,11,79–81]. Table 4.4 gives a short comparison of the most relevant analytical capabilities of the MeV ion beam techniques for elemental analysis with those of X–ray fluorescence (XRF) [17], electron probe microanalysis (EPMA) [19], Auger electron spectroscopy (AES) [21], secondary ion mass spectrometry (SIMS) [82], low energy ion scattering (LEIS) [83], laser microprobe mass analysis (LAMMA) [84], X–ray photoelectron spectroscopy (XPS) [85], electron spectroscopy for chemical analysis (ESCA) [85], and scanning transmission electron microscopy (STEM) [86].

Unlike SIMS, XPS, LAMMA, and AES, MeV ion techniques give information that is independent of the chemical bonding in the sample. This is a weakness in that no chemical information about the sample may be obtained, but it is also the greatest strength of MeV ion techniques, because this insensitivity to chemical effects allows quantitative elemental analysis. EPMA is more quantitative than PIXE for elemental concentrations greater than 500 ppm, but it is much less sensitive. The spatial resolution attainable with PIXE and backscattering spectrometry is comparable with EPMA, but poorer than the 0.05 μm is attainable with high-resolution SIMS. The main advantages of the nuclear microprobe over electron beam methods are its quantitivity for light elements using NRA, its multielemental high sensitivity using PIXE, depth informa-

TABLE 4.4. Comparison of Techniques for Elemental Analysis[a]

Method	Measured Signal	Spatial Resolution (μm)	Depth Resolution (μm)	Detectable Elements (Atomic Number)	Detection Sensitivity (wppm)	Quantitivity (%)
PIXE	X-rays	0.3	5	>11	0.1	5
RBS	Backscattered incident ions	0.5	0.02	>2	1	3
NRA	Charged particle reaction products	1	0.005	all low Z	0.01	3
ERDA	Forward scattered sample ions	>1	0.005	<15	500	3
SIMS	Sample ions	0.05	0.005	all	0.001–10	50
AES	Auger electrons	0.1	0.001	>2	1,000	50
XPS	Photoelectrons	1000	0.002	>2	1,000	50
ESCA	Photoelectrons	500	0.003	>2	1,000	50
LEIS	Backscattered ions	100	0.0003	>2	1,000	
EPMA	X-rays	0.5	1	>6	100	1
LAMMA	Ions	0.5	0.1	all	0.07–20	50
XRF	X-rays	3	5	>11	1	5
STEM	X-rays	0.0005	—	>11	500	5

[a]The values given in this table are only approximate, since the capabilities of each technique vary widely depending on operating conditions and the analytical requirements of different samples.

tion obtained using backscattering spectrometry, and large analytical depth. The advantages of the nuclear microprobe over SIMS are its quantitivity and its depth-profiling capability using backscattering spectrometry and NRA. Analysis with backscattering spectrometry and NRA is possible without the erosion of the sample needed for SIMS, which can introduce problems of diffusion and pitting, as well as there being uncertainties in the erosion rate.

4.6. ION INDUCED ELECTRON IMAGING

A single MeV ion produces many ionized electrons in the sample because of its large electronic energy loss. Most of the ionized electrons have a kinetic energy of less than 100 eV, so that those escaping from the sample surface come from depths of a few nanometers. Because the number of electrons created close to the surface is proportional to the rate of electronic energy loss by the ions, the number of electrons emitted falls with increasing MeV ion energy and rises with sample atomic number.

Ion induced electron emission is a drawback for quantitative elemental analysis using PIXE, backscattering spectrometry, NRA or ERDA, because these techniques require accurate measurement of the incident ion dose, which is affected by electrons leaving the sample surface. This effect is minimized by applying a low positive voltage to the sample to suppress electron emission. Because the electron emission coefficient does not change much for a range of different metals [87,88] for the same ion type and energy, the usual contrast produced in ion induced electron images is from sample topographical variations. Sternglass [89] predicted that the yield from insulators should be larger than from metals, because low-energy electrons generated in insulators do not undergo the inelastic collisions with bound electrons that limit the depth from which electrons can escape from metals. Much work on ion induced electron emission can be found in the proceedings of the Atomic Collisions in Solids conferences listed in Appendix 1. Detailed discussions on the interpretation of backscattered and secondary electron images generated in a scanning electron microscope can be found in Ref. 90.

The generation of images showing variations in the intensity of electrons produced by the ion beam in a nuclear microprobe was first developed by Younger and Cookson [91] using a photomultiplier detector [92,93]. Figure 4.20a shows two types of photomultiplier detector used for this work. In both cases, low-energy electrons emitted from the sample are accelerated and focused onto the detector front face by a large positive voltage on the aluminium layer. The light generated by the keV electrons striking the scintillator is transmitted through the light pipe to the photomultiplier where it is converted into an electrical signal. This is used to modulate the brightness of a cathode ray tube in synchronism with the position of the scanned beam. Detector type A [92], where the high voltage is screened in the chamber, is commonly used on scanning electron microscopes. In detector type B [93], which is smaller and simpler, the high

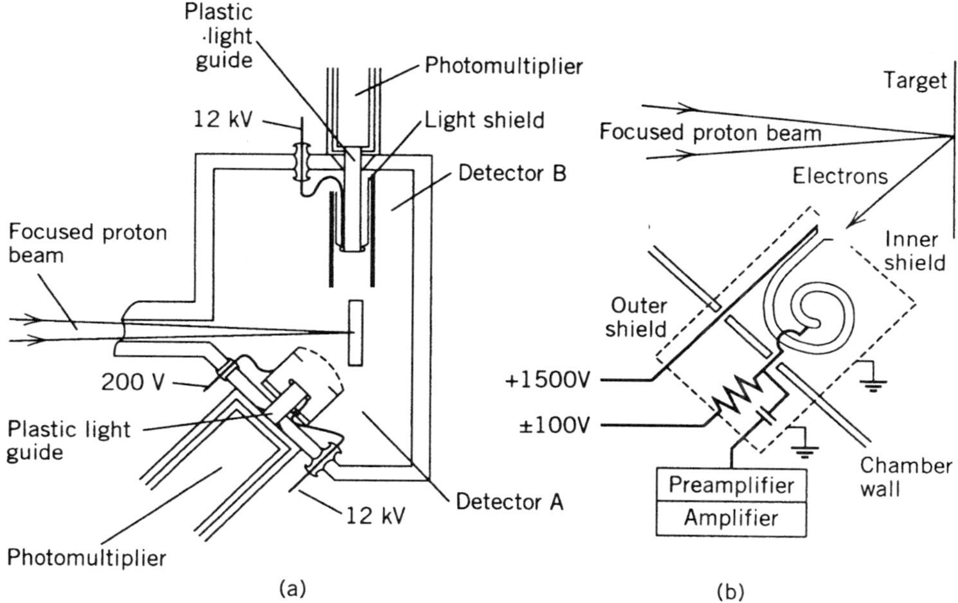

Figure 4.20. Electron imaging systems using (a) a photomultiplier that typically consists of a plastic scintillator attached to the end of a light pipe with a thin evaporated aluminium layer on the front face. Reprinted from Ref. 91 with kind permission from Elsevier Science B.V., Amsterdam, The Netherlands. (b) Channeltron detector where the front end was at a small positive or negative voltage and the center was at a high voltage, such that electrons striking the front were accelerated toward the center. A shield around the channeltron assembly was found to be necessary to reduce the noise level.

voltage is unscreened in the chamber. This presents no problems for MeV ions but would cause unacceptable beam deflection in a scanning electron microscope.

Fluctuations in the ion beam current hitting the sample in the nuclear microprobe, which are unavoidable owing to the slit stabilization method for Van de Graaff accelerators, destroy electron image contrast, but this can be compensated [91] by normalizing the measured electron signal strength to the fluctuating beam current. The effect of beam current fluctuations on the electron image contrast has been recently investigated [94]. The use of a Disctron accelerator, which has no slit stabilization system, has resulted in a constant beam current, which has allowed the generation of ion induced electron images with no compensation required [95].

The spatial resolution of approximately 400 nm, presently attainable in an electron image using a nuclear microprobe [36], is much poorer than the 5 nm resolution attainable using keV electrons in a scanning electron microscope. Because low-energy secondary electrons are only emitted from very close to the sample surface, the spatial resolution is determined by the size of the incident beam spot and not by electron beam spreading in the sample. Ion induced electron imaging using keV heavy ions, in SIMS for example, attains a spatial resolution of approximately 50 nm [96], and again the large ion scattering in the sample does not affect the spatial resolution. In both these other types of microscopy, the beam current stays constant. The simpler process of image formation that this allows has resulted in a widespread use of this imaging method with these types of microscopes.

Another approach [97,98] is to use a channel electron multiplier (channeltron) [99] to generate electron images. Individual electrons emitted from the sample are detected in the form of a charge cascade and then fed into the standard pulse-height-analysis data acquisition system. Figure 4.20b shows a schematic of a typical channeltron system for generating electron images. This is simple and cheap, because it utilizes existing nuclear microprobe electronics and can be used simultaneously with other techniques. However, it limits the maximum number of electrons that contribute to the resultant image because of a limited counting rate, and so results in a statistically noisy electron image. With photomultiplier systems, the variation of a d.c. signal strength is used to modulate the intensity of a cathode ray tube, so a large measured electron flux results in less noisy images. Figure 4.21a compares an optical image of a copper grid with an ion induced electron image measured in a nuclear microprobe using a photomultiplier detector [91]. Figure 4.21b shows an ion induced electron image of a copper grid (top) measured using a channeltron [100]. The lower noise level in the image using a photomultiplier detector compared with the channeltron image is obvious, resulting in much clearer topographical contrast. Figure 4.21b also shows an extracted horizontal linescan across the line indicated in the electron image. The counts measured using the channeltron are compared with the copper X–ray counts measured with PIXE. The enhanced electron yield compared with the X–ray yield from the edges of the copper grid can be seen in the overlaid linescans in Figure 4.21b, demonstrating that care must be taken in interpreting the observed electron image contrast.

Examples of ion induced electron images of microcircuit devices are shown in Figure 4.22, measured with a photomultiplier detector [101,102]. In each case, there is good contrast between the light-colored metallizations and the darker insulating regions of the devices. References 103 and 104 show other examples of ion induced electron imaging of microcircuit devices by the same microprobe group at Albany. An example of an ion induced electron image of a microelectronic device measured using a channeltron is shown in Figure 4.23. Figure 4.23a shows an optical image of a metallization finger, and Figure 4.23b shows an electron image of the area within the dashed box. The arrowed line indicates the position of a linescan used to determine the surface layer composition using backscattering spectrometry both on and away from the contact pad

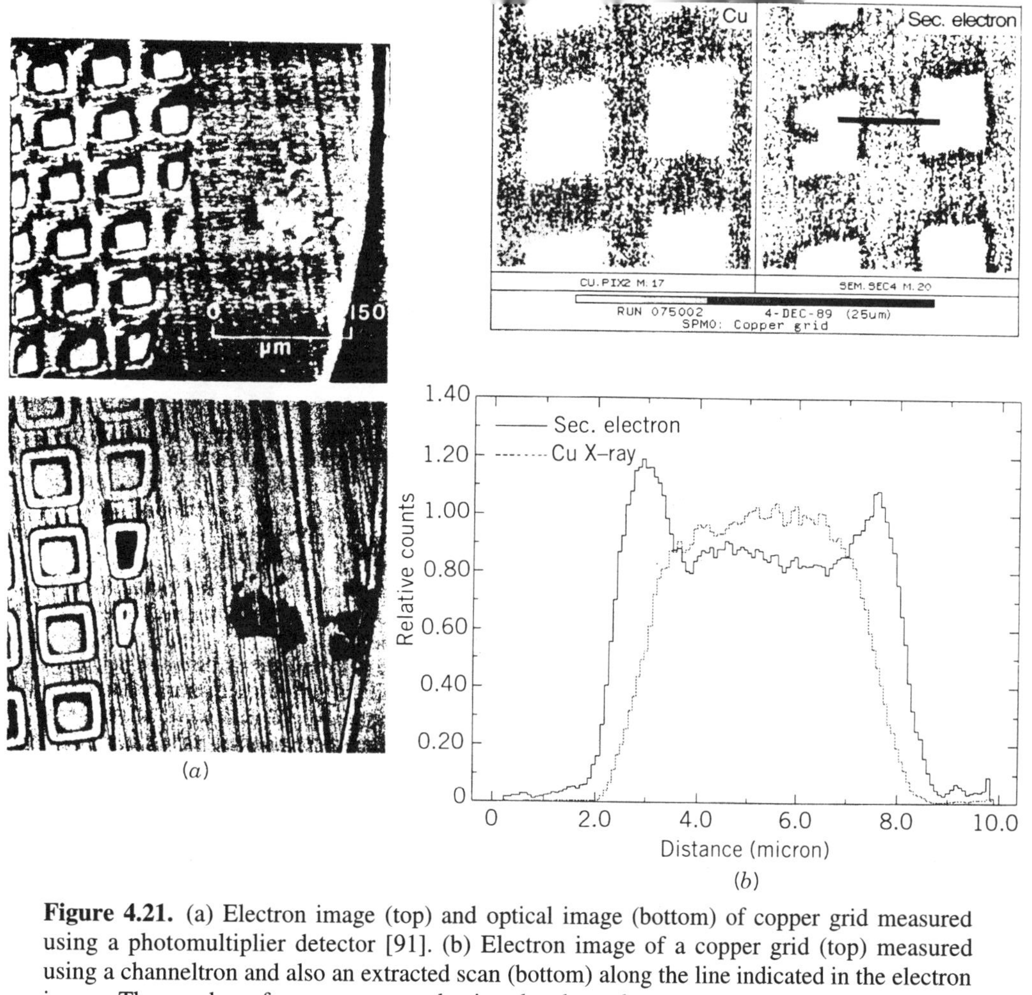

Figure 4.21. (a) Electron image (top) and optical image (bottom) of copper grid measured using a photomultiplier detector [91]. (b) Electron image of a copper grid (top) measured using a channeltron and also an extracted scan (bottom) along the line indicated in the electron image. The number of counts measured using the channeltron are compared with the copper X-ray counts measured with PIXE [100]. Reprinted with kind permission from Elsevier Science B.V., Amsterdam, The Netherlands.

Figure 4.22. Electron images of various device test structures measured over 5-s periods using a beam current of 100 to 200 pA of 2 MeV ^{4}He ions focused to a spot size of 2 μm. Reprinted from Refs. 101 and 102 with kind permission from Elsevier Science B.V., Amsterdam, The Netherlands, and W.G. Morris.

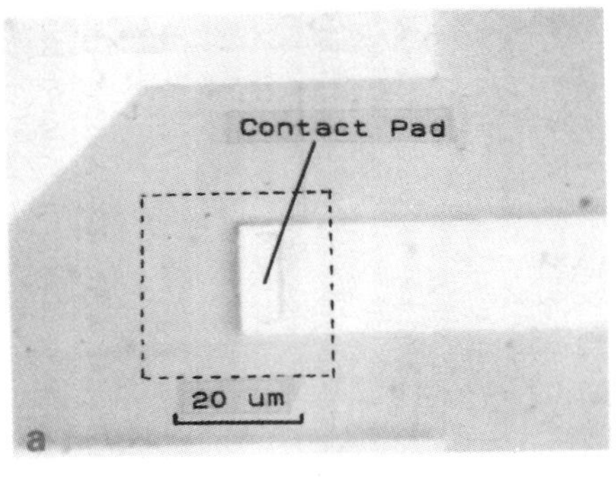

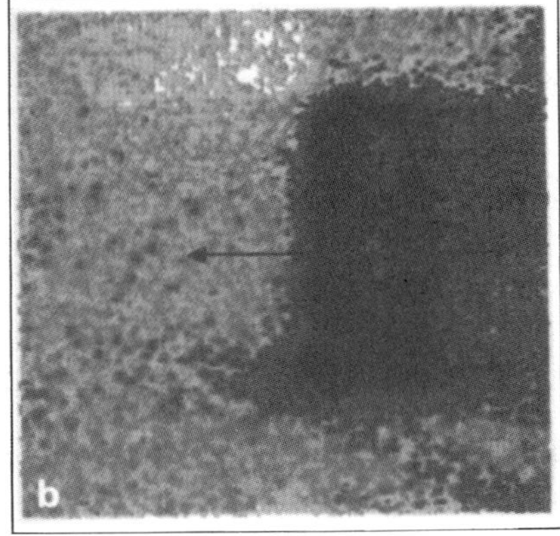

Figure 4.23. (a) Optical image of a metallization finger with a 4×12 μm^2 contact pad. (b) Electron image of this area measured using the channeltron assembly shown in Figure 21b with 10 pA of 2 MeV ^{4}He ions focused to a 1 μm diameter spot over a 10-min period. Reprinted from Ref. 53 with kind permission from Elsevier Science B.V., Amsterdam, The Netherlands.

[52]. The fact that the measurement period for this electron image was much longer than those for the images in Figure 4.22 emphasizes the noisy nature of electron images generated using a channeltron.

In an electron microscope, there are several other modes of generating image contrast with electrons emitted from materials that have not been investigated using a nuclear microprobe [90]. Voltage contrast microscopy, for example, involves detecting changes in the electron image of a microelectronic device due to different bias voltages affecting the number of electrons emitted from different device areas. Images with contrast due to magnetic variations have been produced from samples such as magnetic tapes, where the magnetic field alters the emitted electron trajectories and hence the measured electron intensity.

4.7. SCANNING TRANSMISSION ION MICROSCOPY

The energy loss of an ion transmitted through a thin sample depends on the elemental composition and thickness, that is, the areal density, where

$$\text{areal density} = \int_{E_o}^{E_r} \left[\frac{dE}{d(\rho z)} \right]^{-1} dE \tag{4.14}$$

where E_o is the initial ion energy and E_r is the remaining ion energy after passing through the sample. With STIM, the transmitted ion energies and number of ions at each pixel within the scanned area are measured using a semiconductor detector located behind the sample and used to generate an image showing variations in the areal density. Scanning transmission ion microscopy was developed primarily as a method of quantitatively imaging the areal density distribution of thin biological samples [105–107] and identifying features of interest for subsequent analysis with PIXE or backscattering spectrometry. It has also been used to normalize PIXE images to the measured variation of the sample areal density [108] and to image the density of small living insects in air [109]. An important feature for biomedical research is that STIM can image variations in the areal density of unstained tissue sections for subsequent PIXE analysis, thus avoiding the serious problem of the contamination of samples with the chemical dyes that are normally used to highlight features. Much of the work in the biological field is reviewed in Refs. 107 and 110, and Ref. 111 provides a detailed account of the theory and applications of STIM. This method has also been used to image the distribution of metallization layers on microelectronic devices [112,113] as described in Chapter 7; other materials applications are described in Chapter 9.

Although energy loss images have been generated in a scanning electron microscope with electron-transparent samples, this has not been widely applied. Also, STIM can be likened to X–ray microradiography [114] in which the X–ray attenuation gives a variation in the transmitted X–ray intensity through thin samples, with a spatial resolution of approximately 5 μm. STIM has also been used as an imaging method for other ion energy regimes [115–117], but only the use of MeV ions in conjunction with a nuclear microprobe is described here.

The large number of collisions of the MeV ions in the sample results in energy straggle, lateral spread, and angular spread of the beam emerging from the rear of the sample. Table 4.5 compares the lateral beam spread, rate of energy loss, and range and energy straggle for ions of different masses and energies passing through 1 μm thick layers of silicon and gold [118]. The resolution of small variations in areal density improves with the use of heavier ions, because they have a higher rate of energy loss, but is limited by the ion energy straggle. Areal density resolution increases with increasing ion mass, because the rate of energy loss increases much faster than the energy straggle. The lateral beam spread for the different ions passing through a 1 μm thick layer of a particular material does not vary greatly, but the spread is significantly greater for ions passing through denser material. Although MeV ^{1}H ions can pass through materials more than 100 μm in thickness, the range of heavier MeV ions drastically decreases with increasing ion mass; this limits the use of heavy ions for STIM measurements. Most STIM measurements have consequently been carried out using MeV ^{4}He and ^{1}H ions.

The transmission of 3 MeV ^{1}H ions and 2 MeV ^{4}He ions through carbon are compared in Figure 4.24, using work described in Ref. 111. The measured ion energy spread through a homogeneous material increases as $[E_d^2 + (2.35\Omega_B)^2]^{1/2}$, where E_d is the full-width-at-half-maximum (FWHM) detector resolution and Ω_B is the Bohr energy straggle of the ions after passing through the material. A simplified expression for the energy straggle has been derived [107], which allows simple calculation of the expected energy straggle from the sample density

$$\Omega_B^2 = \Lambda \int_0^T \rho(z)\,dz \tag{4.15}$$

TABLE 4.5. Comparison of Different Ions for STIM Analysis[a]

Sample	Ion	Energy (MeV)	r_e (nm)	dE/dz (keV/μm)	Ω_B^2 (keV)	Range (μm)
Silicon	p	5.5	5.9	13	2.6	251
$\rho = 2.3$ g/cm^3	Li	6.0	6.3	356	6.1	14.5
	C	12.0	6.1	1046	8.2	10.8
Gold	p	3.0	32	714	8.3	27.0
$\rho = 19.3$ g/cm^3	p	5.5	22	502	9.9	67.3
	Li	6.0	34	1050	37	5.7
	C	12.0	35	3120	84	4.2

[a]The table shows the root-mean-square lateral beam spreading r_e rate of energy loss dE/dz, and energy straggling Ω_B^2 for a 1 μm thick sample and three ions. The ion range in each element is also shown.

SOURCE. Reprinted in modified form from Ref. 118 with kind permission from Elsevier Science B.V., Amsterdam, The Netherlands.

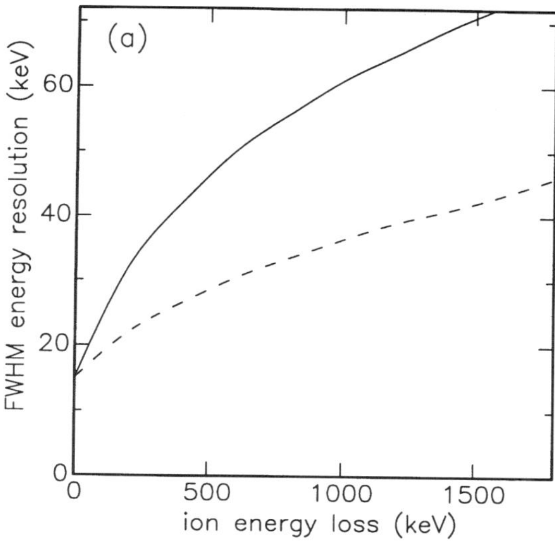

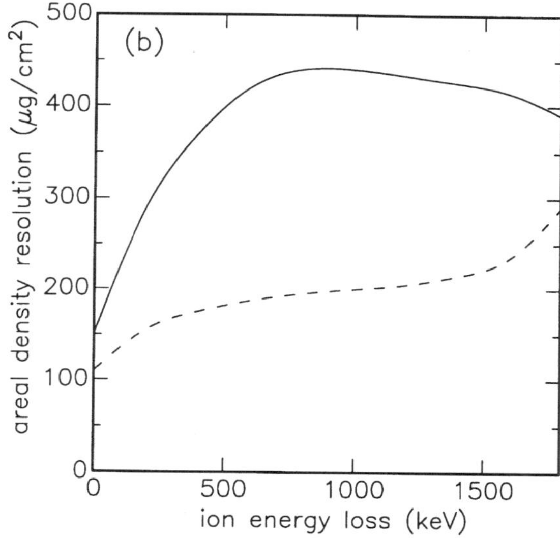

Figure 4.24. The transmission of 3 MeV ^{1}H ions (full lines) and 2 MeV ^{4}He ions (dashed lines) through carbon. (a) FWHM ion energy resolution for a semiconductor detector resolution of 15 keV. (b) FWHM areal density resolution as a function of energy loss assuming one ion per pixel, with the curve for ^{4}He ions multiplied by a factor of 10. (c) Root-mean-square angular spread as a function of ion energy loss. (d) Fraction of transmitted ions measured in a detector with a maximum half-angle of 3.6° about the beam axis. (e) Root-mean-square radial spread of 3 MeV ^{1}H ions. Modified from Ref. 111 with permission of G.S. Bench.

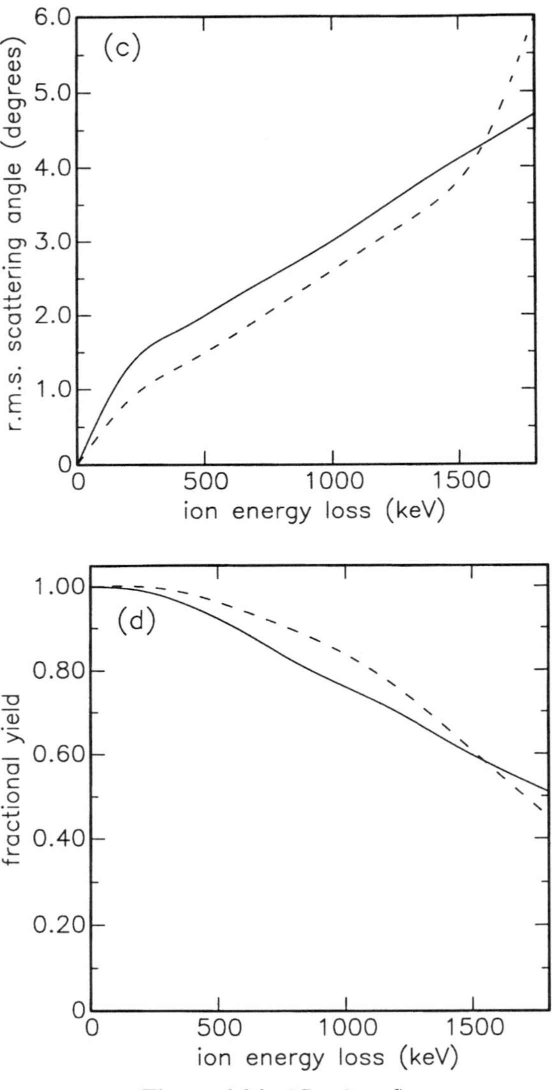

Figure 4.24. (*Continued*)

where T is the sample thickness, and Λ depends on the atomic number of both the incident ion and the particular sample nucleus. Figure 4.24a shows the increase in the measured energy resolution as a function of transmitted ion energy loss. Figure 4.24b shows the resultant areal density resolution as a function of ion energy loss, based on Figure 4.24a and the rate of ion energy loss, assuming one ion per pixel in the measured STIM image. For ^{1}H ions, the areal density resolution first worsens with energy loss, and then improves because the

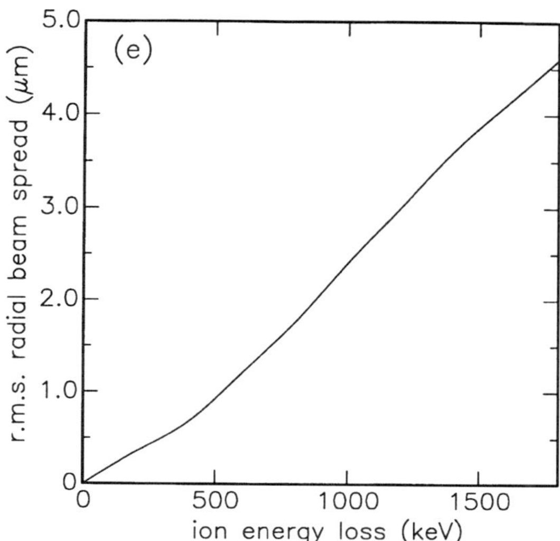

Figure 4.24. (*Continued*)

rate of energy loss increases faster than the energy straggle. For 2 MeV ^{4}He ions, the areal density resolution increases rapidly toward a large energy loss, because the rate of energy loss decreases while the energy straggling continues to increase. For N ions per pixel, the uncertainty in the measurement of the average ion energy and therefore in the areal density decreases by a factor of $1/\sqrt{N}$. Figure 4.24b can be scaled by this; so for five ions per pixel, the areal density resolution is better than 200 μg/cm^2 for 3 MeV ^{1}H ions and better than 10 μg/cm^2 for 2 MeV ^{4}He ions.

The root-mean-square scattering angle as a function of ion energy loss is shown in Figure 4.24c, based on Ref. 119. The angular spread of the transmitted ions increases with the square root of the material thickness and is inversely proportional to the ion energy. This angular spread decreases the number of ions that are measured in a given solid angle behind the sample. Figure 4.24d shows the fraction of ions that are measured in a detector with a maximum half-angle of 3.6° centered about the beam axis, based on Figure 4.24c. In a material containing regions of widely differing areal density, there may be different numbers of ions measured at each pixel by a detector with a small acceptance angle. The measurement of the ion fraction that only undergoes a small angular scatter has been called *bright-field* imaging [120], and the use of an off-axis detector measuring scattered ions has been called *dark-field* imaging. It has been pointed out that this scattering contrast may degrade the accuracy of measuring the ion energy loss [111], and the use of a large acceptance angle detector was recommended to avoid this. With a large acceptance angle detector, a very small beam current of about 1 fA can be used for STIM, enabling a spatial resolution of approximately 100 nm [121]. There is no reason why even higher resolution STIM probes should not be attained when limitations from ions scattered by the microprobe collimators, vibrations, and stray magnetic fields are eliminated.

Lateral scattering, resulting from the ion–electron collisions, causes the transmitted ions to lose energy in regions of the sample outside their defined image pixel. This can result in an incorrect energy loss value and limits the spatial resolution achievable in a STIM image. Figure 4.24e shows the lateral ion spread for 3 MeV ^{1}H ions through carbon. Lateral scattering is typically a few percent of the distance that the ion travels through the material and only becomes a serious limitation to STIM resolution for thick samples. This ability allows fine lateral variations in areal density to be imaged with STIM.

Figure 4.25 shows a STIM image of a carbon replica of a diffraction line grating [122] measured using 2 MeV ^{4}He ions focused to a spot size of approximately 100 nm. From this image, a minimum resolvable thickness variation of approximately 20 nm was deduced.

Figure 4.25. STIM image of a diffraction grating, using five 2 MeV ^{4}H ions per pixel. The line spacing is 463 nm and the grating linewidth is 100 nm. The scan size is approximately $8 \times 7\ \mu m^2$. Reprinted from Ref. 122 with kind permission from Elsevier Science B.V., Amsterdam, The Netherlands.

4.7.1. Methods of Scanning Transmission Ion Microscopy Image Formation

STIM images are generated by measuring either the transmitted ion energy or variations in the number of ions at each image pixel within the scanned area. Only one ion per pixel is required in principle to measure the energy loss, but, in practice, owing to energy straggling, several ions are required to generate low-noise images. There are obviously many similarities between image formation with STIM and CSTIM (described in Chapter 5, where relevant sections should also be consulted).

The simplest method of generating STIM images uses the map mode of data collection described in Chapter 2, whereby the transmitted ion energy spectrum from the entire scanned area is divided up into windows. The number of counts within each window is used to generate images showing areas of different energy loss within the sample. All the measured images can then be combined to generate a single image showing the average transmitted ion energy at each pixel. The effective energy resolution in the resultant image is poor, because the average ion energy at each pixel is partially limited by the width of the energy window on the transmitted energy spectrum.

The best method of generating a STIM image is using an event-by-event data acquisition system, as described in Chapter 2. The measured data set of the ion energy loss values at each pixel can then be manipulated in different ways to give the best image contrast. Average and median processing [123, 124] of event-by-event STIM data improve the accuracy of measuring the ion energy loss at each pixel by utilizing the information from all the measured ions, which reduces the effect of the measured ion energy spread. With average processing, the outlying events in the energy spectrum at each pixel, arising from slit scattering or other forms of noise, have a large effect on the average energy loss and so can introduce noise into the image. This can be avoided by discarding outlying events, but care must be taken that information is not lost by doing this. With median processing, the measured energy loss values in each pixel are ordered by increasing energy loss, and the central value is chosen. This defines the location of edges present to within a pixel and effectively eliminates noise in the image.

A spectrum of measured ion energies at each pixel can be influenced by unresolved spatial structures smaller than the beam spot size and the lateral spread of the beam. With both average and median processing, a single energy loss value is chosen at each image pixel to generate a STIM image, leading to a loss of information from the processed energy spectrum at each pixel. A method for overcoming this has now been developed [124]. First, anomalous ion energy loss values are discarded if they are too far away from the average or median value, and the remaining data is used to construct an average or median energy loss image in the normal manner. Moments of the remaining ion energy loss value at each pixel are then calculated by summing various powers of the energy differences away from the average or median. Images are then constructed from these moments, which emphasize regions of the sample that have spectra with similar deviations from symmetry. The moment of a pixel (x, y) is defined as

$$M(x, y) = \frac{\sum_{i=1}^{N}(E_i - E)^p}{N} \qquad (4.16)$$

where N is the measured number of ions in the image pixel, E is the calculated value of the average or median energy loss in the pixel, E_i is the energy loss of each individual ion, and p is the integer-valued order of the moment. This definition can be refined further by defining an odd and an even moment, which use just those ions with an energy respectively less than or greater than the average or median energy. This process emphasizes outlying values of energy loss, and, in particular, odd moments can provide structural information that may not be obtained from an average or median image. In the presence of noise from slit scattered ions or incomplete charge collection within a detector, the lower second moment was found to give optimum image contrast [124]. Figure 4.26 shows a lower second moment STIM image of a 220 nm diameter latex sphere, measured using 2 MeV ^{4}He ions focused to a spot size of approximately 100 nm. This moment image was more sharply defined than the corresponding median processed image and had a lower noise level in the surrounding area.

Because most work with STIM has been carried out for biomedical analysis, examples from this research field are shown in Figure 4.27 to illustrate the

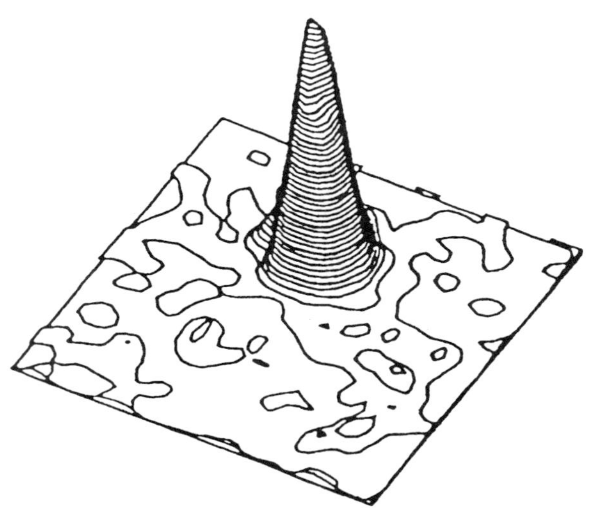

Figure 4.26. Three-dimensional depiction of a 1×1 μm^2 lower second moment STIM image of a 220 nm diameter latex sphere. Reprinted from Ref. 124 with kind permission from Elsevier Science B.V., Amsterdam, The Netherlands.

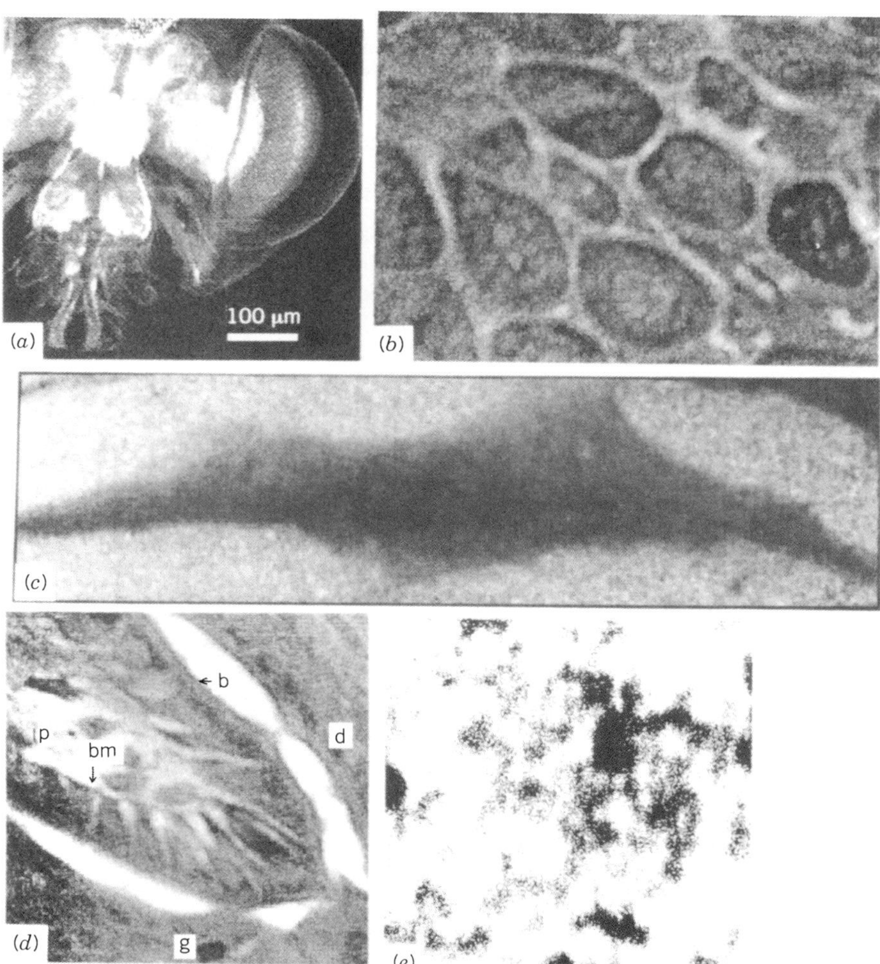

Figure 4.27. (a) 4 MeV ^{1}H ion STIM image of a part of the head of a fruit fly [125], where light areas indicate higher energy loss. In the rest of these images, dark represents high ion energy loss. (b) 90 × 70 μm^2 image of a section of mouse kidney tissue [126], using 2 MeV ^{4}He ions. (c) 90 × 24 μm^2 image of a fibroblast, using 2 MeV ^{4}He ions [127]. The dark area corresponds to the cell nuclear membrane. (d) 95 × 70 μm^2 image of section of mouse kidney tissue, using 2 MeV ^{4}H ions [126]. (e) 100 × 100 μm^2 image using 3 MeV ^{1}H ions of the same area of unstained brain tissue as was shown in Figure 4.7 [113]. All reprinted with kind permission from Elsevier Science B.V., Amsterdam, The Netherlands.

ability of STIM to resolve fine spatial and depth structure. Figure 4.27a shows a 4 MeV ^{1}H ion STIM image of a part of the head of a fruit fly [125]. The mouth parts are at the bottom of the image and an eye is visible in the top right regions. Figure 4.27b shows a 90×70 μm^2 STIM image of mouse ileum tissue, and Figure 4.27d shows a 68×68 μm^2 STIM image of mouse kidney tissue, measured using 2 MeV ^{4}He ions in both cases [126]. The tissue structure and many features can be identified in these images. Figure 4.27c shows a 90×24 μm^2 STIM image of a fibroblast cell, using 2 MeV ^{4}He ions, where the dark area in the center corresponds to the cell nuclear membrane [127]. Figure 4.27e shows a 100×100 μm^2 STIM image using 3 MeV ^{1}H ions of the same area of unstained brain tissue shown in Figure 4.7. This image was used to locate the senile plaque for PIXE and backscattering spectrometry [113]. Figure 4.27a to 4.27d were collected with an event-by-event data acquisition system, whereas Figure 4.27e was collected using the map mode of data collection by which the spectrum was split into windows.

4.7.2. Mixed Beams

Recent work on the use of mixed beams with a nuclear microprobe [128] has increased the range of ion types and energies available for STIM (and also for IBIC microscopy described in Chapter 6). A mixed beam incorporates two different types of ion with closely similar magnetic rigidity. Because the incident ion charge state is important for this work, it is explicitly included in this section.

The field required in magnetic quadrupole lenses to focus a particular ion depends on the ion's magnetic rigidity R_m

$$R_m = \frac{M_1 E_o}{Z^2} \tag{4.17}$$

where M_1, E_o, and Z are the incident ion mass, energy, and charge state, respectively. Heavier or higher energy ions are more difficult to focus because of their greater rigidity. Higher charge state ions are conversely easier to focus. The typical maximum ^{4}He$^+$ ion energy that the magnetic quadrupoles used in nuclear microprobes can focus is about 2.5 MeV. Because their range is less than 10 μm in most materials, this limits their use for STIM analysis. The magnetic rigidity of ^{4}He^{2+} ions is only a quarter that of ^{4}He$^+$ ions according to Eq. (4.17) so they can be focused with four times the incident energy, enabling a greater sample thickness to be analyzed. Two different approaches for producing and focusing ^{4}He^{2+} ions with a nuclear microprobe have recently been developed to extend the range of applications of STIM using a single-ended Van de Graaff accelerator. Both methods rely on using a mixture of hydrogen and helium gas in the source bottle and the use of magnetic quadrupoles as the focusing elements of the nuclear microprobe.

The first method involves initially focusing $^1H_2^+$ ions (molecular hydrogen) onto the sample in the microprobe chamber, produced from hydrogen in the source bottle. The small beam current of $^4He^{2+}$ ions, produced from the helium present, is then extracted by altering the strength of the velocity selector at the exit of the source bottle. As Eq. (4.17) shows, these two ions have nearly the same magnetic rigidity, with a small adjustment to the acceleration voltage, they travel through the accelerator and round the analyzer magnet with the same path to the microprobe object collimator. They focus on the sample surface at the same quadrupole lens strengths, enabling $^4He^{2+}$ ions to be used for STIM analysis even though it is not easily possible to directly focus such a small beam current typically available of this ion type.

The second method developed to increase the ^4He ion energy available for use with a nuclear microprobe involves first focusing $^1H^+$ ions in the microprobe and then extracting $^4He^+$ ions from the source bottle. These ions are stripped to produce $^4He^{2+}$ ions by passing them through a very thin carbon foil just before the analyzer magnet (i.e., post-acceleration stripping). This process gives $^1H^+$ ions and $^4He^{2+}$ ions that are the same energy and same magnetic rigidity. They focus with the same quadrupole lens current but have widely differing ranges, a property that is useful for extending the use of ion microtomography, which is described in the following section.

Figure 4.28 shows examples of energy spectra produced using both these methods of generating $^4He^{2+}$ ions. In both cases, the focused ions are transmitted through a 2 μm thick copper grid. Figure 4.28a shows the transmitted energy

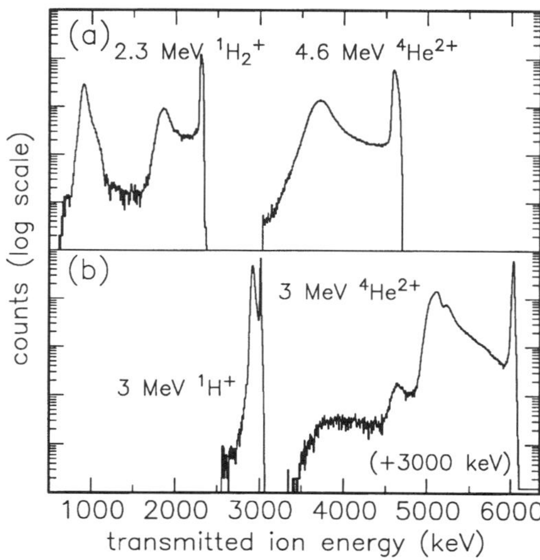

Figure 4.28. Energy spectra for mixed beams transmitted through a 2 μm thick copper grid. (a) 2.3 MeV $^1H_2^+$ ions and 4.6 MeV $^4He^{2+}$ ions. (b) 3 MeV $^1H^+$ ions and 3 MeV $^4He^{2+}$ ions (shown offset by +3,000 keV along the horizontal axis for clarity).

spectra for 2.3 MeV $^1H_2^+$ ions and 4.6 MeV $^4He^{2+}$ ions, which were obtained using the first method described above, with a terminal voltage of 2.3 MV. The lower energy peaks in each spectrum are caused by ions passing through the grid bars, and the full-energy peaks are caused by ions passing through the grid holes. The two lower-energy peaks for the $^1H_2^+$ ions are caused by 2.3 MeV $^1H_2^+$ ions dissociating into two 1.15 MeV $^1H^+$ ions on impact with the grid bars. The middle peak is due to both ions being measured, whereas the low-energy peak is due to just one of the ions being detected. Figure 4.28b shows transmitted energy spectra for 3 MeV $^1H^+$ ions and 3 MeV $^4He^{2+}$ ions, which were produced using the second method described above, with a terminal voltage of 3 MV. The measured spatial resolution in the resultant STIM images of the copper grid using the $^4He^{2+}$ ions was comparable in both cases with that initially measured using (a) 2.3 MeV $^1H_2^+$ ions and (b) 3 MeV $^1H^+$ ions, demonstrating the capability for producing well-focused beams of $^4He^{2+}$ ions for STIM analysis.

4.8. ION MICROTOMOGRAPHY

Scanning transmission ion microscopy uses a focused MeV ion beam to generate images showing variations in material areal density as viewed from a specific orientation, which is usually with the sample surface perpendicular to the incident ion beam. This gives a two-dimensional view of the areal density, and methods of viewing the structure in three dimensions are now described.

The use of stereo imaging to provide information on small density variations by generating two STIM images from different orientations [111,129] only gives qualitative information on the three-dimensional density distribution. If STIM images of a thin sample are measured at many different orientations, then a three-dimensional image of the areal density can be reconstructed. This technique has been called ion microtomography (IMT), and it has been used to resolve submicron density variations. The use of ion energy loss to generate images in conjunction with computed tomography was discussed in the 1960s [130], and demonstrated in the 1970s [131,132]. Two-dimensional ion microtomography was demonstrated in the 1970s using 25 MeV ^1H ions [133] and recently with 1.4 MeV/nucleon argon ions [134], but the majority of development in this field has used ^1H ions of less than 10 MeV [135,136]. Ion microtomography has much in common with other forms of tomography, such as X–ray tomography, but relies on ion energy loss rather than X–ray absorption as the image contrast mechanism. This can be advantageous in giving both a reduced radiation dose, and increased sensitivity and spatial resolution in some cases.

The requirements on the nuclear microprobe hardware and data acquisition system for generating tomographic STIM images have been described [122, 135,136–138]. The regions of interest on the sample are placed as close as possible to the axis of rotation of the goniometer, and then the sample is rotated about one axis in incremental steps of typically 1° through 360°. A STIM image is measured at each orientation, usually by raster scanning the beam over the sample and measuring a predefined number of ions, usually approximately 10, at each pixel. The choice of the type and energy of the ion used for tomography is a compromise between using heavy ions for high energy loss contrast and using light ions to ensure that the ion beam is transmitted through the sample at all the angles. The ideal solution of a high-energy, heavy-ion beam is only possible using large tandem Van de Graaff accelerators, but nuclear microprobes using these beams have not yet achieved the high spatial resolution attainable with light ion beams.

The measured STIM images at the various orientations are median processed to reduce the uncertainty associated with the measured energy at each pixel [139]. Each image is then converted to show the areal density variation of the material, using tabulated stopping powers for individual elements and Bragg's rule for the additivity of energy loss for a mixture of different elements present. Any irregularities in the precession of the rotation axis that result in the features shifting laterally with rotation angle are corrected. All the images are then combined using a filtered back-projection algorithm [140] to give a single object volume containing all the quantitative three-dimensional data associated with the sample. The data set can then be manipulated to show the internal sample structure as viewed from any orientation, and the surface layers can be sliced off to reveal detail below.

The resolvable structure is affected by the reconstructed density determination, the number of orientations used, the number of pixels used in each image, and the form of the reconstruction algorithm. Lateral scattering of the ions through the sample is a major limitation, but an algorithm has been recently developed that compensates for this in certain geometries [119]. With cubic or spherical samples, the total ion energy loss does not greatly change with rotation angle. However with a planar sample the energy loss is low where the sample is perpendicular to the beam and becomes large as the sample rotates to be nearly parallel with the incident beam. Approaches to solving this problem include the use of different energy ions for different rotation angles and the use of mixed beams as described in Section 4.7.2. Algorithms are also being developed that can reconstruct the sample from a small number of orientations, and from a small angular range; these limited angle reconstructions should increase the types of sample that can be analyzed.

The biggest limitation for IMT at present is the data acquisition speed of approximately 10 kHz, resulting in very long measurement times required to generate adequate statistics. In an effort to increase the count rate, a time-of-flight data acquisition system, which can measure more than 50,000 ions/s [141], has been developed for ion microtomography. Figure 4.29 shows a tomographic image of four square silicon pillars analyzed with 3 μm diameter 8 MeV ^{1}H ions [141]. This image involved measuring a total of 90 million ions with the time-of-flight detector in a total time of 45 min, representing an average data acquisition rate of approximately 33 kHz. This seems to be a very promis-

Figure 4.29. Tomographic image of four square silicon pillars using 8 MeV ^{1}H ions measured with a time-of-flight detector. Reprinted from Ref. 141 with kind permission from Elsevier Science B.V., Amsterdam, The Netherlands.

ing approach to solving the problem of obtaining the tomographic data in a reasonable time.

Backscattering spectrometry can also generate three-dimensional tomographic images. Because this method gives depth-resolved information, only a single orientation of the sample is required [142]. A backscattering spectrum is measured at each pixel within the scanned area using the event-by-event data acquisition method. The stored data set can be manipulated to generate three-dimensional backscattering images viewed from different angles. By further manipulation of the data set, two-dimensional images viewed in both cross-section and plan view can be extracted. Work on correcting tomographic backscattering images for effects of scattering kinematics which otherwise cause distorted images have been described [143]. Examples of tomographic backscattering images are shown in Figures 7.33 and 7.34. Tomographic backscattering images of crystal structures can also be generated using a similar procedure, and in this case there is no need for elaborate corrections to compensate for the differing kinematic factors of the elements present, because the contrast depends only on the defect density present in the material [144]. Recently, it has also been demonstrated that tomographic PIXE images can be generated and used to show the different locations of various elements present in rocks [145] and biological tissue [146].

4.9. ION BEAM INDUCED LUMINESCENCE

Ion beam induced luminescence (IBIL) is an emerging technique for use with a nuclear microprobe; it shows considerable promise for the identification of chemical phases, as a check for the presence of certain trace elements, and for monitoring the buildup of ion induced defects in luminescent materials. This technique appears to have particular promise when it is used in conjunction with the trace element sensitivity of PIXE and also the depth profiling capabilities of backscattering spectrometry. At present, IBIL measurements are carried out with a beam current as high as 100 pA because of poor detection efficiency. However, more efficient detectors, currently under construction, will allow beam currents to be considerably reduced.

The term luminescence is used to describe photon radiation in the infrared/visible/ultraviolet region of the spectrum. As such, luminescence usually arises from excited electrons associated with the outer shells of atoms or from the band structure of solids. The study of luminescence excited by photon or electron irradiation has a long history. Techniques such as cathodoluminescence using electrons and photoluminescence using photons for luminescence excitation are well established. The application of these techniques to the characterization of semiconductors and other inorganic solids has been extensively reviewed [147–149].

The literature on IBIL is less extensive; although a recent book by Townsend, Chandler and Zhang [150, chapter 4] provides an excellent overview. Also, the wealth of data already obtained from cathodoluminescence and photoluminescence techniques provides a useful comparison for IBIL studies. This is because, for excitation energies above the band gap energy, the luminescence spectrum is essentially independent of the method of excitation [151]. Luminescence spectra can be influenced by the relative penetration depth of the various excitation methods. The large penetration depth of MeV ions offers an advantage over keV electrons and visible photons in its ability to investigate buried structures. This is seen, for example, in Figure 1.4, where the energy deposited is plotted as a function of depth for 3 MeV ^{1}H ions and 38 keV electrons in silicon. In this example, the ^{1}H ions can produce luminescence from considerably below that of the electrons.

A further application of IBIL has been the study of the Moon. This has been done with natural IBIL produced as a result of the irradiation of the Moon by the solar wind, in particular energetic protons. This luminescence contributes an appreciable fraction to the total light from the Moon [152].

4.9.1. Luminescence Mechanisms

Elemental analysis using a technique such as PIXE is made possible by the fact that the inner shell atomic energy levels which are the source of the X–ray emission are essentially unaffected by the surrounding crystal field. Only in special circumstances is this true for IBIL. The fundamental mechanism for IBIL is where a MeV ion, as with a photon or an electron, creates electron–hole (*eh*) pairs in a solid. A radiative recombination of this *eh* pair then gives rise to luminescence phenomena. Competing with this is nonradiative recombination, which is also possible in many cases.

Several recombination mechanisms are possible. In general the mechanisms that lead to the radiative recombination of valence and of conduction band electrons are dependent on such things as sample temperature, the valence state of any impurity atoms (e.g., Cr^{2+} and Cr^{3+}) and the defect concentration. This makes quantitative luminescence measurements difficult except in special circumstances. A schematic diagram of some of the recombination processes responsible for luminescence in semiconductors is shown in Figure 4.30. Some of these mechanisms are responsible for luminescence from pure, undamaged material. This is *intrinsic* luminescence. Alternatively, the luminescence can be associated with defects, dopants, or other impurities in the material. This is *extrinsic* luminescence.

MeV ions cause more damage than photons and electrons. This results in the production of nonradiative recombination centers that offer energetically more favorable deexcitation pathways. This can result in the degradation of the luminescence yield with ion dose. The reduction of luminescence yield with increasing ion dose has proved to be useful in monitoring the optical effects of ion implantation with unfocused ion beams [150,153].

In most cases, the luminescence signal from the recombination is strongly temperature dependent. In fact, sample cooling brings two major benefits to the practical use of the IBIL signal. Firstly, peaks in the IBIL spectrum become sharper as the sample is cooled. This is because of the reduced thermal motion of the *eh* pairs. Secondly, the signal strength increases as the sample is cooled. For example, Dyer and Mathews [154] found that the light output from type IIa diamonds increased by 170% when the sample was cooled from 290 K to 80 K. Discussion of the mechanisms involved is beyond the scope of this book and the interested reader is referred to Ref. 147 (Chapter 3).

4.9.1.1. Donor–Acceptor Pair Recombination   In some materials, dopants or defect structures can capture electrons or holes. An electron is captured by an acceptor and a hole by a donor. When the captured donor–acceptor pair recombine, luminescence can be produced with an energy that depends on the binding energies of the electron to the donor and the hole to the acceptor, as well as inversely on the pair separation. Because the separation can only have values that are integral multiples of the lattice constant, the luminescence forms a band with fine structure. At room temperature, this is usually not resolved and a broad Gaussian distribution is typically observed. In an early model [155], the so called A-band luminescence from diamond was attributed to donor–acceptor pair recombination, with a nitrogen dimer acting as the donor and substitutional boron as the acceptor. However, later workers proposed alternative models based on an impurity–defect complex or even on other emission mechanisms based on defects [see 156, and references therein]. A typical example of the A-band is shown by the IBIL spectrum from type IIa diamond in Figure 4.31.

4.9.1.2. Recombination of Rare Earth and Transition Elements   As is obvious to anyone who has observed the behavior of geological samples

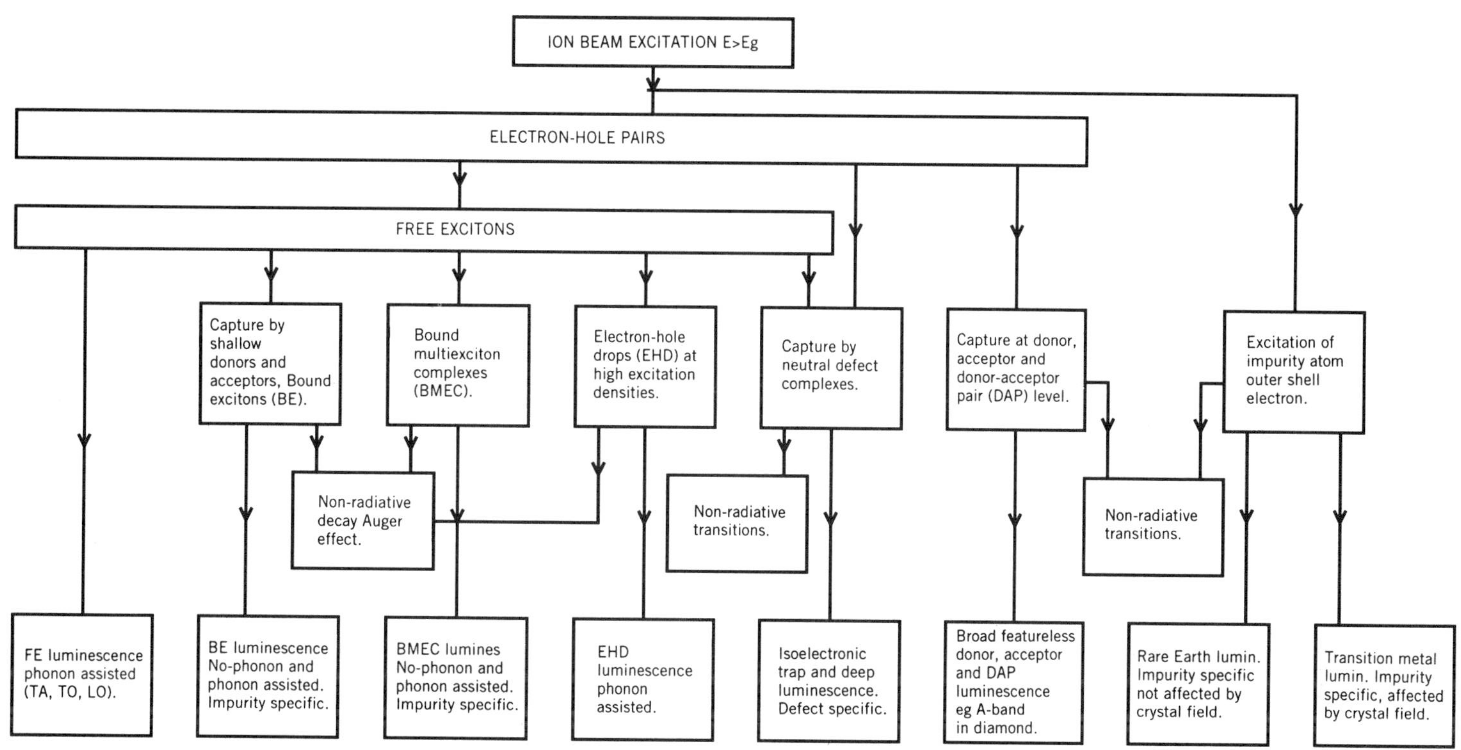

Figure 4.30. Recombination processes in semiconductors, some of which give rise to luminescence.

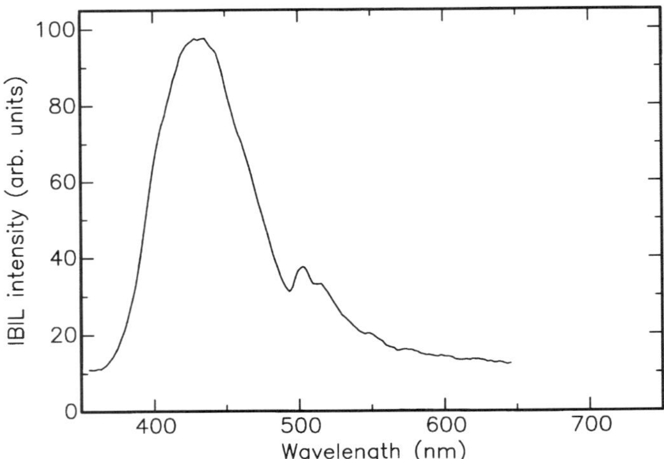

Figure 4.31. IBIL spectrum from type IIa diamond showing the A-band spectrum. 3 MeV ^{1}H ions were used with the sample at room temperature.

being irradiated by a charged particle beam, a colorful range of beautiful luminescence phenomena can be seen [157]. Luminescence from these geological samples does not usually stem from band gap transitions; instead it is the result of electronic transitions across the outer atomic shells with impurity centers, which are usually rare earth or transition metal elements. These are known as *activators*. Other elements can act as *inhibitors* and can quench the luminescence that would otherwise be produced by the activators. Luminescence of this form can be understood with reference to the electron configuration of the activator atom. For example, Mn^{2+} is an activator in a commonly used scintillator screen material $ZnSiO_4$:Mn, also known as the mineral willemite. This is very efficient at producing green light and was used in Chapter 3 as the light-producing medium on the image screen for the grid shadow method of diagnosing nuclear microprobe lens aberrations; it is also used for the ion channeling patterns in Chapter 5.

Transition metals like Mn^{2+} have partially filled 3d outer shells, which make transitions highly sensitive to the surrounding crystal field. The resulting luminescence spectrum is a broad, featureless Gaussian shape, as can be seen from the IBIL spectrum of willemite shown in Figure 4.32. Ions such as Fe^{2+} are common inhibitors of this type of luminescence.

Rare earth metals, on the other hand, have an unfilled 4f shell, which is shielded from the surrounding crystal field by a filled 5s,p,d outer shell. Consequently the transitions that occur at rare earth metal impurities are sharp. This can be seen in the IBIL spectrum from the red phosphor of a video display tube shown in Figure 4.33a. In this example, the IBIL signal arose from europium doping of the phosphor. The europium signal is clearly seen in the

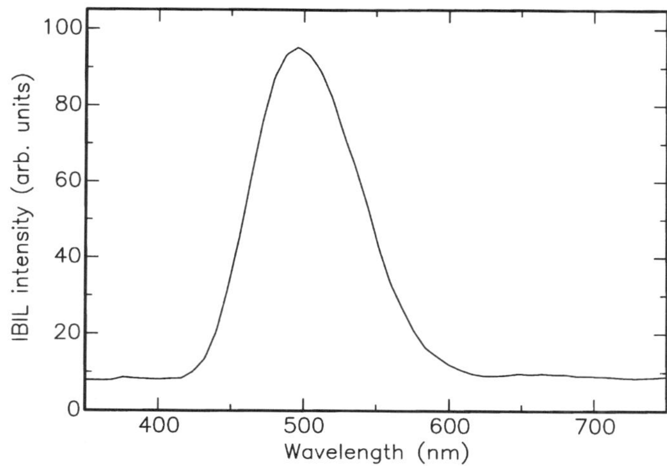

Figure 4.32. IBIL spectrum of willemite ($ZnSiO_4$:Mn). 3 MeV ^{1}H ions were used with the sample at room temperature.

PIXE spectrum, shown in Figure 4.33b, which was collected simultaneously with the IBIL spectrum. Rare earth metals are usually found in the trivalent state with the exception of Eu^{2+} and Sm^{2+}. Generally, the wavelength of transitions that occur at transition metals and rare earth elements are influenced by the valence state of the impurity and its coordination in the crystal.

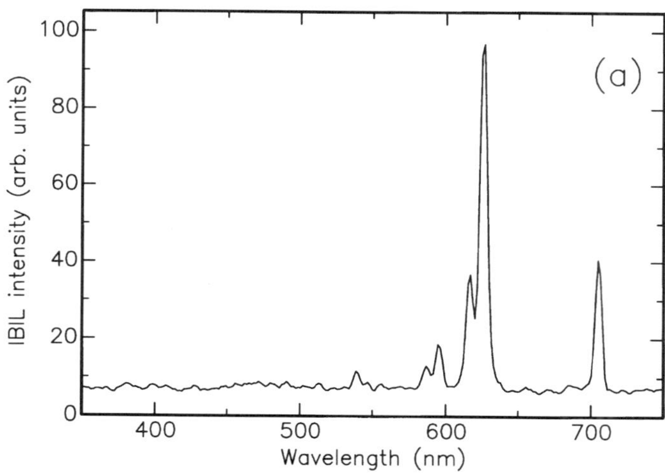

Figure 4.33. (a) IBIL spectrum of the red phosphor from a video display tube showing a number of relatively sharp peaks in the spectrum. 3 MeV ^{1}H ions were used with the sample at room temperature. (b) Corresponding PIXE spectrum, collected simultaneously.

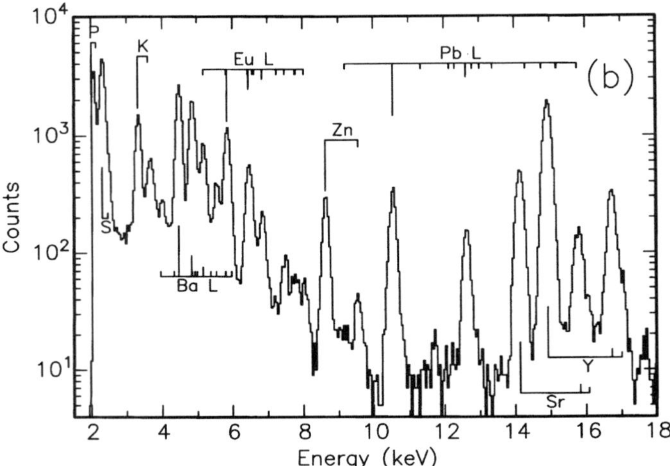

Figure 4.33. (*Continued*)

4.9.1.3. Free and Bound Excitation Recombination   The production of *eh*
pairs in a material can sometimes lead to the formation of an electron-hole
bound state, known as an exciton. The associated energy levels of the exciton lie within the band gap of the material and the level structure is hydro-
genic (i.e., analogous to that of hydrogen). Because the luminescence occurs
at energies close to the band gap energy, this is called *edge emission*. Exci-
tonic recombination radiation produced by cathodoluminescence or photoluminescence has been extensively used to measure characteristics of natural
and synthetic diamonds, both doped and undoped [158,159,160]. As yet these
signals have not yet been exploited for IBIL with a nuclear microprobe, but
such signals are potentially useful because of their sensitivity to electrically
active impurities.

4.9.2. Applications of Ion Beam Induced Luminescence

The first reported use of an IBIL signal from a nuclear microprobe to image structures in a sample was the work of Yang et al. [161], who produced panchromatic
images of the total IBIL signal intensity between 350 and 600 nm. They studied zircons ($ZrSiO_4$), teeth and the effect of ion irradiation damage. In zircon
grains, the IBIL signal revealed considerable inhomogeneities, which were possibly associated with trace impurities of yttrium, iron, or nickel, that were detected
simultaneously with PIXE. The trace impurities were presumably acting as activators and inhibitors. They found that the very high count rate of the IBIL signal
was useful for preliminary assessment of the homogeneity of the grain. This was
also true for the IBIL images produced from the teeth. Thus, IBIL spectra provided evidence for the ionization state of activators in several inorganic samples.

More detailed studies of the IBIL signal from zircons revealed evidence for extrinsic luminescence from Dy^{3+} and Er^{3+} [162]. A very detailed series of studies of IBIL from zircons and other minerals, including some synthetic samples, has been published by the same group [163,164; see also 165].

The strong IBIL signal from the A-band peaks at 415 nm for type IIa diamonds. This has been used in Melbourne [166] to monitor the buildup of damage in microprobe irradiated diamonds. The IBIL signal strength was found to decrease by a factor of two after a dose as low as 2.5×10^{15} 3 MeV ^1H ions/cm^2.

In another study of diamonds, IBIL was used to investigate homoepitaxial diamond films. Evidence was obtained for a correlation between an additional peak in the IBIL spectrum at 540 nm, which may be related to trace element contamination from transition metals acting as activators or it may be from wider donor–acceptor pair separation in the regrown material. In this study, the combined imaging of the green IBIL signal with the PIXE signal from the trace elements was employed [167]. A representative spectra and corresponding images for this work are given in Figure 4.34. Other applications of IBIL, still under development, include the study of optical fibers [168] and trace element activators in video display phosphors [166].

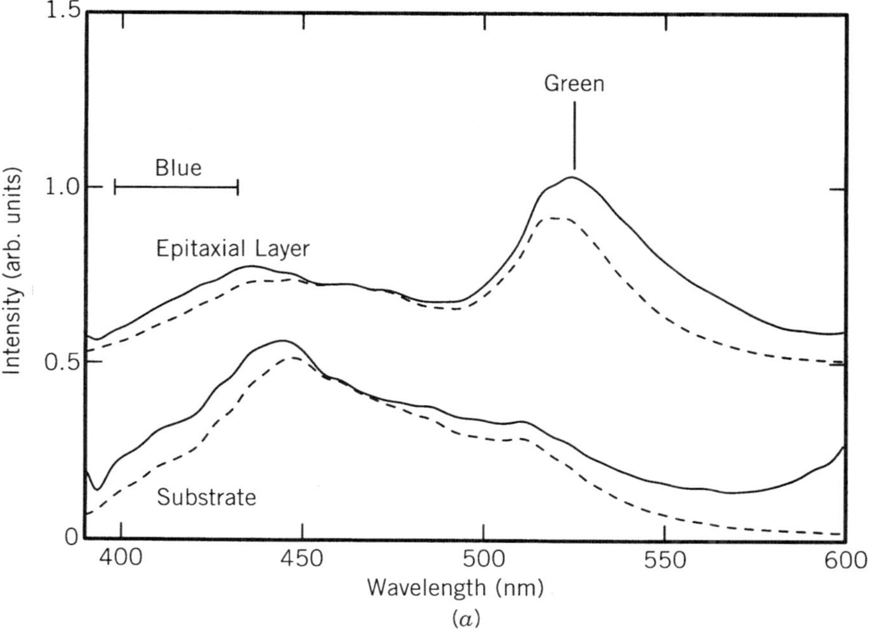

Figure 4.34. IBIL spectra and corresponding 200×200 μm^2 IBIL images from two different areas of CVD homoepitaxial zones on a diamond, showing the green luminescence between 500 and 550 nm. In the spectra, the dashed lines are the raw data and the solid lines are corrected for detector efficiency. In the images, light regions correspond to regions of high signal strength.

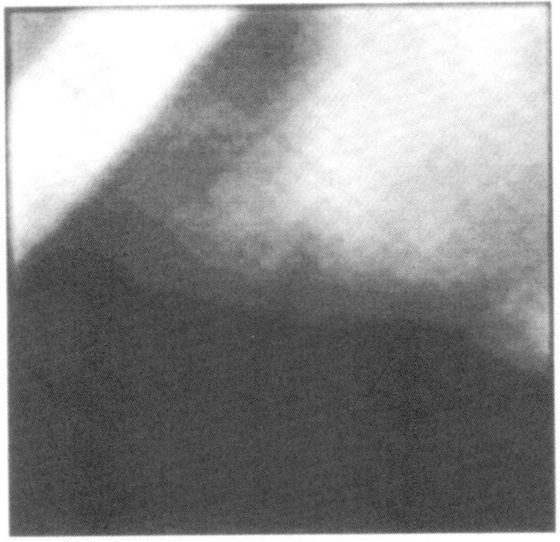 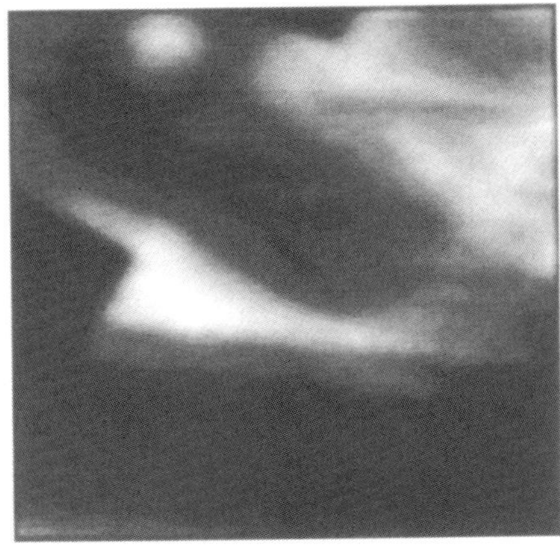

(b)

Figure 4.34. (*Continued*)

REFERENCES

1. J.W. Mayer and E. Rimini eds., *Ion Beam Handbook for Materials Analysis*. Academic Press, New York (1977).

2. W–K. Chu, J.W. Mayer, and M.A. Nicolet, *Backscattering Spectrometry*. Academic Press, New York (1978).

3. J.B. Marion and F.C. Young, *Nuclear Reaction Analysis Graphs and Tables*. North–Holland, Amsterdam (1968).

4. L.C. Feldman and J.W. Mayer, *Fundamentals of Surface and Thin Film Analysis*. North–Holland, Amsterdam (1986).

5. J.R. Bird, P. Duerden, and D.J. Wilson, *Nucl. Sci. Appl.* **B1:**357 (1983).

6. G. Deconninck, *Introduction to Radioanalytical Physics*. Elsevier, Amsterdam (1978).

7. J.F. Ziegler, ed., *New Uses of Ion Accelerators*. Plenum Press, New York (1975).

8. S.A.E. Johansson and J.L. Campbell, *PIXE: A Novel Technique for Elemental Analysis*. Wiley, New York (1988).

9. J.R. Bird and J.S. Williams, eds., *Ion Beams for Materials Analysis*. Academic Press, Orlando (1989).

10. M.B.H. Breese, F. Watt, and G.W. Grime, *Ann. Rev. Nuc. Part. Sci.* **42:**1 (1992).

11. F. Watt and G.W. Grime, eds., *Principles and Applications of High Energy Ion Microbeams*. Adam Hilger, Bristol (1987).

12. S.A.E. Johansson and T.B. Johansson, *Nucl. Instr. Meth.* **137:**473 (1976).

13. M.R. Khan and D. Crumpton, *CRC Crit. Rev. Anal. Chem.* **11:**103, 161 (1981).

14. I.V. Mitchell and K.M. Barfoot, *Nuc. Sci. Appl.* **1:**99 (1981).

15. J.A. Cookson, *Nucl. Instr. Meth.* **181:**115 (1981).

16. S.A.E. Johansson, J.L. Campbell, and K.G. Malmqvist, *Particle Induced X-ray Emission Spectrometry*, Wiley, New York (1995).

17. F.S. Goulding and J.M. Jaklevic, *Nucl. Instr. Meth.* **142**:323 (1981).

18. S.J.B. Reed, *Electron Microprobe Analysis*, Cambridge Monographs on Physics. Cambridge University Press, London (1975).

19. K.F.J. Heinrich, *Electron Beam X–ray Microanalysis*. Van Nostrand, New York (1981).

20. J.D. Garcia, *Phys. Rev.* **A1**:280 (1970).

21. A. van Oostrom, *Surf. Sci.* **89**:615 (1979).

22. S. Raman, In: ed. S. Datz, *Applied Atomic Collision Physics*, Vol. 4. Academic Press, New York (1983).

23. H. Paul, *Nucl. Instr. Meth.* **192**:11 (1982).

24. J.A. Bearden, *Rev. Mod. Phys.* **39**:78 (1967).

25. G.W. Grime and F. Watt, *Beam Optics of Quadrupole Probe–forming Systems*. Adam Hilger, Bristol (1984).

26. E. Clayton, PIXAN: The Lucas Heights PIXE Analysis Computer Package, AAEC/M1, 13 (1986).

27. J.A. Maxwell, J.L. Campbell, and W.J. Teesdale, *Nucl. Instr. Meth.* **B43**:218 (1989).

28. C.G. Ryan, D.R. Cousens, S.H. Sie, W.L. Griffin, G.F. Suter, and E. Clayton, *Nucl. Instr. Meth.* **B47**:55 (1990).

29. G.W. Grime, F. Watt, A.R. Duval, and M. Menu, *Nucl. Instr. Meth.* **B54**:353 (1991).

30. J.L. Campbell and J.A. Cookson, *Nucl. Instr. Meth.* **B3**:185 (1984).

31. J.A. Cookson, *Nucl. Instr. Meth.* **165**:477 (1979).

32. S. Raman and C.R. Vane, *Nucl. Instr. Meth.* **B3**:71 (1984).

33. F. Folkmann, In: ed. O. Meyer, G. Linker, and O. Kappëler, *Ion Beam Surface Layer Analysis*, vol 2. Plenum, New York (1976).

34. F. Bosch, A. El Goresy, W. Herth, W. Martin, R. Nobiling, B. Povh, H.–D. Reiss, and K. Traxel, *Max Planck Institut fur Kernphysik Internal Report*, MPIH–1979–V29 (1979).

35. G.J.F. Legge, C.D. McKenzie, and A.D. Mazzolini, *J. Microsc.* **117**:185 (1979).

36. G.W. Grime and F. Watt, *Nucl. Instr. Meth.* **B75**:495 (1993).

37. J.A. Cookson and F.D. Pilling, *Phys. Med. Biol.* **21**:965 (1976).

38. B.H. Kusko, T.A. Cahill, R.A. Eldred, and R.N. Schwab, *Nucl. Instr. Meth.* **B3**:689 (1984).

39. G. Demortier, *Nucl. Instr. Meth.* **B54**:334 (1991).

40. T.A. Elliott and G.W. Grime, *Nucl. Instr. Meth.* **B77**:537 (1993).

41. A.J. Criddle, H. Tamana, J. Spratt, K.J. Reeson, D. Vaughan, and G.W. Grime, *Nucl. Instr. Meth.* **B77**:444 (1993).

42. J.P. Landsberg, B. McDonald, and F. Watt, *Nature* **360**:65 (1992).

43. J.P. Landsberg, D. Phil. Thesis, University of Oxford (1992).

44. J. Pallon and J. Knox, *Scanning Microsc.* **7**:1207 (1993).

45. G.W. Grime, F. Watt, R.A. Wogelius, and B. Jamtveit, *Nucl. Instr. Meth.* **B77:**410 (1993).

46. C.G. Ryan and D.N. Jamieson, *Nucl. Instr. Meth.* **B77:**203 (1993).

47. C.G. Ryan, D.N. Jamieson, C.L. Churms, and J.V. Pilcher, *Nucl. Inst. Meth.* **B104:**157 (1995).

48. E. Clayton and C.G. Ryan, *Nucl. Instr. Meth.* **B49:**161 (1990).

49. T. Awaya, *Nucl. Instr. Meth.* **165:**449 (1978).

50. L.R. Doolittle, *Nucl. Instr. Meth.* **B9:**344 (1985).

51. J.A. Knapp, J.C. Banks, and B.L. Doyle, *Nucl Instr. Meth.* **B85:**20 (1994).

52. M.B.H. Breese, J.A. Cookson, J.A. Cole, and A.E. Ledbury, *Nucl. Instr. Meth.* **B54:**12 (1991).

53. J.A. Cookson and M.B.H. Breese, *Nucl. Instr. Meth.* **B50:**208 (1990).

54. D.N. Jamieson and C.G. Ryan, *Nucl. Instr. Meth.* **B77:**415 (1993).

55. B.L. Doyle and N.D. Wing, *IEEE Trans. Nuc. Sci.* **30:**1214 (1983).

56. R. Amirikas, D.N. Jamieson, and S.P. Dooley, *Nucl Instr. Meth.* **B77:**110 (1993).

57. J.W. McMillan, P.M. Hirst, F.C.W. Pummery, J. Huddleston, and T.B. Pierce, *Nucl. Instr. Meth.* **149:**83 (1978).

58. D. Dieumegard, B. Maurel, and G. Amsel, *Nucl. Instr. Meth.* **168:**93 (1980).

59. C. Engelmann, *Atom Energy Rev.* (Suppl 2):107 (1981).

60. Y. Llabador, P. Moretto, and H. Guegan, *Nucl Instr. Meth.* **B77:**123 (1993).

61. W. Wang, J. Zhu, A. Le, R. Lu, S. Lin, J. Bao, D. Jiang, J. She, and A. Ji, *Nucl. Instr. Meth.* **B77:**349 (1993).

62. R.A. Jarjis, *Nuclear Cross-section Data for Surface Analysis*, Physics Department, University of Manchester Report, July (1979).

63. J.A. Cookson and F.D. Pilling, Harwell Report AERE–R 6300 (1970); also J.A. Cookson, A.T.G. Ferguson, and F.D. Pilling, *J. Radioanal. Chem.* **12:**39 (1972).

64. N. Toulhoat, P. Trocellier, P. Massiot, J. Gosset, K. Trabelsi, and T. Rouaud, *Nucl. Instr. Meth.* **B54:**312 (1991).

65. P. Courel, P. Trocellier, M. Mosbah, N. Toulhoat, J. Gosset, P. Massiot, and D. Piccot, *Nucl. Instr. Meth.* **B54:**429 (1991).

66. G. Demortier and S. Mathot, *Nucl. Instr. Meth.* **B77:**312 (1993).

67. G. Demortier, ed., Microanalysis of Light Elements (Hydrogen to Neon), *Nucl. Instr. Meth.* **B66:**1–305 (1992).

68. J.W. McMillan, *Nucl. Instr. Meth.* **B30:**474 (1988).

69. T.B. Pierce, J.W. McMillan, P.F. Peck, and I.G. Jones, *Nucl. Instr. Meth.* **118:**115 (1974).

70. J.R. Bird, B.L. Campbell, and P.B. Price, *Atom. Energy Rev.* **12:**275 (1974).

71. D. Heck, *Atomkernenerg.–Kerntech.* **46:**187 (1986).

72. B.L. Doyle, D.S. Walsh, and S.R. Lee, *Nucl Instr. Meth.* **B54:**244 (1991).

73. J.L'Écuyer, C. Brassard, C. Cardinal, J. Chabal, L. Deschênes, J.P. Labrie, B. Terrault, J.G. Martel, and R.St–Jacques, *J. Appl. Phys.* **47:**381 (1976).

74. H.A. Rijken, S.S. Klein, and M.J.A. de Voigt, *Nucl. Instr. Meth.* **B64:**395 (1992).

75. G. Dollinger, T. Faestermann, and P. Maier–Komor, *Nucl. Instr. Meth.* **B64:**422 (1992).

76. C.J. Sofield, L.B. Bridwell, and C.J. Wright, *Nucl. Instr. Meth.* **191:**79 (1981).

77. D. Heck, *Atomkernenerg.–Kerntech.* **46:**187 (1986).

78. M. Mosbah, R. Clochiatti, J. Tirira, J. Gosset, P. Massiot, and P. Trocellier, *Nucl. Instr. Meth.* **B54:**298 (1991).

79. J.A. Cookson, *Nucl. Instr. Meth.* **197:**255 (1982).

80. W. Reuter, *Nucl. Instr. Meth.* **218:**391 (1983).

81. T.J. Shaffner, *Scanning Electr. Microsc.* **1:**11 (1986).

82. K. Wittmaack, *Nucl. Instr. Meth.* **168:**343 (1980).

83. E.N. Haeussler, *Surf. Interface Anal.* **2:**134 (1980).

84. R. Wechsung, F. Hillenkamp, R. Kaufmann, H.J. Heinen, and M. Schurmann, *Scanning Electr. Microsc.* **2:**279 (1979).

85. D. Briggs, *Handbook of X–ray and Ultraviolet Photo–electron Spectroscopy.* Heydrich, London (1977).

86. J.J. Hren, J.I. Goldstein, and D.C. Joy eds. *Analytical Electron Microscopy*, Plenum Press, New York (1979).

87. B. Aarset, R.W. Cloud, and J.G. Trump, *J. Appl. Phys.* **25:**1365 (1954).

88. H.P. Beck and R. Langkau, *Z. Naturforsch.* **A30:**981 (1975).

89. E. Sternglass, *Phys. Rev* **108:**1 (1957).

90. L. Reiner, *Scanning Electron Microscopy*, Springer–Verlag, New York (1985).

91. P.A. Younger and J.A. Cookson, *Nucl. Instr. Meth.* **158:**193 (1979).

92. T. E. Everhart and R.F.M. Thornley, *J. Sci. Instr.* **37:**246 (1960).

93. O.C. Wells and C.G. Bremmer, *Rev. Sci. Instr.* **41:**1034 (1970).

94. M. Takai, Y. Agawa, K. Ishibashi, K. Hirai, S. Namba, K. Inoue, and Y. Kawata, *Nucl. Instr. Meth.* **B54:**279 (1991).

95. K. Ishibashi, K. Inoue, K. Yokoyama, Y. Kawata, H. Fukuyama, and M. Iwasaki, *Nucl. Instr. Meth.* **B75:**526 (1993).

96. P.D Prewett and G.L. Mair, *Focused Ion Beams from Liquid Metal Ion Sources.* Research Studies Press, Britain (1991).

97. K. Traxel and A. Mandel, *Nucl. Instr. Meth.* **B3:**594 (1984).

98. H. Kneis, B. Martin, R. Nobiling, B. Povh, and K. Traxel, *Nucl. Instr. Meth.* **197:**79 (1982).

99. J.G. Timothy and R.L. Bybee, *Rev. Sci. Instr.* **49:**1192 (1978).

100. G.W. Grime, M. Dawson, M. Marsh, I.C. McArthur, and F. Watt, *Nucl. Instr. Meth.* **B54:**52 (1991).

101. W.G. Morris, H. Bakhru, and A.W. Haberl, *Nucl. Instr. Meth.* **B15:**661 (1986).

102. W.G. Morris, W. Katz, H. Bakhru, and A.W. Haberl, *J. Vac. Sci. Technol.* **B3:**391 (1985).

103. H. Bakhru, W.G. Morris, and A. Haberl, *Nucl. Instr. Meth.* **B10/11:**697 (1985).

104. W.G. Morris, S. Fesseha, and H. Bakhru, *Nucl. Instr. Meth.* **B24/25:**635 (1987).

105. J.C. Overley, R.C. Connolly, G.E. Sieger, J.D. MacDonald, and H.W. Lefevre, *Nucl. Instr. Meth.* **218:**39, 43 (1983).

106. R.M. Sealock, A.P. Mazzolini, and G.J.F. Legge, *Nucl. Instr. Meth.* **218:**217 (1983).

107. H.W. Lefevre, R.M.S. Schofield, G.S. Bench, and G.J.F. Legge, *Nucl. Instr. Meth.* **B54:**363 (1991).

108. H.W. Lefevre, R.M.S. Schofield, J.C. Overley, and J.D. MacDonald, *Scanning Microscopy* **1:**879 (1987).

109. H.W. Lefevre, R.M.S. Schofield, and D.R. Ciarlo, *Nucl. Instr. Meth.* **B54:**47 (1991).

110. G.J.F. Legge, *Nucl. Instr. Meth.* **B40/41:**675 (1989).

111. G.S. Bench, PhD thesis, University of Melbourne (1991).

112. M.B.H. Breese, F. Watt, G.W. Grime, and P.J.C. King, *Inst. Phys. Conf. Ser.* **117:**101 (1991).

113. M.B.H. Breese, J.P. Landsberg, P.J.C. King, G.W. Grime, and F. Watt, *Nucl. Instr. Meth.* **B64:**505 (1992).

114. B.P. Richards and P.K. Footner, *Microelectronics* **J15:**5 (1984).

115. Z.H. Cho, M. Singh, and G.C. Huth, *Ann. N.Y. Acad. Sci.* **306:**223 (1978).

116. W.H. Escovitz, T.R. Fox, and R. Levi–Setti, *Ann. N.Y. Acad. Sci.* **306:**183 (1978).

117. R. Levi Seti and T.R. Fox, *Nucl. Instr. Meth.* **168:**139 (1980).

118. A.J. Antolak and A.E. Pontau, *Nucl. Instr. Meth.* **B54:**371 (1991).

119. W.R. Wylie, R.M. Bahnsen, and H.W. Lefevre, *Nucl. Instr. Meth.* **79:**245 (1970).

120. R.M. Sealock, D.N. Jamieson, and G.J.F. Legge, *Nucl. Instr. Meth.* **B29:**557 (1987).

121. G.S. Bench and G.J.F. Legge, *Nucl. Instr. Meth.* **B40/41:**655 (1989).

122. G.S. Bench, A. Saint, G.J.F. Legge, and M. Cholewa, *Nucl. Instr. Meth.* **B77:**175 (1991).

123. J.C. Overley, R.M.S. Schofield, J.D. MacDonald, and H.W. Lefevre, *Nucl. Instr. Meth.* **B30:**337 (1990).

124. G.S. Bench, H.W. Lefevre, and G.J.F. Legge, *Nucl. Instr. Meth.* **B54:**378 (1991).

125. R.M.S Schofield, H.W. Lefevre, J.C. Overley, and J.D. Macdonald, *Nucl. Instr. Meth.* **B30:**398 (1988).

126. B.J. Kirby and G.J.F. Legge, *Nucl. Instr. Meth.* **B77:**268 (1993).

127. G.L. Allan, PhD. Thesis, University of Melbourne (1989); also G.L. Allan, J. Camakaris, and G.J.F. Legge, *Nucl. Instr. Meth.* **B54:**175 (1991).

128. A. Saint, M.B.H. Breese, and G.J.F. Legge, *Nucl. Instr. Meth.* (unpublished).

129. H.W. Lefevre, R.M.S. Schofield, G.S. Bench, and G.J.F. Legge, *Nucl. Instr. Meth.* **B54:**363 (1991).

130. A.M. Cormack, *J. Appl. Phys.* **34:**2722 (1963).

131. K.M. Crowe, T.F. Budinger, J.L. Cahoon, V.P. Elischer, R.H. Huesman, and L.L. Kanstein, Lawrence Berkeley Report LNL–3182 (1975).

132. A.M. Cormack and A.M. Koehler, *Phys. Med. Biol.* **21:**560 (1976).

133. A. Ito, H. Koyama–Ito, *Nucl. Instr. Meth.* **B3:**584 (1984).

134. B.E. Fischer and C. Mühlbauer, *Nucl. Instr. Meth.* **B47:**271 (1990).

135. A.E. Pontau, A.J. Antolak, D.H. Morse, A.A. Berkmoes, J.M. Brase, D.W. Heikkinen, H.E. Martz, and I.D. Proctor, *Nucl. Instr. Meth.* **B40/41**:646 (1989).

136. G.S. Bench, K.A. Nugent, M. Cholewa, A. Saint, and G.J.F. Legge, *Nucl. Instr. Meth.* **B54**:390 (1991).

137. A.E. Pontau, A.J. Antolak, D.H. Morse, and D.L. Weirup, *Nucl. Instr. Meth.* **B54**:383 (1991).

138. A.E. Pontau, A.J. Antolak, and D.H. Morse, *Nucl. Instr. Meth.* **B45**:503 (1990).

139. G.S. Bench, H.W. Lefevre, and G.J.F. Legge, *Nucl. Instr. Meth.* **B54**:378 (1991).

140. R.A. Brooks and G.Di Chiro, *Phys. Med. Biol* **21**:689 (1976).

141. M.L. Roberts, D.W. Heikkinen, I.D. Proctor, A.E. Pontau, G.T. Olona, T.E. Felter, D.H. Morse, and B.V. Hess, *Nucl. Instr. Meth.* **B77**:225 (1993).

142. Y. Mokuno, Y. Horino, A. Chayahara, M. Kiuchi, K. Fujii, M. Satou, and M. Takai, *Nucl. Instr. Meth.* **B77**:373 (1993).

143. M. Takai, Y. Katayama, and A. Kinomura, *Nucl. Instr. Meth.* **B77**:229 (1993).

144. D.N. Jamieson, *Nucl. Instr. Meth.* **B104**:533 (1995).

145. A. Saint, M. Cholewa, and G.J.F. Legge, *Nucl. Instr. Meth.* **B75**:504 (1993).

146. R.M.S. Schofield and H.W. Lefevre, *Nucl. Instr. Meth.* **B77**:217 (1993).

147. B.G. Yacobi and D.B. Holt, *Cathodoluminescence Microscopy of Inorganic Solids.* Plenum Press, New York (1990).

148. B.G. Yacobi and D.B. Holt, *J. Appl. Phys* **59**(4):R1 (1986).

149. R.A. Stradling and P.C. Klipstein, *Growth and Characterisation of Semiconductors.* Adam Hilger, Bristol (1990), p. 135.

150. P.D. Townsend, P.J. Chandler, and L. Zhang, *Optical Effects of Ion Implantation.* Cambridge University Press, London (1994).

151. C.A. Klein, *J. Appl. Phys.* **39**:2029 (1967).

152. Z. Kopal, *Sci. Am.* Volume **212**(5):28 (1965).

153. P.D. Townsend, *Rep. Prog. Phys.* **50**:501 (1987).

154. H.B. Dyer and I.G. Matthews, *Proc. R. Soc. Lond.* **A243**:320 (1957).

155. P.J. Dean, *Phys. Rev.* **139**:A588 (1965).

156. H. Kawarada, Y. Yokota, Y. Mori, K. Nishimura, and A. Hiraki, *J. Appl. Phys.* **67**:983 (1990).

157. D.J. Marshall, *Cathodoluminescence of Geological Materials.* Unwin Hyman, London (1988).

158. A.T. Collins, *Physica B* **185**:284 (1993).

159. L.H. Robins, E.N. Farabaugh, and A. Feldman, *Phys. Rev. B* **48**:19 (1993).

160. H. Kawarada and A. Yamaguchi, *Diamonds Rel. Mater.* **2**:100 (1993).

161. C. Yang, N.P.-O. Larsson, E. Swietlicki, K.G. Malmqvist, D.N. Jamieson, and C.G. Ryan, *Nucl. Instr. Meth.* **B77**:188 (1993).

162. C. Yang, N.P.-O. Homman, L. Johansson, and K.G. Malmqvist, *Nucl. Instr. Meth.* **B85**:808 (1994).

163. C. Yang, N.P.-O. Homman, K.G. Malmqvist, L. Johansson, N.M. Halden, and V. Barbin, Scanning Microscopy **9**:43 (1995).

164. N.P.-O. Homman, C. Yang, K.G. Malmqvist, and K. Hanghöj, presented at the Scanning Microscopy Conference (1994) (unpublished).

165. P. Homman, PhD Thesis, Dept. of Nuclear Physics, Lund, Sweden (1994).

166. A.A. Bettiol and D.N. Jamiesion, *Proceedings of the Second Australian Conference on Compound Optoelectronic Materials and Devices*, COMAD93, ANU, Canberra, 1993.

167. A.A. Bettiol, D.N. Jamieson, S. Prawer, and M.G. Allen, *Nucl. Instr. Meth.* **B85:**775 (1994).

168. D. Redman, H Schöne, B. Doyle, J. Knox, E. Taylor, A. Sanchez, and M. Kelly, Ion Beam Materials Research Lab., Semiannual Report, Sandia National Labs., April 1994.

5

SPATIALLY RESOLVED ION CHANNELING TECHNIQUES

5.1. INTRODUCTION

Ion channeling is an established technique for analysis of crystalline materials. As described in Section 1.4, channeling relies on the regular arrangement of atoms in a crystalline material, so that anything that disrupts this arrangement can in principal be studied using the technique. This includes crystal defects, such as dislocations and stacking faults, which distort or shift the lattice planes, strain in lattice-mismatched epitaxial layers, which can produce abrupt changes in the channeling direction at the strained-layer interfaces, surface effects, such as relaxation or reconstruction, which affect the fraction of the incident ions initially channeled, and the presence of impurity atoms.

Traditional ion channeling analysis of defects, which is discussed in Ref. 1 [for a selection of examples, see 2–9] is not concerned with the production of spatial information. A broad (fraction of a millimeter in diameter), unfocused ion beam, typically 2 MeV ^{4}He ions, is normally used for analysis, and information on the depth distribution of the defects under investigation is deduced from the backscattered, and occasionally transmitted, ion energy spectra. Although it is possible to distinguish between different types of defect, such as whether they are point defects or dislocations, using broad beam channeling analysis, the exact nature of the defects is not determined without recourse to other techniques, such as transmission electron microscopy. The spatial distribution of the defects, and what the defects actually look like, is not observed (an exception to this is described in Section 5.3.1).

The nuclear microprobe offers the capability of studying crystalline materials using traditional ion channeling analysis techniques but with the addition of spatially resolved information from the sample. In particular, the techniques of channeling contrast microscopy (CCM) and channeling scanning transmission ion microscopy (CSTIM), which this chapter introduces, can be used to map variations in quality across a crystalline sample and to produce images of individual crystal defects. Applications of these two techniques are given in Chapters 8 and 9.

The initial section of this chapter provides an account of how spatially resolved information can be produced using ion channeling. This includes a description of the channelograph method of imaging, a forerunner to the microprobe methods, together with accounts of CCM and CSTIM and the differences between these two techniques. Practical aspects of producing channeling images are then discussed, in the order in which they might occur in a typical CCM or CSTIM experiment. Aspects considered are CSTIM sample preparation, sample mountings, location of channeling directions, and image production.

5.2. MECHANISMS OF CHANNELING CONTRAST

All of the methods for producing images of crystal defects with channeling contrast rely on the local deviations of the crystal lattice from the regular atomic arrangement found in perfect crystals. In general, these disruptions of the lattice can lead to a breakdown of the channeling process and an increase in the local probability of ions being dechanneled on passing through the defect (defects can sometimes actually cause channeling rather than producing dechanneling; examples are given in Chapter 8). This dechanneling leads to a local increase in the ion induced X–ray yield and, together with direct scattering of the channeled ions by the defect, an increase in the backscattered ion yield. The CCM technique exploits these yield increases to image local variations in crystal quality. Dechanneling also affects the ions' energy loss rate and range in the material, as described in Section 1.4. The CSTIM and channelography techniques rely on these variations to produce images of crystal defects.

Examples of applications of the CSTIM technique in Chapter 8 involve studies of particular crystal defects, namely dislocations and stacking faults; so it is useful to consider here the effects of these faults on channeled ions.

5.2.1. Dislocations

Only the most basic account of dislocations is given here, and more detailed descriptions can be found in textbooks [e.g., 10,11]. The effect of a dislocation can be visualized by considering the deformation of a crystal, as shown in Figure 5.1a. The right-hand side of the top part of the crystal has been pushed to the left and has slipped by, say, one unit cell length, but the left-hand side, beyond plane ABCD, has not moved yet. The dislocation line (AB) is where the slipped and unslipped parts of the crystal meet on the plane on which the movement is occurring. Figure 5.1b shows the atomic arrangement for the sit-

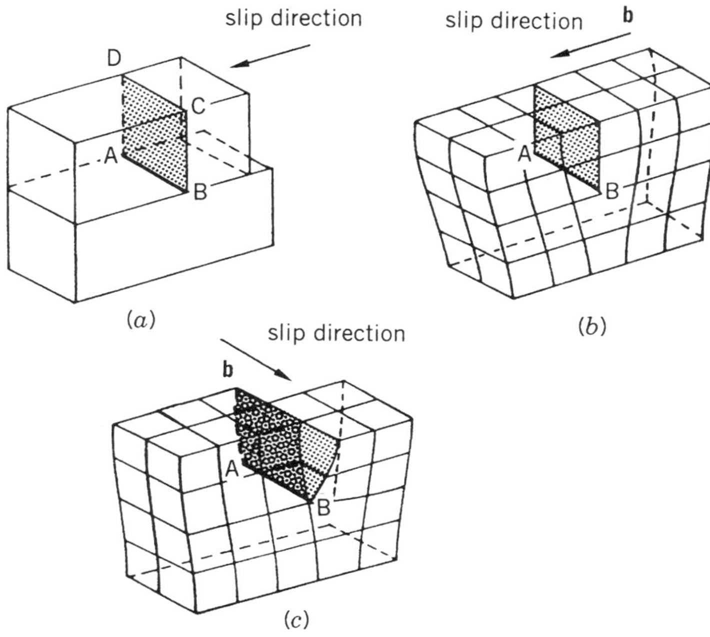

Figure 5.1. (a) Partly slipped crystal, showing location of a dislocation line. (b) The atomic planes around an edge dislocation. (c) The atomic planes around a screw dislocation. **b** indicates the dislocation burgers vector direction.

uation shown in Figure 5.1a. In this case, the direction of slip is perpendicular to the dislocation line, and this is called an edge or 90° dislocation. When the dislocation line is parallel to the direction of slip, a screw or 0° dislocation is generated, as shown in Figure 5.1c. Dislocations can occur that are part edge and part screw in character where the slip direction is at an arbitrary angle to the dislocation line; these are called mixed dislocations.

Dislocations can be characterized by their Burgers vector. Using the idea of the deformation of a crystal again, the Burgers vector represents the amount and direction of slip once the whole crystal has been deformed. Hence, for an edge dislocation, the Burgers vector is at 90° to the dislocation line direction and for a screw dislocation the Burgers vector is parallel to the line direction (see Figures 5.1b,c). Slip of one part of a crystal with respect to another usually occurs on the crystal planes that are the most closely packed in a given crystal structure and along the direction of closest packing within these planes. For semiconductor materials with the silicon crystal structure, such as those studied by CSTIM in Chapter 8, this means that slip occurs on {111} planes (see Figure 1.13), and along $<\bar{1}10>$ directions within these planes. Hence, dislocations in such lattices form with Burgers vectors along $<\bar{1}10>$ directions.

More precisely, dislocations have an associated self-energy from the lattice strain they cause. This strain energy is proportional to the square of the magnitude of the Burgers vector, b^2. Therefore, from energy considerations, dislocations with small Burgers vector are most easily produced. Because the <110> directions in the silicon lattice are the most closely packed, dislocations along these directions have the smallest Burgers vector and so are favored. Such dislocations have the magnitude of a unit lattice vector along this direction; this is $a/\sqrt{2}$, where a is the unit cell size. The Burgers vector is thus written:

$$\mathbf{b} = \frac{a}{\sqrt{2}} \times \frac{1}{\sqrt{2}} <110> = \frac{a}{2} <110> \tag{5.1}$$

Dislocations of the type given by Eq. (5.1) are called perfect dislocations because, as mentioned above, the Burgers vector is equal to a unit lattice vector. However, dislocations can form with Burgers vectors that are shorter than a unit lattice vector, and these are called *partial dislocations*. Partial dislocations are to be found bounding stacking faults (see below) where these defects end inside a crystal.

The primary effect of dislocations on channeled ions is to cause dechanneling owing to the curvature of the lattice planes in the dislocation vicinity, shown in Figure 5.1. It has been assumed [12–14] that a channeled ion would be dechanneled if the axis or plane it was traveling along was curved by more than the channeling critical angle. This assumption was used to calculate the size of the region around a dislocation as viewed by the beam within which a channeled ion would be dechanneled. Outside this region, no dechanneling was assumed to take place. The diameter of the region was given by an expression of the form

$$\sigma_D = k \frac{\sqrt{ba_{tf}}}{\psi} \tag{5.2}$$

where b is the magnitude of the Burgers vector of the dislocation, a_{tf} is the Thomas-Fermi screening radius given by Eq. (1.5) and ψ is the critical angle of the channeling direction given by Eqs. (1.19) or (1.20). The constant k depends on whether axial or planar channeling is used and on the exact nature of the dislocation, and varies between about 0.5 and 2. For 3 MeV ^{1}H ions in silicon, σ_D is approximately 15 nm for channeling along the <100> axis and approximately 50 nm for channeling in the {110} planes when a dislocation Burgers vector of magnitude 3.8 nm is used ($\mathbf{b} = a/2$ <110>).

Computer simulations of the effects of dislocations on channeling have also been performed [15–17]. These show that the above model gives a good indication of the size of the region around a dislocation that will cause dechanneling, but that the assumption of a definite region within which all the ions are dechanneled, and outside of which none are, is an oversimplification. Even very close to the dislocation core, some of the ions can remain channeled, and, further away, the dechanneling was found to decrease approximately exponentially with distance [16]. The simulations show that dechanneling due to dislocations varies as $\sqrt{E}$, where E is the beam energy. This is in agreement with Eq. (5.2) (ψ varies inversely as $\sqrt{E}$), but the simulations have also shown a stronger dependence on b than Eq. (5.2) suggests [15,17].

5.2.2. Stacking Faults

Stacking faults [10,18] are, as their name implies, disruptions in the way that planes of atoms in a crystal are stacked on top of each other. An *extrinsic* fault is formed when an extra plane of atoms, out of sequence with those around it, is added into the crystal, and an *intrinsic* fault is produced when an atomic plane is removed. The effect of a stacking fault on a crystal can be described as a shift of the part of the crystal above the fault with respect to that below the fault. A fault can therefore be described by a translation vector, **R**, which gives the magnitude and direction of the relative shift of the two parts of the crystal. Where a stacking faults ends inside a crystal, it is bounded by a partial dislocation.

For channeling, the effect of a stacking fault is to move rows of atoms below a fault into the channels formed by the rows above it, as shown schematically in Figure 5.2. It is not surprising, therefore, that a stacking fault can cause an initially channeled ion to be dechanneled. Evidence for this dechanneling has been produced using the channelography technique described in Section 5.3.1 for stacking fault tetrahedra in gold crystals [14,19] and from observations of a reduction in the ranges of channeled 40 keV gold ions in silver and gold foils containing planar defects [20]. Backscattering spectrometry has also been used

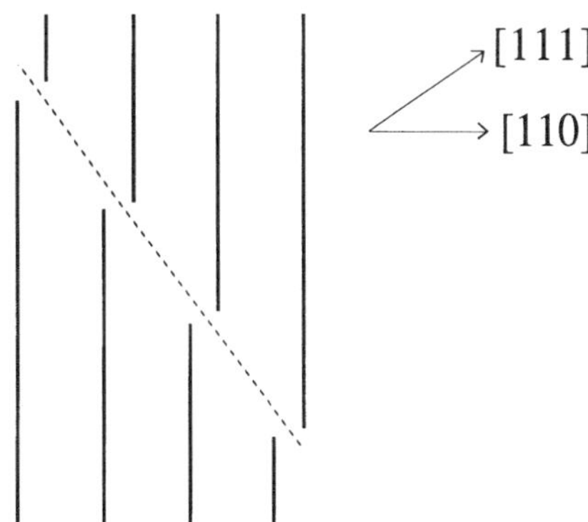

Figure 5.2. Schematic diagram of the effects of a stacking fault in a crystal. The shift in the (110) planes produced by a fault on a (111) plane is shown; this effect can also ben seen in Figure 8.27.

to observe the dechanneling of MeV ^{4}He ions by stacking faults in thin silicon layers grown epitaxially on sapphire crystals [4]. To first order, a stacking fault is expected to dechannel a fraction of the channeled beam equal to the minimum yield for the channeling direction, that is, a few percent for axial directions and approximately 30% for planar directions (Section 1.4). This is because going through a fault, the ions can be considered to encounter a new crystal surface, similar to the crystal entrance surface. Figure 8.27 shows the effects of a stacking fault in a face-centered cubic crystal, and the lattice plane shift produced by a fault can be seen in the atomic resolution electron microscope image of Figure 8.1 (which is best viewed from the right-hand side with the page held close to edge on).

5.3. TECHNIQUES FOR PRODUCING CHANNELING CONTRAST IMAGES

Having looked briefly at how it is possible for defects to affect the channeling process, details of the methods for producing channeling contrast images are now described. The three methods are shown schematically in Figure 5.3. The microprobe techniques of CCM and CSTIM, which involve detection of backscattered ions or ion induced X–rays (CCM), and transmitted ions (CSTIM), evolved in the 1980s and early 1990s. However, the first spatially resolved ion channeling information was produced using the channelography technique in the late 1960s and early 1970s, and this is described first.

5.3.1. Channelography

Some of the first evidence that crystal defects cause dechanneling was produced using a method based on measuring the intensity of α-particles transmitted through thin metal foils [13,14,19,21]. The experimental arrangement for this radiography technique is shown in Figure 5.4. MeV α-particles from an isotropic source were incident on a thin metal foil. Those transmitted through the foil were stopped in a sheet of cellulose nitrate that could be developed to reveal the α-particle tracks, producing a channelograph. An absorber was placed between the source and the metal foil whose thickness was adjusted so that only particles channeled in the metal would get through to the cellulose nitrate owing to their reduced rate of energy loss.

It was found that the presence of stacking faults and dislocations in a metal foil reduced the number of α-particles detected owing to the dechanneling that these defects produced. Grain boundaries at an angle to a foil's surface produced thin ribbons denuded of α-particles in the cellulose nitrate owing to dechanneling. Images of polycrystalline samples were produced, as crystal grains oriented with major channeling directions parallel to the paths of the α-particles allowed many particles to be transmitted, whereas those that permitted slight or no channeling prevented transmission. Channelographs of partly recrystallized

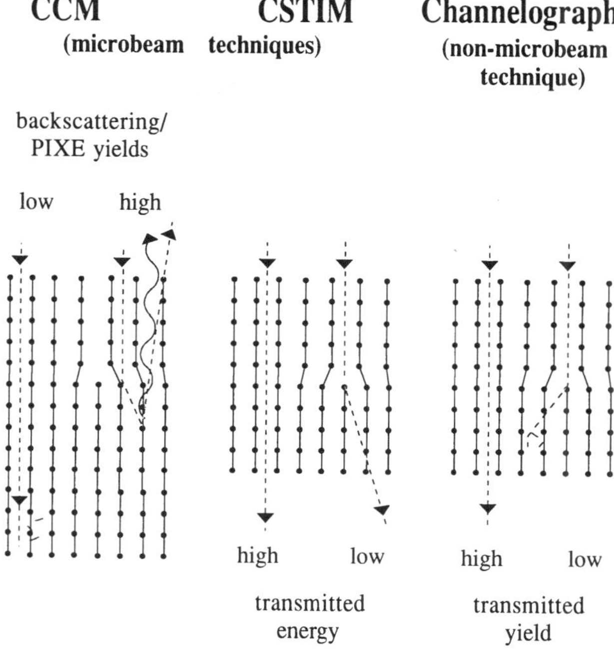

Figure 5.3. Diagram showing the methods of contrast production for the channeling techniques of CCM, CSTIM, and channelography. When an ion beam is channeled, defects in the crystal locally raise the backscattered ion and ion induced X–ray yields, the mechanism behind the CCM technique. The dechanneling a defect produces locally decreases the average energy of transmitted ions, the basis of the CSTIM technique. Channelography relies on the ions dechanneled by a defect having insufficient energy to be transmitted through the crystal, whereas well-channeled ions can be detected owing to their lower energy loss rate.

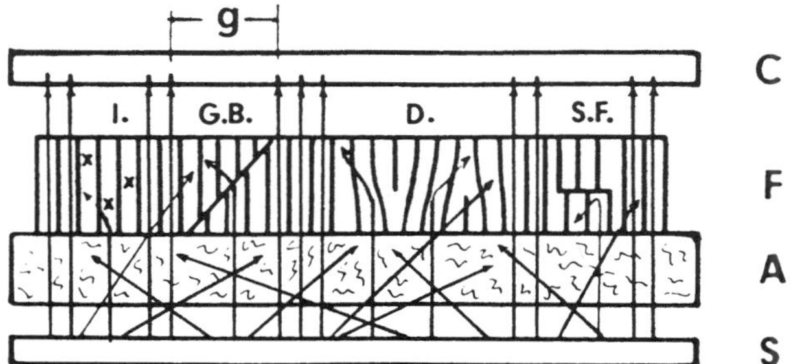

Figure 5.4. Experimental arrangement for the channelography technique. S, source of α-particles; A, amorphous absorber; F, crystalline sample; C, cellulose nitrate. I, G.B., D. and S.F. represent respectively impurity atoms, a grain boundary, dislocations and a stacking fault, with G being the width of the grain boundary image. Reprinted from Ref. 19 with permission (© 1968 American Institute of Physics, Woodbury, NY).

platinum revealed the regions that had not yet crystallized. Images of individual stacking faults or dislocations were not produced, however.

A collimated MeV ^{1}H ion beam from an accelerator was also used with a similar radiography technique to again produce images of grains in polycrystalline metal foils [22]. The images required doses of ~6 × 10^{12} ^{1}H ions/cm^2. An image from this work showing crystal grains and a 20-μm-wide scratch is given in Figure 5.5.

5.3.2. Channeling Contrast Microscopy

The first technique to be developed for the production of channeled images of crystals with a nuclear microprobe was channeling contrast microscopy [23], which refers to production of images based on spatial variations in backscattered ion yield or ion induced X–ray yield using a scanned, focused ion beam, and a beam current of several hundred picoamps. When the incident beam is aligned with a channeling direction of a crystalline material, the yield of backscattered ions from near the surface of the crystal is reduced to a few percent of that for a nonaligned beam for an axial channeling direction and approximately 30% for a planar channeling direction, as described in Section 1.4. Similarly, the ion induced X–ray yield also reduces in channeled alignment. Dechanneling produced by defects in the crystal causes a local increase in the backscattered ion or ion induced X–ray yields, and defects can cause immediate backscattering of channeled ions (called direct scattering), also increasing the backscattered ion yield.

This can be seen, for example, in the backscattered ion energy spectra shown

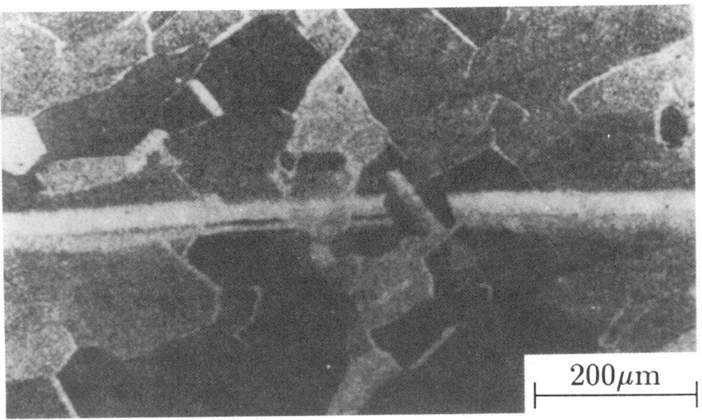

Figure 5.5. Channelograph of grains in a polycrystalline metal foil. Darker corresponds to beter channeling. The bright band running from left to right across the center of the image is a scratch. Reprinted from Ref. 22 with permission (© Taylor & Francis, Basingstoke, U.K.).

in Figure 5.6. These were all taken from crystals that consisted of a layer of $Si_{1-x}Ge_x$ grown on to a silicon substrate [24]. Such crystals are described more fully in Chapter 8. Three spectra measured with 2 MeV ^{1}H ions are shown: one taken from a crystal consisting of a 1 μm layer of $Si_{0.85}Ge_{0.15}$ on silicon with the incident beam not channeled, one taken from the same crystal with the beam aligned with the [001] crystal axis, and one taken with the beam channeled in a different crystal consisting of a 0.3 μm layer of $Si_{0.88}Ge_{0.12}$ on silicon. This latter crystal was expected to have few defects at the layer–substrate interface, and it can be seen that there is a slow increase with depth into the crystal in the backscattered ion yield owing to natural dechanneling. The $Si_{0.85}Ge_{0.15}/Si$ crystal was found to contain significant numbers of dislocations at the layer–substrate interface (Chapter 8). The backscattered spectrum, in this case, undergoes a sharp increase at an energy of approximately 1700 keV, owing to the dechanneling produced by the interface dislocations.

If the yields of backscattered ions or ion induced X–rays are mapped with the incident beam aligned with a channeling direction, defects and regions of poor crystallinity will be revealed by the yield increase they produce, which is the basis of the CCM technique. The first demonstration of CCM [23] [see also 25] used a 2 MeV ^{4}He ion beam focused to a spot size of 12×15 μm^2 to image a region of an ion implanted silicon crystal that had been previously laser annealed. The image was produced by mapping the counts in the backscattered

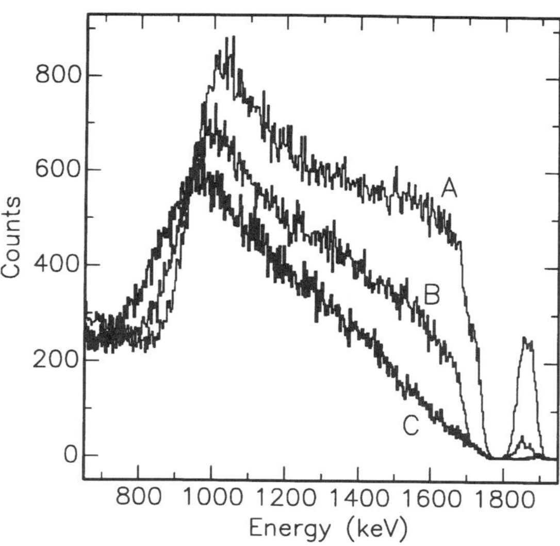

Figure 5.6. MeV ^{1}H ion backscattering spectra. A: $Si_{0.85}Ge_{0.15}/Si$ sample, non-channeled. B: $Si_{0.85}Ge_{0.15}/Si$ sample, beam channeled along the [001] axis. C: $Si_{0.88}Ge_{0.12}/Si$ sample, beam channeled along the [001] axis. Reprinted from Ref. 24 with permission (© 1993 American Institute of Physics, Woodbury, NY).

ion energy spectrum from depths corresponding to the damage layer produced by the implantation. By selecting different regions of a backscattered ion energy spectrum from which to produce images, depth resolved information can be produced using CCM. Depth information can also be generated by using X–ray lines of different energies to produce images. For example, low-energy M shell X–rays for a particular element will have a shallower average production depth than the L shell X–rays from the same element, so that images produced using the latter lines will show deeper features.

The choice of ion beam for CCM analysis has been discussed [26] based on a comparison of the relative advantages and disadvantages of using 3 MeV ^{1}H ions or 2 MeV ^{4}He ions. These differences are summarized in Table 5.1. ^{4}He ions have the advantage of an order of magnitude better depth resolution than ^{1}H ions owing to their greater stopping power. The depth resolution for ^{4}He ions is of the order of 20 nm at a sample surface for a detector at a steep backward angle, although this can be improved further using a grazing exit angle geometry [27]. ^{4}He ions are more sensitive to elements heavier than silicon compared with ^{1}H ions owing to their larger Rutherford scattering cross-sections; for light elements, the scattering cross-sections for ^{1}H ions have nuclear resonances (as described in Sections 1.2 and 4.3, and Ref. 28), increasing the sensitivity but making it harder to produce quantitative measurements. ^{1}H ions also have larger cross-sections for PIXE. Channeling critical angles [29] for 2 MeV ^{4}He ions are approximately 70% larger than for 3 MeV ^{1}H ions, making orientation of

TABLE 5.1. Summary of the Relative Advantages of 2 MeV ^{4}He Ions and 3 MeV ^{1}H Ions for CCM Analysis[a]

2 MeV ^{4}He Ions

+ Good depth resolution, ~20 nm (BS)
+ High sensitivity for depth profiling heavy elements (BS)
+ Wide channeling angles
0 Depth profiles down to ~1 μm (BS)
0 Mean production depth $\leq$1 μm (PIXE)
− Low sensitivity to light elements by BS
− Low sensitivity by PIXE, except for atomic number <12

3 MeV ^{1}H Ions

+ High sensitivity by PIXE
+ High sensitivity to light elements by BS
0 Depth profiles down to ~10 μm (BS)
0 Mean production depth $\leq$10 μm (PIXE)
− Narrow channeling angles
− Low depth resolution, ~200 nm (BS)
− Low mass resolution for depth profiling (BS)

[a]BS refers to backscattering spectrometry. +, −, and 0 signify that the property is respectively beneficial, detrimental, or neutral for analysis.

a crystal for channeling easier. Table 5.2 gives some relevant parameters for analysis of a silicon crystal using backscattering spectrometry or PIXE with 2 MeV ^{4}He ions and 3 MeV ^{1}H ions. Examples of CCM images are given in Chapters 7, 8, and 9.

5.3.3. Channeling Scanning Transmission Ion Microscopy

The STIM technique for producing images was described in Section 4.7, and involves the detection and energy measurement of ions transmitted through thinned samples. Local variations in the average or median ion energy loss, or in

TABLE 5.2. Parameters of 2 MeV ^{4}He and 3 MeV ^{1}H Ion Beams for CCM Analysis of a Silicon Crystal Matrix[a]

	2 MeV ^{4}He	3 MeV ^{1}H
Axial channeling half-angle, $\psi_{\frac{1}{2}}$ (degrees)		
$\langle 100 \rangle$	0.40	0.24
$\langle 111 \rangle$	0.61	0.36
$\langle 110 \rangle$	0.47	0.28
PIXE bulk yield (counts/(μC · msr))		
Si $K_{\overline{\alpha}}$	1.6×10^6	2.3×10^7
Sb $L_{\overline{\alpha}}$ from dopant	2.3×10^2/wt.%	1.0×10^4/wt.%
Sb $K_{\alpha 1}$ from dopant	1.4×10^{-2}/wt.%	1.0×10^2/wt.%
PIXE mean production depth (μm)		
Si $K_{\overline{\alpha}}$	0.61	3.7
Sb $L_{\overline{\alpha}}$ from dopant	0.52	2.1
Sb $K_{\alpha 1}$ from dopant	0.49	9.7
BS matrix scattering height (counts/(μC · msr · keV))		
Si	17	55^b
O dopant	0.088/wt.%	0.98^b/wt.%
Sb dopant	0.53/wt.%	0.68/wt.%
BS characteristic depth scale[c] (nm/keV)		
Si	2.1	24
O dopant	2.2	24
Sb dopant	2.0	24

[a] All quantities are calculated for normal incidence; PIXE quantities calculated for a measurement angle of 135°; $K_{\overline{\alpha}}$, $L_{\overline{\alpha}}$ designate the sum of the contributions from both the α_1 and α_2 X-ray line yields or the weighted average contribution for the mean depth of X-ray production; backscattering quantities calculated for a scattering angle of 160°. BS refers to the backscattering spectrometry. The backscattering matrix scattering height is the height in counts of the signal from an element at its surface energy.
[b] Calculated from the relevant non-Rutherford nuclear elastic cross-section for oxygen and silicon [28].
[c] The depth scale for BS can be reduced by an order of magnitude for glancing exit angle geometry [27].

the number of ions detected with energies in a certain range, or in the number of ions scattered into a given solid angle, can be used to produce images showing the changes in areal density of a sample. When the sample is crystalline, the transmitted ion energy loss and ion scattering can also be affected by whether the ions have been channeled for all or part of their passage through the crystal. The STIM technique is used to produce images of crystals that reveal local variations in crystal quality, with the primary contrast mechanism used to date being variations in the average transmitted ion energy loss.

In Section 1.4, we described how the channeling process affects the ion energy loss rate as the ion travels through the crystal, with channeled ions suffering on the order of 50% of the energy loss rate of nonchanneled ions. This can be seen in Figure 5.7, which shows two transmitted [^1]H ion energy spectra taken through a thinned silicon crystal with the beam channeled in the {111} planes of the crystal and with the beam not channeled [30]. The peak on the high-energy side of the channeled spectrum is from those ions channeled throughout the crystal (this is very prominent in the spectrum shown because a detector with a restricted acceptance angle was used, as described further in Section 5.4.7). Defects in a crystal will locally disrupt the channeling process and lead to a locally enhanced probability of dechanneling. Dechanneled ions revert back to the higher, nonchanneled energy loss rate. This can be seen in Figure 5.8, which shows two transmitted [^1]H ion energy spectra, both taken

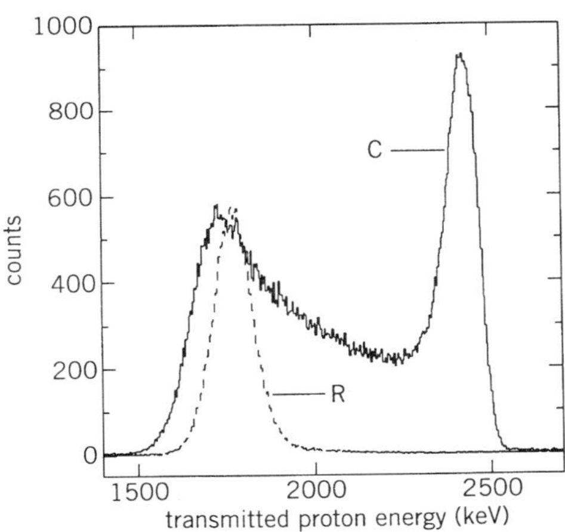

Figure 5.7. Transmitted 3 MeV [^1]H ion energy spectra, normalized to the same peak height, taken through a thinned silicon crystal. R, beam not aligned with a channeling direction; C, beam aligned with a set of the {111} planes of the crystal. High energy peak in spectrum C is due to well-channeled ions; this peak is prominent in the spectrum owing to use of a detector with an acceptance half-angle of only 0.4°. The effective sample thickness for the nonchanneled spectrum was slightly less than that for the channeled spectrum owing to the difference in sample tilt angle. Reprinted from Ref. 30 with kind permission of Elsevier Science B.V., Amsterdam, The Netherlands.

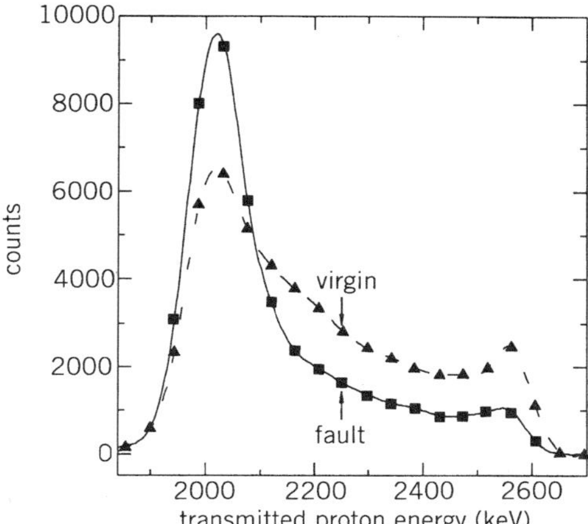

Figure 5.8. Transmitted ^{1}H ion energy spectra from a silicon crystal with the incident beam channeled in {111} planes of the sample. Broken line: spectrum from good crystal. Unbroken line: spectrum from a region of the sample with a stacking fault close to the surface. Spectra extracted from the data used to produce the image shown in Figure 8.35a, so that they are normalized to the same incident beam dose. The detector acceptance half-angle was 1°, so the high energy peak is less pronounced than in Figure 5.7.

with the incident beam channeled, but one with the beam on a defect and one with the beam on virgin (i.e., perfect) crystal. The effect of the defect was to reduce the fraction of ions transmitted in the high-energy peak of the spectrum. The average transmitted ion energy calculated from these two spectra is 2196 keV for the spectrum measured from virgin crystal and 2120 keV for the spectrum taken from the defect. The average energy loss of the ^{1}H ions transmitted through the defect was thus greater than through virgin crystal owing to the dechanneling the defect produced. Average energy loss STIM images of a crystalline sample taken with the incident beam channeled can therefore reveal regions where channeling is disrupted owing to the presence of defects. This is the basis of the CSTIM technique.

The first use of CSTIM employed 1 MeV and 3.9 MeV ^{1}H ion beams channeled along a <111> axis to image regions of 58 μm thick and 4.2 μm thick silicon crystals that had been previously damaged by a high ion beam current [31,32]. The effects of sample damage on the transmitted energy spectra of 1.1 MeV and 2.5 MeV ^{4}He ions passing through a 4.2 μm thick silicon crystal have been observed [33], and CSTIM was again used to image the damaged region. Figure 5.9 shows a three-dimensional contour plot of an average energy loss image of a region of a silicon crystal damaged by a high beam current [31]. Subsequent development of the CSTIM technique at Oxford produced images of individual crystal defects (examples are given in Chapter 8).

Although a ^{4}He ion beam has been used for CSTIM imaging [33], 2 or 3 MeV ^{1}H ions have been used more extensively at Oxford because of their

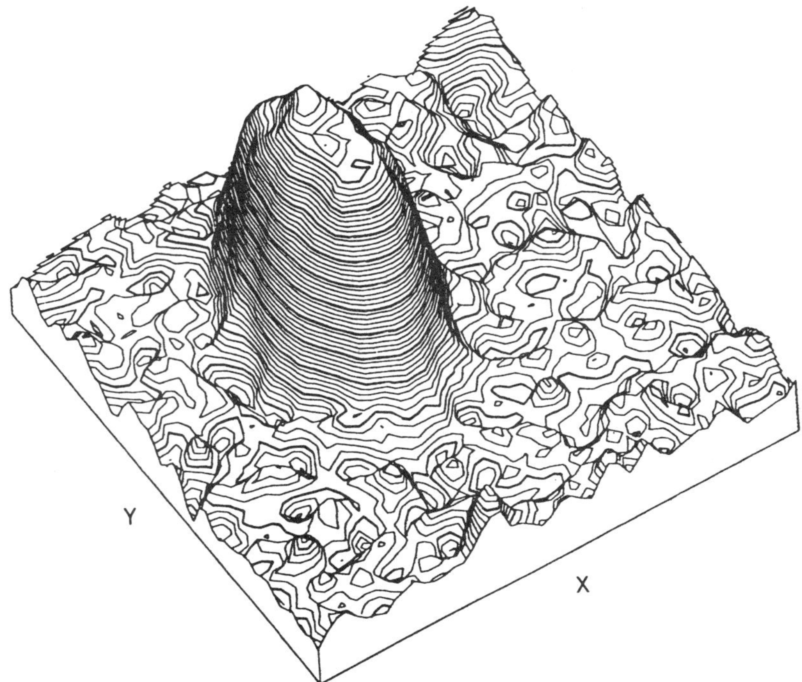

Figure 5.9. 50×50 μm^2 CSTIM average energy loss contour plot produced from a 58 μm thick silicon crystal using a 3.9 MeV ^{1}H ion beam. Region of increased energy loss near the center is caused by lattice damage introduced deliberately using a high beam current. Reprinted from Ref. 31 with permission (© 1990 American Institute of Physics, Woodbury, NY).

greater range than similar energy ^{4}He ions, enabling crystals several tens of microns thick to be analyzed. As mentioned in Section 4.7, the STIM technique (and therefore CSTIM as well) can be very efficient if a detector with a large acceptance angle is used to capture the transmitted ions. Virtually every incident ion is detected; with a count rate normally limited by the data acquisition system to less than 10 kHz, a beam current of 2 fA or smaller is required. Larger beam currents are required if a detector with a restricted acceptance angle is used, as discussed in Section 5.4.7.

5.3.4. Differences between Channeling Contrast Microscopy and Channeling Scanning Transmission Ion Microscopy

The differences between the capabilities of CCM and CSTIM may make one technique more suitable than the other for a particular application. Channeling contrast microscopy can analyze bulk samples without requiring any preparation. If backscattered ions are measured, depth-resolved channeling contrast images can be produced by selection of appropriate regions of the energy spectrum for image production; depth-resolved CCM images from high T_C superconductor crystals are given in Section 9.1. On the other hand, CSTIM requires samples to be thinned to a few tens of microns in thickness, and does not in general produce depth-resolved information. However, the sample preparation required for CSTIM is much less than that required for transmission electron microscopy analysis of crystal defects.

However, CSTIM has a significant advantage over CCM because it requires a much lower beam current for analysis. Reduction of the beam current to approximately 1 fA for CSTIM is achieved by decreasing the size of the microprobe object and collimating apertures, as described in Section 5.4.3. This has the effect of decreasing the geometrical and lens-aberrated beam spot size at the sample, so that spot sizes of the order 100 to 300 nm are routinely achievable. Therefore the spatial resolution of CSTIM is better than that of CCM, in general. The count rate for CCM tends to be limited by the amount of beam current that can be focused into a small beam spot with a low beam convergence, whereas the count rate for CSTIM is limited by the speed of the data acquisition system. Because a much higher count rate is achievable with CSTIM, images generally have much better statistics and are less noisy than those produced by CCM. The relative merits of CSTIM and CCM images are well illustrated in Section 8.5.1, which compares images from a SiGe/Si crystal produced with both methods.

Another major advantage of the low beam current required for CSTIM analysis is the reduction in sample damage owing to the smaller dose needed to produce an image compared with CCM. Beam induced damage in crystalline materials is discussed more fully in Section 1.5. One effect of the low beam dose required for CSTIM analysis is that it is possible to locate channeling directions at the same region of the crystal as required for subsequent channeling analysis. This is particularly important if the angular location of the channeling direction varies with position over the crystal. In contrast, for CCM analysis, the channeling direction often has to be found in a different region of the crystal than where the measurement is to be performed in order to prevent damage to the region of interest from the beam during the alignment.

It has also been observed [31,32] that the change in shape of the CSTIM energy spectrum on moving from good crystal to a region where defects are present can be more noticeable than the corresponding change in the backscattered energy spectrum, possibly making CSTIM a more sensitive technique for detection of low defect concentrations than CCM.

5.3.4.1. Ion Optical Considerations

Ion channeling measurements are usually carried out using an analysis beam that is nearly parallel (i.e. the beam convergence half-angle at the sample is much less than the channeling critical angle). This results in a large fraction of the incident ions becoming channeled, at least in the surface region of the crystal. However, an ion beam that has been focused on to a crystal sample by a probe-forming lens system of a nuclear microprobe can have a relatively steep convergence angle, determined by the desired beam spot size, the brightness of the ion source of the accelerator, the demagnification factor of the probe-forming lens system, and the minimum beam current required to obtain statistically significant data in a reasonable time without damaging the sample excessively.

The low current required for CSTIM analysis means that the incident beam naturally has a low convergence because the collimating aperture is reduced to a very small size. On the Oxford microprobe, the collimating aperture size is typically $50 \times 130~\mu m^2$ for CSTIM analysis. This means that the beam half-divergence into the lenses is $0.0004° \times 0.001°$, for a distance of 6.8 m between the object aperture and collimating aperture. The convergence half-angle at the sample is $0.04° \times 0.024°$ as the lens demagnifications of 80 and 23 are asymmetrical in the horizontal and vertical planes on the high excitation triplet lens configuration of the Oxford system. The beam convergence used for CSTIM analysis is therefore much less than channeling critical angles, which are of the order of $0.1°$ for 3 MeV ^1H ions and $0.2°$ for 2 MeV ^4He ions planar channeled in silicon.

For nonchanneled backscattering and PIXE measurements, which require a beam current of at least 100 pA, the collimating aperture for normal operation is of the order of $600 \times 600~\mu m^2$ at Oxford. This produces a beam convergence half-angle at the sample greater than $0.8°$ in the horizontal plane and greater than $0.2°$ in the vertical plane, which are far too large to be used for channeling measurements. To perform CCM analysis, the collimating aperture must be reduced to allow the beam to channel. To produce a large enough beam current, however, it is necessary to increase the size of the object aperture, with a resulting detrimental affect on the beam spot size. This high excitation triplet configuration is thus not ideal for CCM measurements because of the large lens demagnifications. A quadruplet lens configuration is more suitable for CCM analysis because of the lower lens demagnifications. This means that larger collimating apertures can be used with a consequent increase in the beam current available for analysis. Typical parameters for the operation of the Melbourne microprobe for CCM measurements are shown in Table 5.3. Using this lower demagnification system, it is easy to carry out CCM measurements with beam convergence angles of considerably less than the channeling critical angle.

The effect of a very large beam convergence produced by the Melbourne microprobe quadruplet lens configuration, aligned with axial and planar channels used for CCM measurements, can be demonstrated by observations of the transmission channeling patterns produced from thin crystals (described more fully in Section 5.4.5). For the geometry of this experiment, the convergence angle of the incident beam was approximately twice the axial channeling critical angle (such a beam is not normally used for CCM measurements). Figure 5.10a shows a transmission channeling pattern from a silicon crystal irradiated with a convergent ^1H ion beam along the <100> axis. The pattern shows two distinct regions. In the central region, there is a bright cross produced by ions

TABLE 5.3. Ion Optical Properties of the Melbourne Nuclear Microprobe for CCM Measurements[a]

Typical Parameters for CCM Imaging	Sample Material	
	Si <100>	C <110>
Microprobe beam	2 MeV ^{4}He	1.4 MeV ^{1}H
Object aperture size	20 μm	20 μm
Beam spot diameter	1 μm	1 μm
Beam convergence angle, $2\theta_i$	2.2 mrad	2.8 mrad
Channeling half-angle $\psi_{\frac{1}{2}}$[b]	0.40	0.44
Beam current on sample	100 pA	100 pA
Matrix scattering height (counts/μC/keV/msr)[c]	17.1	820
Time for 3% statistics[d]	1.2 hours	1.5 min
Dose in a 100×100 μm^2 scan[a]	3×10^{16}/cm^2 (13% of $\mathbf{D}_s$)	4×10^{14}/cm^2 (0.5% of $\mathbf{D}_s$)

[a]The critical dose for swelling induced dechanneling D_s was discussed in Section 1.5.

Probe-forming lens system demagnification factor	-20
Maximum beam divergence from accelerator	0.35 mrad
Maximum convergence angle on sample	6.8 mrad (0.39°)

[b]$\psi_{\frac{1}{2}}$ values were taken from Ref. 29.

[c]Sample in random orientation, using the measured non-Rutherford cross-section [28] for carbon.

[d]3% statistics in the total energy spectrum from the sample regardless of the (x, y) coordinate. Detector solid angle of 35 msr.

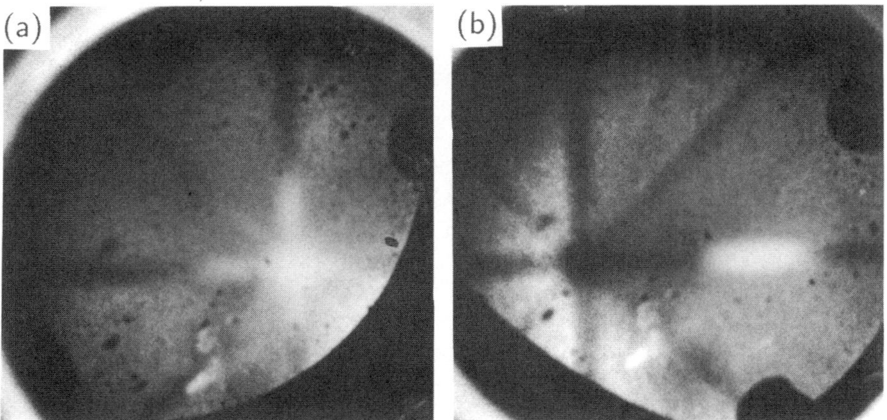

Figure 5.10. (a) Transmission axial channeling pattern from a 7 μm thick <001> silicon crystal. The crystal was irradiated with a 4 MeV ^{1}H beam that had a convergence angle of 0.4°; i.e., approximately twice the channeling critical angle. (b) As for (a) except showing (110) planar channeling.

that are channeled through the crystal within the <100> axis, as well as through the lowest order planes around this axis. The fraction of the incident beam with a convergence angle larger than the axial and planar channeling critical angles was not channeled through the crystal and was responsible for a low-intensity halo around the central bright region. Within the halo, a dark blocking pattern is evident; the shadows of the crystal planes can be seen extending from the bright cross pattern of the channeled beam, because the scattered ions have a low probability of emerging from the crystal parallel to a plane or axis.

The effect of increasing incident beam convergence angle on a backscattering spectrum for 2 MeV ^{1}H ions aligned with the silicon <001> axis is shown in Figure 5.11. These backscattering spectra are all measured at axial channeling alignment but with increasing beam convergence angle on the sample surface. The beam convergence angle ranges from half the axial channeling critical angle for the best channeled spectrum to the channeling critical angle for the worst channeled spectrum. The main effect of increasing the beam convergence angle is to increase the dechanneling rate of the initially channeled ion beam.

Planar channeling through the same thin crystal is shown in Figure 5.10b. Here the crystal has been rotated about the vertical axis from the alignment shown in Figure 5.10a, such that the same incident beam is no longer aligned with the <100> axis. The convergence angle of the beam is generally too steep to allow planar channeling to be used analytically for CCM, since too little beam current is available with sufficiently small convergence angle, even with this low demagnification microprobe.

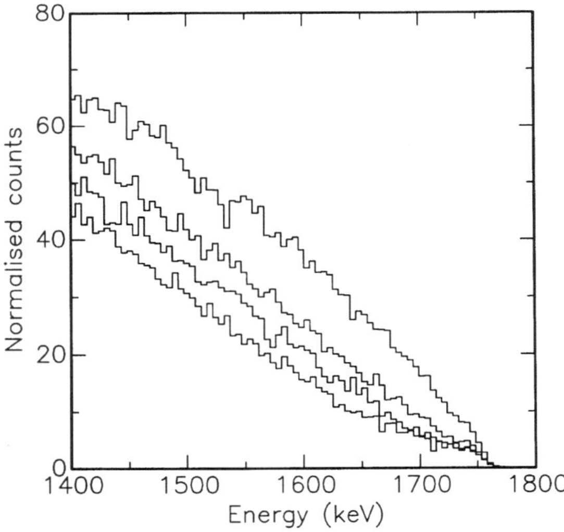

Figure 5.11. Axial dechanneling as a function of increasing 2 MeV ^{1}H ion beam convergence angle on a <001> silicon surface. The dechanneling rate increases with increasing beam convergence angle.

5.4. PRACTICAL ASPECTS OF ION CHANNELING IMAGING

This section gives some practical details of CCM and CSTIM experiments. The order of the discussion is much the same as the order in which a typical experiment might proceed. Sample preparation for CSTIM is described, followed by the mounting of samples in the microprobe chamber and the setting of the microprobe apertures for analysis. Experimental details for the location of channeling directions and the production of CCM and CSTIM images are also given.

5.4.1. Sample Preparation

For CCM analysis, bulk crystalline samples can be studied and little or no preparation is needed. For CSTIM analysis, however, the samples must be thin enough to transmit the incident beam. A 3 MeV ^{1}H ion beam, the beam used to obtain the majority of the CSTIM results presented in Chapter 8, has a range of approximately 90 μm in silicon. In practice, silicon crystals used for CSTIM analysis with 3 MeV ^{1}H ions on the Oxford microprobe were thinned to 20 to 40 μm in thickness.

The samples studied in Chapter 8 were all semiconductor crystals grown in the form of wafers (~500 μm in thickness initially) with the [001] crystal axis normal to the wafer surface. The method which was used for thinning these crystals, and the method used in general to prepare crystals for CSTIM analysis on the Oxford microprobe, is given here. Small pieces, approximately 2×2 mm^2, were taken for analysis by cleaving the wafers. Silicon cleaves along $\{111\}$ planes that meet the (001) surface along [110] and [$\bar{1}$10] directions; so the material was cleaved by scoring it with a diamond scribe along one of these directions and gently applying pressure until the wafer snapped along the scored line.

For the thinning process, the samples were fixed on to a 30 mm diameter optical flat with a glue melting just above 70°C and soluble in acetone. Two samples of the same material were often thinned simultaneously. They were stuck on opposite sides of the face of the flat to enable the uniformity of the thinning to be monitored. To thin the samples and remove large scratches from the thinned surface, a lapping and polishing machine was used. The samples were thinned from their back (unpolished) surface with wet and dry paper until they were approximately 60 μm thick; the flat was held by hand against the paper and periodically the thickness of the samples was measured with a micrometer. When the required thickness was reached, the wet and dry paper was replaced with a polishing pad and the thinned surfaces polished with diamond lapping compound, starting with 25 μm particle size and finishing with 1 μm particle size. This process removed the large scratches that were produced by thinning, so that they did not show up in CSTIM energy loss images. This thinning and polishing process usually took about an hour. The optical flat was then left in acetone to dissolve the glue. When the samples had come away from the flat, they were washed in acetone and then in alcohol. This relatively simple method of producing thin crystals often caused the crystals to have a gradual bend, so that only small areas (<150 × 150 μm^2 in size) could be analyzed in a single CSTIM image owing to the change in channeling orientation across the crystal.

Features of interest on the samples were located before microprobe analysis using optical microscopy. A small drop of silver paint placed on to the sample using a hair acted as a marker for location of particular sample regions and so that regions could be located again in experiments performed at different times.

The minimum sample thickness that this method could achieve was approximately 20 μm, and production of high-contrast transmission channeling patterns (Section 5.4.5) requires samples thinner than this. To produce very thin crystals with low curvature an alternative thinning technique can be used. Large areas of silicon (~1 cm across) can be thinned to less than 1 μm in thickness and left supported by surrounding thick crystal using a method involving boron doping and chemical etching. Crystals produced by this process are not curved and are very uniform in thickness [34]. Such very thin silicon crystals have been used for studies of the energy loss of channeled ions [35] and for observations of transmission channeling patterns [for example, see 36,37]. An example of such a pattern produced from an approximately 1.4 μm thick silicon crystal prepared using this method is given below in Figure 5.20.

5.4.2. Sample Mountings and Goniometers

For channeling analysis, it is necessary to rotate and translate the sample in the microprobe chamber for positioning and to align the channeling direction with the beam. This is usually achieved by mounting the sample on a eucentric goniometer which allows it to be both translated along two orthogonal directions in its plane and rotated about two axes, either both in its plane or with one normal to the sample surface. It is important for the goniometer to have sufficient angular resolution for alignment of channeling directions: the critical angle for channeling in major planes of silicon with 3 MeV ^{1}H ions is approximately 0.1°, so that a goniometer with an angular resolution of at least 0.05° is required. The eucentricity of the goniometer is important for two reasons: first, to prevent lateral movements of the sample with tilting, which would cause different regions of the sample to be analyzed, and, second, so that the sample's position along the beam axis is not changed, which might cause the sample to move out of the plane of the beam focus.

A goniometer used on microprobes in Melbourne, Oxford, and Singapore [38] is shown in Figure 5.12a. It enables the sample to be rotated about the x and y axes, and to be translated parallel to these two axes. The smallest angular movement that can be made with the goniometer is 0.025° and the angular range of the axes is ±20° around the vertical (y) axis and ±15° around the horizontal (x) axis. The smallest sample translation that can be made is 5 μm, and both translation directions have ranges of ±12.5 mm. The sample is mounted as close

Figure 5.12. (a) The four-axis eucentric goniometer used for channeling experiments on nuclear microprobes in Melbourne, Oxford, and Singapore [38]. (b) Close-up of the goniometer sample stage used on the Oxford microprobe, showing the particle detector (A), the shutter used to protect the detector from high beam currents (B), and the arm used to open and close the shutter (C). Reprinted from Ref. 39 with the kind permission of Elsevier Science, B.V., Amsterdam, The Netherlands.

as possible to the eucentric position of the goniometer, although it is common in CSTIM analyses at Oxford to find that the sample undergoes small lateral movements, of the order of 1 μm per 0.1° of rotation, when tilted.

For CSTIM analysis, a semiconductor charged particle detector is positioned behind the sample. Such a detector cannot be mounted behind the goniometer shown in Figure 5.12a, so at Oxford it is mounted on the sample stage itself. Figure 5.12b shows a close-up of the goniometer stage onto which the sample and detector are mounted [39]. The detector-sample distance is approximately 2 mm, so that the detector subtends a solid angle of 2.6 steradians; this means that virtually all the transmitted ions are detected and their energies measured. The

Figure 5.12. (*Continued*)

detector is surrounded by a plastic cover to insulate it from the rest of the sample stage so that the beam current can be measured from the stage. To protect the detector from damage when high beam currents are used for beam focusing or backscattering measurements, a shutter is mounted between the detector and the sample that can be opened and closed by means of an arm that extends outside the microprobe chamber. The shutter is shown open in Figure 5.12b, exposing the detector to the beam, and the arm is shown in position to close the shutter.

The sample is fixed to the goniometer stage, over the active region of the detector for CSTIM analysis, often using conducting silver paint. Commonly mounted with the sample are a piece of quartz to enable the beam to be focused, a piece of 2,000-lines-per-inch copper grid for checking the beam focus and suitable calibration samples for the backscattered ion detector (such as a piece of silicon with a thin gold layer on its surface) and CSTIM detector (such as a thin aluminium foil of known thickness).

The properties and suitability of the semiconductor detectors used for backscattering and transmission measurements have been discussed previously [40]. For CSTIM analysis on the Oxford microprobe, a semiconductor detector with an active area of 25 mm^2 and an energy resolution (FWHM) of approximately 20 keV is used. This detector has a low outer casing thickness of 7 mm, which enables it to be mounted on the goniometer sample holder behind the sample. For backscattering measurements on the Oxford microprobe, a semiconductor detector subtending a solid angle of 80 msr is positioned at an angle of 155° to the incident beam direction. On the Melbourne microprobe, the detector solid angle is between 40 and 250 msr at an angle of 150° to the incident beam. It is also possible to use unencapsulated photodiodes, which are much cheaper than charged particle semiconductor detectors, for charged particle detection [41].

5.4.3. Production of Channeling Scanning Transmission Ion Microscopy Beam Current

For backscattering or PIXE measurements, a minimum beam current of approximately 100 pA is required. For CSTIM analysis, particularly when a large acceptance angle detector is employed, the current must be several orders of magnitude less than this. A suitable procedure for production of a beam current of about 1 fA is as follows:

1. The transmitted ion detector is shielded from the beam (by moving it off the beam axis or using a shutter arrangement as described above) and a beam current of ~100 pA is used to focus the beam by observation of the beam spot on a fluorescent material with a stereo zoom optical microscope, in a standard manner.
2. The object aperture is closed to reduce the beam current from 100 pA to ~1 pA.
3. The collimating aperture is closed so that no beam at all is entering the microprobe chamber.
4. The detector is placed on the beam axis, and any shutter arrangement is opened.
5. The collimating aperture is slowly opened until ~2,000 ions per second (0.3 fA) are being recorded by the detector by observing the count rate on a rate meter connected to the amplifier.
6. The spatial resolution of the focused beam is then checked using STIM images of a copper grid.
7. The energy scale of the detector is calibrated, using the spectrum peaks produced by the straight-through beam and by the beam that has passed through a sample of known thickness.

For CSTIM analysis on the Oxford microprobe, it is important that the collimating aperture is increased slowly, because once it is open wide enough to allow some ions through, small increases in the aperture size result in large increases in the beam current. The collimating aperture should be increased symmetrically about the beam axis to reduce scattering of the beam; scattering results in a low-energy tail on the CSTIM spectrum and adds to the noise level on the CSTIM images. If slit jaws are used for the object aperture, care must be taken when reducing the aperture size to prevent too much beam being stopped by any one set of slit jaws. On occasions when too much beam is stopped in this way, it has been found that the set of jaws heats up. This results in the jaws slowly closing in and reducing the small beam current to zero after a short time. In the absence of slit heating effects, a suitably small beam current can be produced that is stable for several hours.

5.4.4. Choice of Channeling Direction

Channeling analysis begins with location of a suitable channeling direction for the incident beam. The particular direction chosen can significantly affect interpretation of the results of the experiment. This is particularly true when images of individual defects are produced using CSTIM, and there are several factors to be considered when deciding on the channeling direction to be used:

1. It is often more beneficial to use a planar channeling direction than an axial direction, as defects can cause a greater fraction of planar channeled ions to be dechanneled than axially channeled ions. For example, it was described in Section 5.2.1 how the amount of dechanneling produced by a dislocation is inversely proportional to the critical angle of the channeling direction used, so that dechanneling is greater for planar channeling (narrow critical angles) than for axial (wider critical angles). Dechanneling caused by dislocations is seen much more prominently in backscattering spectra taken in planar alignment than in those taken with the beam axially aligned [7,42]. It has also been shown that CSTIM images reveal stacking faults with greater contrast when the incident beam is planar channeled than when it is axially channeled (Section 8.6.2).

2. Planar channeling, rather than axial, can also give information on the nature of a particular defect under investigation. In particular, a defect may affect some sets of lattice planes but not others, so that the defect is invisible for certain channeling directions. This is again demonstrated by stacking faults as described in Section 8.6.2. Invisibility of a defect for some planar channeling directions provides information on the way the defect is disrupting the crystal lattice (a similar technique is used in transmission electron microscopy to characterize individual defects [43]). It also means that individual defects could be missed if only a single channeling direction was used for analysis.

3. In semiconductor crystals with the silicon structure, the {111} planes have the lowest ratio of channeled to random energy loss and the lowest natural dechanneling rate (Section 1.4). This means that defects can be revealed most quickly and with the greatest contrast for {111} planar channeling, and often a {111} direction is the best to use for a first look at a crystal.

4. For observation of deep defects, {111} planes are again the best to use for crystals with the silicon structure because of their low natural dechanneling rate (see Section 8.6.2 for CSTIM images of defects 50 μm below the surface of a sample taken using {111} channeling).

It is therefore often beneficial to produce images of a region of a crystal using a variety of channeling directions for the incident beam, in order to fully characterize the defects under investigation. The directions used may be limited by the range of tilt angles of the goniometer. Lower index directions are much easier to locate than those with high indices, as they have larger critical angles and produce more stable channeling. For thin crystals, however, quite high-index planes and axes can be observed in transmission channeling patterns, as described in Section 5.4.5.

An example of the use of multiple channeling directions for analysis with a microprobe is given in Section 8.6.2 and Ref. 44. In the latter study, a region of a silicon crystal was imaged at nine different planar channeling directions. The 40 μm thick sample was fixed to a stage allowing rotation about a single axis in the sample plane, and a detector with a restricted acceptance angle was used (see Section 5.4.7). Figure 5.13a is a stereographic projection showing the angular path of the beam as the sample was tilted from $-45°$ to $+45°$ with respect to the sample surface. The nine channeling directions used were the four $\{111\}$ directions, four $\{110\}$ directions, and the (010) direction, as marked in Figure 5.13a.

Figure 5.13b shows the average transmitted ^{1}H ion energy as a function of sample tilt angle for the path shown in Figure 5.13a, with $0°$ being the angle at which the beam was aligned with the (010) channeling direction. As the tilt angle was increased from zero, the average transmitted ion energy decreased owing to the effective increase in sample thickness (which varied inversely as the cosine of the tilt angle). However, the average transmitted energy was increased every time the beam was aligned with one of the nine planar channeling directions. The peaks in the transmitted energy are numbered in Figure 5.13b according to their position on Figure 5.13a, and their angular distance from the (010) direction is given. The largest increase in average transmitted energy was produced by the $\{111\}$ planes, as they produce the most stable channeling and have the lowest ratio of channeled to random energy loss mentioned above. The $\{110\}$ planes produced the next strongest effect and the (010) plane produced the weakest effect of those used. It was found that the image contrast for the defects being studied (stacking faults) was greatest for $\{111\}$ channeling, followed by $\{110\}$ channeling, with the (010) planes producing the weakest contrast. However, only the (010) channeled image revealed all of the defects present in the region of the sample imaged, as, for the other directions, the nature of the defects and the beam-sample geometry meant that some or all of the defects could not be seen.

5.4.5. Location of Channeling Directions

Alignment of the beam with a channeling direction is normally achieved by mounting the crystal on a goniometer, which allows rotation about two orthogonal axes, finding the amount of misalignment between the incident beam direction and the channeling direction required, and then tilting the crystal about one

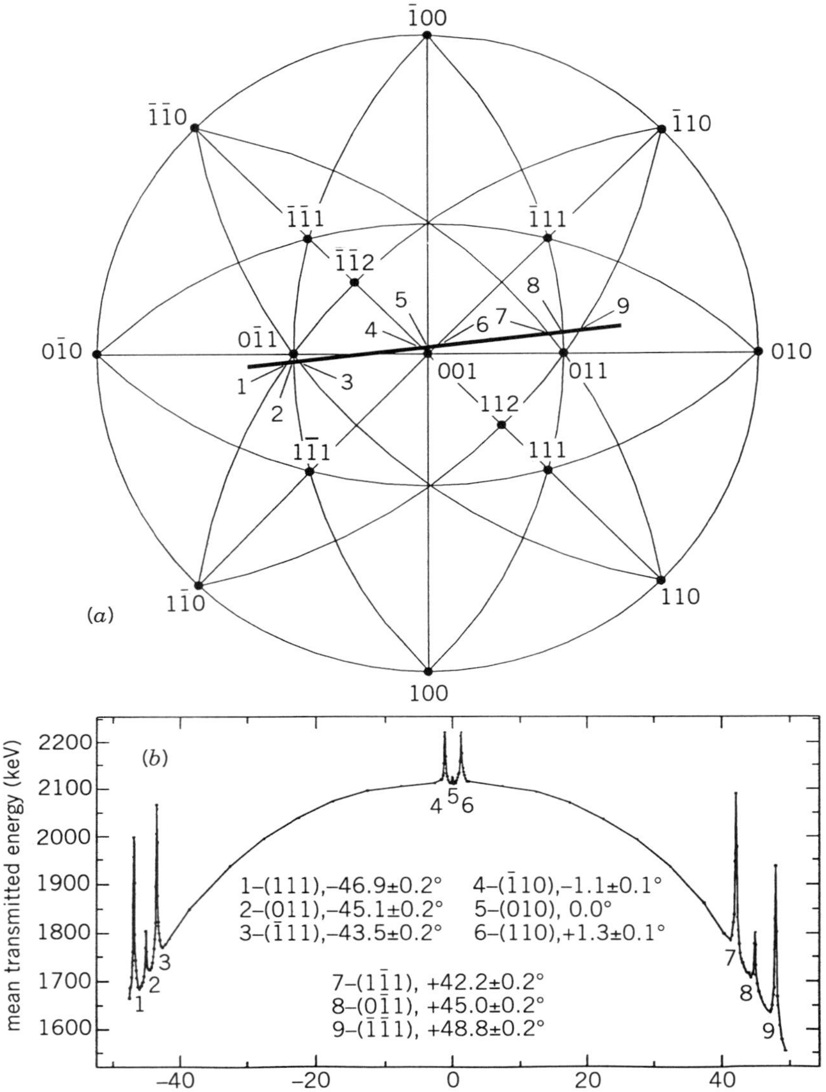

Figure 5.13. (a) Stereographic projection showing the angular path (dark line passing close to the center of the plot) of a ^{1}H ion beam as the silicon sample was tilted. The positions of intersection of nine major planar channeling directions with the beam path are marked. (b) Average transmitted energy of the 3 MeV ^{1}H ions as a function of sample tilt angle away from beam alignment with the (010) planes. The peaks in the curve occurred when the beam was aligned with the nine planar directions indicated in (a), and their angle from (010) channeling is given. The curve was produced by recording transmitted proton energy spectra over the angular range and calculating the average transmitted energy from these. The points have been joined by a straight line to show the curve's shape more clearly. Reprinted from Ref. 44 with permission (© 1995 American Physical Society, Woodbury, NY).

or both axes so that the incident beam is channeled. The sample tilt axes can either both be in the same plane as the sample or one can be parallel to the sample normal direction. Two methods for aligning the crystal are discussed below: the first involves observation of the shape of transmitted or backscattered ion energy spectra as the crystal is tilted, the second involves observations of the channeling patterns produced by transmitted or backscattered ions.

5.4.5.1. Observation of Transmitted or Backscattered Ion Energy Spectra

The channeling process can dramatically affect the shape of backscattered or transmitted ion energy spectra. In the former case, the yield of backscattered ions from the surface of the crystal is reduced, and, in the latter case, a fraction of the ions is transmitted with a much lower than normal energy loss. Observation of the shape of the spectra as the crystal is tilted allows channeling directions to be located. Typically, two or more planar channeling directions are found, and, from their positions, an axial channeling direction can be aligned with the incident beam, if required.

In the case of a goniometer allowing rotation about the sample normal direction (referred to as the *rotation axis*) and about an axis in the plane of the sample (*tilt axis*), the procedure for aligning a near-normal axis has been described [45, see also 40, Section 8.2]. Changing the rotation angle describes a circle with angular radius given by the tilt angle. This arrangement has the advantage for transmission channeling that the same crystal thickness is presented to the beam at all rotation angles for a given tilt [46]. To locate a major axial channeling direction, the crystal is turned through 360° about the rotation axis with the tilt angle at a value large enough so that the cone traced out includes the desired axis. The rotation angle values corresponding to alignment of the beam with major planar directions that intersect at the axis are found, with each plane being located at two rotation angle values. The locations of the planar directions are plotted on a stereographic projection drawn with the incident beam direction at its center and with radius given by the tilt angle. The two rotation angle values for each planar direction are joined; the point on the plot where all the lines meet gives the location of the axis.

An example of this alignment procedure is shown in Figure 5.14. Here an 0.9 MeV ^{4}He ion beam was focused onto a nickel crystal with the <100> axis near to alignment with the beam direction. The yield of backscattered ions from the top 100 nm of the nickel crystal is shown plotted on a polar diagram as a function of rotation angle. The precise location of the axis is determined by linking the planar minima, as shown.

For a goniometer with two orthogonal rotation axes both in the plane of the sample, axial alignment is again produced by location of planar channeling directions. In this case, the above method can be used if both axes are tilted simultaneously to produce a rotation scan [47,48]. Alternatively, one axis at a time can be used for rotation. For some crystal alignments, location of an axis is relatively straightforward with the latter method; the silicon crystals studied in Chapter 8 all had their [001] axis normal to the sample surface and were

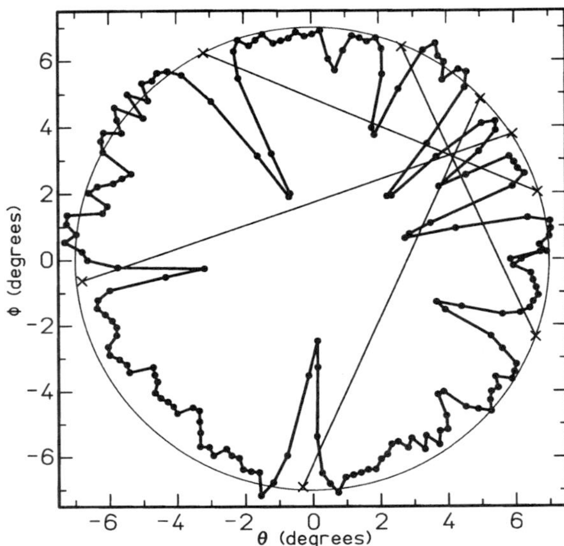

Figure 5.14. Example of the channeling alignment technique using azimuthal plots. The yield of backscattered ^{4}He ions from the top 100 nm of a nickel crystal is shown plotted on a polar diagram as a function of rotation angle. The precise location of the axis is determined by linking the planar minima, as shown. Data obtained using the Melbourne nuclear microprobe.

mounted so that their (110) and ($\bar{1}$10) planes were running approximately horizontally and vertically, respectively. In this case, rotation about one axis allowed location of the (110) planar direction and rotation about the other allowed location of the ($\bar{1}$10) planes. Setting the crystal so that the beam was channeled in both sets of planes corresponded to [001] axial alignment.

5.4.5.2. Change in Shape of the CSTIM Spectrum with Angle from Channeling
This section gives a more detailed description of the effects of channeling on transmitted ^{1}H ion energy spectra.

As indicated in Section 1.4, channeled ions have a lower energy loss rate than that of nonchanneled ions, so they produce a high-energy tail on the transmitted ion energy spectrum. Such a spectrum, measured with a large acceptance angle detector, is shown in Figure 5.15. It was taken from a 31 μm thick piece of a $Si_{0.95}Ge_{0.05}$/Si sample (studies of which are presented in Chapter 8) with a 3 MeV ^{1}H ion beam channeled in the (110) planes. As the sample was tilted from a nonaligned direction toward the channeling orientation, more ^{1}H ions were channeled, so the high-energy tail on the spectrum increased. Figure 5.15 also shows a nonchanneled spectrum, and a semichanneled spectrum taken with the sample 0.08° from channeling alignment (the channeling critical angle was ~0.1°).

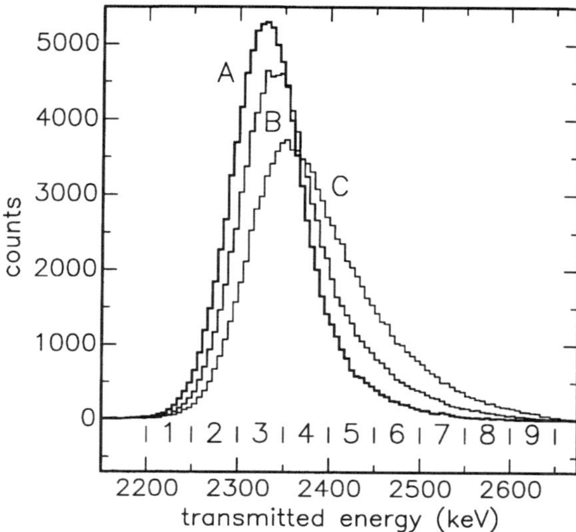

Figure 5.15. Three transmitted ^{1}H ion energy spectra, normalized to the same number of counts. A, beam not aligned with a channeling direction. B, beam 0.08° from (110) planar channeling. C, beam within 0.05° of (110) planar channeling. Marked along the x axis are the windows used to produce Figure 5.16.

To observe the change in the fraction of ^{1}H ions transmitted in the high-energy tail as the sample was tilted through a (110) planar channeling orientation, a set of eleven energy spectra was taken (Figure 5.15 shows only three of these spectra) at different sample tilt angles from the channeling orientation. The spectra were all normalized to contain the same numbers of counts and were divided up into nine energy windows, which are marked on Figure 5.15. The number of counts in each of these energy windows is shown as a function of tilt angle from the channeling orientation in Figure 5.16, with the counts for each window expressed as a percentage of the counts in that window when the crystal was −0.4° from the channeling orientation. As Figure 5.16 reveals, windows 5 to 9 on the high-energy side of the spectrum show a dramatic increase in the number of counts at the channeled position by up to two orders of magnitude, whereas the number of counts at the channeled position for windows 1 to 3 on the low-energy side can be less than half the number recorded away from the channeling direction. The number of counts in window 4 shows little change with crystal orientation. This procedure shows how an uncollimated detector can be used for channeling alignment; the procedure for a collimated detector is described in Section 5.4.7.

For the measurement of backscattered ions, there is a large reduction in the yield from close to the sample surface when the beam is channeled. This can be seen in the two channeled spectra shown in Figure 5.6. The variation with angle from the channeling axis of the yield of ^{1}H ions scattered from silicon

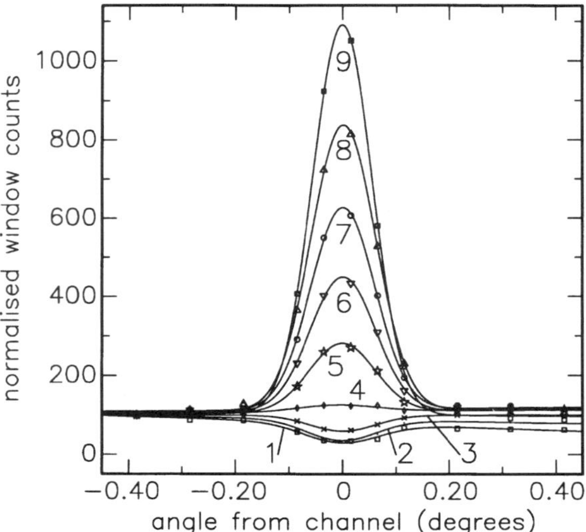

Figure 5.16. Variation with angle from (110) alignment of counts in each of the windows set on the spectra of Figure 5.15. For each window, the points have been fitted with Gaussian curves.

atoms in the layer of the $Si_{0.85}Ge_{0.15}/Si$ sample, whose channeled spectrum is shown in Figure 5.6, can be seen in Figure 1.15.

5.4.5.3. The Alignment Procedure The changes in the ion energy spectrum on advancing through the channeling direction can be used to align the beam for channeling analysis. Initial alignment can be performed using a rate meter linked to the number of transmitted or backscattered ions being detected. The lower level on the detector amplifier is increased, so that only ions transmitted with a low energy loss, or ions backscattered from near the sample surface, are recorded. The crystal is then slowly tilted while the rate meter is observed. For the measurement of transmitted ions, alignment with a channeling direction produces a large increase in the rate meter reading, whereas, for the measurement of backscattered ions, alignment produces a drop in the count rate.

The channeling direction can then be found more accurately by observing the full energy spectrum on the computer as the crystal is tilted slightly about the angles found from observing the rate meter, as shown in Figures 5.15 and 5.16. For transmitted ions, energy spectra can be recorded for the same incident beam dose (equivalent to the same number of transmitted ion counts for a detector with a large acceptance angle), and the number of counts falling in a window located on the high-energy part of the spectrum maximized as a function of sample tilt angle. However, the beam current required for CSTIM analysis is usually too low to be measured, so that it is not possible to record spectra for the same incident dose, and the facility to record spectra with the same number of counts may not be available. In this case, two energy windows are chosen, one near the peak of the nonaligned spectrum and one half way along the high-energy tail of the spectrum (for example windows 3 and 6 in Figure 5.16). The ratio of counts in these two windows is then measured as the crystal is tilted, and the channeling direction is found from the angle at which the ratio is at a minimum (for counts in window 3 divided by counts in window 6). The spectra used for Figures 5.15 and 5.16 were normalized off-line.

For a good-quality crystal mounted so that the channeling direction is close to the beam axis, the rough alignment takes only a few minutes and the full alignment described above takes less than 30 min. For transmitted ions, the alignment process is complicated if the sample varies considerably in thickness, for example, if it has been prepared initially for transmission electron microscopy analysis and is not at the exact eucentric point of the goniometer. In this case, tilting the sample causes it to undergo small translations as well. This results in the ions encountering different sample thicknesses and leads to changes in the transmitted energy spectrum that can impair attempts to find the channeling direction.

For backscattering measurements, a window is set on the part of the ion energy spectrum corresponding to scattering from near the sample surface. Energy spectra are recorded as a function of tilt angle, each for the same incident beam dose, and the number of counts falling in the window is plotted versus tilt angle. The minimum of the dip reveals the angular location of the channeling direction.

Automatic methods for crystal alignment have been developed based on the above techniques [48–50]. These involve computer-controlled rotation of the sample using stepper motors to drive the goniometer. An azimuthal scan can be performed with the computer recording the backscattered yield at each point on the scan, finding the major planar dips and calculating from them the position of the axis of interest. Alternatively, the sample can be tilted about two axes in its plane through a fixed range of angles with a preset step size to produce an angular image of the backscattered ion yield. The location of the axis can be found as the point on the image where the yield is lowest. Figure 5.17a shows such a contour plot produced by an automatic alignment system for a nickel crystal using a 0.9 MeV ^{4}He ion beam and an angular range of $\pm7°$ about each tilt axis. Figure 5.17b shows a similar image produced using the same incident ion beam from a BISCO superconductor crystal [48], over an angular range of $\pm2°$.

5.4.5.4. *Direct Observation of Ion Channeling Patterns*

The orientation of a single crystal can also be obtained from observations of the patterns produced by ions scattered from the crystal or transmitted through a thin crystal. Such blocking or transmission star patterns provide a simple projection of the planes and axes of the crystal on to the recording medium [51], so that the crystal orientation can be determined and alignment of a particular channeling direction performed.

The patterns of ions leaving a crystal can be recorded in a number of ways. To produce rapid, continuous patterns suitable for orienting a crystal, a fluorescent screen coated with, for example, zinc sulfide can be used [52,53] and the patterns photographed as required [54]. Patterns can also be recorded using pho-

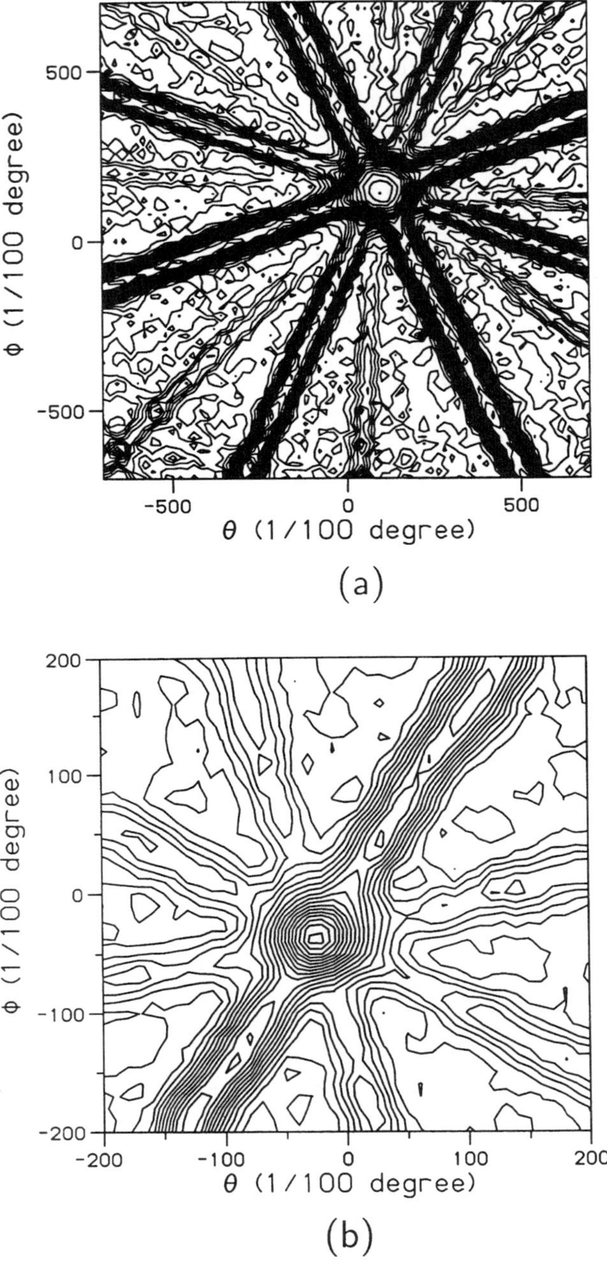

Figure 5.17. Channeling angular scans about the <100> axis of (a) nickel crystal and (b) BISCO. These contour plots were produced from variations in the signal intensity of 0.9 MeV ^{4}He ions backscattering from the top of 100 nm of the samples, which were tilted through an angular range of (a) ±7° and (b) ±2°. Adapted from Ref. 48.

tographic film or cellulose nitrate foils (see Ref. 55 and Section 5.3.1), which both need developing to reveal the ion distributions. Position-sensitive charged particle detectors [46] or a detector scanned across the emerging particle distribution [56] can be used to record both the angular distribution and the energies of the emerging ions.

If the recording material is placed at a backward scattering angle for the beam, blocking patterns are produced [51]. *Blocking* [57] refers to the process by which ions emitted from close to atom sites in the lattice (such as those scattered through large angles owing to close nuclear collisions) have a decreased probability of being emitted along channeling directions. The ion paths are steered away from channeling directions by the same correlated series of small-angle scatterings that produces channeling. An example of a blocking pattern obtained by bombarding a silicon single crystal with 150 keV ^{1}H ions is shown in Figure 5.18 [51]. The pattern of planes about the [001] axis can be compared with Figure 1.16. Measurement of the position on a pattern of an axis of interest with respect to the incident beam directions can be used to align the crystal for channeling [55].

If a thin crystal is being studied and the recording medium is placed behind

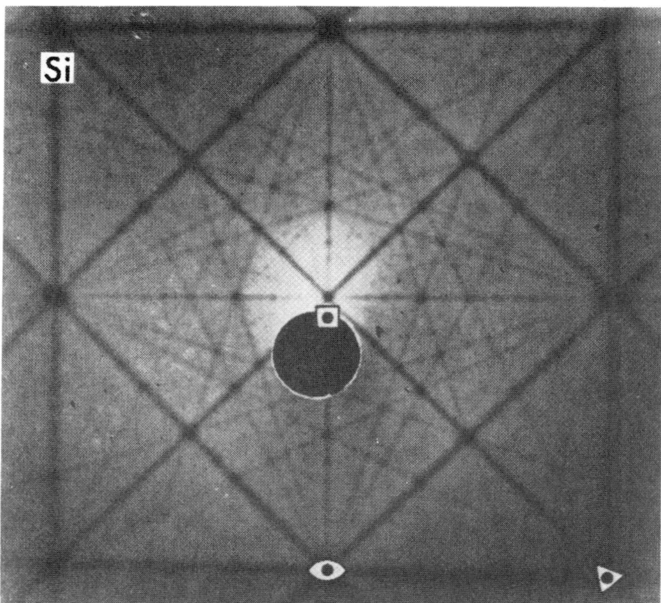

Figure 5.18. [001] blocking pattern of 150 keV ^{1}H ions incident on a silicon crystal. Black lines and spots represent ^{1}H ion-deficient regions. The corners of the square near the edge of the pattern correspond to <100> directions 45° from the central <110> axis. Reprinted from Ref. 51 with permission (© American Institute of Physics, Woodbury, NY).

the crystal, the angular distribution of the ions transmitted through the crystal can be displayed and used for alignment [53,58]. In this case, the intensity of the transmitted ions is increased where the incident beam is aligned with a channeling direction. Planes and axes away from the incident beam direction produce lower-than-background intensity owing to blocking. A photograph of the experimental set-up on the Oxford microprobe used to produce transmission channeling patterns is given in Figure 5.19. A fluorescent screen, taken from an old oscilloscope, is mounted 660 mm behind the sample at the end of a pipe attached to the back port of the microprobe chamber. Patterns produced on this screen by the passage of 3 MeV ^{1}H ions through a thin silicon crystal are shown in Figure 5.20 (and also in Figure 5.10). The photographs in Figure 5.20 show the effect on the pattern of tilting the crystal about the vertical axis so that the incident ^{1}H ion beam went through alignment with a (110) planar channeling direction. The appearance of such patterns depends on the beam energy, and the orientation and thickness of the crystal [59]. The patterns are very useful for alignment of CSTIM samples.

Use of a nuclear microprobe to produce transmission channeling patterns has two advantages over use of a broad beam: firstly, patterns can be produced from

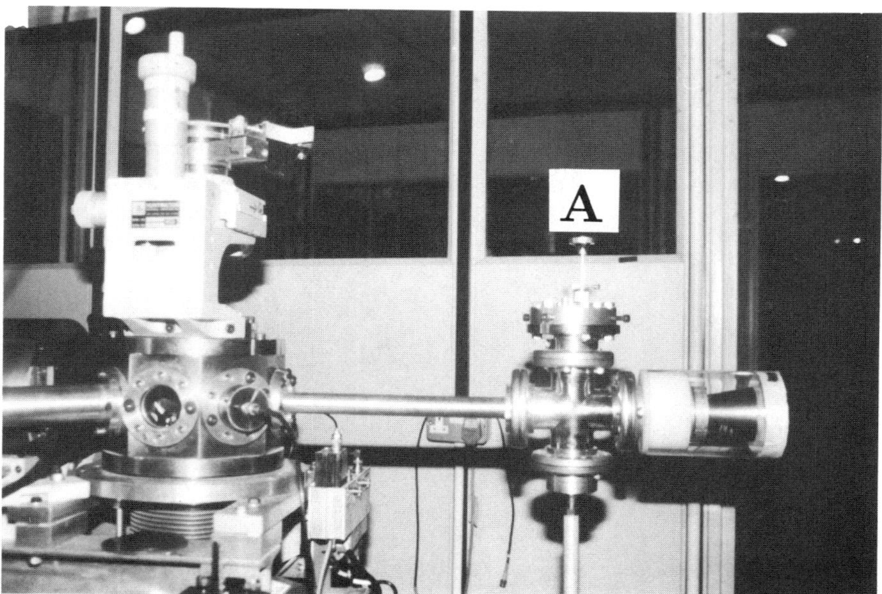

Figure 5.19. Photograph of the experimental arrangement for the production of transmission channeling patterns on the Oxford microprobe. A fluorescent screen is mounted on the beam axis 660 mm behind the sample. The arm (A) allows a particle detector to be moved on to the beam axis in front of the screen to record the energy spectrum of the transmitted ions.

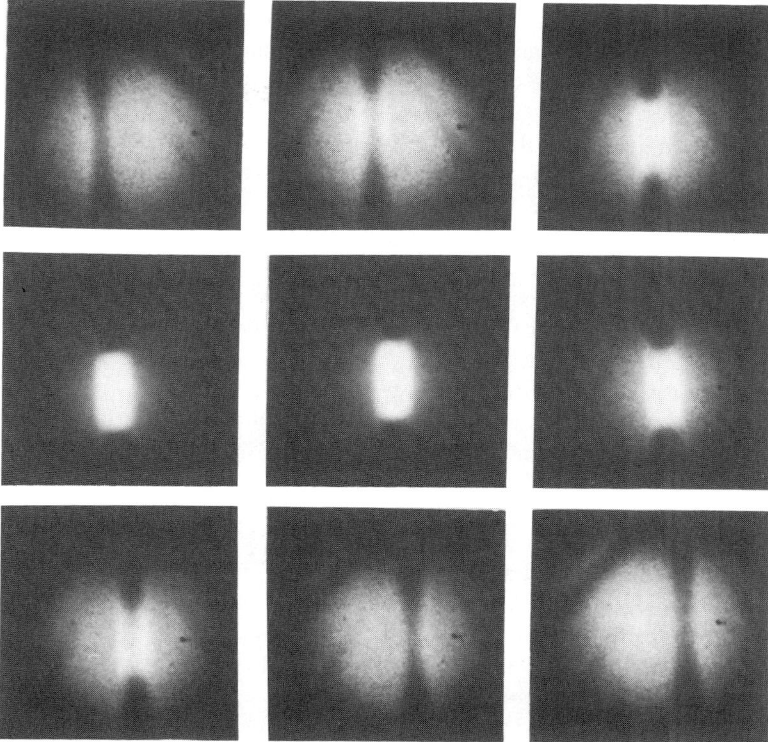

Figure 5.20. Example of transmission channeling patterns produced using 3 MeV ^{1}H ions on the Oxford microprobe. A silicon crystal with a thickness of approximately 1.4 μm was used, and the beam was incident close to the sample [001] axis. Shown are nine patterns taken with the sample tilt about the vertical axis progressively changed (0.06° between each photograph) so that the incident beam went through alignment with the (110) planes. This planar direction is revealed as a dark band (low transmitted ion intensity) with the beam away from alignment owing to blocking (photographs in the top left and bottom right) and as a region of very high transmitted ion intensity when the incident beam was channeled.

very small crystals (or small areas of a large crystal) and secondly the angular resolution of the images is not limited by the size of the beam spot when it hits the recording material. Care must be taken, however, to ensure that the crystal is not damaged by the analyzing beam, as currents similar to those required for backscattering measurements can be necessary to produce patterns.

5.4.6. Image Production

Once the crystal is aligned and the region of interest located, images of the sample can be produced. For CCM, this involves acquiring a backscattered ion or ion induced X–ray spectrum as the beam is scanned over the sample. Either before data collection is started, or during off-line analysis if the data is taken in event-by-event format (see Section 2.3), regions of the spectra are selected and images produced based on the number of counts falling in each energy region at each point in the scanned area. The windows are selected based on particular X–ray lines or scattering from particular elements at particular depths.

For CSTIM analysis, a similar procedure is followed. A transmitted ion energy spectrum is recorded, and, if the data is taken in event-by-event format, a map showing the average ion energy loss can be produced off-line. Quantities other than the average energy can also be used to produce images if the data is taken in event-by-event format, and this is described in Section 4.7. If the data is recorded in windowed format, as for the CSTIM images of Chapter 8, the procedure is slightly different. A diagram of the method of CSTIM image production as used on the Oxford nuclear microprobe, where data is taken in windowed format, is given in Figure 5.21 taken from Ref. 60. The figure is based on the production of an average energy loss ^{1}H ion image of a sample of the $Si_{0.95}Ge_{0.05}$/Si. The imaged region was quite large ($920 \times 920~\mu m^2$) and the crystal had been curved by the thinning process. This meant that if the incident beam was aligned with a channeling direction for one part of the crystal, it was not necessarily channeled at another part owing to the bend of the crystal tilting the lattice planes away from alignment.

Figure 5.21a shows a schematic of the typical set-up for a CSTIM experiment. As the beam is scanned over the sample, a transmitted ion energy spectrum is built up (Figure 5.21b). Images of the sample are produced by setting energy windows on this spectrum (labeled 1–11 in Figure 5.21b). Images can be produced from a single energy window, in which case contrast results from variations in the number of ions falling in the window at each pixel. Figure 5.21c shows two such images of the $Si_{0.95}Ge_{0.05}$/Si sample. Image 1 is from window 2 on the spectrum, that is, a window where the ^{1}H ions had quite a high energy loss, and image 2 is from window 5 in the lower energy loss part of the spectrum. These two images are therefore almost the inverse of each other. To produce a final image of the sample showing the average ^{1}H ion energy loss, $\overline{E}$, the information in all of the individual energy window images is combined together according to the formula:

$$\overline{E}(x, y) = \frac{\Sigma_{w=1}^{W} n_w (E_b - E_w)}{\Sigma_{w=1}^{W} n_w} \tag{5.3}$$

where x, y refers to a pixel position on the map, n_w is the number of ^{1}H ions at this pixel that have fallen in window w, E_w is the energy in the middle of window w, E_b is the beam energy, and W is the total number of windows. Image 3 of Figure 5.21c shows the average energy loss image for the $Si_{0.95}Ge_{0.05}$/Si sample. In this, and the CSTIM images shown in Chapter 8, darker greys represent higher average transmitted energy loss. In image 3 of Figure 5.21c, the light

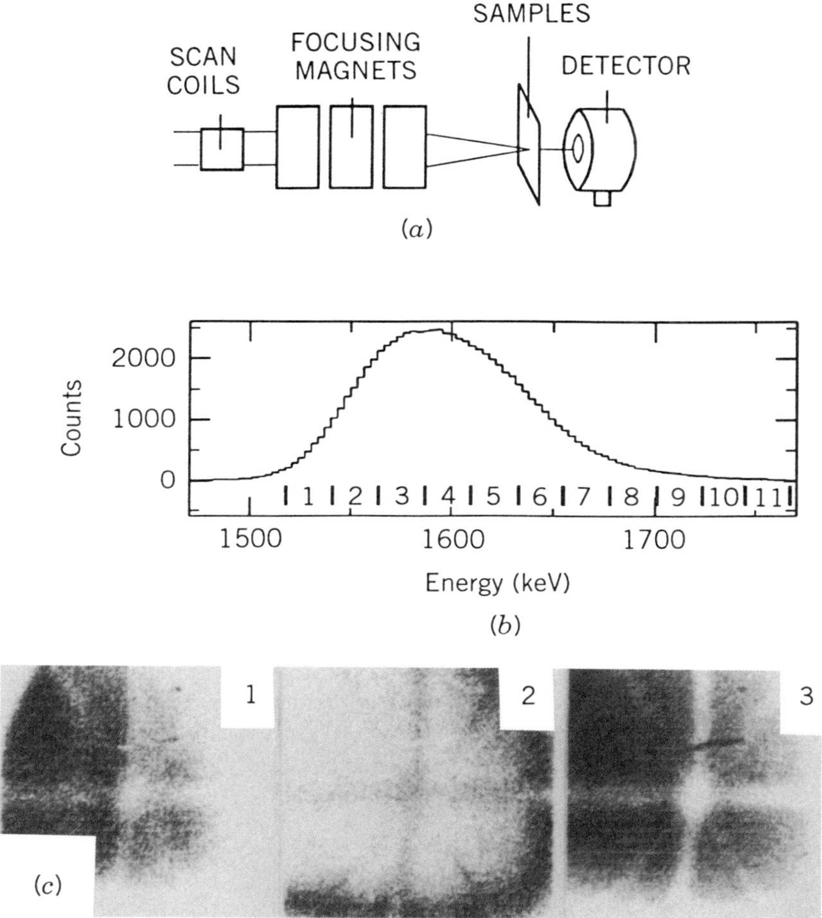

Figure 5.21. CSTIM image production when the data is recorded in windowed format. (a) Schematic of the experimental arrangement. (b) Energy spectrum recorded for the production of the images in (c). Numbers 1–11 below the spectrum mark the positions of the energy windows. (c) 920 μm wide images of a curved crystal. Image 1: map of the intensity of counts falling in spectrum window 2. Image 2: Intensity map from window 5. Image 3: Average energy loss image produced by combining the images from all the slices set on the spectrum. Darker greys represent higher energy loss. Reprinted from Ref. 60 with kind permission of Elsevier Science B.V., Amsterdam, The Netherlands.

region at the center of the image was where the incident beam was channeled along the [001] axis of the crystal. The horizontal and vertical arms coming from the central region correspond to where the beam was channeled in {110} planes of the sample, and the diagonal arms are where the beam was {100} channeled. Therefore, this image is effectively a map of the crystal bend.

Average energy loss images can often benefit from some simple image processing techniques. Smoothing can be used to reduce the image noise level, as was applied, for example, to the images shown in Figure 8.35. In this case, the energy loss value at each pixel was convoluted with those of its eight nearest neighbors, weighted by a factor depending on their distance. Histogram equalization [61] can also be used to increase the image contrast. This technique assigns equal numbers of pixels to each grey level of the final displayed image, spreading the energy loss values over the whole range. For samples that vary in thickness within the imaged region, the resulting background energy loss variations can dominate the contrast so that features of interest cannot be seen. It has been found useful in such cases to remove the background caused by the thickness variations. This can be done by fitting a polynomial surface to the background and then subtracting this from the whole image. An example of this process is given in Figure 5.22, and the procedure was also used on CSTIM images of a wedge-shaped sample (Figure 8.34). Methods of data collection and image production for STIM are also discussed in Section 4.7, and much is also applicable to CSTIM.

5.4.6.1. Time for Data Collection   The time period for which data is collected is determined by two competing factors: the time required to gather sufficient image statistics and the amount of damage done to the sample by the analysis beam. The low beam current required for CSTIM analysis means that even for a small scan (e.g., $30 \times 30 \ \mu m^2$) a relatively long measurement period (e.g., 1 hr) would cause the sample to receive only 4×10^{12} ^{1}H ions/cm^2. As discussed in Chapter 1, lattice damage sufficient to be detectable with ion channeling begins

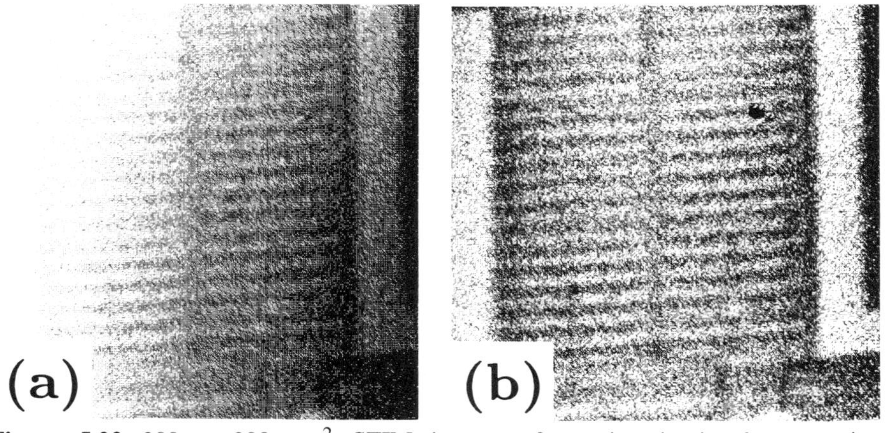

Figure 5.22. $200 \times 200 \ \mu m^2$ STIM images of a microcircuit, demonstrating nonuniform-background subtraction. (a) Original image, with a background produced by sample thickness variations that obscures some of the image details. (b) Image after a background was fitted and subtracted. This is the same microcircuit device area as that described in Section 7.1 for IBIC microscopy and shown in Figure 7.3.

to occur for doses of approximately 10^{16} ions/cm^2 for silicon, so that detectable sample damage is rarely a worry in CSTIM measurements with large acceptance angle detectors. Larger incident beam currents are necessary if a detector with a restricted acceptance angle is used, as described in Section 5.4.7. Even in this case, the current is unlikely to exceed 1 pA, and the dose delivered to the sample in the CSTIM measurement given as an example above would still be less than that required to cause detectable damage using ion channeling.

Of more concern in a CSTIM experiment is the ion dose received by the semiconductor detector. Damage sufficient to cause a degradation in the electrical properties of a semiconductor detector begins to occur at dose levels of 10^9 to 10^{11} ions/cm^2 (see Section 1.5). Prolonged use of the same detector for CSTIM experiments can result in energy calibration changes and a loss of energy resolution, particularly if the detector is collimated so that a small area receives a large dose. These effects can be minimized by changing the region of the detector that receives most of the dose, avoiding runs where the beam is stopped at a point on the sample and ensuring that the detector is not accidentally exposed to high-beam currents during the setting up of the experiment (see Section 5.4.3). However, experience on the Oxford microprobe suggests that replacement of the detector used for CSTIM measurements every few months, equivalent to approximately 200 hr of CSTIM data collection, is necessary.

The data acquisition time for CSTIM analysis is then primarily limited by the time required to gather sufficient statistics. This is determined by such factors as the choice of channeling direction, sample thickness, and the particular defects being studied, and is discussed again in Section 8.3.2. However, typical recording times as used for most of the CSTIM images in Chapter 8 are of the order of 20 to 30 min for count rates of approximately 2,000 ^{1}H ions/s.

For CCM measurements, the damage done to the sample by the analyzing beam is more significant. Figure 5.23 shows the time to deliver a dose of 10^{16} ions/cm^2 to a sample as a function of scan size, assuming a beam current of 100 pA. For a data acquisition time of half an hour, the use of scan sizes below approximately $100 \times 100 \ \mu$m^2 will cause the sample to receive doses greater than this value, and sample damage becomes a serious consideration. Analysis of crystal defects requiring smaller area scans may thus be prohibited by sample damage for CCM.

5.4.7. Use of a Restricted Acceptance Angle Detector for Channeling Scanning Transmission Ion Microscopy

In Section 5.4.2, the sample holder containing the CSTIM detector used with the goniometer on the Oxford microprobe is described. This detector subtends a large solid angle so that virtually all of the incident ions are detected. However, use of a transmitted ion detector with a restricted acceptance angle can be beneficial to CSTIM analysis. A second sample mounting, shown schematically in Figure 5.24, is also available for use on the Oxford microprobe. It allows the sample to be translated along two orthogonal directions and rotated about

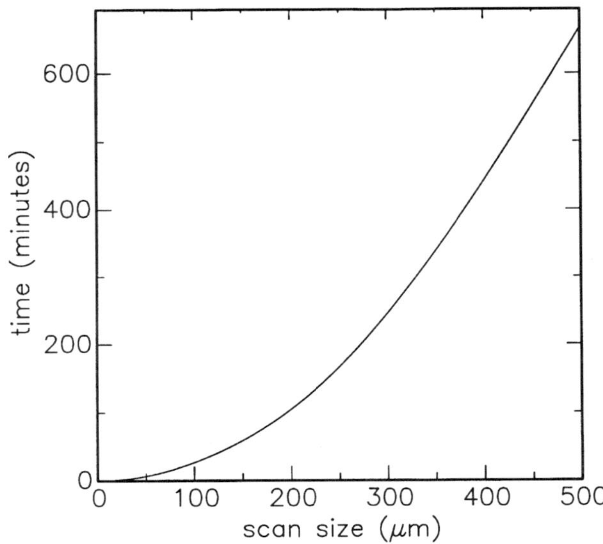

Figure 5.23. Graph showing the time necessary to deliver a dose of 10^{16} ions/cm^2 as a function of scan size for an incident beam current of 100 pA.

a single axis lying in the sample plane. The angular range is much larger than that allowed by the goniometer mounting, and allows planar channeling directions at large angles to the surface normal direction to be reached. This stage also enables a particle detector to be placed 40 mm behind the sample that can be collimated to have a restricted acceptance angle. The collimator, which con-

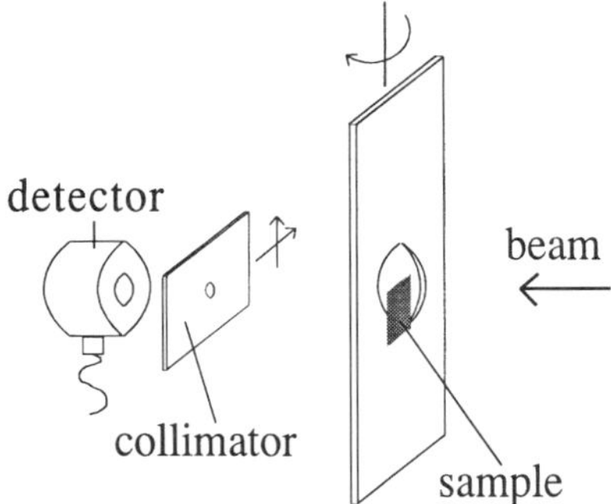

Figure 5.24. Schematic of the sample mounting allowing use of a restricted acceptance angle detector on the Oxford microprobe. The sample holder allows rotation about a single (vertical) axis through an unlimited range of angles. The detector is positioned approximately 40 mm behind the sample. Its acceptance angle can be reduced by placing a moveable collimator in front of it.

sists of a small hole drilled in a piece of aluminium, is mounted in front of the detector and can be moved laterally small distances to enable it to be positioned on the beam axis. The collimator is aligned by scanning an area larger than the collimating aperture without a sample in position. This produces an image of the collimating aperture as seen by the detector and enables the aperture to be shifted laterally until its image is centered.

As discussed in Section 1.4, channeled ions are transmitted through a crystal within a narrow angular range (of the order of the channeling critical angle) about the incident channeling direction, whereas nonchanneled ions are transmitted with a larger angular distribution. Collimation of the transmitted ion detector thus results in preferential detection of the channeled ions. This can be seen in Figure 5.25, which shows two transmitted ^{1}H ion energy spectra taken with the beam channeled in the (110) planes of a silicon crystal. For one, a large acceptance angle detector was used and for the other, the detector had an acceptance half-angle of $0.4°$. This was still greater than the channeling critical angle (~$0.1°$) yet it can be seen that the collimation caused exclusion of many ions from the low-energy part of the spectrum, emphasising those that were well channeled.

Detector collimation can enable defects to be observed more quickly in a CSTIM experiment, and observations of deep defects can require a collimated detector to be used (see Section 8.6.2). Detector collimation also makes the location of channeling directions using transmitted ions much easier. The overall count-rate into the detector is increased significantly at alignment for a constant incident beam current, as is the number of high-energy pulses. This can be seen in Figure 5.26, which shows two photographs of an oscilloscope screen, displaying pulses from the amplifier output of a detector receiving ^{1}H ions that had been transmitted through a silicon crystal. The detector had an acceptance half-angle of $0.4°$. Also shown by each photograph is the resultant energy spectrum of the transmitted protons. For Figure 5.26a, the incident ^{1}H ion beam was $1°$ away from alignment with a planar channeling direction. The count rate into the detector was 540 counts/s on average for the 2 min for which the spectrum was recorded. For Figure 5.26b, the beam was aligned with a {111} planar channeling direction of the crystal. In this case, the overall count rate into the detector was increased, the count rate during recording of the spectrum (also taken for 2 min) being 1,600 counts/s on average. The change in the height of the pulses displayed on the oscilloscope screen can also be easily seen, allowing major channeling directions to be aligned by observing the oscilloscope as the sample is tilted. With a large acceptance angle detector, the change in the oscilloscope pulses is too small to be seen from the oscilloscope screen on going through channeling alignment, and there is no observable change in the count rate into the detector.

Collimation of the detector acceptance angle necessitates the use of a higher incident beam current on the sample than for a large acceptance angle detector. This is further exacerbated as the sample thickness is increased, yet the beam current required is still normally two orders of magnitude less than that required for backscattering measurements. Care must be taken when using a collimated detector if a relatively thick sample is wound off the beam axis to be replaced by

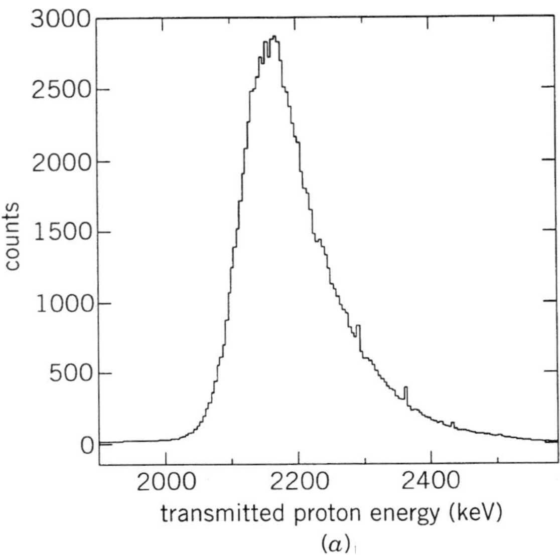

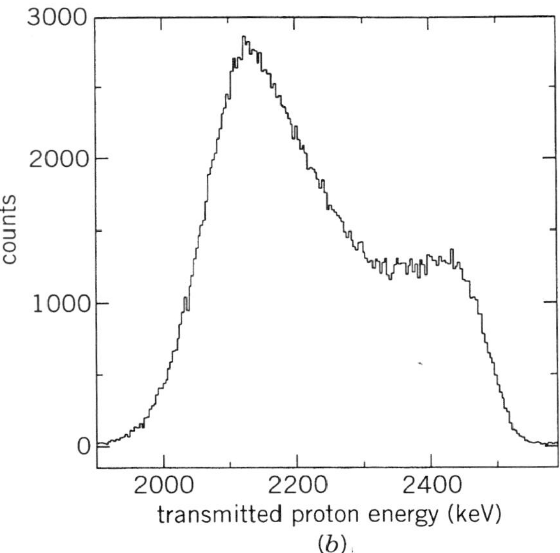

Figure 5.25. Transmitted 3 MeV ^{1}H ion energy spectra from a silicon crystal. Both taken with the incident beam channeled in the (110) planes of the sample, but (a) using a detector with a large acceptance half-angle of 45° and (b) using a detector with a small acceptance half-angle of 0.4°. Reprinted from Ref. 30 with kind permission of Elsevier Science B.V., Amsterdam, The Netherlands.

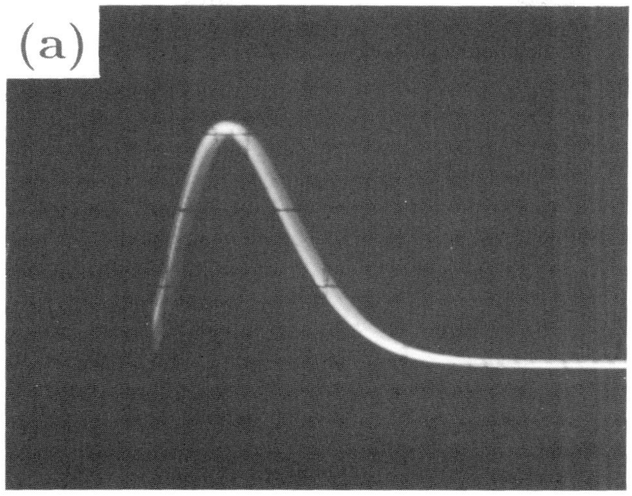

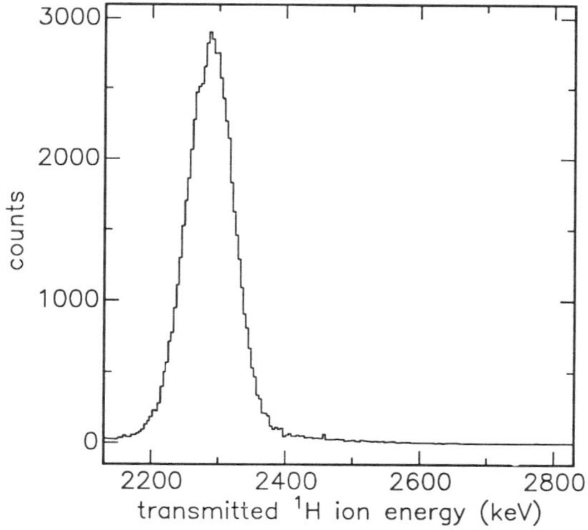

Figure 5.26. (a) Photograph of an oscilloscope screen showing pulses from the amplifier of a collimated detector receiving ^{1}H ions transmitted through a silicon crystal. Beam 1° from channeling. Shown beneath the photograph is the corresponding transmitted ^{1}H ion energy spectrum. (b) Same as (a), but with the beam aligned with a set of {111} planes of the crystal. Photographs in (a) and (b) taken with the same exposure time, and both spectra were recorded for 2 min each.

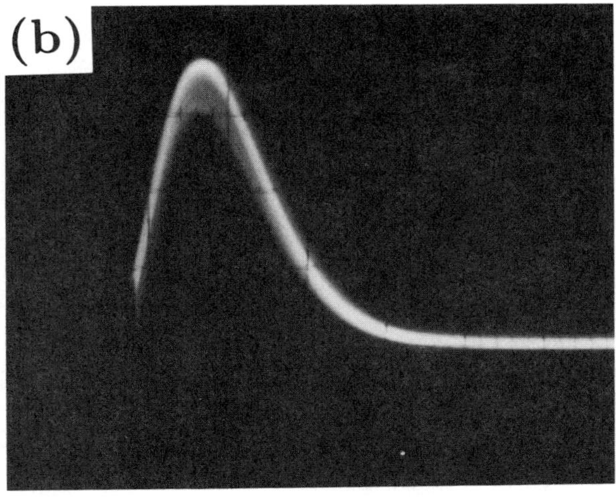

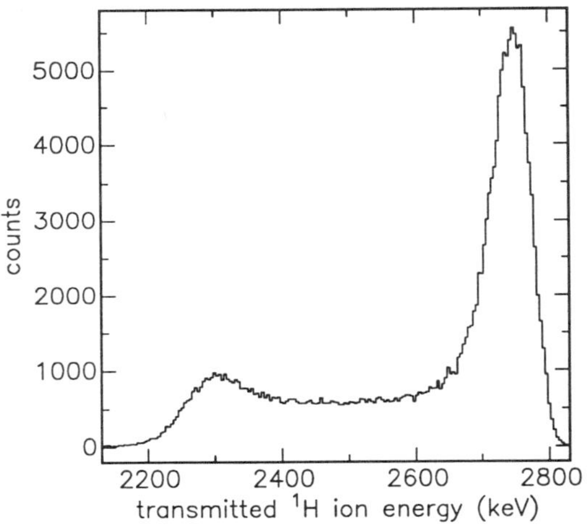

Figure 5.26. (*Continued*)

a thinner sample (or no sample at all). In this case, the reduction in the scattering of the transmitted beam by the sample can cause the detector to suddenly receive a high beam dose and therefore suffer damage.

REFERENCES

1. L.C. Feldman, J.W. Mayer, and S.T. Picraux, *Materials Analysis By Ion Channeling*. Academic Press, San Diego (1982).
2. K.L. Merkle, P.P. Pronko, D.S. Gemmell, R.C. Mikkelson, and J.R. Wrobel, *Phys. Rev.* **B8**(3):1002 (1973).

3. S.T. Picraux, E. Rimini, G. Foti, and S.U. Campisano, *Phys. Rev.* **18**(5):2078 (1978).

4. S.U. Campisano, G. Foti, E. Rimini, and S.T. Picraux. *Nucl. Instr. Meth.* **149**:371 (1978).

5. W.F. Tseng, J. Gyulai, T. Koji, S.S. Lau, J. Roth, and J.W. Mayer. *Nucl. Instr. Meth.* **149**:615 (1978).

6. N.W. Cheung, R.J. Culbertson, L.C. Feldman, P.J. Silverman, K.W. West, and J.W. Mayer. *Phys. Rev. Lett.* **45**(2):120 (1980).

7. A. Turos and D. Wielunska. *Nucl. Instr. Meth.* **201**:481 (1982).

8. K. Kimura. *J. Phys. Soc. Jap.* **52**(3):895 (1983).

9. S.T. Picraux, B.L. Doyle, and J.Y. Tsao. *Semiconductors and Metals*, Vol. 33. AT&T Bell Labs, New Jersey (1991).

10. D. Hull, *Introduction to Dislocations*. Pergamon Press, Oxford (1975).

11. J. Weertman and J.R. Weertman, *Elementary Dislocation Theory*. Macmillan, New York (1966).

12. Y. Quéré. *Phys. Stat. Sol.* **30**:713 (1968).

13. Y. Quéré. *Ann Phys.* **5**:105 (1970).

14. J. Mory and Y. Quéré, *Rad. Effects* **13**:57 (1972).

15. D. van Vliet, *Phys. Stat. Sol.* **2**:521 (1970).

16. H. Kudo. *J. Phys. Soc. Jap.* **40**(6):1645 (1976).

17. J.A. Ellison and S.T. Picraux, *Phys. Lett.* **83A**(6):271 (1981).

18. G.R. Booker and W.J. Tunstall, *Phil. Mag.* **13**:71 (1966).

19. Y. Quéré and H. Couve, *J. Appl. Phys.* **39**(8):4012 (1968).

20. T. Schober and R.W. Baluffi, *Phys. Stat. Sol.* **27**:195 (1968).

21. Y. Quéré, J.-C. Resneau, and J. Mory. *C. R. Acad. Sci. Paris* **B262**:1528 (1966).

22. Y. Quéré and E. Uggerhøj, *Phil. Mag.* **34**(6):1197 (1976).

23. J.C. McCallum, C.D. Mckenzie, M.A. Lucas, K.G. Rossiter, K.T. Short, and J.S. Williams, *Appl. Phys. Lett.* **42**(9):827 (1983).

24. M.B.H. Breese, P.J.C. King, J. Whitehurst, G.R. Booker, G.W. Grime, F. Watt, L.T. Romano, and E.H.C. Parker, *J. Appl. Phys.* **73**(6):2640 (1993).

25. J.S. Williams, J.C. McCallum, and R.A. Brown. *Nucl. Instr. Meth.* **B30**:480 (1988).

26. D.N. Jamieson, R.A. Brown, C.G. Ryan, and J.S. Williams, *Nucl. Instr. Meth.* **B54**:213 (1991).

27. J.S. Williams, J.C. McCallum, and R.A. Brown, *Principles and Applications of High-Energy Ion Microbeams*, eds. F. Watt and G.W. Grime. Adam Hilger, Bristol (1987), Chapter 9.

28. R. Amirikas, D.N. Jamieson, and S.P. Dooley. *Nucl. Instr. Meth.* **B77**:110 (1993).

29. J.W. Mayer and E. Rimini, eds., *Ion Beam Handbook For Material Analysis*. Academic Press, New York (1977).

30. P.J.C. King, M.B.H. Breese, P.R. Wilshaw, and G.W. Grime. *Nucl. Instr. Meth.* **B104**:233 (1995).

31. M. Cholewa, G.S. Bench, G.J.F. Legge, and A. Saint, *Appl. Phys. Lett.* **56**(13):1236 (1990).

32. M. Cholewa, G.S. Bench, A. Saint, and G.J.F. Legge, *Nucl. Instr. Meth.* **B54**:397 (1991).

33. M. Cholewa, G.S. Bench, A. Saint, G.J.F. Legge, and L. Wielunski, *Nucl. Instr. Meth.* **B56/57**:795 (1991).

34. N.W. Cheung. *Rev. Sci. Instr.* **51**(9):1212 (1980).

35. A. Dygo, M.A. Boshart, M.W. Grant, and L.E. Seiberling, *Nucl. Instr. Meth.* **B93**:117 (1994).

36. D.D. Armstrong, W.M. Gibson, and H.E. Wegner. *Rad. Effects* **11**:241 (1971).

37. J.S. Rosner, W.M. Gibson, J.A. Golovchenko, A.N. Goland, and H.E. Wegner, *Phys. Rev.* **B18**(3):1066 (1978).

38. J.C. Hodges. Mark II goniometer. Richmond, CA. U.S.A.

39. P.J.C. King, M.B.H. Breese, G.R. Booker, J. Whitehurst, P.R. Wilshaw, G.W. Grime, F. Watt, and M.J. Goringe, *Nucl. Instr. Meth.* **B77**:320 (1993).

40. W.K. Chu, J.W. Mayer, and M.A. Nicolet. *Backscattering Spectrometry*. Academic Press, New York (1978).

41. G.S. Bench, PhD thesis, University of Melbourne (1991).

42. P. Baeri, S.U. Campisano, G. Foti, E. Rimini, and S.T. Picraux, *Phys. Lett.* **68A**(2):244 (1978).

43. G. Thomas and M.J. Goringe, *Transmission Electron Microscopy of Materials*. Wiley-Interscience, New York (1979).

44. P.J.C. King, M.B.H. Breese, P.R. Wilshaw, and G.W. Grime, *Phys. Rev.* **B51**(5):2732 (1995).

45. J.U. Anderson, J.A. Davies, K.O. Nielson, and S.L. Anderson, *Nucl. Instr. Meth.* **38**:210 (1965).

46. B.R. Appleton, C. Erginsoy, and W.M. Gibson, *Phys. Rev.* **161**(2):330 (1967).

47. J.A. Borders and S.T. Picraux, *Rev. Sci. Instr.* **41**:1230 (1970).

48. D.N. Jamieson and W.B. Belcher, *Nucl. Instr. Meth.* **B104**:124 (1995).

49. D.J. Diskett, A.J. Avery, and R.E.T. Marshall, *Nucl. Instr. Meth.* **B64**:836 (1992).

50. M.D. Strathman and S. Bauman, *Nucl. Instr. Meth.* **B64**:840 (1992).

51. C.S. Barrett, R.M. Mueller, and W. White, *J. Appl. Phys.* **39**(10):4695 (1968).

52. G.O. Engelmohr, R.M. Mueller, and W. White, *Nucl. Instr. Meth.* **83**:160 (1970).

53. D.N. Jamieson, M.B.H. Breese, and A. Saint, *Nucl. Instr. Meth.* **B85**:676 (1994).

54. R.S. Nelson. *Phil. Mag.* **15**:845 (1967).

55. D.A. Marsden, N.G.E. Johansson, and G.R. Bellavance, *Nucl. Instr. Meth.* **70**:291 (1969).

56. R.E. Holland and D.S. Gemmell, *Phys. Rev.* **173**(2):344 (1968).

57. D.S. Gemmell and R.E. Holland, *Phys. Rev. Lett.* **14**(23):945 (1965).

58. G. Dearnaley, I.V. Mitchell, R.S. Nelson, B.W. Farmery, and M.W. Thompson, *Phil. Mag.* **18**:985 (1968).

59. B.R. Appleton and L.C. Feldman, *Rad. Effects*, **2**:65 (1969).

60. P.J.C. King, M.B.H. Breese, P.R. Wilshaw, P.J.M. Smulders, and G.W. Grime, *Nucl. Instr. Meth.* **B99**:419 (1995).

61. R.C. Gonzalez and P. Wintz, *Digital Image Processing*. Addison-Wesley, London (1977).

6

ION BEAM INDUCED CHARGE MICROSCOPY

The main uses for this technique are as a means of imaging the distribution of buried junctions in semiconductor devices as described in Chapter 7 and as a means of imaging deep dislocation networks in semiconductors, which is described in Chapter 8. This chapter begins by reviewing the semiconductor theory relevant to ion beam induced charge (IBIC) microscopy. The theory used to calculate the ion induced charge pulse height as a function of ion type and energy, minority carrier diffusion length, and surface and depletion layer thickness is then described. Analogies with electron beam induced current (EBIC) microscopy are used to show the strengths and weaknesses of these two methods. Ion induced damage is the main limitation of IBIC microscopy, because this limits the maximum number of ions that can be used to form an IBIC image, so that images can be statistically noisy. The effects of ion induced damage, both in the substrate and in the depletion layer, on the measured charge pulse height is outlined, and the resulting charge pulse height reduction is characterized. Methods of compensating for the effects of ion induced damage so that a higher ion dose can be used to generate an IBIC image are outlined. Finally, it is shown how the damage produced by a focused MeV ion beam can be imaged with IBIC microscopy to reveal the extent of the beam halo.

The recent development of IBIC microscopy [1] has come from several branches of research. The large amount of experience gained in the last 30 years in the production and use of semiconductor charged particle detectors [2–5] has been very important, because the methods used for noise reduction, measurement and calibration of the measured charge pulse height, and ion induced damage are all relevant. The ability to measure small ion induced charge pulses also depends crucially on the availability of high-sensitivity and low-noise electronics [2]. Work on other beam induced current types of microscopy such as EBIC, which uses a keV electron beam [6], and optical beam induced current (OBIC) which uses a laser beam [7] have also been important in providing a relevant technical and theoretical background. A description of different EBIC imaging modes, as well as more detailed accounts of the theory and technical aspects of image generation, is given in Ref. 8. Both EBIC and OBIC have been used to image dislocations [9] and inversion layers in semiconducting materials and depletion regions in devices [10]; examples are shown in Chapter 7. The main advantages of IBIC are its larger analytical depth and lower scattering through surface layers, enabling higher spatial resolution in buried layers, and also the choice of being sensitive or insensitive to topographical information.

However, there is no displacement damage in silicon using keV electrons, and a low-power laser beam such as argon or helium–neon used for OBIC also produces no material damage, whereas MeV light ions generate defects in the semiconductor; this is a major limitation for IBIC. Work on ion and electron induced damage in semiconductors [11–13] provided an important basis for characterizing the charge pulse height reduction produced by a decrease in the semiconductor diffusion length.

The first use of charge collection microscopy with a nuclear microprobe was to image a region of ion induced damage in a *pn* junction [14]. The different types of image contrast achievable with an event-by-event data acquisition system were described. Research into the ability of MeV ions to cause single-event upsets (SEUs) in microelectronic device memories by causing the device state to alter was another line of research [15,16] that helped the development of IBIC as an imaging technique, because SEU studies can locate the device regions most susceptible to upsets. Another useful topic is backscattering spectrometry [17], because concepts such as the rate of MeV ion energy loss and depth resolution are relevant to IBIC microscopy.

6.1. SEMICONDUCTOR THEORY

Only basic semiconductor theory relevant to IBIC is reviewed here. Other sources are recommended for a more in-depth discussion of these and other theoretical aspects [18–20]. Ionizing radiation such as MeV light ions, keV electrons or laser light can create mobile charge carriers in semiconducting materials by transferring enough energy to the atoms to move valence electrons to the conduction band, leaving behind a positively charged hole. The average energy, E_{eh}, needed to create this electron-hole pair does not depend on the type of ionizing radiation and is constant for a given material [21]. E_{eh} is typically three times larger than the energy band gap; values for a range of commonly encountered semiconductors have been listed [2,3,18]. Charge carriers produced by ionizing radiation diffuse randomly through the semiconductor lattice, and if it contains no electric field they become trapped and recombine, until all excess carriers are removed.

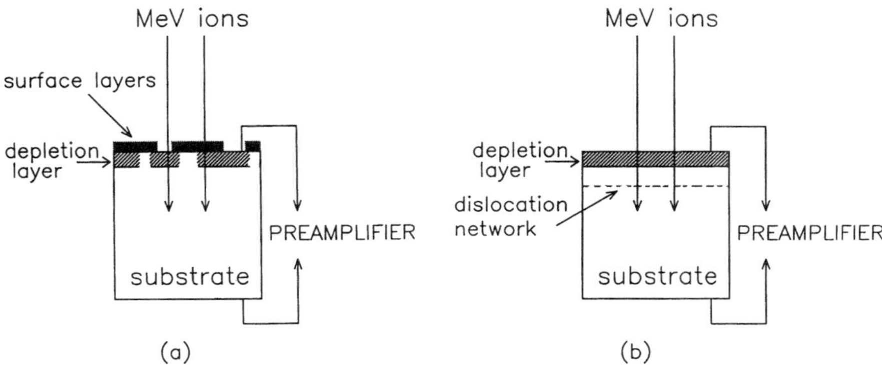

Figure 6.1. Schematic of geometry used for IBIC analysis of (a) microcircuits and (b) semiconductor slices containing dislocations.

If the semiconductor contains an electric field, the charge carriers are separated in the field region and this charge flow can be measured in an external circuit. Typical sample geometries used for IBIC microscopy are shown in Figure 6.1; the electric field can be provided by device *pn* junctions (Figure 6.1a) or by a Schottky barrier manufactured on the sample surface (Figure 6.1b). Microelectronic devices (Figure 6.1a) typically consist of a semiconductor substrate with a patterned array of *pn* junctions at the semiconductor surface, which comprise the different transistors and other electronic components present [22,23]. Above the semiconductor surface there are usually thick, patterned layers of insulating and metal tracks that make up the interconnecting device layers, with a thick passivation layer over the device surface. The total thickness of all the surface layers present can be several microns. For the analysis of dislocations in semiconductor slices (Figure 6.1b), a metal layer is deposited over the surface that forms a Schottky barrier and provides the depletion region. A beam current of less than 1 fA of MeV ions is incident on the front face of the sample, and each ion generates charge carriers all along its trajectory. The number of charge carriers generated by each incident ion is measured for IBIC microscopy using a contact from the depletion layer at the front face and a contact to the rear surface, which are connected to a charge-sensitive preamplifier. This gives an output voltage that is amplified and measured by the data acquisition computer.

6.1.1. Charge Diffusion and Drift

In *n*-type silicon, the majority of equilibrium charge carriers are electrons, and holes are the minority carriers. MeV ions generate equal numbers of excess majority and minority charge carriers. If the excess majority carrier density Δn is smaller than the equilibrium density n, the excess majority carriers follow the diffusive motion of the injected minority carriers Δp to preserve charge neutrality. The motion of both excess carrier types is thus determined by the excess minority carriers. If a minority carrier is trapped at a defect, a majority carrier might also make the transition to the defect level, depending on the nature of the defect. If this happens, then recombination can occur, which leads to a reduction in the measured charge pulse height. The average time a minority carrier is free to move before being trapped is called the minority carrier lifetime τ

$$\tau = \frac{1}{N_d \sigma_d v_{th}} \tag{6.1}$$

where N_d is the defect density (cm^{-3}), σ_d is the trapping cross-section of the initial defects present (cm^2), and v_{th} is the carrier thermal velocity ($\sim 10^7$ cm/s). Values of the minority carrier lifetime can range from a few milliseconds for very pure silicon and germanium to a few nanoseconds in heavily doped materials or those with a high defect density. For example, for a material with a defect density of 10^{15}/cm^3 and a typical defect trap cross-section of 10^{-15} cm^2, Eq. (6.1) gives $\tau \sim 100$ ns.

In material that contains a homogeneous defect distribution, excess carriers are continually removed through recombination at a rate of $\Delta p / \tau$ for a small excess minority carrier concentration. Excess carriers may be lost from a given region by diffusion or drift, and in equilibrium they are replaced by more ion induced excess charge carriers, according to the minority carrier continuity equation

$$g = \frac{\nabla J}{e} - \frac{\Delta p}{\tau} \tag{6.2}$$

where g is the ion induced charge carrier generation rate, J is the minority carrier flux, and the recombination rate is controlled by minority carrier trapping.

The range of MeV ^{1}H ions is considerably greater than the depletion layer thickness encountered in samples used for IBIC microscopy, so most charge carriers are generated in the semiconductor substrate beneath the depletion layer. In the case that the flux J in Eq. (6.2) is determined solely by charge diffusion, then

$$J = -eD_l \frac{\partial \Delta p}{\partial r} \tag{6.3}$$

where D_l is the minority carrier diffusion coefficient (equal to 12 cm^2/s for holes in silicon). After the passage of an ion, the solution for $g = 0$ has the form

$$\Delta p \propto e^{-r/L} \tag{6.4}$$

where r is the distance from the point of creation of the charge carriers and L is the minority carrier diffusion length. Charge carriers created in the substrate diffuse a distance L in a time τ, where $L = \sqrt{D_l \tau}$. The measured carrier intensity thus decreases exponentially with lateral distance from the point of generation, so the measured charge pulse height decreases away from a depletion region. A specific three-dimensional solution to Eq. (6.2) for a real sample structure means that the boundary conditions must be defined to quantitatively determine the diffusive flux at each point. This has been only partially solved for specific dislocation and device geometries, and the theory used in Section 6.2 relies on a one-dimensional solution to Eq. (6.2), such that radial symmetry about the ion trajectory is assumed. Complete three-dimensional solutions are briefly discussed in Section 6.1.4.

6.1.2. Semiconductor Depletion Region

The two most common methods of forming a depletion region in a semiconductor are with a *pn* junction and a Schottky barrier. Figure 6.2 shows a schematic of the characteristics of a *pn* junction, which is usually formed by selectively doping different regions of the same semiconductor wafer. Because of the difference in charge carrier concentrations on either side of the junction, electrons from the *n*-type material diffuse into the *p*-type material and recombine with

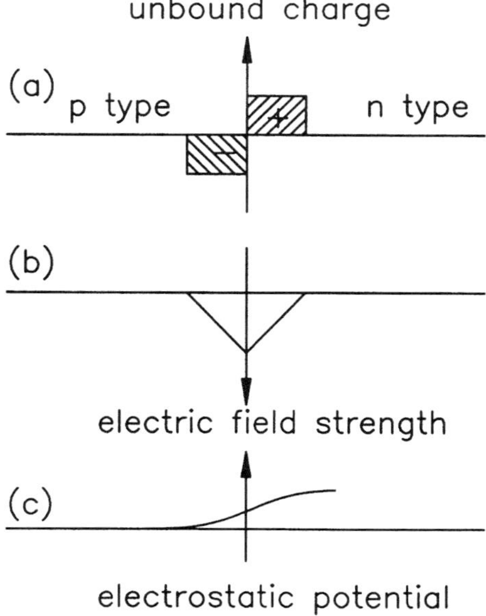

Figure 6.2. Schematic of a *pn* junction showing (a) the distribution of unbound fixed charge, (b) electrostatic field distribution, and (c) electrostatic potential distribution.

some of the majority holes, and vice versa. This uncovers positive charge in the n-type material and negative charge in the p-type material, resulting in a region on either side of the junction that is depleted of unbound charge carriers. This imbalance results in an electric field that has a maximum strength at the junction and extends into the material on either side. If the material on one side of the junction is heavily doped and the other side is lightly doped, the depletion region extends much further on one side of the junction than the other, and the thickness z_d of the depletion region is

$$z_d = \sqrt{\frac{2\xi(V_b + V_e)}{ne}} \tag{6.5}$$

Here ξ is the dielectric constant, n is the majority carrier density, V_b is the built-in voltage across the junction, and the n-type material is at a higher potential than the p-type material. V_e is an externally applied bias voltage that is defined as being positive for reverse bias, that is, positive n region compared with the p region. The depletion width in silicon as a function of doping density and voltage $(V_b + V_e)$ is shown in Figure 6.3 for an abrupt one-sided pn junction. For a doping density of $10^{15}/\text{cm}^3$ and a built-in junction voltage of $V_b = 0.5$ V, the depletion width is of the order of 1 μm with no external bias applied. Applying a reverse bias voltage, as shown above the line for $V_e = 0$ V, uncovers

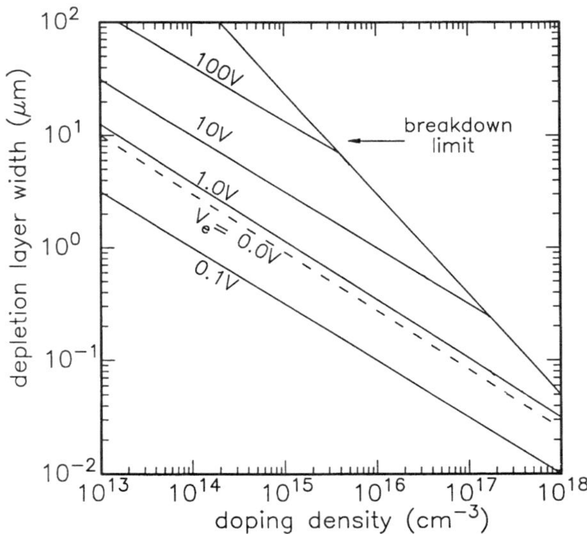

Figure 6.3. Depletion layer width as a function of doping density and voltage $(V_b + V_e)$ for an abrupt one-sided silicon pn junction.

more bound charge on either side of the junction, which increases the depletion layer width. Conversely applying a forward bias, as shown below the line for $V_e = 0$ V, covers up more bound charge on either side of the junction, which decreases the depletion layer width.

When a metal and a semiconductor surface are brought into contact, then charge flows to equalize the Fermi levels in the two materials. If the work function W of the metal is greater than the electron affinity of the semiconductor A, a depletion region is established at the semiconductor surface. The potential height, called the Schottky barrier, $\Phi \approx (W - A)$ eV, ignoring the presence of interface states at the semiconductor surface. According to the model for the barrier height proposed by Schottky, most metals form a Schottky barrier on n-type silicon because the metal's work function is usually greater than the electron affinity of the silicon. The actual barrier height depends strongly on the methods of preparing the surface and depositing the metal layer, but typical values for the barrier height on n-type silicon are $\Phi = 0.72$ eV and 0.8 eV for Al and Au metal layers, respectively. For p-type silicon, the barrier height is typically 0.3 eV less than for n-type silicon. The thickness of a depletion region formed by a Schottky barrier is similar to that for an abrupt, one-sided pn junction, and may be calculated using Eq. (6.5) with V_b replaced with the Schottky barrier height. The theory and preparation of Schottky and ohmic contacts is discussed in more detail in Refs. 24 and 25.

Ion induced charge carriers created in the depletion region are separated under the influence of the electric field, and move with a drift velocity $v = \mu\epsilon$, where μ is the carrier mobility and ϵ is the electric field strength. For a 1 μm thick depletion layer with 1 V across it, and a mobility of $\mu = 1,350$ cm^2/V/s, carriers take about 10 ps to cross the depletion layer. This is much shorter than the carrier lifetime, so charge carriers have a low probability of being trapped at any defects present in the depletion layer. The IBIC theory in Section 6.2 assumes 100% charge collection in the depletion layer. It is shown in Section 6.3 that ions stopped in the depletion region result in both the maximum topographical IBIC image contrast and also the minimum sensitivity to ion induced damage.

6.1.3. Carrier Trapping and Recombination

Carrier recombination may occur directly between the conduction band and the valence band or via defect states in the semiconductor band gap. In indirect band gap semiconductors, such as silicon, the probability of band-to-band recombination depends on the square of the majority carrier concentration. It is thus only significant at a high carrier density, as there are usually enough mid-band defect states so that recombination by these states is more likely. In direct band-gap semiconductors, such as gallium arsenide, band-to-band recombination with subsequent photon emission is an important process, and this is the basis of cathodoluminescence [26] and ion beam induced luminescence (IBIL) [27] described in Section 4.9.

A trapping process occurs when a charge carrier is captured by a defect and then has a high probability of being re-emitted into the conduction or valence band (i.e., high detrapping probability). The detrapping rate is proportional to $\exp(-E_d/kT)$, where E_d is the energy necessary to raise a carrier from the trap level to the band edge (here called the defect energy level). This process is shown schematically in Figure 6.4a for a shallow trap that is located close to the edge of the conduction band (small E_d, labeled 3), an intermediate trap level (labeled 2), and a deep trap that has an energy level close to the center of the band gap (large E_d, labeled 1). Figure 6.4b shows the measured transient response at a fixed temperature for the three trap levels, and Figure 6.4c shows how the transient response changes with temperature for one of these trap levels, on a longer timescale. For the deep trap, there is a low probability that carriers will be reemitted, so there is a higher probability of recombination. The transient response is fast, but the measured charge pulse is small because of loss of charge carriers by recombination. For the shallow traps, there is a high probability of carrier emission and the carriers might be trapped and then detrap several times before being measured. The transient response is slower because of the time the carriers have spent in the traps, but a bigger charge pulse is measured. If the semiconductor is heated then the probability of detrapping increases so a larger charge pulse is measured and the transient response is faster because of the shorter detrapping time, as shown in Figure 6.4c.

MeV ions create additional defects in semiconductors as described in Section 1.5, and the resultant increase in defect density increases the recombination rate and so reduces the measured charge pulse height. It is important to characterize this for IBIC microscopy to understand how damage affects the measured charge pulse height spectrum and also to minimize its effects. Ion induced defects can have a range of energy levels within the semiconductor band gap, and techniques such as deep level transient spectroscopy (DLTS) [28] can measure the density and energy levels of the defects. It is therefore possible, in principle, to calculate the effects of each ion induced defect level on carrier trapping and recombination to determine the charge pulse height reduction. However, the effect of a particular trap level depends on its occupation, which in turn depends on the position of the defect energy level in the band gap relative to the Fermi level. It would thus be very difficult to derive a meaningful relationship for the effect of all of the different defect energy levels on the measured charge pulse height, and a simpler approach is developed in Section 6.2.

6.1.4. Charge Funneling

The passage of a high-energy, heavy ion through a semiconductor can produce a region with a very high density of excess charge carriers along the ion path because of the ion's high rate of electronic energy loss. If a heavy ion passes through a depletion region of an integrated circuit memory, the resultant plasma formation can distort the shape of the depletion region, giving rise to charge

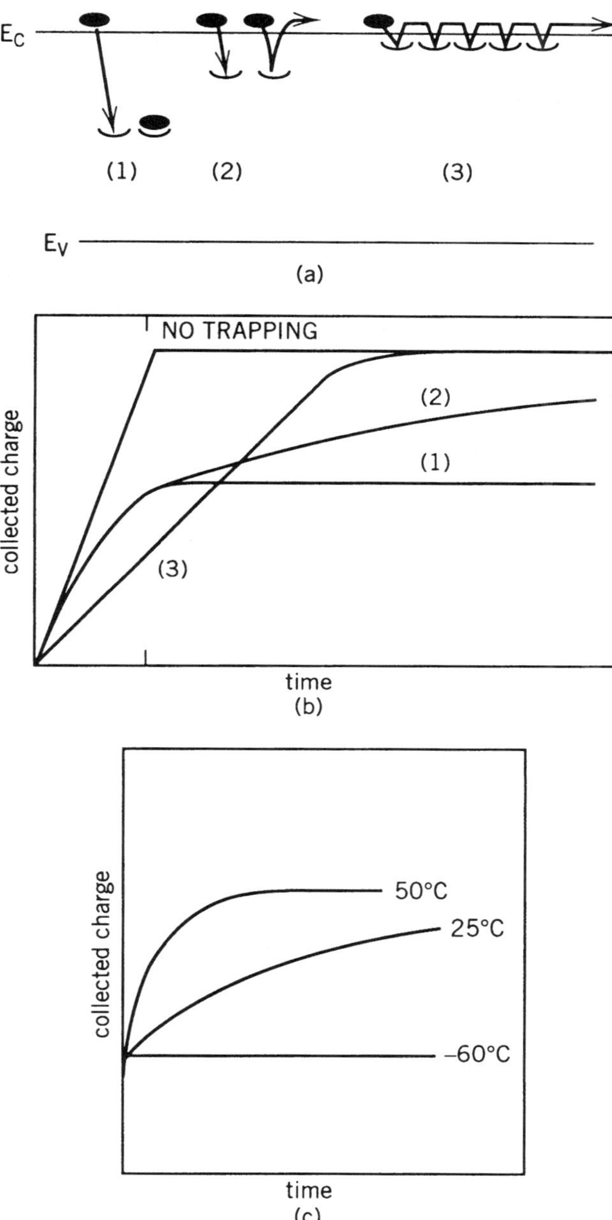

Figure 6.4. (a) Schematic of trapping and detrapping processes for a deep trap (1), intermediate trap level (2), and shallow level (3). (b) Transient response for these three trap levels. (c) Transient response for a trap level for three different temperatures. Modified from Ref. 2 with kind permission from Elsevier Science B.V., Amsterdam, The Netherlands.

funneling [15]. This increases the vulnerability of integrated circuit memories to single event upsets (SEUs), which means that the information stored in the memory is altered. The dense plasma created around high-energy heavy ions is a reasonable experimental analogy for the passage of strongly ionizing cosmic radiation passing through integrated circuit memories in satellite-based micro-electronic devices, which is why this has become a popular research area.

Figure 6.5 shows the passage of a heavy ion through an np junction of a memory device (the substrate is p-type). The carrier density can reach up to $10^{20}/cm^3$ along the path of a heavy ion, which is much greater than the typical device substrate doping density. The junction depletion layer in the vicinity of the ion path is quickly neutralized, and the high electric field in the junction is screened by electrons being drawn off at the electrode. The electric field associated with the junction is reduced in size and becomes elongated in the direction of the ion path, and Figure 6.5c shows the distorted equipotential line associated with this effect. The effect of this charge funnel is to give a much larger measured charge pulse than would otherwise occur. The charge is also much more localized around the struck np junction, as less charge diffuses away to be absorbed by other junctions (Figure 6.5d). A large amount of charge can arrive at the struck junction and results in a much increased susceptibility of the device to change its existing memory state, resulting in a corruption of the stored information. An analytical model describing the effective length of the charge funnel, so that the additional increase in charge arriving at the struck junction can be calculated, is described in Ref. 15.

Charge funneling is important to IBIC microscopy, because it should be taken into account when calculating the charge pulse height due to heavy ions if the ionization carrier density is much greater than the substrate density. It is also important because SEUs provide a considerable impetus for the continuing development of IBIC as a method of determining which device areas are subject to upsets, as described in Section 7.1.4.

Sophisticated two- and three-dimensional computer codes [29] can be used to model the effects of charge funneling in order to gain a detailed understanding of the basic mechanisms by which this effect can upset devices and to distinguish funneling from the effects of charge drift and diffusion. An example of a system that has been modeled is a heavily doped $(5 \times 10^{20}/cm^3)$ n^+ diffusion in a p-type silicon substrate with different doping concentrations [30,31]. The passage of a 100 MeV iron ion, with a linear energy transfer (see Section 1.1) of about 28 MeV/(mg $\cdot$ cm^{-2}) through this system was modeled to investigate under what conditions it could upset the device. Figure 6.6 shows the simulated charge collection transients from one quadrant of the symmetrical device area for three different substrate doping densities [30]. The total amount of charge collected is 1.06 pC (or 97.2% the total), 1.01 pC (92.7%), and 0.74 pC (67.9%) for the substrate doping concentrations of $1.5 \times 10^{14}/cm^3$, $1.5 \times 10^{15}/cm^3$ and $1.5 \times 10^{16}/cm^3$, respectively, so the measured amount of charge decreases with increasing substrate doping density. Whereas the two lightly-doped substrates show only one collection regime, the higher doped substrate clearly shows two

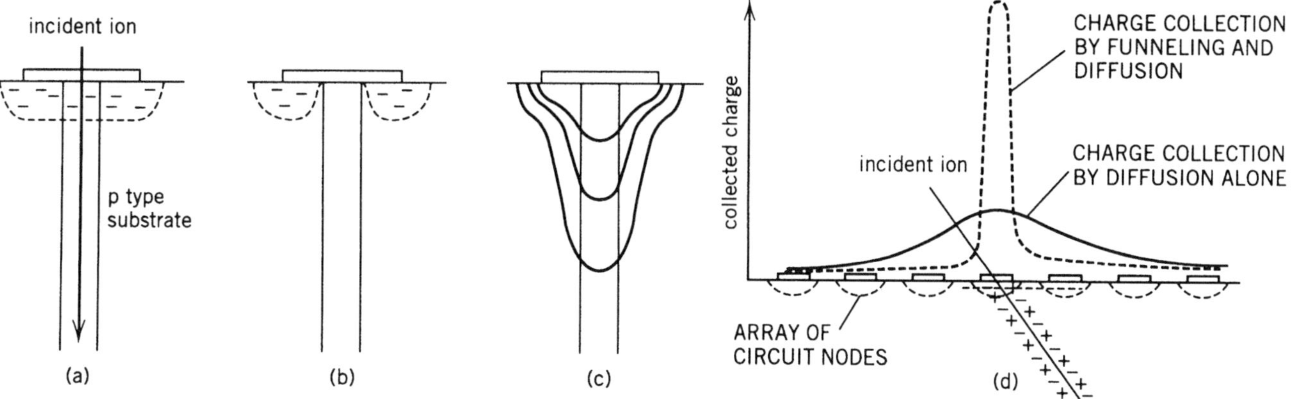

Figure 6.5. Schematic of charge funneling showing (a) a heavy ion hitting a junction, (b) the depletion region being neutralized by the resultant plasma column, (c) equipotential lines from the junction being extended down. (d) Shows the resultant charge collection profile over a circuit array, with enhanced charge collection due to charge funneling. Modified from Ref. 15 (©1982 IEEE).

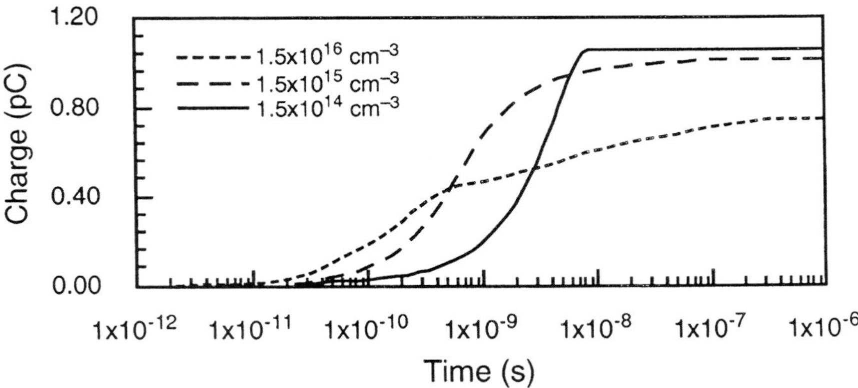

Figure 6.6. Charge transients generated by a 100 MeV Fe ion in three n^+/p silicon diodes with substrate doping densities shown in the top left. Reprinted from Ref. 30 ©1994 IEEE and with permission of P.E. Dodd, Sandia National Laboratories.

regimes, with the breakpoint at approximately 400 ps. For the lightly-doped substrates, nearly all the charge carriers are collected by the funnel, so there is little extra charge to be collected by diffusion after the funnel collapses some 10 ns after the passage of the ion through the device. For the more heavily doped substrate of $1.5 \times 10^{16}/cm^3$, the funnel collapses 400 ps after ion impact, leaving a considerable fraction of charge carriers free to diffuse through the device. Some carriers diffuse to the collecting junction and others recombine, so less charge is measured in total.

The above shows the insight that can be gained into the charge collection mechanism by detailed computer simulations, and this also seems to be the best path to gaining a better quantitative understanding of three-dimensional charge collection mechanisms with IBIC. However, reasonable results can be obtained using a basic one-dimensional interpretation which as described below.

6.2. QUANTITATIVE INTERPRETATION OF THE ION BEAM INDUCED CHARGE

This section considers the relationship between the amount of charge measured with different energy ^{1}H ions and ^{4}He ions passing through variable thickness surface and depletion layers into a substrate with a variable diffusion length. The use of ^{1}H$_2$ ions (molecular hydrogen) for IBIC microscopy is also considered at the end of this section. Long range ions are primarily considered here (i.e., ions that come to rest deep in the substrate), whereas Section 6.3 considers the effects of short range ions (i.e., ions that are stopped in the device depletion layers).

The ion induced charge pulse height is given throughout in units of keV so that the process is conceptually similar to backscattering spectrometry. For example, a measured charge pulse height of 250 keV from a 2 MeV ion means that 12.5% of the beam energy has generated charge carriers that contribute to the charge pulse height, and the remaining 1750 keV has been lost through the surface layers or through carrier recombination. Sections 6.2 and 6.3 are based on work described elsewhere [1,32–34].

6.2.1. Carrier Generation Volume

An important asset of IBIC microscopy is that it can image deeply buried device layers because of the long ion range. The ability to image small, deeply buried regions also crucially depends on the low lateral scattering of the focused ion beam. The shape of the charge carrier generation volumes for MeV ^{1}H ions and keV electrons is compared in Figure 6.7, assuming that both are incident on a single point on a silicon surface. The measured carrier generation volume for keV electrons [similar to that in Ref. 35], is shown in Figure 6.7a in the form of generated carrier concentration contours, and calculation has also shown a similar effect [36]. The axes are normalized to the electron range R_e and the generation volume is approximately spherical. The lateral extent of the generation volume can be reduced by lowering the electron beam energy, but at the expense of further decreasing the analytical depth. It has been calculated that the effective spatial resolution in EBIC is highly dependent on the shape of the generation volume rather than the diffusion length [37]. Because of the large lateral extent of the generation volume of the electron beam within the

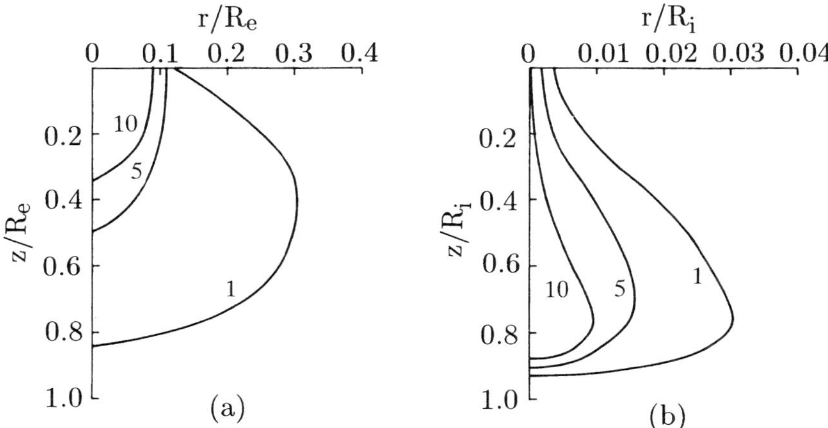

Figure 6.7. (a) Carrier generation contours for 10 keV electrons and 3 MeV ^{1}H ions in silicon. The lateral distance r and the depth z are plotted as fractions of (a) the electron range R_e and (b) the ion range R_i. The numbers 10, 5, and 1 refer to the intensity of the individual generation contour. Reprinted from Ref. 1 with permission (©1992 American Institute of Physics, Woodbury, NY).

silicon, the high spatial resolution of the electron beam on the sample surface is degraded, so deeply buried areas cannot be imaged with high spatial resolution.

Figure 6.7b shows a similar plot of the calculated generation volume for 3 MeV ^{1}H ions, based on values for the ion range, lateral scattering, and rate of energy loss using values given in Ref. 38. There is very little lateral scattering of the 3 MeV ^{1}H ions in the top few microns, and most occurs close to the end of range, resulting in a teardrop shape carrier generation volume. If the ion range is much greater than both the depletion depth and the diffusion length of the sample, carriers generated deep within the sample where there is significant scattering will not be measured. At a depth of 1 μm, for example, the degradation of spatial resolution with IBIC using 3 MeV ^{1}H ions is typically 10 times less than that with EBIC using 10 keV electrons.

6.2.2. Calculation of the Charge Pulse Height

It is assumed throughout this section that all the carriers generated in the depletion region and those that diffuse to its edges are measured, and there is no dependence on the electric field strength. Long range ions are mainly considered here, whereas Section 6.3 considers the effects of short range ions that are stopped in the depletion layers of the device. Low excess carrier conditions are assumed throughout, so that charge funneling is ignored. A semi-infinite lateral extent of the depletion layer is assumed, such that it is much greater than the ion range and the diffusion length. The three-dimensional carrier generation volume shown in Figure 6.7 can thus be represented by a one-dimensional depth distribution.

The total measured charge pulse height of P keV can be calculated by integrating the ion electronic energy loss dE/dz (as shown in Figure 1.4) between the semiconductor surface ($z = 0$) and the ion range ($z = R_i$):

$$P = \int_0^{z_d} \frac{dE}{dz}\, dz + \int_{z_d}^{R_i} \frac{dE}{dz} \exp-\left(\frac{z - z_d}{L} \right) dz \qquad (6.6)$$

where L is the diffusion length in the sample. The first term is the contribution from the charge carriers generated within the depletion region, which is z_d μm thick. The second term is the contribution from the charge carriers diffusing to the depletion region from the substrate. Equation (6.6) can be numerically integrated to calculate the measured charge pulse height for different energy MeV light ions.

Figure 6.8 shows the calculated charge pulse height for ^{1}H ions and ^{4}He ions for different diffusion lengths. The three-dimensional plots in Figure 6.9 show the calculated variation of the charge pulse height with both diffusion length and ^{1}H ion and ^{4}He ion energy. Both Figures 6.8 and 6.9 are based on the assumption of negligibly thin surface and depletion layers. For a short

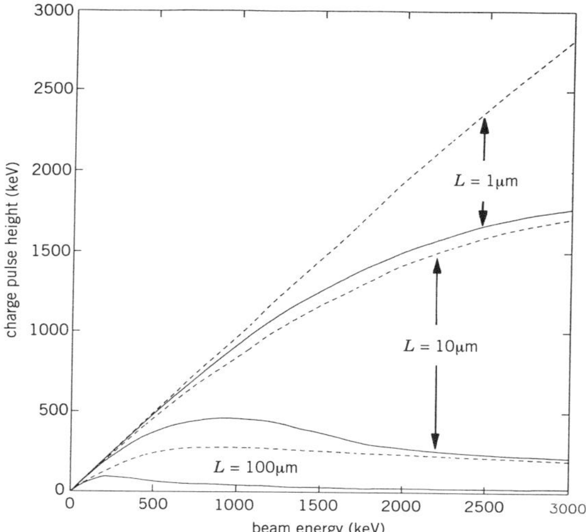

Figure 6.8. Variation of the charge pulse height with ^{1}H ions (solid lines) and ^{4}He ions (dashed lines) versus energy, for diffusion lengths of L = 1, 10, and 100 μm. Reprinted from Ref. 32 with permission (©1993 American Institute of Physics, Woodbury, NY).

diffusion length, the charge pulse height reaches a maximum value at a certain ^{1}H ion energy and then decreases as the ^{1}H ion energy is raised further. This is because at a low beam energy there is a high rate of energy loss close to the surface where the generated charge carriers can be measured. As the beam energy is raised further, there is a gradual reduction in the rate of energy loss at the surface and most carriers are generated too deep to be measured. A similar maximum is also found for keV electrons [39,40].

Figure 6.10a shows the increase in the charge pulse height for 1 and 2 MeV ^{1}H ions and 1 MeV ^{4}He ions with diffusion length with a 1 μm thick depletion layer, relative to the charge pulse height with a negligibly thin depletion layer. Each of the curves has a maximum at a diffusion length that increases with the ion range. This is due to the high rate of energy loss at the end of ion range, and occurs because the presence of the 1 μm thick depletion layer enables carriers to be measured from this region that are not measured with a negligibly thin depletion layer. Figure 6.10b shows the variation in the charge pulse height for 1 and 2 MeV ^{1}H ions and 1 MeV ^{4}He ions with diffusion length with a 1 μm thick surface layer relative to the charge pulse height for a negligibly thin surface layer. For a long diffusion length, the charge pulse height from a sample with a 1 μm thick surface layer is smaller than that measured from the same sample with no surface layer. This is because the energy lost in passing through the surface layer results in fewer carriers generated in the substrate. For a shorter diffusion length, the charge pulse height difference decreases, and at a short diffusion length the charge pulse height is larger with a 1 μm surface layer than for no surface layer.

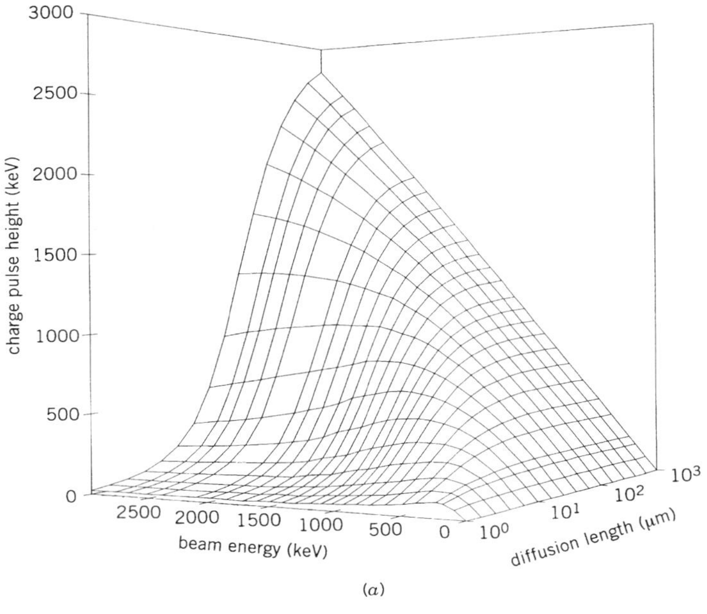

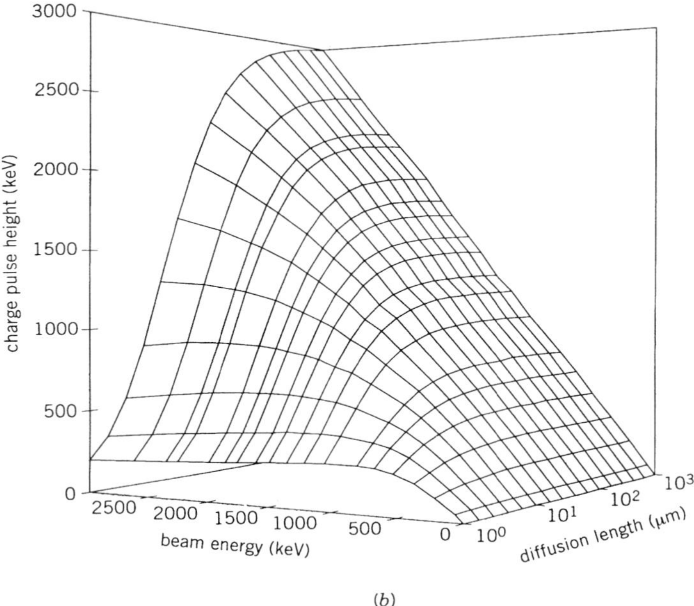

Figure 6.9. Three-dimensional representations of the charge pulse height variation with diffusion length and (a) ^{1}H ion energy, and (b) ^{4}He ion energy. Reprinted from Ref. 32 with permission (©1993 American Institute of Physics, Woodbury, NY).

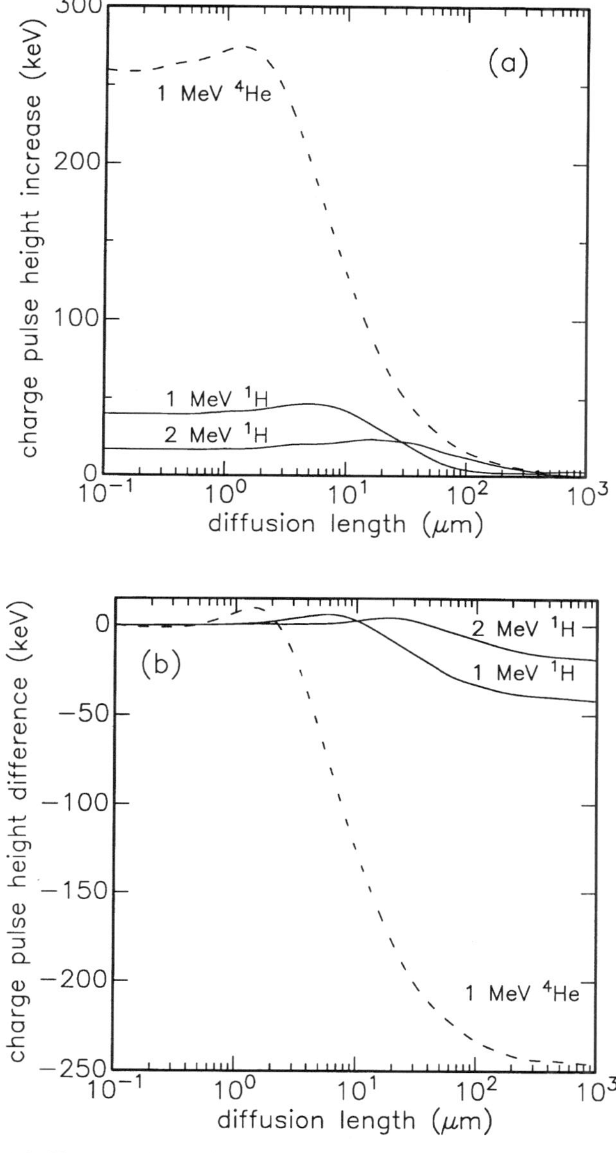

Figure 6.10. (a) Charge pulse height increase with diffusion length for 1 and 2 MeV ^{1}H ions (solid lines), and 1 MeV ^{4}He ions (dashed line) for a depletion layer 1 μm thick compared with a negligibly thin depletion layer. (b) Charge pulse height difference with diffusion length for 1 and 2 MeV ^{1}H ions (solid lines), and 1 MeV ^{4}He ions (dashed lines), for a surface layer 1 μm thick compared with a negligibly thin surface layer. Reprinted from Ref. 32 with permission (©1993 American Institute of Physics, Woodbury, NY).

In a microcircuit device with a typical diffusion length of 1 to 10 μm, it can be expected that the charge pulse height for ^{4}He ions changes rapidly with surface layer thickness, whereas the charge pulse height with ^{1}H ions does not. IBIC image contrast from microcircuit structures, where there is a lot of topographical variation, will be sensitive to topography with ^{4}He ions but much less sensitive with ^{1}H ions. This point is discussed again in Section 6.3 after the use of molecular hydrogen for IBIC has been described.

6.2.3. Effect of Ion Induced Damage in the Substrate

The main damage mechanism of MeV light ions in semiconductors is the creation of vacancy/interstitial pairs, called Frenkel defects, as described in Section 1.5. These can occur singly or in clusters, and they exhibit donor/acceptor characteristics. The primary effect for IBIC microscopy of MeV light ion induced damage in semiconductors is a reduction in the minority carrier diffusion length, because the defects act as trapping and recombination centers. A low ion irradiation dose is assumed, such that only a small fraction of the semiconductor lattice atoms are displaced and the ion induced defects do not change the doping concentration of the semiconductor substrate. It is assumed here that MeV ^{1}H ions and ^{4}He ions produce the same types of simple defects, and the effects of defect clusters are ignored. No account of the effect of defect mobility, thermal annealing, or self-annealing on the charge pulse height is taken. The defects present before the creation of any ion induced defects are called here the initial defects to distinguish them from the ion induced defects. The minority carrier diffusion length before ion irradiation can be expressed (in cm) as

$$L = \sqrt{\frac{D_l}{N_d \sigma_d v_{th}}} \tag{6.7}$$

Published values for the defect trap cross-sections [41] vary from 10^{-13} to 10^{-17} cm^2. A trap cross-section of the initial defects of $\sigma_d = 10^{-15}$ cm^2 is thus chosen to be a representative value throughout, with $D_l = 12$ cm^2/s in silicon. This approach gives an arbitrary value of the initial defect density N_d in Eq. (6.7) for a given diffusion length.

Figure 1.18 showed the ion induced defect depth distribution in units of defects/ion/μm for 3 MeV ^{1}H ions and 3 MeV ^{4}He ions in amorphous silicon. Since neither the defect energy levels or the occupation statistics of the ion induced defects are known, the introduced defects are characterized here by a trap cross-section $\gamma \sigma_d$; that is, their cross-section is a factor of γ different from the trap cross-section of the initial defects. This γ factor now takes into account the energy level dependence of the detrapping time and occupation statistics at a given temperature. The assumption is also made that the same type of defects (i.e., they have the same γ factor) are created along the full ion range. The effect of ion induced damage on the diffusion length can now be expressed as

$$L = \sqrt{\frac{D_l}{(N_d + N_{dd}\gamma)\sigma_d v_{th}}} \tag{6.8}$$

where N_{dd} is the ion induced defect density. The change in diffusion length can be found using Figure 1.18, and Eq. (6.8) used in Eq. (6.6). Although this approach is very basic, it does enable the effects of ion induced damage to be simply characterized and understood. The results are summarized in Figure 6.11.

For all the curves, there is a region at low ion dose where the charge pulse height does not decrease with ion dose, and the charge pulse height depends only on the initial defects present. At a higher ion dose, the ion induced defects start to modify the charge pulse height, and there is then a region where the charge pulse height decreases logarithmically with dose.

The rate of charge pulse height reduction for 1 and 2 MeV ^{1}H ions, for the same initial diffusion length $L = 5$ μm is compared in Figure 6.11. The charge pulse height reduces faster for 1 MeV ^{1}H ions because the end-of-range region, where most of the ion induced defects are created, is closer to the sample

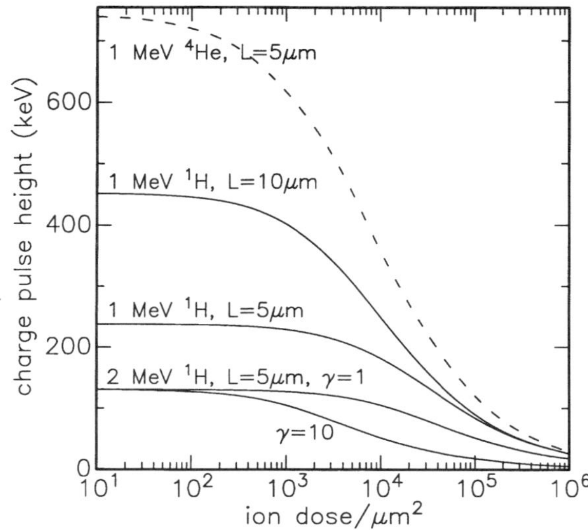

Figure 6.11. Charge pulse height with cumulative ion irradiation with 1 and 2 MeV ^{1}H ions (solid lines), and 1 MeV ^{4}He ions (dashed lines). The diffusion length is shown in each case, and $\gamma = 1$ except where indicated. Reprinted from Ref. 32 with permission (©1993 American Institute of Physics, Woodbury, NY.)

surface than for 2 MeV ^{1}H ions. The rate of charge pulse height reduction for initial diffusion lengths of $L = 5$ and 10 μm for 1 MeV ^{1}H ions is also shown. The charge pulse generated with the longer diffusion length is larger, but it also starts to reduce after a smaller dose and decreases faster than for a shorter diffusion length. Although MeV ^{4}He ions give a larger measured charge pulse than similar energy ^{1}H ions, the charge pulse height begins to decrease after a much lower ion dose. It also decreases considerably faster since ^{4}He ions create more defects close to the surface.

6.2.4. Ion Channeling in the Substrate

The charge pulse height decreases when a silicon slice with a thin Schottky barrier at the surface is brought into channeling alignment with ^{1}H ions [1]. This sensitivity of charge pulses to changes in the sample crystallographic alignment is interesting in that the similar effect in EBIC is very small [42]. This is because keV electrons lose their well-collimated profile in the thin Schottky barrier layer and so cannot channel well in the underlying semiconductor. Because of their high penetration and low scattering, the MeV ions are relatively unperturbed by a thin metal layer, and the parallel ion beam stays tightly collimated through this layer. The effect of ^{1}H ion channeling on the IBIC image contrast measured from bunches of 60° dislocations is described in Chapter 8.

Ion channeling is described briefly in Section 1.4 and Chapter 5 and in detail elsewhere [43,44]. Only those aspects relevant to ion induced charge collection from the semiconductor substrate are discussed here. It is assumed that the energy needed to create an electron-hole pair in channeled alignment is the same as in nonchanneled alignment for MeV light ions, which may not be the case for channeled heavy MeV ions [45].

The minimum yield $\chi_{\min}$ just below the sample surface (Eqs. 1.12 and 1.13) gives the fraction of the beam that is not channeled as the beam enters the semiconductor surface. The channeled beam fraction is assumed to lose energy at a reduced rate of $\eta(dE/dz)$ compared with the nonchanneled beam fraction. The amount of beam that is not channeled increases with depth in the sample due to dechanneling (Section 1.4.4), and the axial dechanneling rate per micron is represented as a linear term Bz. The average dechanneled beam fraction with depth is therefore $D_C = \chi_{\min} + Bz$, and the total energy loss per micron (dE_{av}/dz) in channeled alignment is

$$\frac{dE_{\mathrm{av}}}{dz} = [(1 - D_C)\eta + D_C]\frac{dE}{dz} \tag{6.9}$$

The charge pulse height in channeled alignment is found by using the channeled rate of energy loss from Eq. (6.9) in Eq. (6.6). Figure 6.12 shows the decrease of the channeled charge pulse height compared with nonchanneled alignment for 1, 2, and 3 MeV ^{1}H ions and 2 MeV ^{4}He ions with diffusion length. There

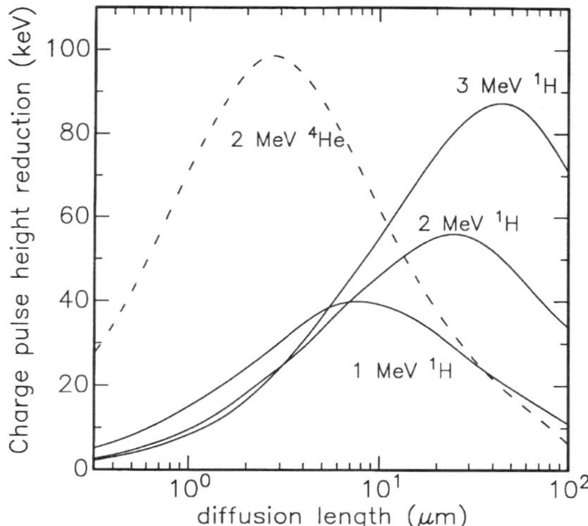

Figure 6.12. Charge pulse height reduction in channeled alignment compared with nonchanneled alignment with diffusion length for ^{1}H ions (solid lines), and ^{4}He ions (dashed lines). In each case $\gamma = 1$, $\eta = 0.5$ and $\chi_{min} = 0.05$. Reprinted from Ref. 32 with permission (©1993 American Institute of Physics, Woodbury, NY).

is typically a maximum of 30 to 100 keV ($\simeq 10$ to 20%) decrease in the channeled charge pulse height compared with nonchanneled alignment.

The lower rate of charge pulse height reduction in channeled alignment owing to a lower defect production rate is demonstrated in Figure 6.13. Here a thin gold Schottky barrier layer was deposited over a p-type silicon slice to create a depletion layer approximately 1 μm thick. The sample was irradiated with 2 MeV ^{1}H ions and the charge pulse height was measured with cumulative dose in separate areas in channeled and nonchanneled alignment. This was simulated and the effects of ^{1}H ion induced damage were modeled by assigning the ^{1}H ion induced defects a trap cross-section that was a factor of γ different to that of the initial defects. The good agreement between the measured and simulated rates of charge pulse height reduction shows that the simple model described above is able to give a reasonable insight into this effect.

6.2.5. Molecular Hydrogen for Ion Beam Induced Charge Microscopy

Below an energy of 1 MeV, ^{1}H ions are not routinely available for use with a nuclear microprobe owing to accelerator instability at a low terminal voltage. However, a ^{1}H$_2$ ion dissociates on impact with the sample into two ^{1}H ions, each with half the energy of the original ^{1}H$_2$ ion, and it has recently been shown [47] that this enables the use of lower energy ^{1}H ions for IBIC microscopy.

There are two ways in which the use of ^{1}H$_2$ ions differs from the use of ^{1}H ions for IBIC microscopy. Firstly, the charge pulses from both lower energy ^{1}H

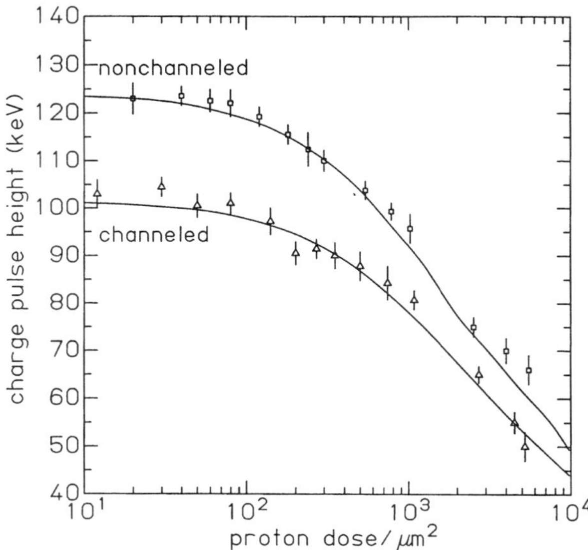

Figure 6.13. Variation of the charge pulse height with cumulative dose of 2 MeV ^{1}H ions in channeled and nonchanneled alignment [46]. The measured results are shown as points and the simulated results as solid lines, with $\gamma = 32$ and an initial diffusion length of 4.4 μm.

ions produced when the ^{1}H$_2$ ion dissociates on impact with the sample surface are measured simultaneously, as described in Section 4.6.2. In addition to this, lower energy ^{1}H ions produce larger charge pulses than higher energy ^{1}H ions, because the rate of electronic energy loss close to the surface is higher, as shown in Figure 6.8. The charge pulse height measured with ^{1}H$_2$ ions is thus considerably larger than using the same energy ^{1}H ions, which is important, because, with ^{1}H ions, the charge pulses may be too small to be detected against the noise level in some cases. Secondly the two ^{1}H ions produced when the ^{1}H$_2$ ion dissociates have a shorter range than the same energy ^{1}H ion, which makes ^{1}H$_2$ ions more sensitive to device topography than the same energy ^{1}H ions, as described in the following section. Further work on the use of molecular ions for IBIC microscopy has been described [34,47].

6.3. INCORPORATING THE EFFECTS OF THE DEPLETION LAYER

The theory used in Section 6.2 to calculate and interpret the ion induced charge pulse height was primarily for long range ions that come to rest deep in the substrate. This theory is now extended to incorporate the effects of short range ions that are stopped in the depletion layer. The conditions necessary for obtaining the maximum topographical image contrast and also the necessary condition for the maximum insensitivity to ion induced damage are discussed.

6.3.1. Maximizing the Topographical Contrast

The topographical contrast sensitivity is now reconsidered for ^{1}H ions, ^{4}He ions, and molecular hydrogen, and also for keV electrons as used in EBIC microscopy. Figure 6.14a shows the rate of electronic energy loss with distance traveled for 2 MeV ^{1}H ions, 3 MeV ^{4}He ions, 1,660 keV ^{1}H$_2$ ions (which are considered to have the same range as 830 keV ^{1}H ions, and twice their rate of energy loss), and 38 keV electrons. The 1,660 keV ^{1}H$_2$ ions and 38 keV electrons have the same range of 11.6 μm as the 3 MeV ^{4}He ions, so that the charge pulse height from these different charged particles with the same range can be compared.

Figure 6.14b shows the charge pulse height resulting from a depletion layer thickness of 1 μm and a substrate diffusion length of $L = 6$ μm with increasing surface layer thickness for these same charged particles. The ions have the most rapid variation in the resultant charge pulse height when the surface layer thickness is about the same as the ion range. The maximum topographical contrast thus occurs when the ions are stopped in or just beneath the depletion layer. For 3 MeV ^{4}He ions, the maximum slope of the charge pulse height variation with surface layer thickness occurs close to the end of range and is equal to 24 keV per 100 nm, whereas the charge pulse height for 2 MeV ^{1}H ions with surface layers less than 10 μm thick only changes at a rate of 0.2 keV per 100 nm. This factor of 100 difference in the sensitivity thus demonstrates why MeV ^{4}He ions are so much more sensitive to topography than similar-energy ^{1}H ions. The maximum slope of the charge pulse height variation for the 1,660 keV ^{1}H$_2$ ions is 21 keV per 100 nm. This is comparable with the MeV ^{4}He ions, which demonstrates that, with lower energy ^{1}H ions produced by the break up of ^{1}H$_2$ ions, IBIC images can be made sensitive to topographical contrast.

With EBIC, a current induced by a steady beam current of keV electrons is measured rather than individual charge pulses, because these would not be resolved from the noise level, which is typically 50 to 100 keV. However, in Figure 6.14, the charge pulse height from single 38 keV electrons is shown on the same scale as for MeV ions to compare the sensitivity of EBIC to changes in surface-layer thickness. There is no sharp variation in the resultant charge pulse height because the keV electrons have no well-defined range. For 38 keV electrons the maximum slope of the charge pulse height variation is between that measured with ^{1}H ions and with MeV ^{4}He ions, emphasizing that IBIC can be made either sensitive or insensitive to topographical contrast, whereas EBIC cannot.

6.3.2. Minimizing the Effects of Ion Induced Damage

Ion induced damage of the semiconductor substrate decreases the measured amount of charge because the increased recombination of the slowly diffusing

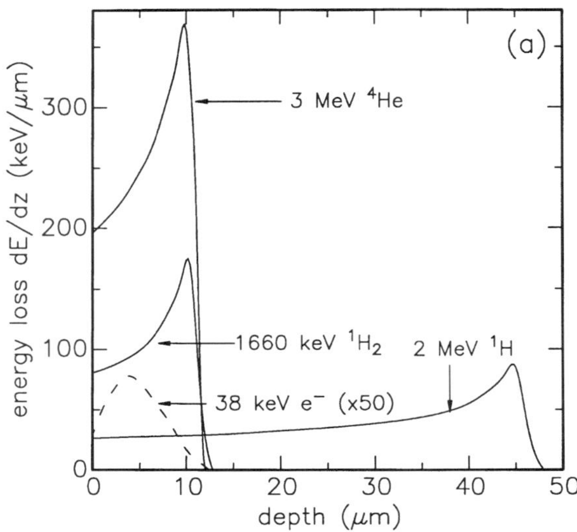

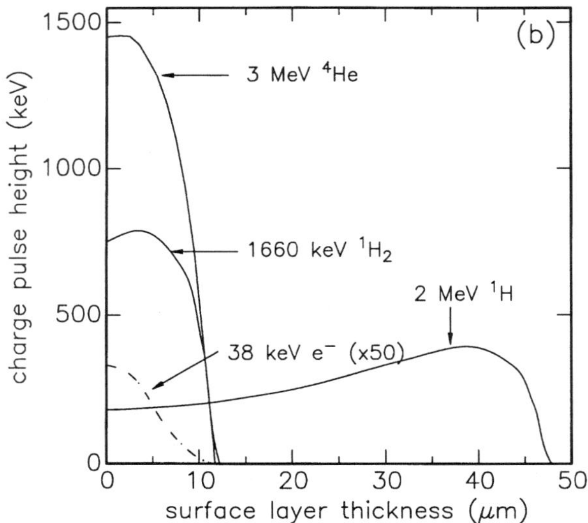

Figure 6.14. (a) Electronic energy loss with depth for 2 MeV ^{1}H ions, 3 MeV He ions, 1,660 keV ^{1}H$_2$ ions and 38 keV electrons (dashed line with the energy loss is multiplied by 50). (b) Charge pulse height resulting from a depletion layer thickness of 1 μm and a diffusion length of 6 μm as a function of increasing surface layer thickness for the same charged particles as (a). Reprinted from Ref. 34 with permission (©1995 American Institute of Physics, Woodbury, NY).

charge carriers causes a reduction in the carrier diffusion length. The mechanism by which the amount of charge measured from the depletion region reduces with ion induced damage is more complex [11–13]. Charge carriers generated in the depletion region have a much lower recombination probability than those generated in the substrate because of the associated electric field, and, in Section 6.2, it was assumed that all the charge carriers generated in the depletion layer were measured because of this. It was shown in Figure 6.14 that ions that were stopped in the vicinity of the depletion layer maximized the topographical contrast, and this same criteria is now discussed as a method of minimizing the effects of ion induced damage.

Figure 6.15 shows a schematic of the final locations of ions with a fixed energy penetrating through an increasingly thick surface layer into a narrow depletion region and then into the substrate. This is similar to considering ions with decreasing energy penetrating through a constant-thickness surface layer. For a thin surface layer (or a high ion energy, shown in Figure 6.15a), the ions penetrate into the substrate and most of the measured charge is due to diffusion from the substrate. Because the distribution of the nuclear energy loss along the

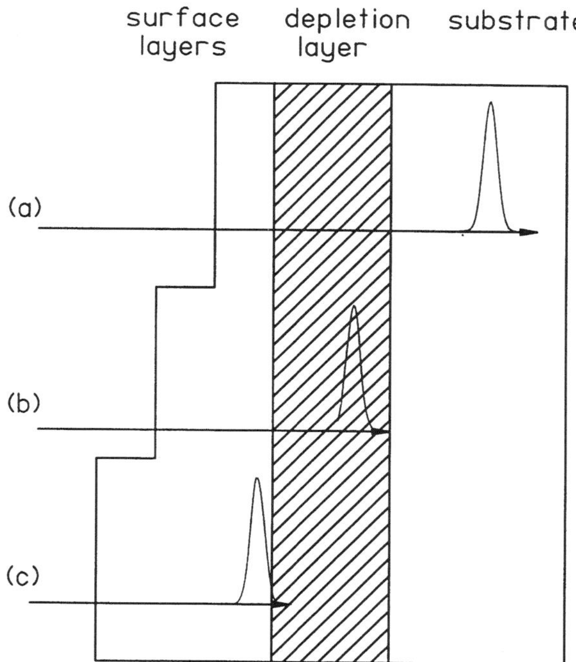

Figure 6.15. Schematic of constant energy ions penetrating through increasingly thick surface layers from (a) to (c). In (a) ions are stopped in the substrate, in (b) they are stopped in the depletion layer, and in (c) they are stopped in the surface layers. Reprinted from Ref. 34 with permission (©1995 American Institute of Physics, Woodbury, NY).

ion trajectory follows a similar trend as the electronic energy loss, it also reaches a maximum close to the end of the ion range, so this is where the maximum ion induced defect density occurs. In this case, most of the defects are created in the substrate where they have a large effect on the measured charge pulse height. For a thicker surface layer (or a lower ion energy, shown in Figure 6.15b), all of the ions are stopped in the depletion layer. The maximum defect generation rate also occurs here, where the ions have a much smaller effect on the measured charge pulse height. For a still thicker surface layer (or a still lower ion energy) as shown in Figure 6.15c, most of the ions are stopped in the surface layer and only a few penetrate into the depletion region. If the ion does not generate enough charge carriers to be measured above the noise level, then it will not be detected.

In summary, short range ions that are stopped in the depletion layer result in the least ion induced damage and the greatest topography contrast, whereas long range ions result in substantial ion induced damage and weak topography contrast.

6.4. EXPERIMENTAL PROCEDURE

The basic layout used for IBIC analysis of microcircuits and Schottky barrier samples is shown in Figure 6.1. The ion induced charge pulses are measured using standard charged particle detection electronics, similar to those used for backscattering spectrometry, NRA, and ERDA. Each pulse is fed to a low-noise charge-sensitive preamplifier whose output voltage is proportional to the measured number of carriers generated by individual ions. Charge-sensitive preamplifiers are ideal for use in IBIC experiments because they integrate the induced charge on a feedback capacitor C. A charge-sensitive preamplifier typically has an open loop gain of approximately 10^4, so that it appears as a large capacitance to the sample, rendering the gain insensitive to changes in the sample capacitance. Details of preamplifier design, pulse shaping, and methods of noise reduction have been described in [2] and many other texts. The preamplifier output voltage V_o is

$$V_o = \frac{1000Pe}{E_{eh}C} \tag{6.10}$$

where P is the measured charge pulse height (in keV) given by Eq. (6.6). With a feedback capacitance of $C = 1$ pF, a typical preamplifier gain is 44 mV/MeV of energy of the incident ion. The small output pulses from the preamplifier are fed into an amplifier that gives an output voltage of approximately 1 V. This is then fed into the data acquisition computer, and an image is generated showing either variations in the average measured charge pulse height or the intensity of counts from different windows of the spectrum at each pixel within the scanned area.

A maximum beam current of approximately 2 fA should be used for IBIC microscopy, because the maximum data acquisition rate available with most microprobes is less than 10 kHz. To produce a focused spot containing such a small current, a much larger beam current of 100 pA is first focused in a conventional manner, as described in Chapter 2. The object and divergence slits are then closed until the remaining current is a few thousand ions per second, as measured by a semiconductor detector placed in the path of the ion beam. This procedure ensures that the sample is not irradiated with a high ion dose prior to analysis. A similar method for production of a low beam current for CSTIM is outlined in Section 5.4.3. The sample is then moved onto the beam axis and a large area IBIC image is used to identify and position the region of interest. The scan size is reduced and the process repeated until the required feature for analysis is in the middle of the image.

The surface of a packaged device must be exposed for IBIC microscopy, and it is very important that the unpacking process should not damage the wiring or layer structure. Metal can packages are opened by gently sawing or grinding away the can top, and ceramic packages can be opened with a knife. In polymer potted devices, the polymeric material generally fills the entire internal volume above the device and between the bond wires, so mechanical removal methods cannot be used. A jet of sulfuric acid will dissolve most polymer cases but leave the device passivation bond wires intact, leaving the device surface exposed for analysis.

It is important to test the electrical contacts to the sample before irradiating the sample in the microprobe; otherwise, there is a high probability of unintentionally damaging the sample by mistaking the absence of measured pulses for the absence of beam irradiating the sample. First check that the sample is sensitive to light by measuring the current flowing through it. Light generates charge carriers just as the ions do, so there should be an increase in the measured current with light shining on the sample. Next, connect the sample in the shielded microprobe chamber and check that the noise level measured with an oscilloscope (with the chamber lights off) is small enough to resolve charge pulses. If the noise level is excessively large, this might be due to poor barrier preparation or wrong connections to the microelectronic device. A further wise step is to check that charge pulses generated by ^{4}He ions from a test source can be resolved from the noise level. Also, check that the amplifier polarity is set to give positive pulses into the data acquisition system.

A typical charge pulse height spectrum measured from a microcircuit device using MeV ^{1}H ions is shown in Figure 6.16. The charge pulses in this case are typically two to five times larger than the noise level that is shown at the lower end of the spectrum. The lower input threshold of the amplifier should be raised to a level indicated in Figure 6.16, such that very few noise pulses are measured or else the high noise level will saturate the data acquisition system and the ion induced charge pulses will not be measured. However, if the threshold is raised too far, then the smallest charge pulses will not be measured.

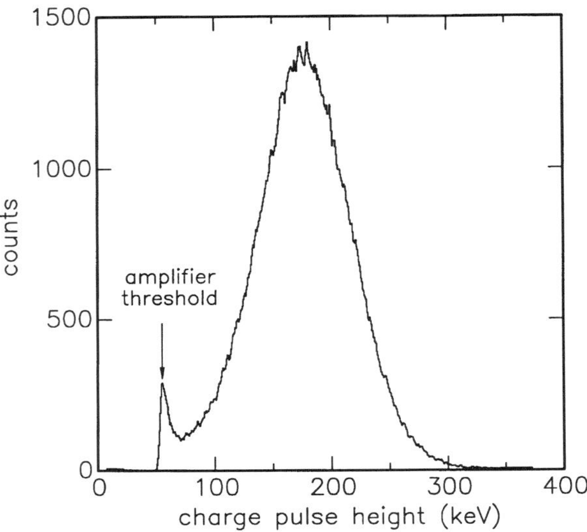

Figure 6.16. Typical charge pulse height spectrum from a microcircuit using MeV ^{1}H ions.

6.4.1. Methods of Noise Reduction

The noise level typically encountered with IBIC samples is 50 to 100 keV for microcircuit devices. When the charge pulse height reduces below the noise level, it cannot be accurately measured, which imposes a limitation on the ion dose that can be used to generate an IBIC image. Reducing the noise level increases the ability to measure charge pulses from material with a short diffusion length and also increases the ion dose that can be used to generate an IBIC image.

Noise in charge-sensitive preamplifiers has three main sources; the input field effect transistor, the input capacitance, and the preamplifier resistance. The noise contribution from the input capacitance increases at a typical rate of 15 to 20 eV/pF, so any excess capacitance should be removed. Leads between the preamplifier and the sample should be as short as possible, because the capacitance increases with lead length and ideally the preamplifier should be mounted inside the target chamber to be as close as possible to the sample. The area of the Schottky barrier, or the number of pins connected to the preamplifier, should be as small as possible to minimize the total active area, as this reduces the capacitance. The leads should be well screened, and the sample should be isolated from earth loops, circulating currents, and radio frequency pickup from other components of the microprobe electronics.

The noise level in the measured charge pulse height spectrum also depends on the amplifier time constant. A typical value of 1 μs is used for the best signal-to-noise ratio. The optimum value depends on both the current flowing in the sample and its capacitance. Decreasing the time constant too much can decrease the measured charge pulse height, because not all the trapped carriers may have detrapped, so it may be necessary to make a compromise between the best signal-to-noise ratio and quantifying the diffusion length from the measured charge pulse height. Once all excess input capacitance to the preamplifier has been eliminated, the remaining measured noise level is dominated by thermal generation of charge carriers. Cooling the sample reduces the thermally generated noise level. Where thermally generated noise is the major noise contribution, the total level typically reduces to 30% of its room temperature level at liquid nitrogen temperature, which allows smaller ion induced charge pulses to be resolved. Cooling the sample may also cause the measured charge pulse height to decrease, because the detrapping time becomes longer.

The charge pulse height spectra measured from Schottky barriers for IBIC microscopy have been very noisy to date, because large area barriers have been used in order to manually attach a thin wire to the front surface using silver paint. Because both the capacitance and the thermally generated noise contributions increase with the barrier area, this should ideally be as small as possible. An approach that has been used to minimize the Schottky barrier area with EBIC microscopy [48] has been to fabricate a large number of smaller barriers by depositing the metal layer through a mask, such as a coarse mesh grid. A contact is then made to the small area barrier within the electron microscope using a thin wire mounted on a micromanipulator and positioned with the aid of secondary electron images. This approach cannot be used for IBIC microscopy, because the dose required to generate an ion induced electron image would cause a large amount of damage to the material. The approach being developed for IBIC microscopy is to deposit a number of small barriers through a mask as described above and then mount the sample in a standard dual-in-line pin package. Connecting wires from the small metal barriers to the pins of the dip package are then made using a wire-bonding machine. This approach also solves another practical problem of reducing the chances of the bond wires breaking or dangling in front of the ion beam, obscuring the region required for analysis.

Another interesting development, which may further the capabilities of IBIC, is the production of time-resolved, or transient, IBIC images [49]. This may enable the charge components measured directly from the depletion region and from the substrate to be distinguished owing to the different charge collection timescales.

6.5. MEASUREMENT AND COMPENSATION OF ION INDUCED DAMAGE

Many charge pulses should be measured at each pixel to reduce the statistical noise in an IBIC image. Ion induced damage is the major drawback with IBIC microscopy, because it limits the number of charge pulses that can be measured in an image. The effects of ion induced damage and methods of compensating for its effects are described here for both methods of collecting data described in Chapter 2, which are the simple map mode and the more complex event-by-event mode.

6.5.1. Compensation of Damage Using the Map Mode of Data Collection

The device used here to demonstrate the effects of ion induced damage on IBIC image contrast is a 4 Mbit dynamic random access memory (DRAM) with 1 μm wide trench cells [50]. This makes it ideal for studying the effects of ion induced damage as a small scan size must be used to resolve the small features. The doping concentration of the p-type silicon substrate was 1 to $2 \times 10^{15}/\text{cm}^3$, and the area approximately 100 nm around the 1 μm holes was heavily n-doped. To prepare a sample suitable for IBIC microscopy, all the device surface layers were removed using hydrofluoric acid, leaving just the p-type substrate and the n-doped trench walls. Electrical contacts were then produced by depositing a thin gold layer onto the back surface to form an ohmic contact and a thin aluminium layer onto the front surface to form a Schottky contact.

Figure 6.17a shows the measured charge pulse spectrum from a 25 × 25 μm^2 area after a dose of fifty 3 MeV ^{1}H ions/μm^2, and the measured spectrum

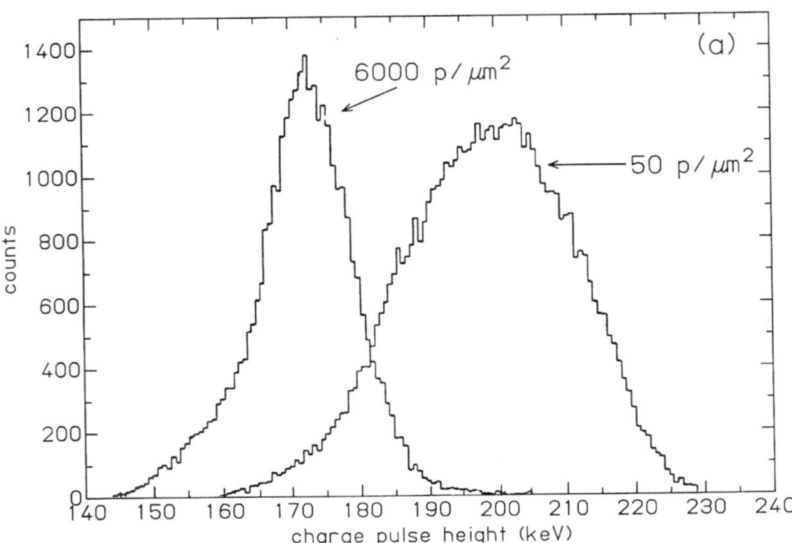

Figure 6.17. (a) Measured charge pulse height spectrum from a DRAM sample after a dose of 50 ^{1}H ions/μm^2, and after 6,000 ^{1}H ions/μm^2. (b) Measured variation of the average charge pulse height, and (c) measured variation of the charge pulse FWHM with cumulative beam dose. Reprinted from Ref. 50 with kind permission from Elsevier Science B.V., Amsterdam, The Netherlands.

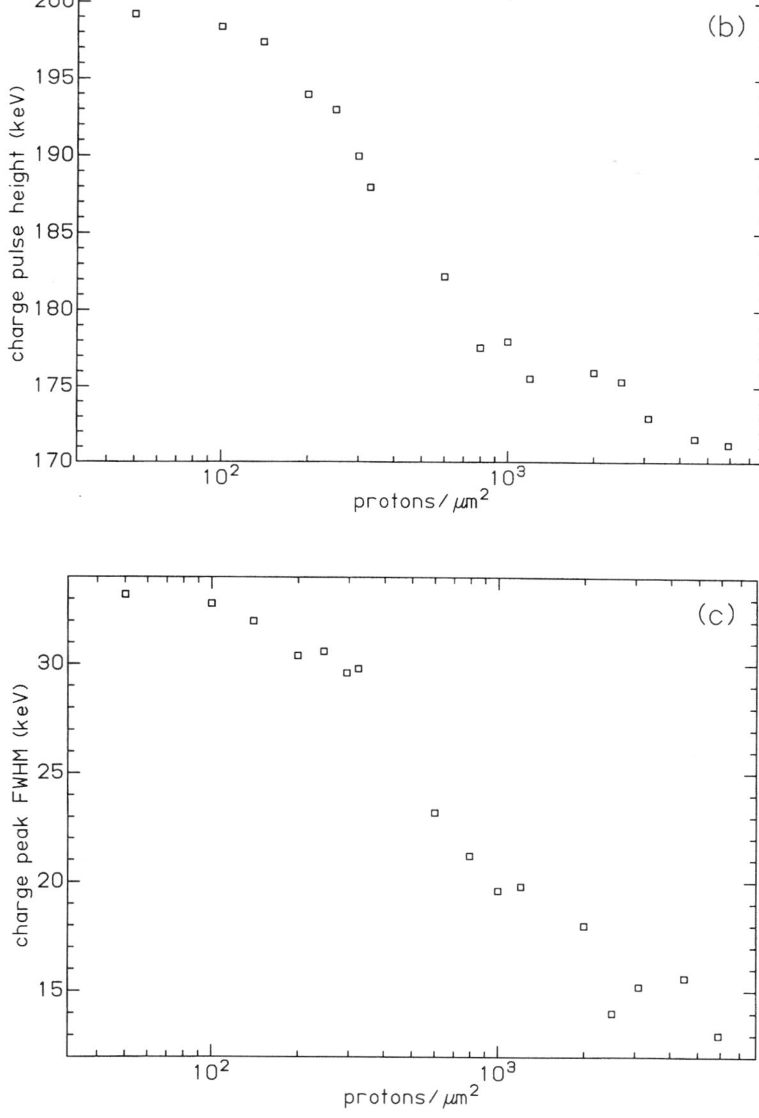

Figure 6.17. (*Continued*)

after a dose of 6,000 ^{1}H ions/μm^2. The charge pulse height and peak width start to decrease after a dose of ~100 ions/μm^2, as shown in Figure 6.17b,c. The mechanism for the charge pulse height reduction due to ion induced damage in the semiconductor substrate was discussed in Section 6.2. With N^* ^{1}H ions/μm^2 in a scanned area of X by X μm^2, using 256 $\times$ 256 image pixels, the number of ions measured at each image pixel is $N = N^*(X/256)^2$. For $N^* = 100$ ions/μm^2 and for a scan size of 100×100 μm^2, a total of 16 ions can be measured at each pixel. If the scan size is reduced to 25×25 μm^2, then only one ion can be used at each pixel for the same beam dose.

Mapping mode was used here for data collection whereby the measured charge pulse height spectrum was divided into windows as described in Chapter 2. Images were produced showing the number of pulses falling in a window at each pixel; these were then combined to form a single image showing the average measured charge pulse height within the scanned area. If the charge pulse height decreases during data collection, then the image contrast is destroyed. To solve this problem, a peak-following routine was included in the data acquisition software that checked at given time intervals during data collection whether the average measured charge pulse height had decreased [50]. If it had, then the position of each window on the charge pulse height spectrum was decreased by the same amount, such that the windows were always located on the same part of the measured charge pulse height distribution.

The left image in Figure 6.18 shows a 20×20 μm^2 IBIC image of the DRAM sample, measured with a dose of 200 ^{1}H ions/μm^2 (i.e., 1 ^{1}H ion/pixel). This was the maximum dose that could be used before the image

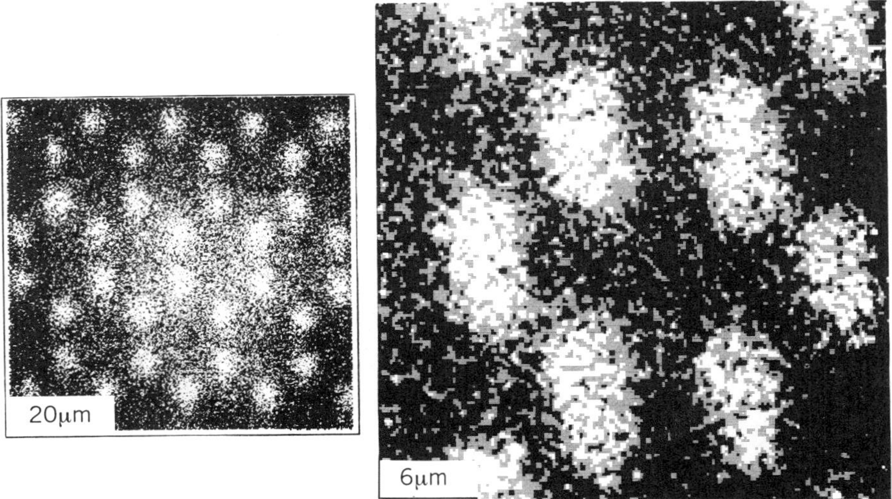

Figure 6.18. 20×20 μm^2 IBIC image measured using 200 ^{1}H ions/μm^2, and not compensated for ion induced damage (left), and 6×6 μm^2 IBIC image measured using 16,000 ^{1}H ions/μm^2 and compensated for the decreasing pulse height (right). The degradation of the spatial resolution in the vertical plane in the compensated image was not connected with the compensation method. Reprinted from Ref. 50 with kind permission from Elsevier Science B.V., Amsterdam, The Netherlands.

contrast was destroyed by the ion induced damage without any compensation. The light-colored honeycomb structure of the 1 μm wide trench cells can be seen. The right image in Figure 6.18 shows a 6×6 μm^2 IBIC image with the effects of damage compensated as described above, measured using a dose of 16,000 ^{1}H ions/μm^2 (i.e., 9 ^{1}H ions/pixel). The dose/μm^2 used in the compensated image is 80 times greater than that in the uncompensated image, which demonstrates that compensation of the decreasing charge pulse height is possible. However, it is not possible to compensate for the changing shape of the charge pulse height spectrum with cumulative dose using this method, because it relies on altering the measured data at each pixel by the same amount, even though different regions may be affected differently by damage.

Another problem observed with such samples was an uneven distribution of the average charge pulse height measured from different parts of the scanned area. The left image in Figure 6.19 shows a 65×65 μm^2 image of the DRAM sample, and the uneven contrast across the image is evident (ignore the top left corner). An image of the same area is shown on the right-hand side of Figure 6.19, which was generated from the same data set, but the uneven contrast effect has been filtered out by displaying the variance of the different charge pulse heights measured at each pixel. Here the image contrast is uniform across the scanned area even though the average charge pulse height was not. This uneven contrast is thought to be due to either an ion induced charging effect or an uneven depletion field width caused by poor sample preparation or nonuniform deposition of the Schottky barrier metal layer.

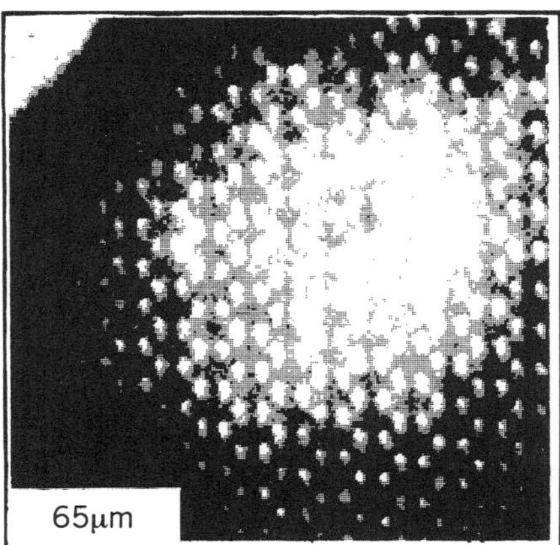

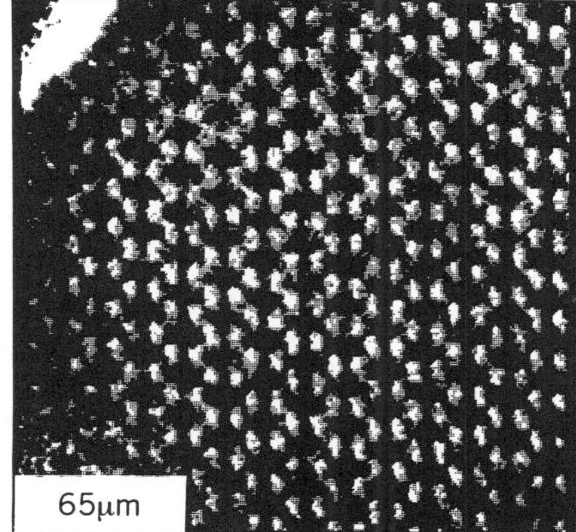

Figure 6.19. 65×65 μm^2 IBIC image of the DRAM, showing (left) the effects of nonuniform charge collection across the image, and (right) this effect filtered out. Reprinted from Ref. 50 with kind permission from Elsevier Science B.V., Amsterdam, The Netherlands.

6.5.2. Compensation of Damage Using Event-by-event of Data Collection

An event-by-event data acquisition system, as described in Chapter 2, stores the measured charge pulses along with their position coordinates for subsequent processing and analysis. This is ideal for generating images in which the contrast arises from a variation of the average charge pulse height and for low statistics images, as discussed in Section 4.7 for STIM analysis. The specific advantage for IBIC microscopy, however, is that the measured charge pulse height data set can be sliced up into sequential dose increments so that the evolution of the charge pulse height spectrum and image contrast from different parts of the scanned area can be examined with a cumulative ion dose.

In the previous section, a method of compensating for the effects of ion induced damage was used in which the charge pulse reduction was compensated for by increasing each subsequently measured pulse by the same amount at each pixel. This process does not work if the charge pulse height measured from different regions within the scanned area varies in a different manner, and so was only partially successful. With an event-by-event data acquisition system, the choice of how to tackle this problem is more flexible, but the best method used so far is to use the average value of the measured charge pulse heights at each pixel. Even though the charge pulse height in adjacent pixels may vary with cumulative ion dose, i.e., the contrast may change, the charge pulses can still be used to generate an IBIC image and so increase the ion dose that can be used. The resultant image is not quantitative, because there is no definite relationship between the average charge pulse heights at different pixels. This can be solved by slicing the data set up in specific regions to measure the charge pulse height at the start of analysis (i.e. before the affects of damage are detectable). This is essential for quantitative analysis to correctly interpret the measured charge pulse height in terms of the surface and depletion layer thicknesses present.

It is strongly recommended that an event-by-event data acquisition system be used for all IBIC measurements and that the measured data sets be sliced up to check for damage related effects. Examples of characterizing the damage with an event-by-event acquisition system and constructing IBIC images in the presence of damage are given in Chapter 7. There is a strong possibility that the best method of reconstructing IBIC images that suffer from ion induced damage has not yet been developed; more work is required to assess different reconstruction methods.

6.6. STUDY OF NUCLEAR MICROPROBE BEAM HALO USING ION BEAM INDUCED CHARGE

Figure 6.20 shows a 125×125 μm^2 image of the same DRAM sample discussed in the previous section, where the central 25×25 μm^2 region had been

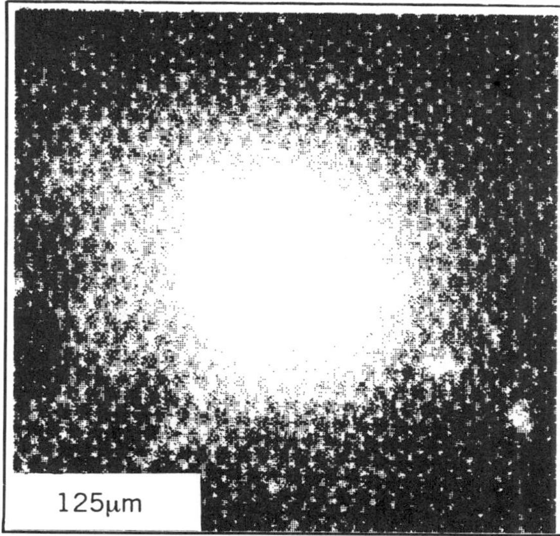

Figure 6.20. 125×125 μm^2 IBIC image of the DRAM, where the central 25×25 μm^2 region has been irradiated with 6,000 ^{1}H ions/μm^2. The honeycomb network of the trench holes can be seen around the edges. Reprinted from Ref. 50 with kind permission from Elsevier Science B.V., Amsterdam, The Netherlands.

irradiated with a dose of 6,000 ^{1}H ions/μm^2. A large area around the central irradiated area has also been damaged by the beam halo, which limits the spatial resolution attainable with the nuclear microprobe. This section demonstrates how the ion induced damage produced in a blank silicon slice by a focused MeV ion beam can be imaged with IBIC to reveal the extent of the focused beam halo. This method can be used as a diagnostic tool [51] for analyzing the factors limiting spatial resolution with the nuclear microprobe.

A three-dimensional representation of an IBIC image of the damage distribution caused by a current of 200 pA focused into an area approximately 1 μm across, with a total dose of 20 nC, is shown in Figure 6.21. The horizontal plane is the x direction and the vertical plane is the y direction. The charge pulse height from the central 1 μm wide region is lost in the noise level, because this region has received a very high dose and has been extensively damaged. The resultant IBIC image shows the damage distribution from just the beam halo. The larger extent of the beam in the vertical plane than the horizontal plane is immediately obvious. By measuring the reduction in the charge pulse height from a given increase in the total dose, the vertical scale in Figure 6.21 can be calibrated in terms of dose received from the beam halo. These dose contours can then be integrated over the full area in which damage is detectable to calculate the total beam fraction in the halo. In Figure 6.21, approximately 1% of the total beam current is outside a central area of 3 μm in diameter, so

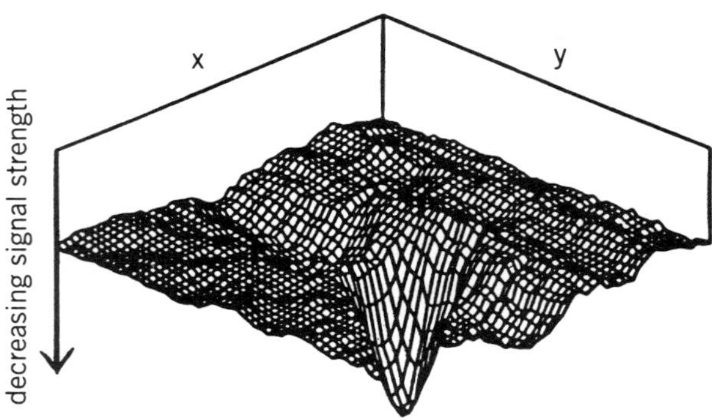

Figure 6.21. 80×80 μm^2 three-dimensional representation of an IBIC image of the damage distribution caused by a dose of 20 nC of 3 MeV ^{1}H ions. The measured charge pulse height is plotted on the vertical axis. Reprinted from Ref. 51 with kind permission from Elsevier Science B.V., Amsterdam, The Netherlands.

this small fraction is unlikely to limit spatial resolution at this beam current of 200 pA.

The chromatic aberration coefficients for the Oxford nuclear microprobe used for this experiment are $(x/\theta\delta) = 390$ μm/mrad% in the horizontal plane, and $(y/\phi\delta) = 970$ μm/mrad% in the vertical plane, where δ is the percentage momentum spread of the beam passing through the quadrupole lenses (aberration coefficients are described in depth in Chapter 3). The effect of any changes in the beam energy passing through the quadrupole lenses will thus be approximately 2.5 times greater in the vertical plane than in the horizontal plane for the same divergence angle in the two planes. From Figure 6.21, the ratio of the extent of the vertical to horizontal beam halo is 2.4, which is in good agreement with the ratio of the vertical to horizontal plane chromatic aberration coefficients. This supports the hypothesis that the measured halo is caused by a momentum spread of the beam fraction in the halo.

For a beam current of 1 fA the object and divergence slit sizes are much less than for a beam current of 200 pA, so the beam fraction in the halo is not necessarily the same. Figure 6.22 shows contour plots for the variation of the measured charge pulse height due to point irradiations at different divergences with a beam current of 0.3 fA. The different shape of the beam halo is immediately apparent. By integrating the damage contours, the amount of beam current outside the central 3 μm area was found to be 10% to 15%. This large fraction of the beam current in the halo at a low beam current is thus likely to limit the attainable spatial resolution at a low beam current.

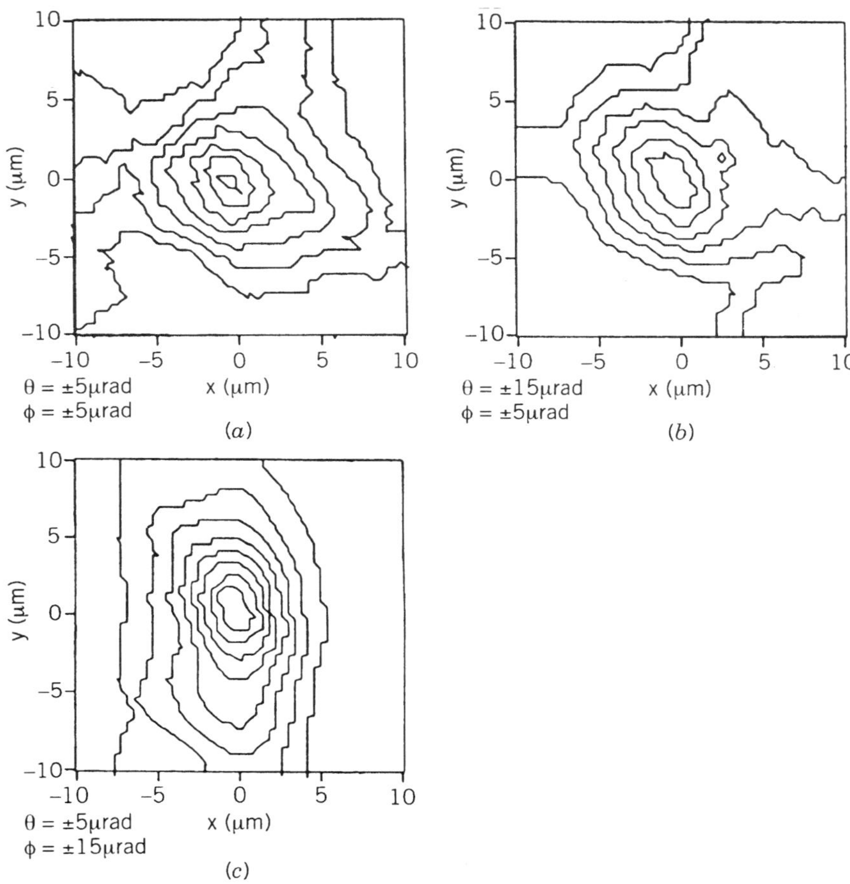

Figure 6.22. IBIC contour plots showing the variation of the measured charge pulse height due to point irradiations at different divergences shown in the bottom left corner. The outermost contour shows the area that has received a dose of 2,000 ^{1}H ions/μm^2, and the contours inside this have received higher doses. Reprinted from Ref. 51 with kind permission from Elsevier Science B.V., Amsterdam, The Netherlands.

REFERENCES

1. M.B.H. Breese, G.W. Grime, and F. Watt, Oxford Nuclear Physics Rept. OUNP–91–33 (1991); also M.B.H. Breese, P.J.C. King, G.W. Grime, and F. Watt, *J. Appl. Phys.* **72**(6):2097 (1992).

2. G. Bertolini and A. Coche, eds., *Semiconductor Detectors*, North–Holland, Amsterdam (1968).

3. J.B.A. England, *Techniques in Nuclear Structure Physics*. Macmillan Press, London (1974).

4. G. Dearnaley and D.C. Northrop, *Semiconductor Counters for Nuclear Radiations*. E. & F.N. Spon Ltd, London (1963).

5. F.S. Goulding, *Nucl. Instr. Meth.* **43**:1 (1966).

6. H.J. Leamy, *J. Appl. Phys.* **53**:R51 (1982).

7. C.J.R. Sheppard, *Scanning Microsc.* **3**:15 (1989).

8. D.B. Holt, M.D. Muir, P.R. Grant, and I.M. Boswarva, *Quantitative Scanning Electron Microscopy*, Academic Press, London (1974).

9. C. Donolato, *Appl. Phys. Lett.* **34**:80 (1979).

10. K.F. Galloway, K.O. Leedy, and W.J. Keery, *IEEE Trans. Parts, Hybrids, Packag.* **12**:231 (1976).

11. J.W. Corbett, *Electron Radiation Damage in Semiconductors and Metals*. Academic Press, New York (1966).

12. V.A.J. van Lint, T.M. Flanagan, R.E. Leadon, J.A. Naber, and V.C. Rogers, *Mechansims of Radiation Effects in Electronic Materials*. Wiley, New York (1980).

13. R. Grube, E. Fretwurst, and G. Lindström, *Nucl. Instr. Meth.* **101**:97 (1972).

14. D. Angell, B.B. Marsh, N. Cue, and J.-W. Miao, *Nucl. Instr. Meth.* **B44**:172 (1989).

15. F.B. McLean and T.R. Oldham, *IEEE Trans. Nuc. Sci.* **29**:2018 (1982).

16. A.R. Knudson and A.B. Campbell, *Nucl. Instr. Meth.* **218**:625 (1983).

17. W.K. Chu, J.W. Mayer, and M.A. Nicolet, *Backscattering Spectrometry*. Academic Press, New York (1978).

18. S.M. Sze, *Physics of Semiconductor Devices*. Wiley-Interscience, New York (1981).

19. C. Kittel, *Introduction to Solid State Physics*. Wiley, New York (1986).

20. D.A. Frazer, *The Physics of Semiconductor Devices*. Oxford University Press, New York (1983).

21. C.A. Klein, *J. Appl. Phys.* **39**:2029 (1968).

22. S.M. Sze, ed., *VLSI Technology*. McGraw–Hill, New York (1983).

23. D.F. Horne, *Microcircuit Production Technology*. Adam Hilger, Bristol (1986).

24. E.H. Rhoderick, *Metal–Semiconductor Contacts*. Oxford University Press, New York (1978).

25. C.R.M. Grovenor, *Microelectronic Materials*. Adam Hilger, Bristol (1989).

26. M.D. Lumb, *Luminescence Spectroscopy*. Academic Press, London (1978).

27. C. Yang, N.P.O. Larsson, E. Swietlicki, K.G. Malmqvist, D.N. Jamieson, and C.G. Ryan, *Nucl. Instr. Meth.* **B77**:188 (1993).

28. D.V. Lang, *J. Appl. Phys.* **45**(7):3023 (1974).

29. H.L. Grubin, J.P. Keskovsky, and B.C. Weinerg, *IEEE Trans. Nuc. Sci.* **31**:1161 (1986).

30. P.E. Dodd, F.W. Sexton, and P.S. Winokur, *IEEE Trans. Nuc. Sci.* **41**(6):2005 (1994).

31. P.E. Dodd and F.W. Sexton, Sandia National Laboratories Semi-annual report on 'Radiation Assurance,' (1994).

32. M.B.H. Breese, *J. Appl. Phys.* **74**(6):3789 (1993).

33. M.B.H. Breese, J.S. Laird, G.R. Moloney, A. Saint, and D.N. Jamieson, *Appl. Phys. Lett.* **64**(15):1962 (1994).

34. M.B.H. Breese, A. Saint, F.W. Sexton, H.A. Schöne, K.M. Horn, B.L. Doyle, J.S. Laird, and G.J.F. Legge, *J. Appl. Phys.* **77**(8):3734 (1995).

35. G.E. Possin and J.F. Norton, In: O. Johari., ed. *Scanning Electron Microscopy.* Illinois Institute of Technology Research Institute, Chicago, IL (1975), p. 457.

36. R. Shimizu and T.E. Everhart, *Optik* **36**:59 (1972).

37. C. Donolato, *Appl. Phys. Lett.* **34**:80 (1979).

38. J.F. Ziegler, J.P. Biersack, and U. Littmark, *The Stopping and Range of Ions in Solids.* Pergamon, New York (1985).

39. J.C. Chi and H.C. Gatos, *J. Appl. Phys.* **50**(5):3433 (1979).

40. C.J. Wu and D.B. Wittry, *J. Appl. Phys.* **49**(5):2827 (1978).

41. G.L. Miller, D.V. Lang, and L.C. Kimmerling, *Ann. Rev. Mater. Sci.* 377 (1977).

42. J. Hjelen and B. Tolleshaug, *Micron Microsc. Acta* **23**:179 (1992).

43. C.V. Morgan, ed., *Channeling Theory, Observations and Applications.* Wiley, New York (1973).

44. L.C. Feldman, J.W. Mayer, and S.T. Picraux, *Materials Analysis by Ion Channeling.* Academic, New York (1982).

45. C.D. Moak, J.W.T. Dabbs, and W.W. Walker, *Bull. Am. Phys. Soc.* **11**:101 (1966).

46. M.B.H. Breese, unpublished work.

47. A. Saint, M.B.H. Breese, and G.J.F. Legge, unpublished.

48. T. Fell, PhD Thesis, University of Oxford (1992).

49. J.S. Laird, A. Saint, and G.J.F. Legge, unpublished.

50. M.B.H. Breese, G.W. Grime, and M. Dellith, *Nucl. Instr. Meth.* **B77**:332 (1993).

51. M.B.H. Breese, G.W. Grime, and F. Watt, *Nucl. Instr. Meth.* **B77**:243 (1993).

7

MICROELECTRONICS ANALYSIS

The trend in modern microelectronic device fabrication is to stack layers on top of each other and to decrease the feature size to less than 0.5 μm in order to achieve a greater feature density. References 1–4 discuss many aspects of device fabrication and technology. These trends place very stringent requirements on the analytical techniques used to study the thickness, composition and distribution of the device layer structure and on the methods used to analyze the electrically active regions of the device. Common device failure mechanisms [5], and the analytical steps typically used in device failure analysis [6,7] have been described.

There are a wide range of established methods used to analyze the structure of devices. Optical microscopy has a spatial resolution of approximately 0.5 μm and is ideal for surface examination but gives no information from beneath metallization layers because of the high attenuation of light through metal. The scanning electron microscope [8–10] can achieve a spatial resolution of less than 10 nm by measuring secondary electrons emitted from the surface and is an excellent tool for imaging device topography. Other electron beam related signals, such as X–rays or backscattered electrons, can analyze subsurface device layers up to a depth of approximately 1 μm. A keV electron beam cannot pass through thick metallizations or passivation layers on devices without suffering serious degradation of spatial resolution because of scattering in the sample, so etching and sectioning techniques must be used for analyzing deeper layers. Microradiography [11] can give subsurface information on devices with a spatial resolution of approximately 5 μm, and is ideal for studying macroscopic faults. Because silicon is virtually transparent to infrared light, an infrared microscope [12] can image silicon devices through the substrate from the rear side with a spatial resolution of approximately 1 μm. All these methods are nondestructive, and the device still functions after analysis, but none of them can give high spatial resolution information on buried device layers.

Other methods that are capable of analyzing subsurface layers with higher spatial resolution involve etching or sputtering the surface layers. Transmission electron microscopy (TEM) [13,14] is a high-resolution technique for detailed imaging of the device structure but involves difficult sample preparation [15] and the analytical volume is small. It is, however, a very important device analysis technique, because it can resolve features a few nanometers across, so that the thinnest layers present in device structures can be imaged. Secondary ion mass spectrometry (SIMS) [16] uses a focused keV heavy ion beam, and achieves a spatial resolution of 50 nm. It has high elemental sensitivity but poor elemental quantitivity. Depth profiles of the elements present can be obtained with a depth resolution of 5 nm, by continually sputtering the surface layers. Auger electron spectroscopy [17] is capable of surface analysis of the chemical composition and can measure depth profiles, but is difficult to quantify. The atomic force microscope is also being used in the analysis of microelectronic devices for such applications as profilng the device morphology. Despite all these methods, there are still no easy methods of obtaining quantitative elemental depth distributions and imaging the distribution of deeply buried subsurface layers within intact devices with high spatial resolution.

Examples of some of the different techniques used for the analysis of the physical structure of microelectronic devices are shown in Figure 7.1. Figure 7.1a shows a cross-section TEM image of a processed silicon wafer used to fabricate device structures on. Stacking faults and twins can be seen beneath the surface in this 100 nm thick section. Figure 7.1b shows a plan view TEM image of a 4 Mbit dynamic-random-access-memory (DRAM) field. The light areas are the trench holes, and four dislocations are arrowed in this image. Figure 7.1c shows a cross-section SEM image of a silicon-on-insulator device structure that shows the recrystallization of the silicon epilayer with the silicon substrate. Figure 7.1d shows an atomic force microscope (AFM) image of the word lines on a 16 Mbit DRAM device where the image contrast is due to the device topography. Example of SIMS images of a device structure are given in Ref. 21, and also in Section 7.2.1. Reference 22 shows examples of scanning acoustic microscopy images of semiconductor devices.

Imaging methods showing the distribution of the electrically active device areas include electron beam induced current (EBIC) [23] using a keV electron beam, and optical beam induced current (OBIC) [24] using a focused laser beam. Typical uses of such techniques are the measurement of the semiconductor diffusion length [25,26], imaging the formation of microplasmas and leakage channels, testing the gain and logic states of devices, and in tracing faulty device interconnections that create open circuits. Analysis is again limited by thick surface and metallization layers, and the image contrast is very sensitive to energy losses and attenuation in the surface layers, making interpretation of

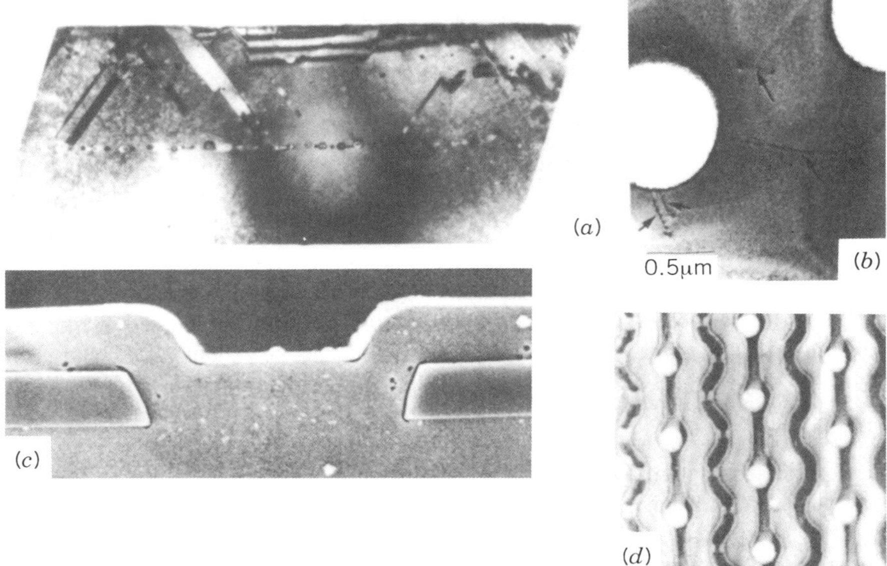

Figure 7.1. (a) Cross-section TEM image of processed silicon with a 1.6 μm wide field of view, reprinted courtesy of C.R. Marsh, Materials Department, University of Oxford. (b) Plan view TEM image of a 4 Mbit DRAM memory field, reprinted from Ref. 18 with permission (© 1993 IOP Publishing Ltd., Bristol, U.K.). (c) Cross-section SEM image of a silicon-on-insulator device structure, reprinted with permission from Ref. 19, (d) AFM image of the word lines on a 16 Mbit DRAM device, with a 10 μm wide box length. Reprinted from Ref. 20 with permission (© 1993 IOP Publishing Ltd., Bristol, U.K.).

image contrast difficult [27]. Figure 7.2 shows examples of images of devices measured with EBIC and OBIC. In both cases, the bright regions correspond to a large induced current and the dark regions show the device topography. Figure 7.2a shows an EBIC image of bipolar transistors in a logic array where the bright regions represent regions of diffusion isolation of the transistor collectors. Figure 7.2b shows an OBIC image of a transistor structure, and it can be seen that no charge was induced beneath the metallized regions due to the high attenuation of the incident laser light through the metal layers.

Voltage contrast microscopy [28] is a method of secondary electron imaging with a scanning electron microscope, whereby the image contrast is varied by applying a positive or negative voltage to certain device pins. Charging of the insulating surface layers either locally enhances or reduces the measured secondary electron yield, but it is difficult to relate precisely the observed image contrast to the underlying device structure. These and other methods used for the analysis of surfaces, thin films, and microelectronic devices are discussed further elsewhere [29,30].

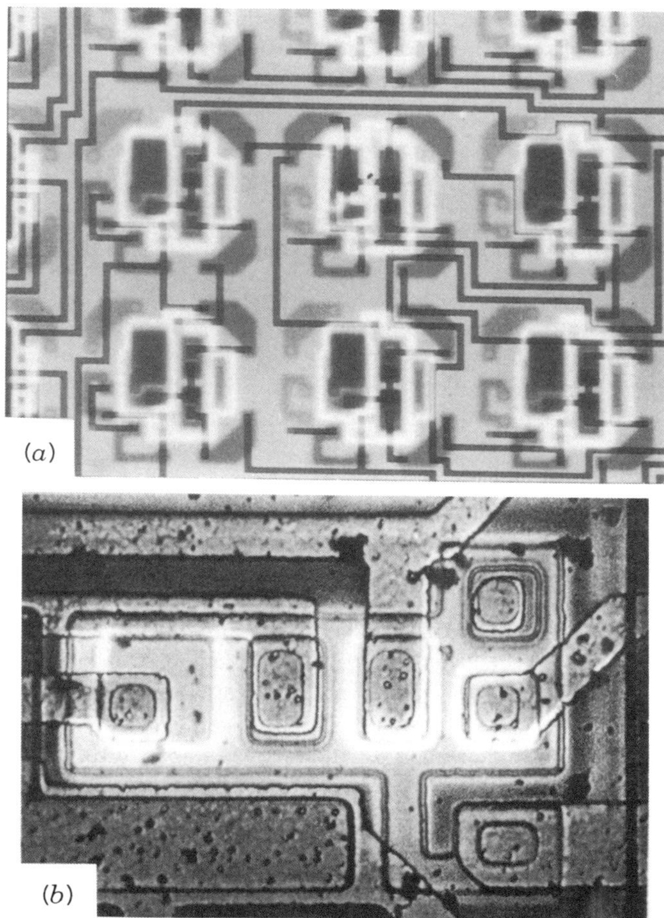

Figure 7.2. (a) EBIC image of bipolar transistors. Reprinted with permission from Matelect Limited, London, (b) OBIC image of a transistor structure. Reprinted with permission from Lasertec Corporation, London.

MeV light ions have a high penetrating power in semiconductors, so they can generate charge carriers from under the surface layers of fully intact devices for analysis of the buried depletion regions with IBIC microscopy [31]. The high penetrating power of MeV ions allows the full thickness to be imaged and quantified with PIXE and backscattering spectrometry. If the device substrate is mechanically thinned and polished to approximately 30 μm, then MeV ^{1}H ions have sufficient energy to travel through the remaining layers, and the variation in the transmitted ion energy loss can be measured with STIM. This chapter reviews the use of the nuclear microprobe in conjunction with all these MeV ion analytical techniques to characterize both the electrical and structural properties of microelectronic devices. Work related to the analysis of the crystalline perfection of semiconductors and microelectronic devices using channeling techniques is described in Chapters 5, 8 and 9.

7.1. ANALYSIS OF DEVICE ACTIVE REGIONS

7.1.1. Ion Beam Induced Charge Microscopy of a EPROM Memory Device

An n-type metal oxide semiconductor (nmos) memory device was the first structure used to demonstrate image formation with IBIC microscopy and the capability to image buried depletion regions within intact microelectronic devices. Different sensitivities to both the device topography and to ion induced damage by ^{1}H and ^{4}He ions has been shown, this work is based on Refs. 31–34.

Figure 7.3a shows an optical image of the device area analyzed, and Figure 7.3b shows the area inside the dashed box in Figure 7.3a at a higher magnification. This area, shown in Figure 7.3a, contains two output driver field effect transistors (FETs) for the data pin shown on the left, and also two transistors comprising the input buffer. The metallization, which is the light-colored area of Figure 7.3b, is a 1 μm thick layer of aluminum (1%Si). There is an approximately 1.5 μm thick SiO_2 passivation layer over the surface. The small circles along the width of the metallized areas are the contacts to the transistor n-type drains and sources underneath. These can also be seen in Figure 7.4, which shows low and higher magnification SEM images of this same area. Figure 7.3c shows the layout of two transistors within this area, and Figure 7.3d shows how their drains and sources relate to the optical images shown in Figure 7.3a,b. Separate preamplifiers were connected to the different device pins being studied so that separate IBIC images could be measured simultaneously to show the connections between different parts of the device.

Figure 7.5 shows the measured charge pulse height spectrum from the area shown in Figure 7.3, generated using 3 MeV ^{1}H ions, and with a charge sensitive preamplifier connected between the data pin and ground. By dividing the charge pulse height spectrum into individual 'windows' and mapping out the number of counts in each window, IBIC images can be formed to shows regions of equal measured pulse height within the scanned area. All these individual images can then be combined to give a single image showing the variation of the average charge pulse height. Figure 7.6 shows the individual and average pulse height (labeled EBAR in the bottom right of the figure) IBIC images generated from this charge pulse height spectrum. Images 1 to 3 correspond to areas away from the depletion regions. Only a few charge carriers generated by the MeV ^{1}H ions passing through the areas shown in images 1 to 3 have laterally diffused into the depletion regions. The interdigitated structure of depletion regions is shown in image 4, and image 5 shows a square depletion region at the top left of the scanned area. The average charge pulse height image shown in the

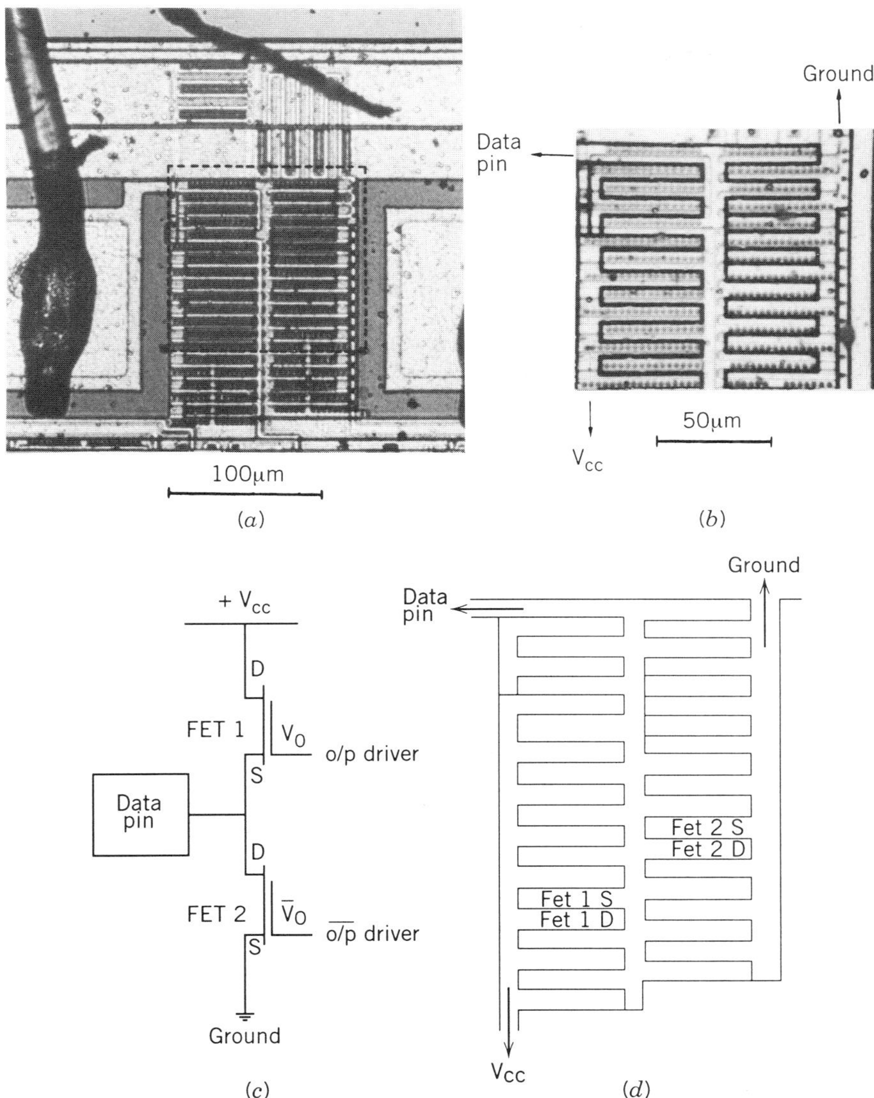

Figure 7.3. (a) 300×300 μm^2 optical image of an area of the memory device. (b) Higher magnification optical image showing the gaps in the metallization together with the metal to drain and metal to source contacts. (c) Schematic device layout for the two output transistors. (d) shows how their drains and sources relate to Figure 7.3a,b. Reprinted from Ref. 31 with permission (© 1992 American Institute of Physics, Woodbury, NY).

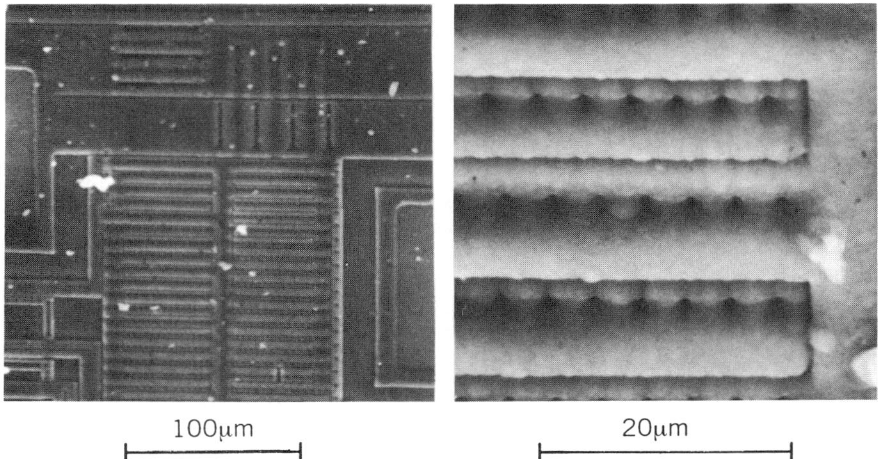

Figure 7.4. Low- and high-magnification SEM images of the device area. Reprinted from Ref. 34 with permission (© 1994 American Institute of Physics, Woodbury, NY).

bottom right of Figure 7.6 displays all this information in a single image. In all the average pulse height IBIC images shown in this book, the darker the region the larger the average charge pulse height measured from it, unless otherwise stated. Figure 7.7 is a 75×75 μm^2 IBIC image of the central region from Figure 7.6 under the same imaging conditions. The circles along the lengths of the source and drain regions that can be seen in Figures 7.3b and 7.4 are not detectable in this image, demonstrating the insensitivity of IBIC using MeV ^{1}H ions, to device topography as described in Section 6.2.

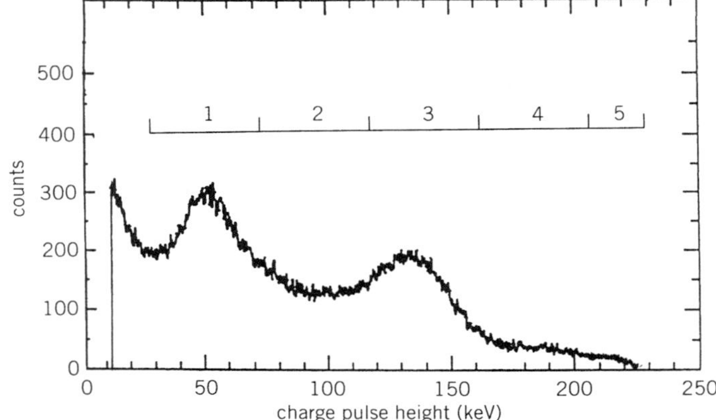

Figure 7.5. 3 MeV ^{1}H ion induced charge pulse height spectrum from the 300×300 μm^2 device area shown in Figure 7.3a. The preamplifier was connected between the data pin and ground.

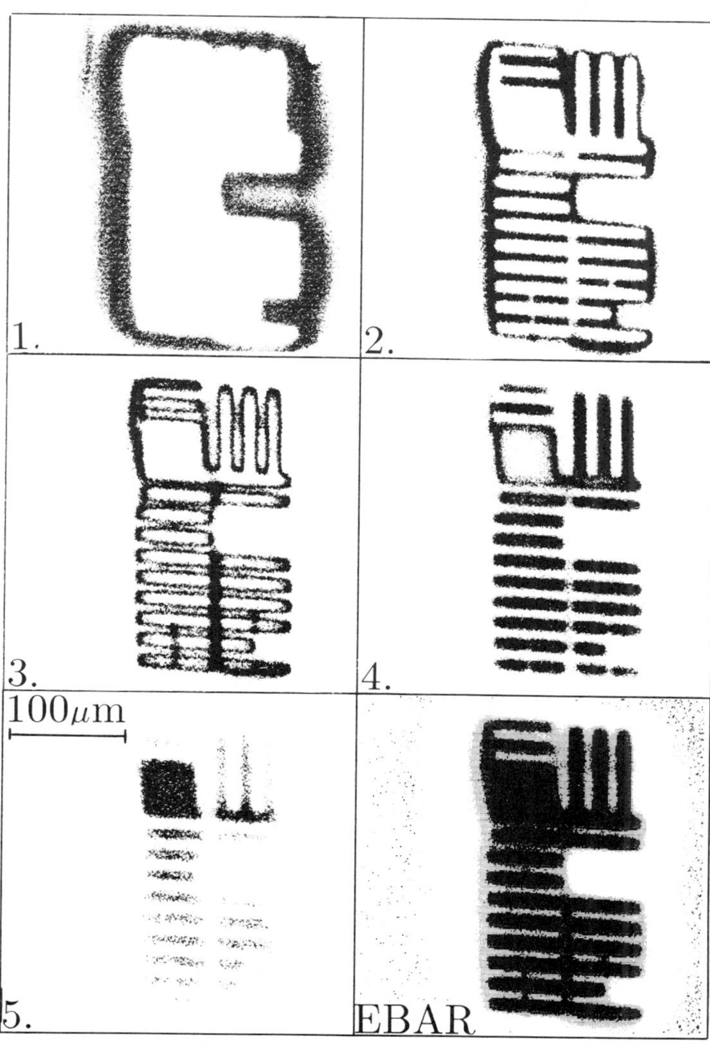

Figure 7.6. $300 \times 300 \ \mu\text{m}^2$ IBIC images obtained from the memory device. The first five images show regions of different measured pulse height from windows numbered in Figure 7.5, and the bottom right image shows the information from the first five images combined to give a single image showing the average charge pulse height within the scanned area. Reprinted from Ref. 33 with kind permission from Elsevier Science B.V., Amsterdam, The Netherlands.

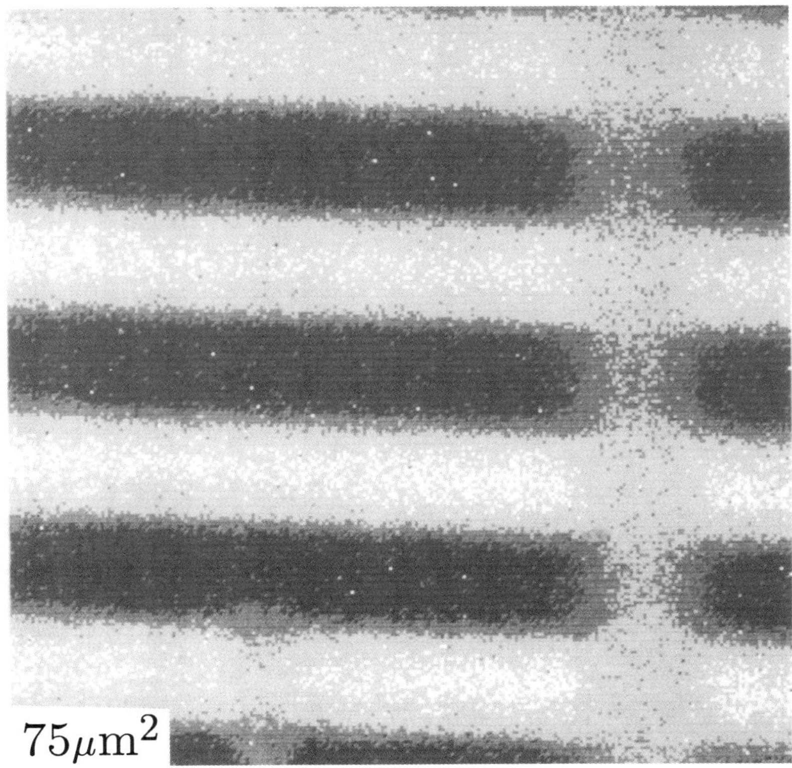

Figure 7.7. 75 × 75 μm² average pulse height IBIC image of the central region in Figure 7.6, under the same conditions. Reprinted from Ref. 31 with permission (© 1992 American Institute of Physics, Woodbury, NY).

The effect of lateral charge diffusion on the spatial resolution attainable with IBIC can be seen in Figure 7.8, which shows a vertical line scan extracted from across the middle of Figure 7.7. The large pulse height areas correspond to the dark regions in Figure 7.7 and the smallest charge pulse height regions correspond to the noise level. The sloping edges of the depletion regions are caused by lateral charge diffusion, not by the beam spot size. By measuring the decrease in the charge pulse height away from the edge of the depletion layer in Figure 7.7 the sample diffusion length is estimated to be approximately 4 μm in this region.

The three IBIC images in Figure 7.9 show the average charge pulse height in this same 300 × 300 μm² area, but with different preamplifier connections in each case. In Figure 7.9a the chip supply voltage pin, V_{CC} = +5 V, and the ground pin are connected to the preamplifier, and the data pin is floating. There is a channel between the drain and source regions in transistor 1, but

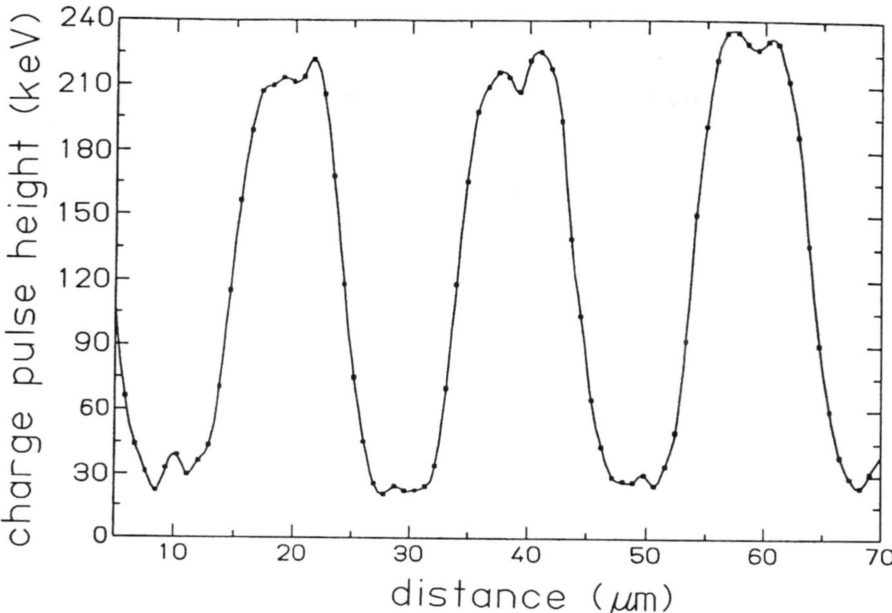

Figure 7.8. Vertical line scan extracted from the middle of Figure 7.7. The vertical scale shows the variation of the average charge pulse height with distance across the scan. Reprinted from Ref. 31 with permission (© 1992 American Institute of Physics, Woodbury, NY).

not in transistor 2. Figure 7.9b also shows the IBIC image measured with the preamplifier across the supply voltage pin and ground under the same conditions as Figure 7.9a, except that the data pin is also now connected to a different preamplifier, such that the images in Figures 7.9b and 7.6 were measured simultaneously. Figure 7.9c was measured between the data pin and ground as in Figure 7.6, but, in this image, the contrast is different than in Figure 7.6, as the other two transistors present in this area are now operating.

Figure 7.10 shows $200 \times 200 \ \mu m^2$ and $40 \times 40 \ \mu m^2$ IBIC images of this same area generated with 2.3 MeV ^{4}He ions. The contrast in these images is so strong that they are not shown at their best in black and white, so color plate 1 shows this same $200 \times 200 \ \mu m^2$ image and a similar $60 \times 60 \ \mu m^2$ IBIC image. These color images allow considerably better appreciation of the strong topographical contrast obtainable with IBIC microscopy. The maximum signal-to-noise level in these IBIC images is approximately 50, whereas in the ^{1}H IBIC images of this same area it was approximately 5.

The gaps in the metallization layer and the holes along the horizontal fingers that can be seen in the SEM images of Figure 7.4 are regions of thinner surface layer coverage. These regions appear dark in the IBIC images because the charge pulses are larger owing to the lower ion energy loss in the surface lay-

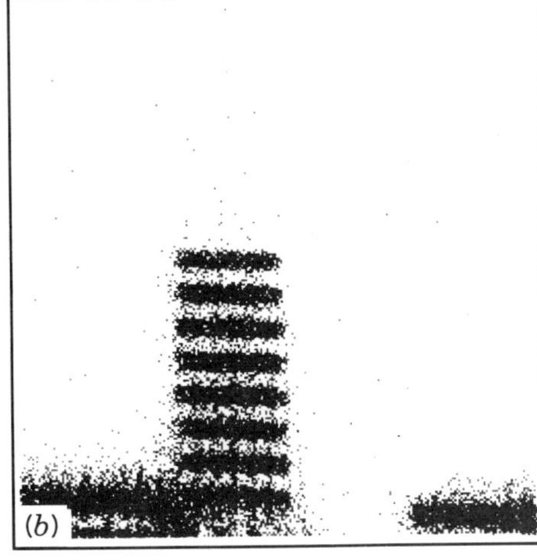
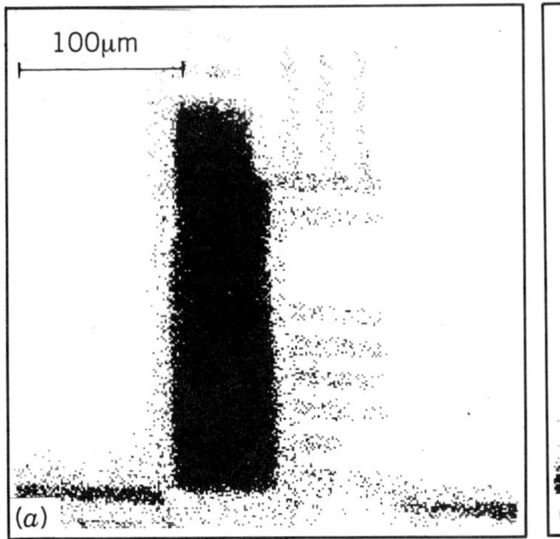

Figure 7.9. Three 300×300 μm^2 IBIC images of the same device area. The preamplifier connections are (a) between the supply voltage pin, with $V_{CC} = +5$ V and ground. The output driver voltage $V_0 = +5$ V, and the data pin is not connected here. (b) Same as (a) except the data pin is now also connected to a different preamplifier. (c) Measured between the data pin and ground but with other transistors on.

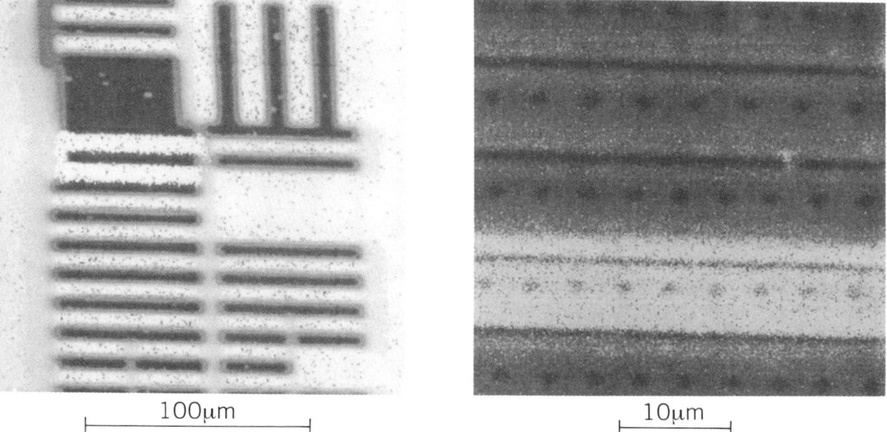

Figure 7.10. 2.3 MeV ^{4}He IBIC images of this same device area with a scan size of 200×200 μm^2 using an ion dose of 16 ions/μm^2, and 40×40 μm^2 scan size using a dose of 176 ions/μm^2. Reprinted from Ref. 34 with permission (© 1994 American Institute of Physics, Woodbury, NY).

ers. In the 60×60 μm^2 IBIC image in Color Plate 1, the effects of both charge diffusion into the depletion layers (the red regions) and also contrast from the device topography (in shades of blue) can be clearly seen (see color insert). The holes along the horizontal metal fingers can be seen even in regions where the measured charge pulses arise only from lateral charge diffusion into the connected pn junctions, such as the lighter grey horizontal strip in the lower half of the 40×40 μm^2 IBIC image in Figure 7.10.

The charge pulse height data sets used to generate the images in Figure 7.10 were then sliced up into sequential dose increments, so that the evolution of the charge pulse height spectrum and image contrast could be examined with cumulative ion dose. There was no detectable evidence of damage at a dose of less than 18 ions/μm^2, so no changes were observed in the data set of the 200×200 μm^2 IBIC image where a dose of 16 ions/μm^2 was used. The 40×40 μm^2 IBIC image in Figure 7.10 was generated using a dose of 176 ions/μm^2, which made it more susceptible to ion induced damage, because the ion dose used was a factor of ten greater than that at which damage was detectable. Figure 7.11a shows the charge pulse height spectra from the entire area of the 40×40 μm^2 IBIC image as a function of eight sequential dose increments of 22 ions/μm^2. The sliced-up spectra continually change shape from cumulative ion induced damage, so that the two peaks visible at the start of data collection have merged into a single peak by the end of data collection. This causes the full spectrum to be broader than the sliced-up spectra, because it is the average of the sliced-up spectra.

Figure 7.11a also shows four selected-area charge pulse height spectra

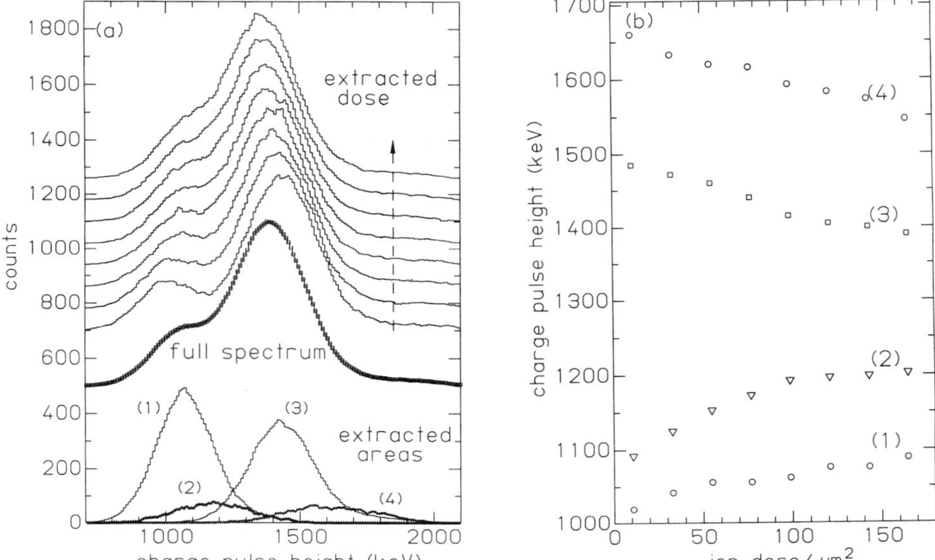

Figure 7.11. (a) Charge pulse height spectra from the $40 \times 40\ \mu\text{m}^2$ IBIC image. The full spectrum from the whole area using the total ion dose of 176 ions/μm^2, is vertically compressed in the middle, and the charge pulse height spectra with cumulative ion dose increments of 22 ions/μm^2 are shown above this, and each is vertically offset by an additional 100 counts. Below the full spectrum are shown four separate spectra extracted away from a depletion region (1), from the holes inside this area (2), at a depletion region (3), and from the holes inside this area (4). The vertical height of these four spectra has been adjusted for clarity. (b) Variation in the average charge pulse height with cumulative ion dose from regions 1 to 4 described in Figure 7.11a. Reprinted from Ref. 34 with permission (© 1994 American Institute of Physics, Woodbury, NY).

beneath the full-dose spectrum. Figure 7.11b shows the variation of the average charge pulse height from these same four areas with cumulative ion dose increments of 22 ions/μm^2. The average charge pulse height measured at a depletion region (3 and 4) gradually decreased, as would be expected assuming that the main effect of ion induced damage is to decrease the diffusion length of the semiconductor substrate beneath the depletion region. However, the average charge pulse height measured away from the depletion region (1 and 2) increased with cumulative ion dose. The most likely cause for this unexpected behavior is ion induced charging of the insulating surface layers. This may have induced an electric field in the substrate leading to enhanced charge collection, as has also been observed with EBIC microscopy.

To ensure that ion induced damage did not affect the image contrast in the $40 \times 40\ \mu\text{m}^2$ IBIC image, a dose of only 18 ions/μm^2 would have to be used, which would result in a statistically noisy image. The approach used to generate these IBIC images was to determine the average charge pulse height at each individual pixel from the stored data set, as described in Section 6.5.2. Even though the charge pulse height in adjacent pixels varies differently with cumulative ion dose, the charge pulses can still be used to generate an IBIC image using an event-by-event data acquisition system.

Figure 7.12 shows the average charge pulse height with cumulative ion dose, measured by positioning the focused ion beam within the square depletion region shown in image 5 of Figure 7.6 for 2 MeV ^{1}H ions and 1.8 MeV ^{4}He ions. Separate areas were used for the different ions to independently study the effects of ion induced damage. The 2 MeV ^{1}H ions penetrate through the depletion region, and charge carriers were generated in the substrate; therefore, charge diffusion contributes to the charge pulse height. The modeling of this reduction is shown for ^{1}H ions as the continuous line in Figure 7.12, using the theory described in Section 6.2 and the assumption that the ion induced trap cross-section γ is a factor of 370 larger than the trap cross-section of the initial defects present, which are given an arbitrary size of 10^{-15} cm^2. A diffusion length of 4 μm, a surface layer thickness of 4 μm, and a depletion layer thickness of 5 μm are assumed. This demonstrates how the charge pulse height decrease due to a reduction in the diffusion length can be modeled. This is crucial, because it shows how the interpretation of the measured charge pulse heights from different regions can remain quantitative even though the pulses are changing in height with cumulative ion dose.

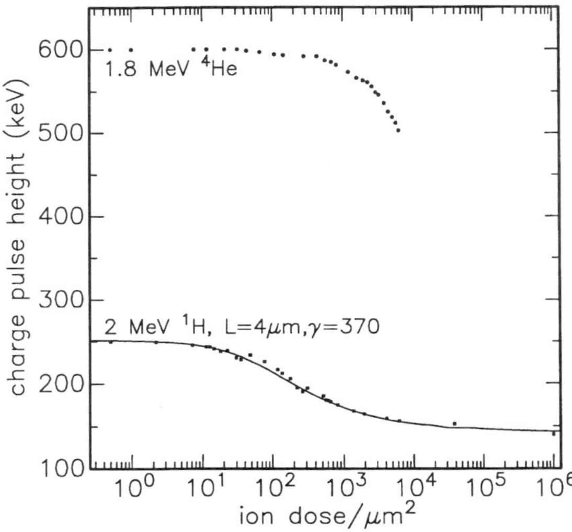

Figure 7.12. Variation of the average charge pulse height with cumulative dose for 2 MeV ^{1}H ions and 1.8 MeV ^{4}He ions from the depletion region shown in window 5 of Figure 7.6, measured between the data pin and ground. Reprinted from Ref. 33 with kind permission from Elsevier Science B.V., Amsterdam, The Netherlands.

The surface layers strongly reduced the measured charge pulse height for 1.8 MeV ^{4}He ions as most of the ion energy was lost in the surface layers. 1.8 MeV ^{4}He ions have a range of 6.2 μm in this region and so did not penetrate through the depletion layer; therefore, there were no charge carriers created beneath the depletion layer. The reduction in the measured charge pulse height for the 1.8 MeV ^{4}He ions is thus caused by recombination within the depletion layer. The different effects of irradiation with ^{1}H ions and ^{4}He ions is further shown by the charge pulse height spectra in Figure 7.13 from these similar areas. The width of the ^{4}He ion induced charge peak increases with cumulative dose and displays

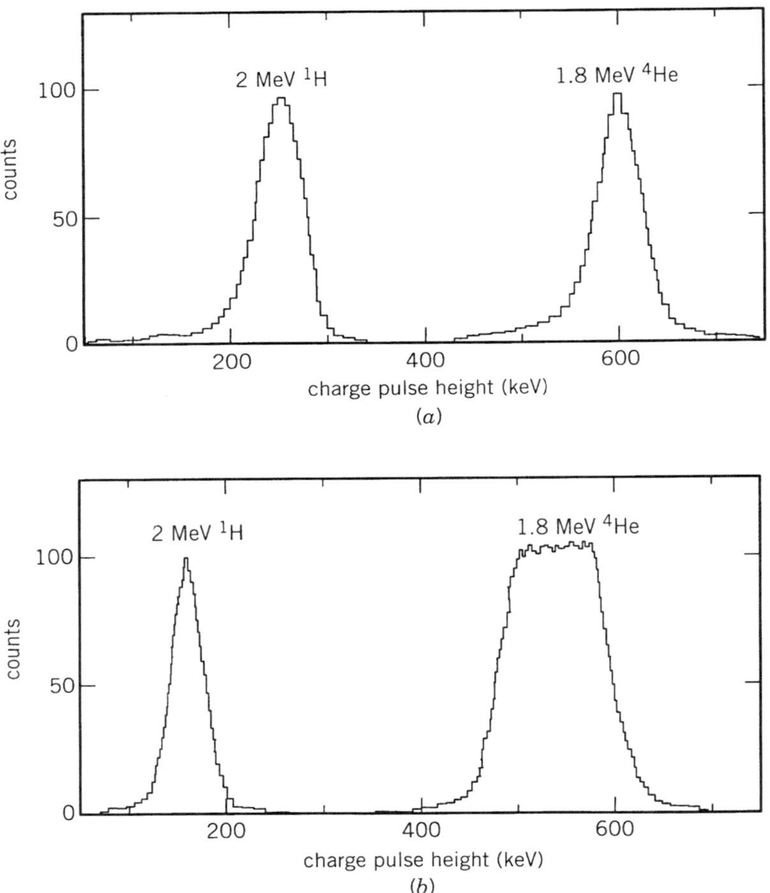

Figure 7.13. Charge pulse height spectra for 2 MeV ^{1}H ions and 1.8 MeV ^{4}He ions using a focused beam positioned in the middle of the square depletion regions shown in image 5 of Figure 7.6. The spectra are shown (a) at the start of irradiation and (b) after a dose of 6,500 ions/μm^2 in each case. Reprinted from Ref. 33 with kind permission from Elsevier Science B.V., Amsterdam, The Netherlands.

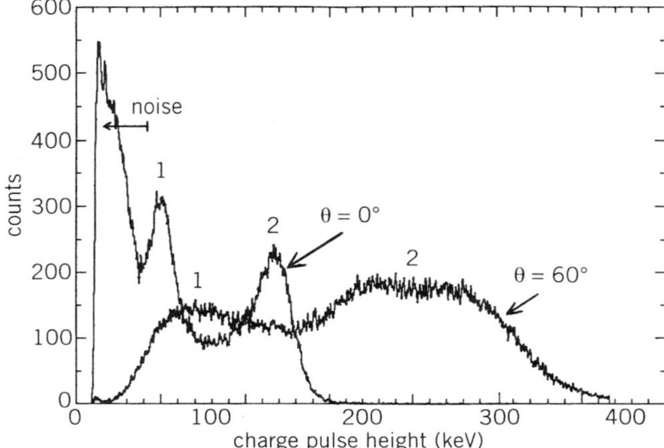

Figure 7.14. Two charge pulse height spectra measured from the same device area shown in Figure 7.6, between the data pin and ground. One curve was measured with the sample at $\theta = 0°$, and the other with the sample at $\theta = 60°$. The increase in the charge pulse peaks marked 1 and 2 is indicated. The corresponding IBIC image for a tilt angle of 60° is shown in Figure 7.15. Reprinted from Ref. 32 with kind permission from Elsevier Science B.V., Amsterdam, The Netherlands.

the characteristic effect of ion induced damage in semiconductor detectors [35]. The rate of charge pulse height reduction here is a thousand times less than that predicted if the ions were stopped in the substrate instead of the depletion layer. This demonstrates the ability to generate IBIC images that are almost insensitive to the effects of ion induced damage if the ions are stopped in the depletion layer, as was described in Section 6.3.

Another method used to increase the size of the charge pulses measured with MeV ^{1}H ions was to rotate this device through an angle of 60° about the vertical axis with respect to the beam direction, so that the rate of ^{1}H ion energy loss perpendicular to the device surface increased [32]. The charge pulse height spectra measured at normal incidence ($\theta = 0°$) and rotated through 60° are shown in Figure 7.14, and the IBIC image obtained from the same 300 × 300 μm^2 area under the same conditions as in Figure 7.6, but rotated through 60° is shown in Figure 7.15. The curved features across the top of the image was caused by ions scattering off the device bond wires into the measured area. The resultant charge pulses measured with the device rotated through 60° are larger, but the contrast as measured at the three vertically running arrows was a little weaker than in the normal incidence IBIC image.

7.1.2. Ion Beam Induced Charge Analysis of a SA3002 Memory Device

Results from a Sandia SA3002 ROM (read only memory) device are now described and interpreted using the theory described in Chapter 6. All aspects of

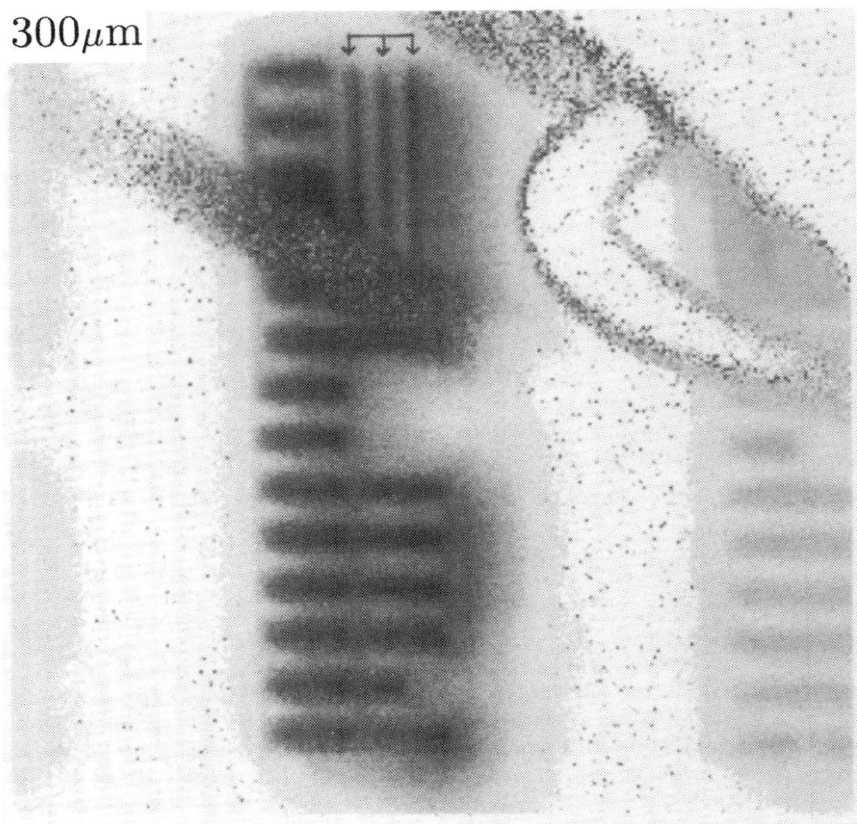

Figure 7.15. 300×300 μm^2 IBIC image of the device area shown in Figure 7.6, tilted through $60°$. The curved feature across the upper half of the image is a bond wire. Reprinted from Ref. 32 with kind permission from Elsevier Science B.V., Amsterdam, The Netherlands.

the measured results have been correlated with the device structure [as described in 36], and only a summary is given here. This work made use of the recently developed mixed-beam facility developed in Melbourne [37], as discussed in Section 4.7.2.

Figure 7.16 shows a schematic of a region of the device memory field, and also a cross-section through the line labeled AA'. Figure 7.17a,b shows two SEM images of the memory field and Figure 7.17c,d shows two cross-section SEM images also through the memory field. This area has a 1 μm thick glassy passivation layer over the surface. The hexagons shown in Figures 7.16 and 7.17a,b are 600 nm thick polysilicon gate regions of the field effect transistors, which make up the memory elements: the drains are inside the hexagons and the source regions are outside them. The vertically-runnning strips are 1 μm thick aluminium met-

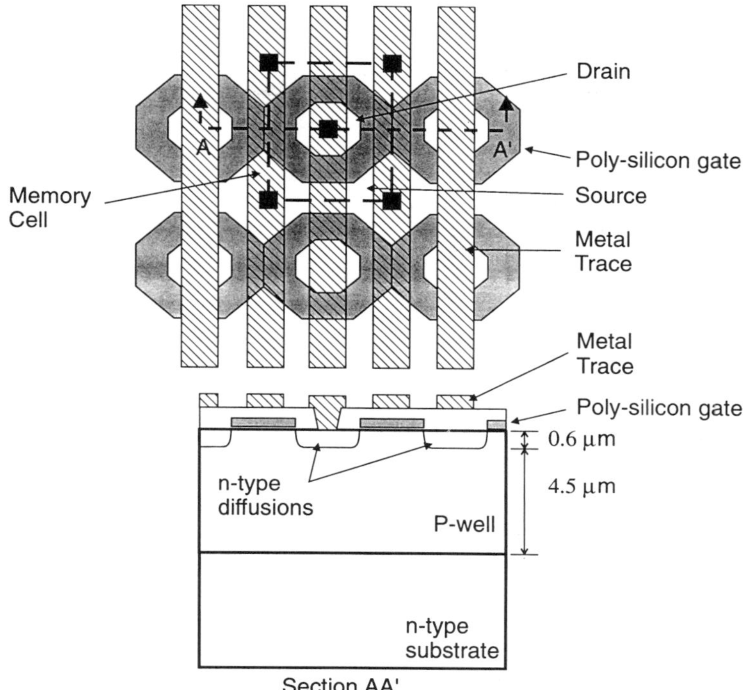

Figure 7.16. Schematic plan view of the memory field and cross-section through the line labeled AA'. Reprinted from Ref. 36 with permission (© 1995 American Institute of Physics, Woodbury, NY).

allizations that are alternately connected to the device source regions (and thence to ground) and drain regions. The metal drain lines run through the centers of the hexagonal gates and the source metal lines are located between the hexagons.

The n^+ drain and source regions are 0.5 μm deep and these lie within a P well depletion region that extends to a depth of 4 to 5 μm beneath the semiconductor surface, as can be seen on the cross-section schematic in Figure 7.16 and the cross-section SEM images in Figure 7.17c,d. There is a field-free region below the P well in the substrate and also in the narrow space above the P well. The charge preamplifier used to make these IBIC measurements was connected between the transistor drains and ground, so it was expected to detect charge from the drain regions of the device but not from the source regions, which were at ground potential.

The six ions used here to analyze this device are listed in Table 7.1, along with the energy, rate of energy loss close to the surface and the ion range. Color Plate 2 shows large area IBIC images measured with (a) 1 MeV ^{1}H$_2$ ions showing the central device logic and the memory fields on either side and (b) a smaller area 2 MeV ^{4}He IBIC image (see color insert). IBIC images showing

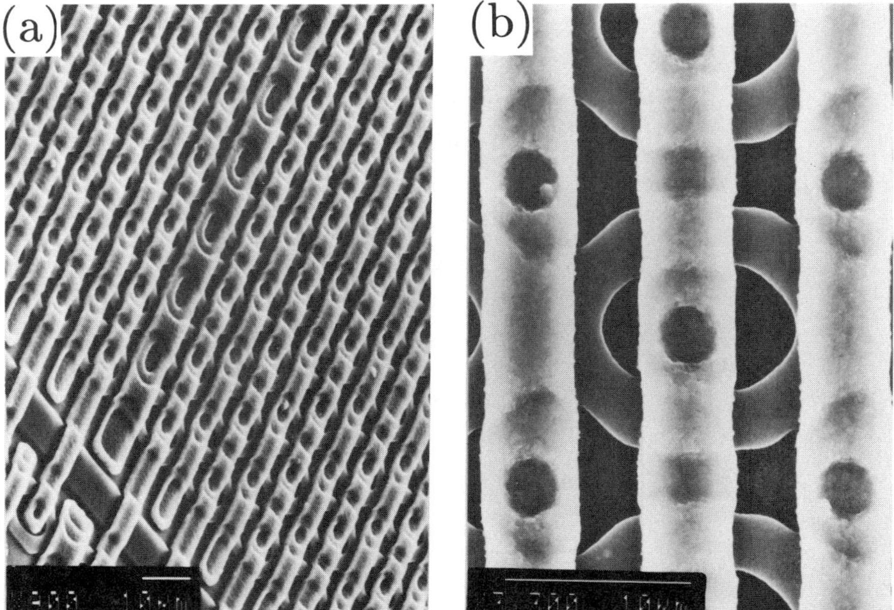

Figure 7.17. SEM images of the memory field. (a) Plan view at an oblique angle show-ing the device topography. (b) View from directly above the memory field. (c) Cross-section showing the n-type diffusions located at a depth of 0.5 μm within the P well which is 4.5 μm deep. (d) Higher magnification cross-section showing a contacted drain cell, and the P well at the bottom of the image. Reprinted from Ref. 36 with permission (© 1995 American Institute of Physics, Woodbury, NY).

the variation in the average charge pulse height across a 80×80 μm^2 region of the device memory field on the right side in Color Plate 2, for each of these six ions, are shown in Figure 7.18. Figure 7.19 shows the charge pulse height spectra from these 80×80 μm^2 areas for 1 MeV ^{1}H ions, 1 MeV ^{1}H$_2$ ions, and 2 MeV ^{4}He ions at the start of data collection and also after fixed sequential dose increments. Figure 7.20 shows the variation of the average charge pulse height with cumulative ion dose during data collection for each of the six different ions.

The low rate of energy loss by the ^{1}H ions resulted in small charge pulses, because most charge carriers were generated too deep in the substrate to diffuse back to the P well. The ^{1}H IBIC images show barely any contrast because the resultant charge pulses from these long range ions were insensitive to variations in the surface layer thickness. Because most of the charge carriers are generated in the substrate, the resultant charge pulse height was easily affected by damage.

The data for 3 MeV ^{1}H$_2$ ions and 6 MeV ^{4}He ions, which have a similar range, are now considered. The vertically-running aluminium metal lines which

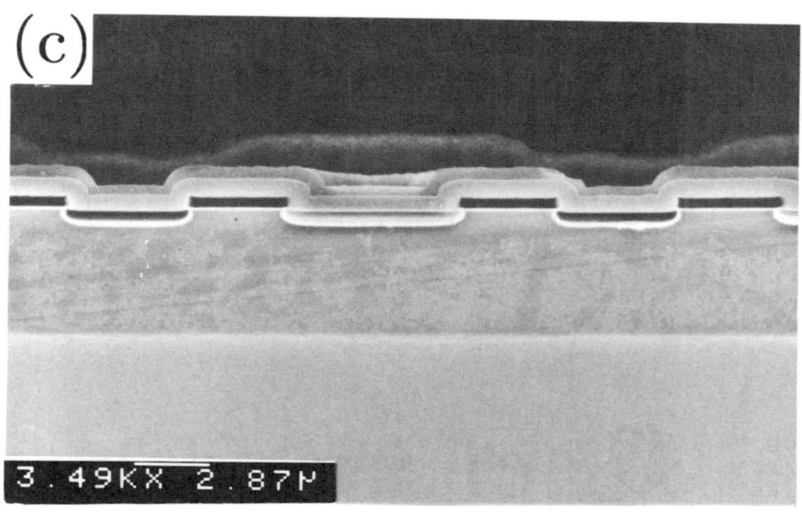

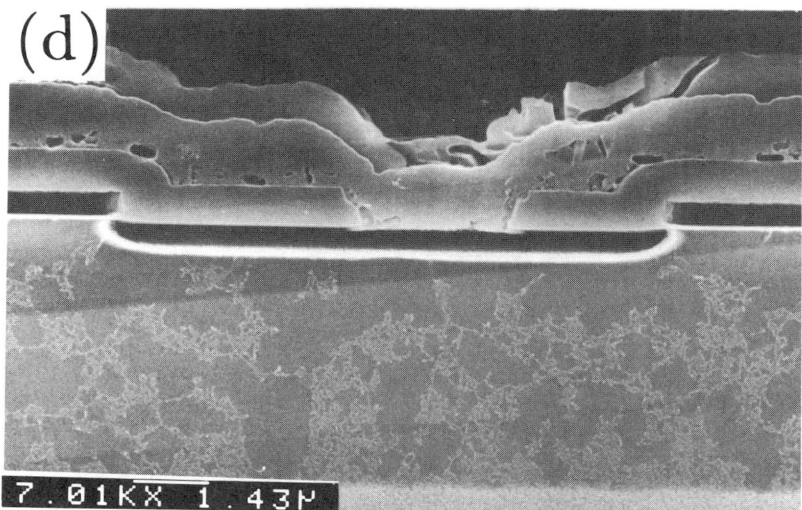

Figure 7.17. (*Continued*)

TABLE 7.1. Ion Beam Parameters in Silicon

Ion	Energy (MeV)	Range (μm)	dE/dz (keV/μm)
H	1	16.0	40
	2	47.3	25
H_2	1	5.7	124
	3	29.8	66
He	2	7.0	240
	6	31.2	140

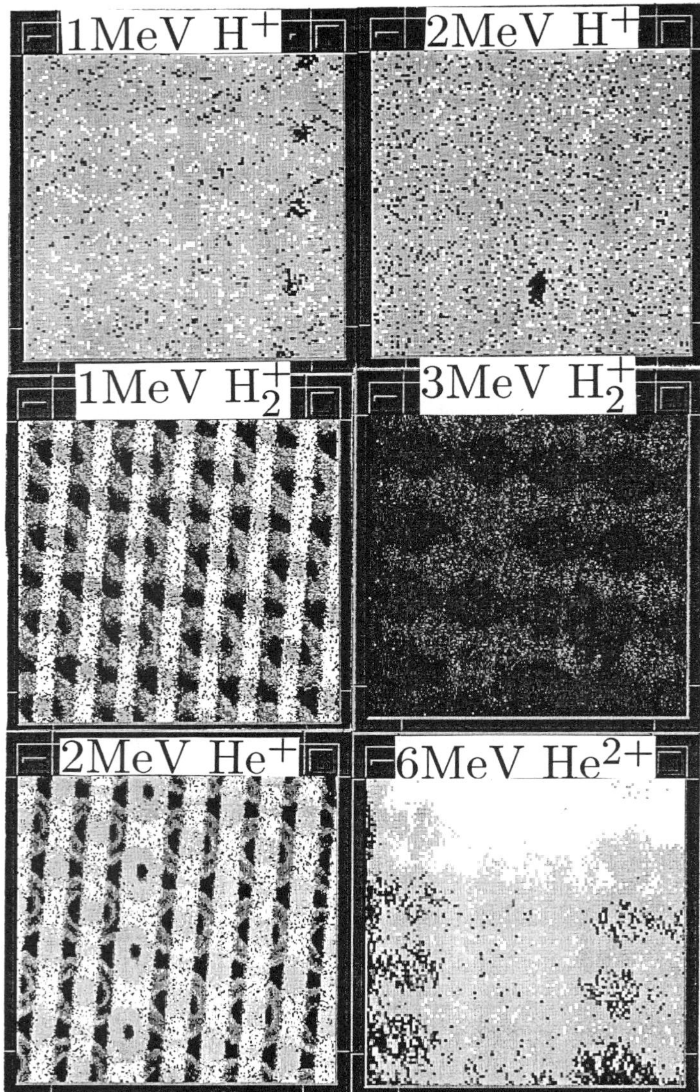

Figure 7.18. IBIC images showing the average charge pulse height of a $80 \times 80 \ \mu m^2$ region of the device memory field for each of the six ions. A different region was used in each case to study separately the effects of damage. The upper portion of the 6 MeV ^{4}He image overlaps a previously damaged region and should be ignored. Reprinted from Ref. 36 with permission (© 1995 American Institute of Physics, Woodbury, NY).

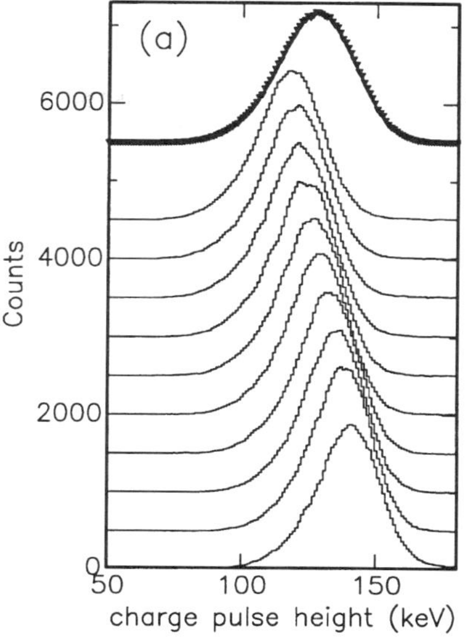

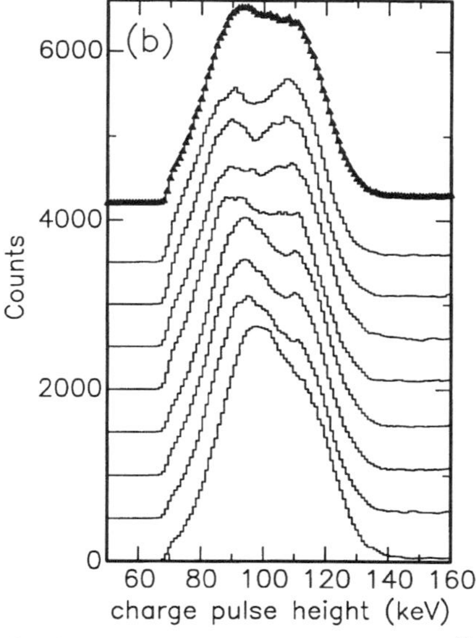

Figure 7.19. Charge pulse height spectra from a 80×80 μm^2 area for (a) 1 MeV ^{1}H ions, (b) 1 MeV ^{1}H$_2$ ions, and (c) 2 MeV ^{4}He ions. The measured charge pulse height spectra with cumulative sequential dose increments of (a) 8 ions/μm^2, (b) 16 ions/μm^2, (c) 9 ions/μm^2 are offset vertically by a fixed arbitrary amount for clarity. The uppermost spectrum in bolder characters in each case is the cumulative spectrum from which the sequential spectra were extracted. Reprinted from Ref. 36 with permission (© 1995 American Institute of Physics, Woodbury, NY).

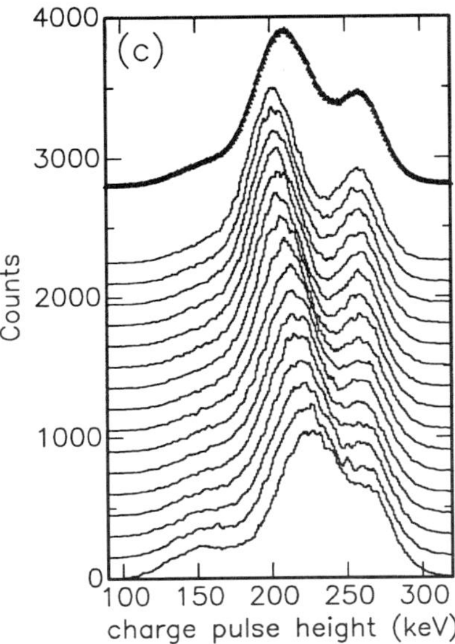

Figure 7.19. (*Continued*)

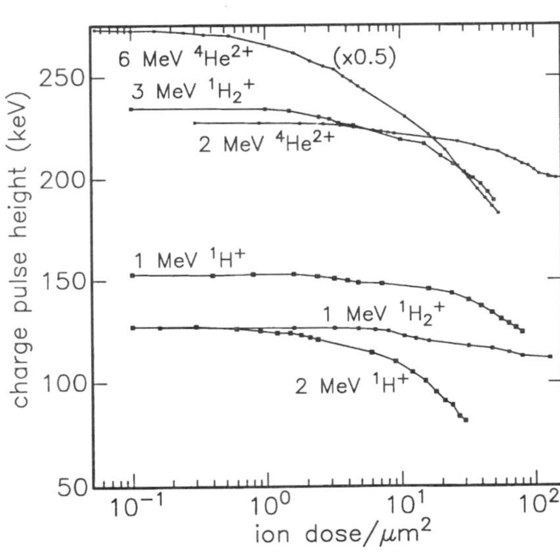

Figure 7.20. Decrease in the average charge pulse height for all six ions with cumulative dose. The data points for each type of ion are joined together for clarity. Reprinted from Ref. 36 with permission (© 1995 American Institute of Physics, Woodbury, NY).

can be faintly seen in the 3 MeV $^1H_2$ IBIC image demonstrate the low topographical contrast obtainable with this ion. The dark regions are the outline of the hexagonal gates and the light regions are the surrounding sources. These are also long range ions with most of the charge generated in the substrate, so the resultant charge pulse height was again easily affected by damage. The charge pulse height from the 6 MeV ^4He ions decreased faster than from the 3 MeV $^1H_2$ ions because of the higher rate of defect generation in the substrate.

The data for the 1 MeV $^1H_2$ and 2 MeV ^4He ions, which have a similar range, are now considered. Neither of the charge pulse height spectra show strong effects of ion induced damage with cumulative irradiation. These two IBIC images shown in Figure 7.18 have very strong topographical contrast and are also shown in Color Plate 3 (see color insert), along with a 40×40 μm^2 IBIC image generated with 2 MeV ^4He ions. The image contrast is so strong for the 2 MeV ^4He ions that it is difficult to appreciate the small variations present, and Figure 7.21 shows separate IBIC images displaying the measured number of counts within the scanned area for dif-

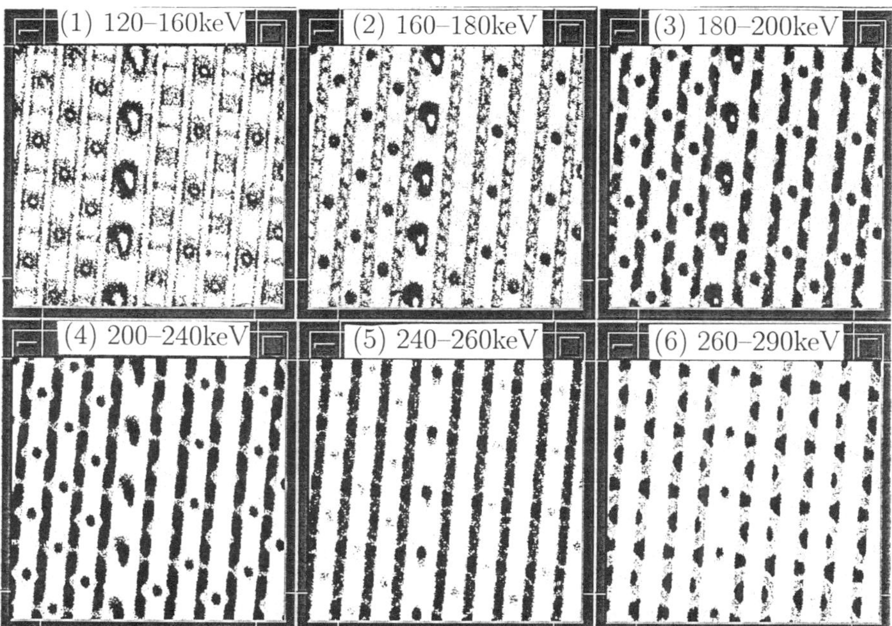

Figure 7.21. IBIC images showing the measured charge pulse intensity from a 80×80 μm^2 region for 2 MeV ^4He ions for different windows on the measured charge pulse height spectrum. The pulse height windows shown at the top correspond to the charge pulse height spectrum shown in Figure 7.19c. Dark represents a high measured intensity and light a low measured intensity. Reprinted from Ref. 36 with permission (© 1995 American Institute of Physics, Woodbury, NY).

ferent windows on the measured charge pulse height spectrum shown in Figure 7.19.

The vertically-running metal strips are regions where the surface layer coverage was thickest and no counts were detected from the 2 MeV ^{4}He ions passing through these regions except at the drain and source contact holes where the surface layers were thinner. The 2 MeV ^{4}He ions are stopped in the field-free region above the P well at these thick metal strips. Since the diffusion length was less than 3 μm in this region, few carriers were able to travel to the collecting junction at the P well, so no pulse was measured above the noise level. These metal strips were slightly thinner at their edges than at the middle, so the ions passing through these regions generated just enough charge carriers in the P well to be detected above the noise level, as shown in window 1 in Figure 7.21.

The gaps between the aluminium metallization correspond to the vertically-running regions shown in windows 4 to 6 in Figure 7.21. The charge pulses measured from those regions shown in window 6 did not decrease with cumulative ion dose, whereas the smaller pulses measured from those regions shown in window 4 decreased slightly. This was because the ions passing through those regions shown in window 6 were stopped in the P well, which was at a depth of 7 μm below the device surface. This was consequently where the maximum defect generation occurred, resulting in the minimum observable effects of ion induced damage. The regions shown in window 4, which correspond to parts of the hexagonal polysilicon gates, are thicker than those regions shown in window 6, so the ions were stopped just before the top of the P-well depletion layer. This resulted in smaller charge pulses and also generated a high defect density in the field-free region above the P well. This caused a decreasing charge pulse height from these regions, because there was a larger component of diffused charge in the resultant pulse.

The above again shows that the ion range can be tuned to give both the maximum topographical IBIC image contrast and also the maximum insensitivity to ion induced damage by stopping the ions in the device depletion region.

7.1.3. High Spatial Resolution Ion Beam Induced Charge Images of Buried Junctions

Figure 7.22 shows IBIC images of a gallium arsenide high-electron-mobility transistor [38]. The sample was fabricated as an n-type GaAs/AlGaAs heterostructure [39], which acts as a two-dimensional electron gas. The surface of the device was patterned with a dose of 1.5×10^{14}/cm^2 20 keV beryllium ions, which formed p-type material and resulted in the formation of a field effect transistor with a very narrow gate depletion region that was confined both laterally and in depth. The IBIC images were generated using contacts between the p- and n-type material, with 10 μm thick aluminum foil placed between the focused 3 MeV ^{1}H ions and the device surface. On the 20×20 μm^2 IBIC image a 0.8 μm wide depletion region is arrowed, and this high spatial resolution was possible due to the low lateral straggling of MeV ^{1}H ions through thick surface layers.

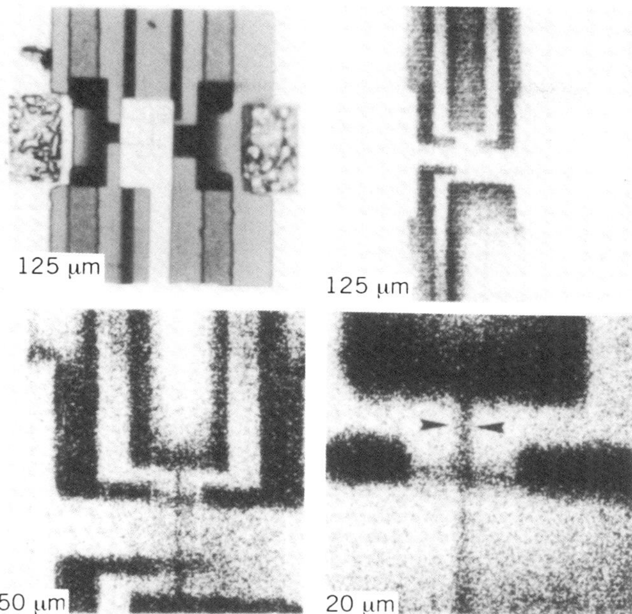

Figure 7.22. Optical image of the gallium arsenic device, and three IBIC images between the *p*- and *n*-type contacts, with scan lengths of 125, 50, and 20 μm. A 0.8 μm wide depletion region is indicated in the 20 × 20 μm^2 image. Reprinted from Ref. 38 with kind permission from Elsevier Science B.V., Amsterdam, The Netherlands.

7.1.4. Single Event Upset Analysis

As microelectronic device features continually shrink, the amount of charge that defines the different logic levels becomes smaller. This means that the charge generated by the passage of ionizing radiation through the device is more likely to cause a change in the device state; this is called a soft upset or a single event upset (SEU). This phenomenon is particularly important for satellite-based devices, which are exposed to a high flux of ionizing cosmic radiation. It is a serious problem in the design of high-density semiconductor memories [40], and considerable effort is currently being placed on developing SEU imaging for device analysis. Charge funneling, which was discussed in Section 6.1.4, increases the upset probability, because the radiation-induced charge becomes localized at the junction that was struck. Much work has been done using unfocused, high-energy, heavy ion beams from large tandem accelerators [41–43], pulsed lasers [44], and pulsed electron beams [45] as sources of ionizing radiation to study upset mechanisms, but the lack of spatially resolved information has led to difficulties in determining which part of the device caused the upsets. The technique of SEU imaging [46–48], developed for the analysis of static random access memory (SRAM) devices, uses a focused heavy ion beam from a nuclear microprobe scanning over the device surface. The information stored within the device is sampled at each beam position within the scanned area, so that an image can be constructed showing the upset probability at each location. Other work on complementary metal oxide semiconductor (CMOS) SRAM devices [49,50] was also able to locate those areas most susceptible to memory corruption using SEU images.

This section describes work [47,48] on radiation-hardened SRAM devices [51]. A SEU image generated using a low-intensity beam of 24 MeV Si^{6+} ions scanning over a 40×40 μm^2 area of the device is shown in Figure 7.23a, and the schematic of the device layout within this area is shown in Figure 7.23b.

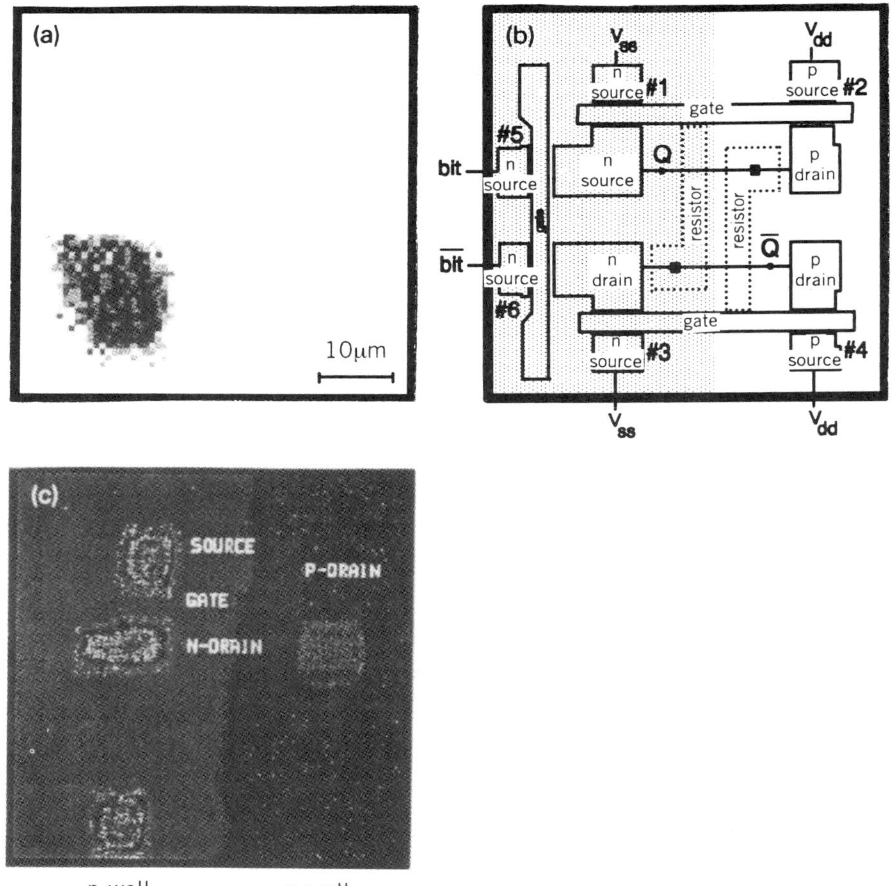

Figure 7.23. (a) 40×40 μm^2 SEU image of the SRAM device. (b) Schematic of the device layout in this area and (c) IBIC image from this same area. Reprinted from Ref. 48 with kind permission from Elsevier Science B.V., Amsterdam, The Netherlands.

The generation of this image required the use of such high-energy heavy ions to achieve the high rate of energy loss required to cause an upset within the device. The SEU image shows variations in the upset probability within this area as darker shading at the center and lighter shading toward the edges of the upset generating region. An IBIC image of this same area, generated using 2.4 MeV ^{4}He ions, is shown in Figure 7.23c. The ion impact locations that give rise to upsets within this scanned area can now be directly identified by comparing the SEU and IBIC images with the device layout, which is an important analytical advance. In this case the upset area was localized around an n-channel transistor drain. In the IBIC image, a large charge pulse was measured from the n drains within the p well of the device. Within the p well, a large charge pulse was also measured from the n source, but this is known from the SEU image not to be sensitive to upsets. There is a weaker region of charge collection within the n well region on the right of the IBIC image, corresponding to one of the device p drains. A detailed explanation of the observed IBIC image contrast in terms of the device layout is given in Refs. 47 and 48, and the SEU image contrast was correlated with the observed IBIC contrast. It was postulated that regions of larger than expected charge pulses could arise from a shunt effect [52] caused by the high ion induced carrier density along the ion path.

7.1.5. Ion Beam Induced Charge Microscopy of a Segmented Pad Detector

The charge collection efficiency in the interpad regions of a segmented pad detector, used to detect subatomic particles in a high-energy physics position-sensitive detector array, has been investigated with IBIC [53]. This is an important topic, because a knowledge of the way in which charge is shared between the different pads present is vital to determine the nature of the particles detected.

Figure 7.24 shows several 4 MeV ^{1}H IBIC images of a segmented pad p^+n 5 × 5 detector, with each pad 900 × 900 μm^2, fabricated on 310 μm thick 13 kΩ cm silicon, with an interpad spacing of 100 μm, and aluminized layers on the surface [53]. In Figure 7.24a, the detector bias was 20 V so that the depletion layer thickness in the individual pads was greater than the ^{1}H ion range of ~150 μm. Five pads were connected to preamplifiers and this IBIC image shows the presence of charge pulses measured using any of these. The wire bonds to the pads can be seen as regions of lower measured charge pulse height.

Figure 7.24b shows the area corresponding to the square superimposed on Figure 7.24a, with no bias voltage applied. The upper left pad 1 has been disconnected, so was not seen in this or subsequent images, in which pads 3 and 4 were connected to a single preamplifier and pad 2 connected to a separate preamplifier. The borders of the complete charge collection outlined in white coincide with the overlying metallization borders. Figures 7.24c,d show IBIC images of this same region, with bias voltages of 2 V and 20 V, respectively. In both cases, the bottom images show the charge collected in pads 3 and 4, while the top images show the charge collected on pad 2 measured by trigger-

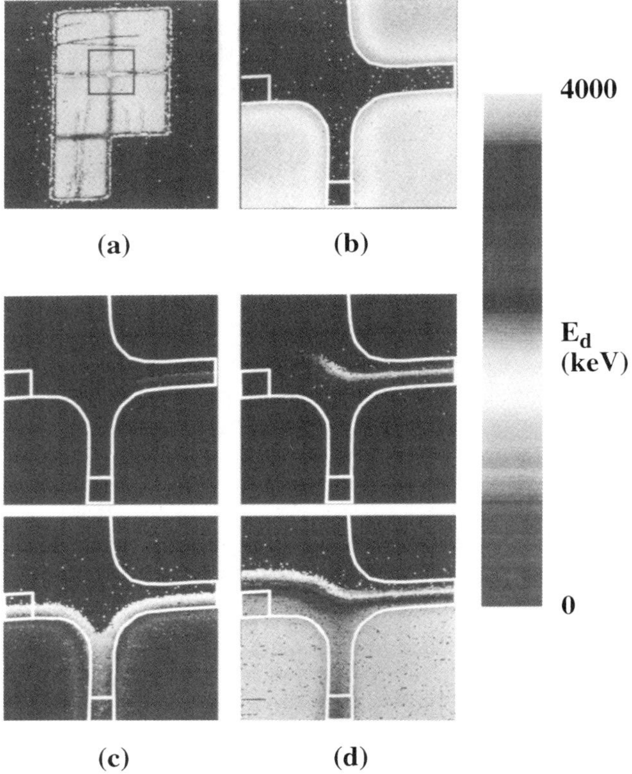

Figure 7.24. (a) IBIC images of segmented detector showing a 3.5 mm^2 area of pads 1–5, numbered from the top left corner. (b) 700 × 700 μm^2 image at intersection of pads 1–4, corresponding to the outlined square shown in (a), with pad 1 disconnected and a bias of 0 V. (c) Same region obtained by triggering station 2 from a pulse in station 1. (d) Same as (c) except with a bias voltage of 20 V. In (c) and (d), the bottom row shows only the charge measured on pads 3 and 4 and the top row shows the charge measured on pad 2, with the data acquisition system triggered from pads 3 and 4. Reprinted with permission of R.A. Bardos, University of Melbourne.

ing the data acquisition from a pulse measured on pads 3 and 4. This top row thus shows the amount of charge that has leaked away from pad 3, and these charge pulses thus have been measured on pad 3 with the wrong height. The pad outline from Figure 7.24b has been superimposed on these images. In Figure 7.24c, the charge collection regions from the three biased pads have extended partially into the corner region, the interpad region, and into the region of the unbiased pad. Complete charge collection was not observed between the pads and falls off with distance away from the metallization in the other regions. In Figure 7.24d, the charge collection between pads has increased. A comparison of the two white outlined boxes adjacent to pad 3, which are both located in the interpad (oxide) region, showed that the charge collection was more efficient in this region when only one of the two pads was biased. This may be from the relatively large electric field component parallel to the surface under the oxide in the former case, compared with a dead region in the latter case, where the parallel field component was small and so the charge drift to either electrode was slower.

Measurement of the depletion layer thickness with bias voltage, showed reasonable agreement with capacitance–voltage measurements of these pad detectors, so long as the depletion thickness was less than the ion range. IBIC thus gives a new method of quantifying and imaging the lateral and depth variation of the depletion region in semiconductor detectors. It avoids problems associated with other methods, such as the frequency dependence of the diode capacitance in the presence of traps for capacitance–voltage measurements, and the presence of currents due to surface states and diffusion that can complicate current–voltage measurements.

7.2. ANALYSIS OF DEVICE PHYSICAL STRUCTURE

In this section, we discuss the use of the nuclear microprobe in conjunction with STIM, PIXE, and backscattering spectrometry for imaging the distribution of buried device metallization layers, and measuring the layer thickness and composition. Much of the earlier work of nuclear microprobe analysis of microcircuit device structure has been described [54] and [55]. Examples are PIXE and ion induced electron images of an integrated circuits, and PIXE and backscattering images of a tin implanted polysilicon resistor are given in [56]. In the latter case, it was also demonstrated how backscattering spectra could be extracted from regions of interest within the scanned area and used to study the tin depth distribution. Another example is of a tin–gallium arsenide contact, where a PIXE image showed strong localization of the tin, and backscattering spectra from these areas showed evidence of the alloying between the tin and the gallium arsenide [54]. The microprobe group at Albany [57–59] studied the metallization thickness and incomplete etching of devices, and used electron images generated by a beam of MeV ^{4}He ions to identify regions of the device required for subsequent backscattering analysis. Doyle studied the tungsten metallization of porous silicon devices [60], and more recently buried oxide layers of a device have been imaged using focused 1.5 MeV ^{4}He ions [61]. If the device substrate is mechanically thinned and polished to a thickness of less than 30 μm, then STIM can be used as a rapid method of determining the uniformity and distribution of metallization layers, even if they are buried under other metal layers [62,63], so it is a higher spatial resolution equivalent of microradiography.

7.2.1. Analysis of a Multilayer Metallization Defect

An optical image of a microcircuit device area [62,63] for which microprobe results are described in this section is shown in Figure 7.25, along with the distribution of the metallization layers present in the central area. There is a TiSi$_2$ layer under the first metallization to assist adhesion to the underlying

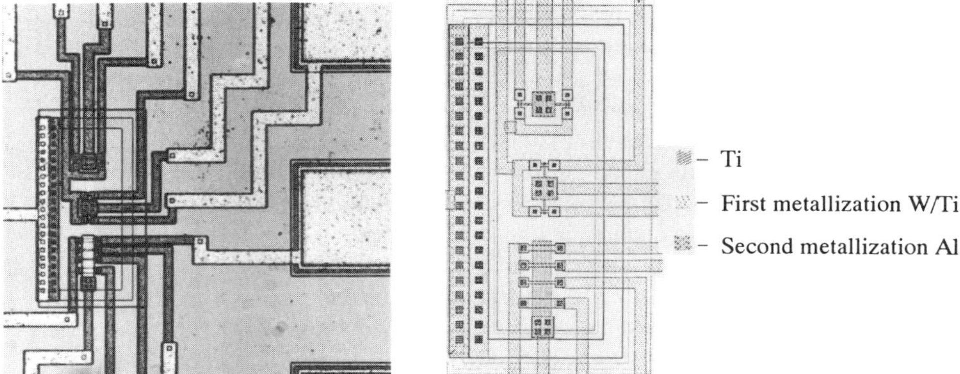

Figure 7.25. Optical image of a 400 × 400 μm^2 device area, and schematic of the metallization scheme for the central region. Reprinted from Ref. 62 with permission (© 1991 IOP Publishing Ltd., Bristol, U.K.).

borophosphosilicate glass. Overlaid on parts of the first metallization layer, and separated by another glassy layer, is the second metallization of aluminium. There is a thick SiO$_2$ passivation layer over the surface except on the contact pads, which can be seen on the far right. The device was thinned and polished to a thickness of 20 μm for STIM imaging, and 3 MeV ^{1}H ions were used for all the nuclear microprobe analysis described below.

This area was first examined with PIXE. Energy windows were set on the aluminium K, silicon K, titanium K, and tungsten L X–ray lines, and the distributions of these elements is shown in the top row of Figure 7.26, measured with 100 pA for 30 min. The silicon X–ray image shows low intensity from the metallized regions, because silicon X–rays generated in deeper levels were absorbed by the overlying metal. There was even lower silicon X–ray intensity from the contact pads because there is no passivation layer over these. The aluminum X–ray image shows that it occupied only the second metallization regions. There is a stronger aluminum signal from the large contact pads than from the aluminum tracks, because there was no absorption from the passivation layer. The titanium and tungsten X–ray images show similar distributions corresponding to the first and second metallization stages.

Images using STIM showing the average energy loss in the same region are shown in the second row of Figure 7.26, with decreasing scan sizes homing in on a small contact hole through the first metallization to the silicon substrate. Dark regions of the images indicates high ^{1}H ion energy loss, and light areas indicates low ^{1}H ion energy loss. The thick tracks of tungsten produce the greatest energy loss and thus the strongest contrast in these images. The additional second metallization over the tungsten tracks is discernible on the 200 × 200 μm^2 image as the darker regions, indicating even higher energy loss. The contact holes visible on the 60 × 60, 20 × 20, and 7 × 7 μm^2 images have higher energy loss around their edges, because the tungsten thickness increases in this

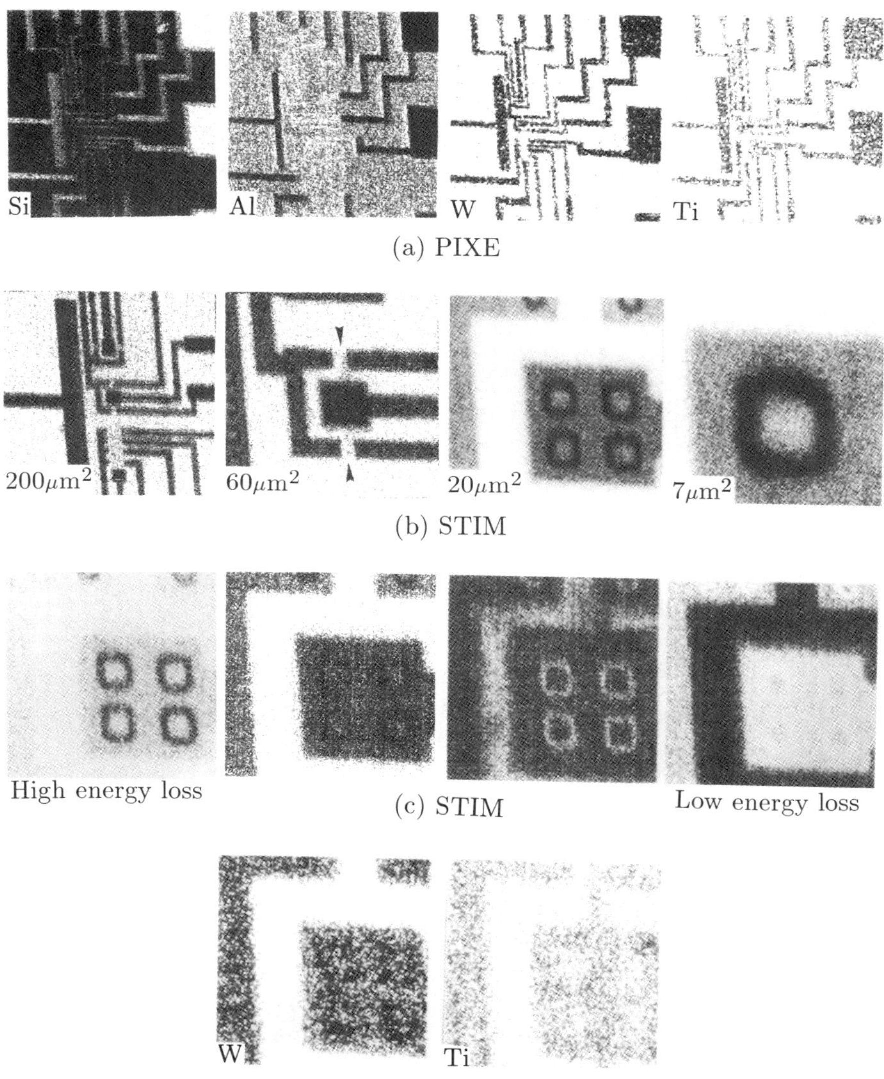

Figure 7.26. Top row: 400×400 μm² PIXE images for silicon K, aluminium K, tungsten L, and titanium K X–ray lines. Second row: STIM images of the same area with the scan size shown in the bottom left corner. Third row: 20×20 μm² STIM energy slices. Fourth row: 20×20 μm² PIXE images of the tungsten L and titanium K X–rays in the same area. Reprinted from Ref. 62 with permission (© 1991 IOP Publishing Ltd., Bristol, U.K.).

region. On the 60×60 and 20×20 μm^2 STIM images there is an indication of a fault in the metal distribution. The upper of the two faint lines indicated by the arrows on the 60×60 μm^2 image is slightly displaced from the center of the gap in the first metallization layer. This defect is more obvious on the STIM images for the separate energy windows of the 20×20 μm^2 area, which are shown in the third row of Figure 7.26. The image on the left has the highest energy loss and corresponds to the thickest regions of tungsten, whereas the right-hand image has the least energy loss and corresponds to regions with no overlying tungsten. On the third image, there is clear evidence that the track is misplaced. To determine the nature of fault giving rise to the contrast variation in the STIM images, PIXE was needed. The bottom row of Figure 7.26 shows the distribution of tungsten and titanium in the same 20×20 μm^2 area. The distribution of tungsten is the same as the layout of Figure 7.25, but the titanium distribution in the upper 1 μm wide track is displaced.

Figure 7.27 shows four images of a 25×25 μm^2 area, measured using PIXE,

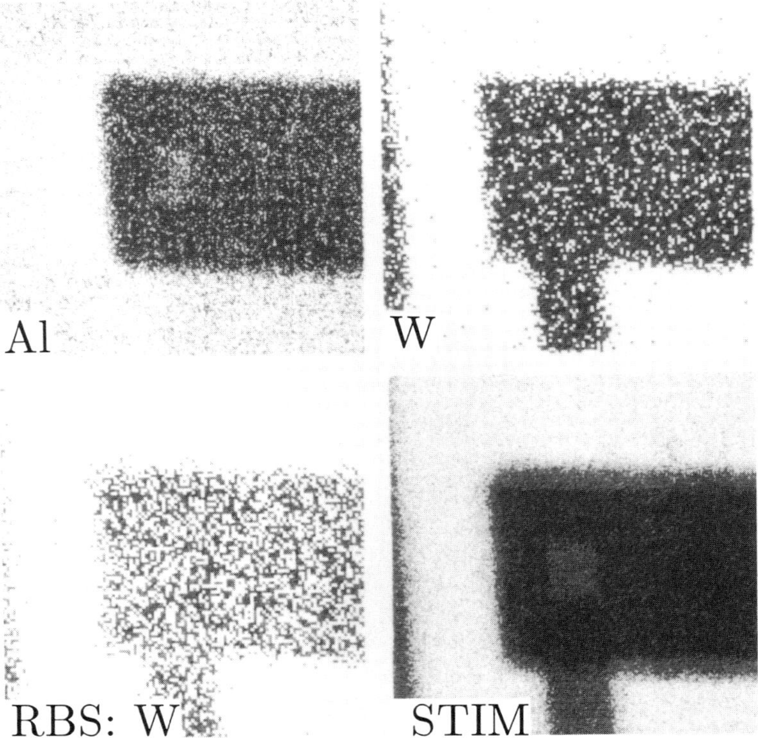

Figure 7.27. 25×25 μm^2 area of a second metallization contact hole showing the aluminum and tungsten PIXE images, backscattering tungsten image, and STIM image. Reprinted from Ref. 63 with kind permission from Elsevier Science B.V., Amsterdam, The Netherlands.

backscattering spectrometry, and STIM, of a second metallization contact hole that was 80 μm to the right of the previous area. Backscattering spectrometry was used to give the composition and depth distribution of the microcircuit layer structure at specific areas as small as 0.5 μm across using STIM images to locate the required areas [62]. Three examples of such backscattering spectra are shown in Figure 7.28 using 3 MeV ^{1}H ions [65]. The major peaks present are identified, and the large oxygen peaks arise from resonant scattering of 3 MeV ^{1}H ions from oxygen present within SiO_2 layers. In spectrum A, only the surface passivation layer and a deeper silicon layer are detectable. In spectrums

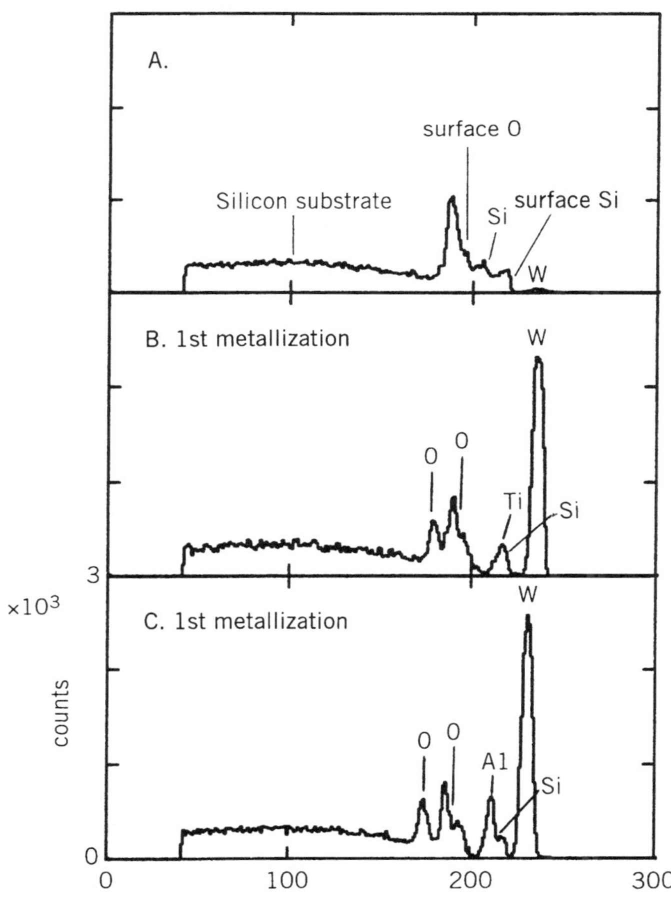

Figure 7.28. Three backscattering spectra using 3 MeV ^{1}H ions at (A) a region with no metallization, (B) a first metallization region, and (C) a second metallization region. The particle detector resolution was 17 keV and was positioned at a backward angle of 170° to the beam. Reprinted from Ref. 65 with permission (© 1991 IOP Publishing Ltd., Bristol, U.K.).

B and C, the metallization layers are indicated. The measured spectra were simulated to determine the composition and depth distribution at each position. Spectra measured from inside first- and second-metallization contact holes in Figures 7.26 and 7.27 indicated that the first metallization layer was 0.7 μm thick and the second metallization aluminium layer was 0.9 μm thick.

A combination of PIXE, backscattering spectrometry, and STIM was able to determine the thickness and spatial distribution of sub-surface metal layers without the need to use etching or sectioning techniques. The rapid imaging capability of STIM allowed it to be used as a method for locating micron-sized areas for subsequent backscattering spectrometry, and to locate a misplaced feature, which was subsequently identified with PIXE.

This same device structure was also analyzed with SIMS [65], and the complementary nature of the results obtained is now briefly discussed. The left side of Figure 7.29 shows the method of SIMS analysis used here. A large beam current of 30 keV gallium ions was used to dig a pit through a metal track where there was a contact between the first and second metallizations, which is also shown in Figure 7.27. The sample was then rotated, and the exposed layer

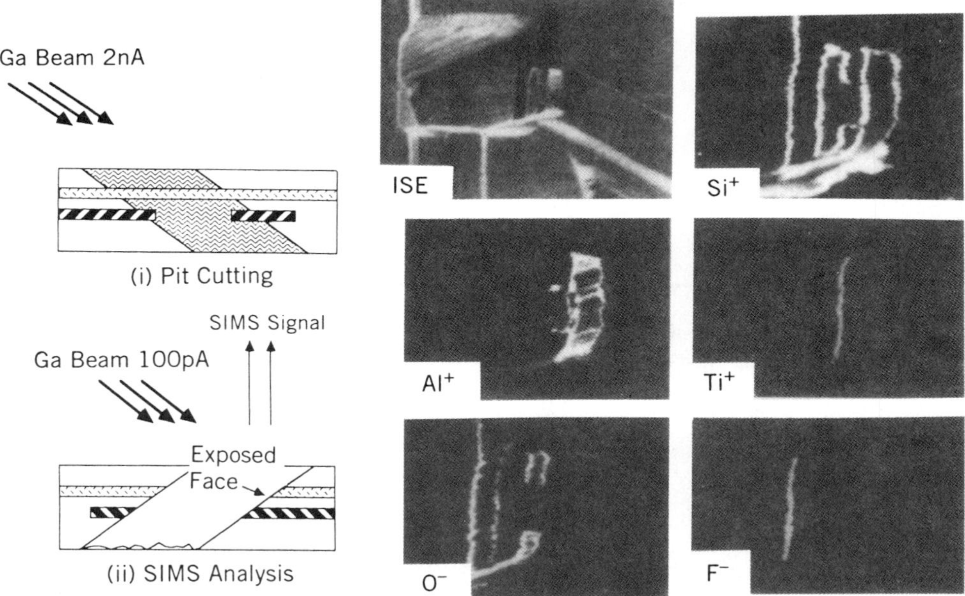

Figure 7.29. SIMS images of the same device structure. The left half shows the method of analysis and the right half shows a 50×50 μm^2 region imaged with ion induced electrons (ISE) and five ion images from the central 20×20 μm^2 area. Reprinted from Ref. 65 with permission (© 1991 IOP Publishing Ltd., Bristol, U.K.).

structure was imaged using the same ion beam but at a lower current. Figure 7.29 shows an ion induced electron image (labeled ISE) of this area and also five SIMS images at higher resolution showing the different elements present. Bright regions correspond to a large measured signal and dark to a small measured signal. The silicon image shows a low signal from the silicon substrate and bright contrast from the substrate surface and around the metal tracks where they were in contact with the insulating oxide. The aluminium image shows the second metallization together with a contact through to the lower tungsten layer. The yield from the tungsten present was too low to generate an image of the first metallization. The insulating oxide layers are clearly shown in the oxygen image, and a strong fluorine signal was found, resulting from a problem with the process used to deposit the tungsten layer. These SIMS images highlight a very powerful means of imaging the layer structure with high spatial resolution. However, great care must be taken with the interpretation of such SIMS results, since the sputtering yield varies widely between elements, resulting in the poor detection of tungsten, for example. In comparison, tungsten is detected very clearly using PIXE and backscattering spectrometry, emphasizing the complementary nature of these two forms of elemental analysis using ion beams.

7.2.2. Backscattering Spectrometry of a Silicon-on-Insulator Device

This device structure was fabricated by zone melt recrystallization of the surface silicon layer using selective epitaxial growth from seed windows in the SiO_2 layer. The seed windows are typically 2 to 8 μm wide with spacings of 20 to 100 μm. The recrystallized surface layer had the same <100> orientation as the substrate, and the device structure was similar to that shown in the cross-section SEM image shown in Figure 7.1c. The recrystallized surface layer was oxidized down to the original buried SiO_2 layer. A schematic cross-section of the structure and a plan view of a 25×25 μm^2 area are shown in Figure 7.30.

Three backscattering spectra [66] corresponding to the regions A, B, and C are also shown in Figure 7.30 for 3 MeV ^{1}H ions. Spectrum A shows an aluminum surface peak from the metal interconnects and a peak from oxygen associated with the buried SiO_2 layer. However the signal from the oxygen in the buried SiO_2 layer exists in all the spectra, because it underlies the entire analysis region. Spectrum B shows surface oxygen extending down to the bottom of the original buried SiO_2 layer, indicating that the original surface silicon layer has been fully oxidized. Spectrum C shows a surface oxygen peak from the passivating oxide, as well as an oxygen peak from the buried SiO_2 isolation layer. Spectrum C has been simulated for the nominal device structure, shown as the smooth curve. The small discrepancy between the simulated and experimental spectra is due to the slightly different detector geometry used between the simulation and the empirical elastic scattering cross-sections used

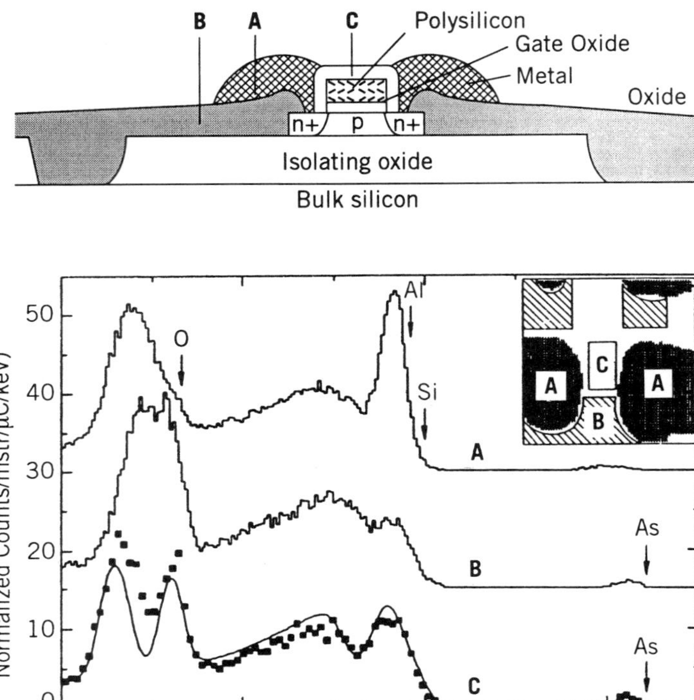

Figure 7.30. Cross-section of the device structure (upper schematic). Three backscattering spectra from the positions indicated in the inset plan view of a 25×25 μm^2 area of this device. Reprinted from Ref. 66 with permission (© 1991 IOP Publishing Ltd., Bristol, U.K.)

for silicon and oxygen. Nevertheless, it was concluded from these spectra that the nominal device structure was in good agreement with the measured structure.

7.2.3. Other Scanning Transmission Ion Microscopy Images of Device Structures

Particle induced X–ray emission has poor elemental sensitivity to aluminum in the presence of a large amount of silicon, since the aluminium K and silicon K X–rays energies are close together. This results in noisy aluminum PIXE images, as can be seen in Figure 7.26. Figure 7.31 shows three STIM images of a silicon-on-insulator device [64], essentially consisting of a silicon substrate

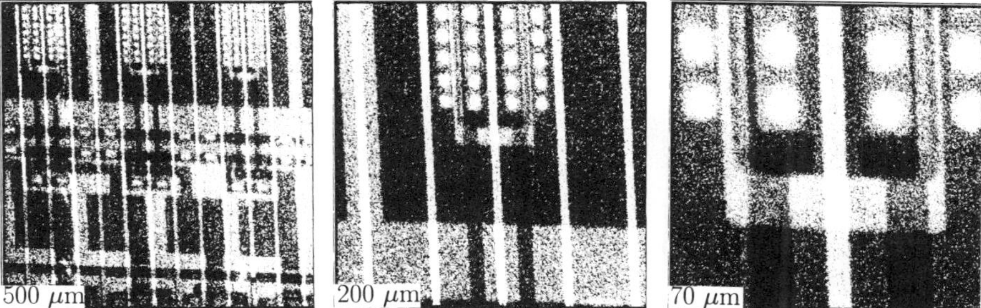

Figure 7.31. STIM images of a device with a single aluminium metallization, with scan sizes shown in the bottom left of each image. Reprinted from Ref. 64 with kind permission from Elsevier Science B.V., Amsterdam, The Netherlands.

with a 1 μm thick aluminum metallization layer, with a thick SiO_2 passivation layer on the surface. The device was thinned and polished from the rear to a thickness of 20 μm. These images were measured using 3 MeV ^{1}H ions with a data collection time of 10 min at a beam current of 0.3 fA. The dark regions correspond to the distribution of the aluminium metallization. A longer data collection time was required for these STIM images than those in the previous section because of the lower energy loss contrast due to the presence of only a single metallization. However, this does show the ability of STIM to image a 1 μm thick subsurface aluminium metallization layer, which PIXE is not good at doing.

Figure 7.32 shows STIM images of a 4 Mbit DRAM with 1 μm diameter trench holes in a silicon substrate [64]. The surface layers were removed prior to analysis using hydrogen fluoride, and the trench holes have been etched to a depth of 4 μm. The device was then mechanically thinned from the rear surface and polished to a thickness of 20 μm, and the STIM images were collected using a beam current of 0.3 fA of 3 MeV ^{1}H ions for 5 minutes. Figure 7.32a shows a 20 $\times$ 20 μm^2 STIM image in which the circular areas are the DRAM trench holes that appear light because the ^{1}H ions lose less energy in passing through these regions. Figure 7.32a also shows a three-dimensional projection of this image. Figure 7.32b shows a 10 $\times$ 10 μm^2 STIM image of this same area and also a three-dimensional projection viewed from the side. This demonstrates the ability of STIM to produce images showing the interior profile of the trench walls without the need to produce cross-sectional slices of the sample. Section 6.5 shows IBIC images of similar DRAM devices.

7.2.4. Three-Dimensional Backscattering Images of Buried Layers

A nuclear microprobe capable of focusing a variety of ions with an energy between 0.5 to 6 MeV has been constructed [67]. It is used for maskless ion implantation, material modification [68], and backscattering spectrometry, uti-

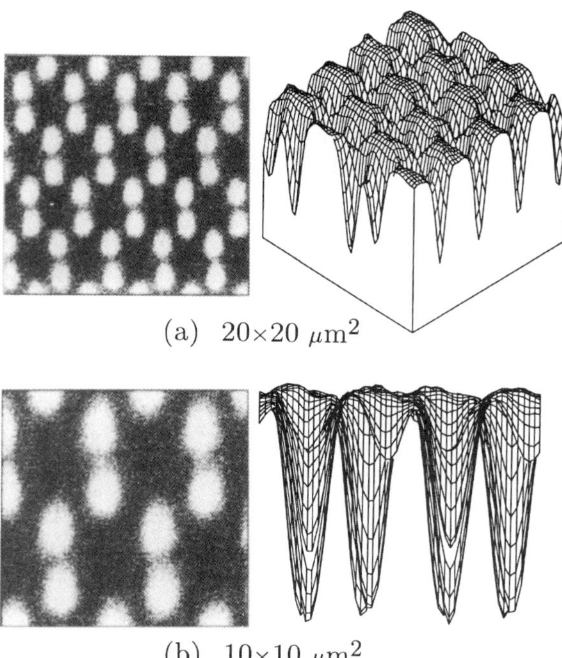

Figure 7.32. (a) 20×20 μm^2 STIM image of a thinned DRAM, together with a three-dimensional representation, which is viewed partially from above. (b) 10×10 μm^2 STIM image and a three-dimensional representation viewed in cross-section. Reprinted from Ref. 64 with kind permission from Elsevier Science B.V., Amsterdam, The Netherlands.

lizing the higher elemental sensitivity of heavy-ion beams compared with that of light-ion beams. Three-dimensional backscattering images of buried device layers obtained using this microprobe are shown here.

For the first example, the silicon substrate was masklessly implanted with 5.6×10^{16} 0.6 MeV gold ions and 10^{17} 1.8 MeV gold ions through a 20 nm thick gold surface pad into an area approximately 25×5 μm^2, as shown in Figure 7.33a. A focused 3 MeV carbon beam was then used to generate a backscattering spectrum at each pixel within the scanned area, which was measured using an event-by-event data acquisition method. Figure 7.33b shows a three-dimensional backscattering image of a 50×50 μm^2 area over the device surface. The vertical axis represents the number of counts in the energy channel, corresponding to the number of backscattered carbon ions from the implanted gold layers. The scale by the right side shows the backscattered carbon intensity at each measured energy channel. Channel 210 is equivalent to carbon backscattered from gold on the device surface, and lower channel numbers represent the implanted gold in the substrate. By further manipulating the data set, two-

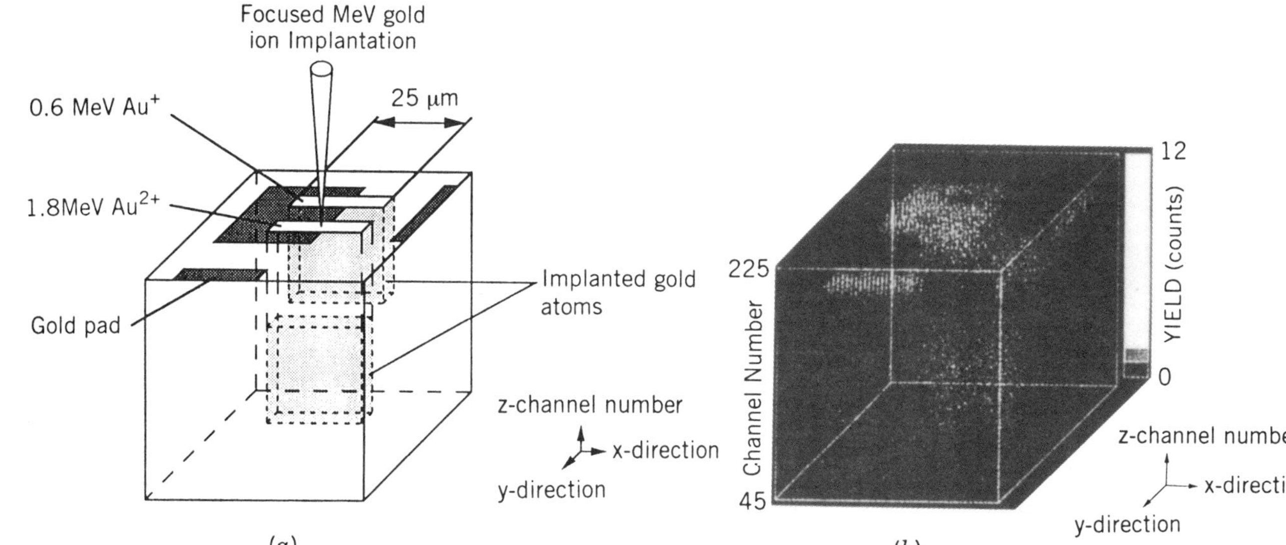

Figure 7.33. (a) Schematic illustration of the masklessly implanted sites A and B implanted at 1.8 and 0.6 MeV, respectively. The implanted area was $25 \times 25 \ \mu m^2$. (b) three-dimensional backscattering images of the implanted gold atoms. The location of the implanted and surface gold atoms is shown as the light-colored dots against a black background. Reprinted from Ref. 69 with kind permission from Elsevier Science B.V., Amsterdam, The Netherlands.

dimensional images viewed in cross-section and in plan view can be extracted, which show the amounts of gold implanted to different depths [69].

Figure 7.34a shows a schematic of a different test structure for multilayered wiring with a gold stripe width of 10 μm [70]. Figure 7.34b shows a three-dimensional backscattering image of this structure obtained with an ion energy of 400 keV and a beam spot diameter of 3 μm. The image has been corrected for the lower energy of the ions penetrating through the top gold layer into the underlying gold layer. Figure 7.34c shows the second gold-stripe layer after removal of the top layer by software, indicating a gap in the gold metallization caused by a process failure. Another example of this process of generating three-dimensional backscattering images from devices is given in Ref. 71.

In conclusion, a nuclear microprobe can image the depletion regions of microelectronic devices using IBIC, and investigate the susceptibility of devices

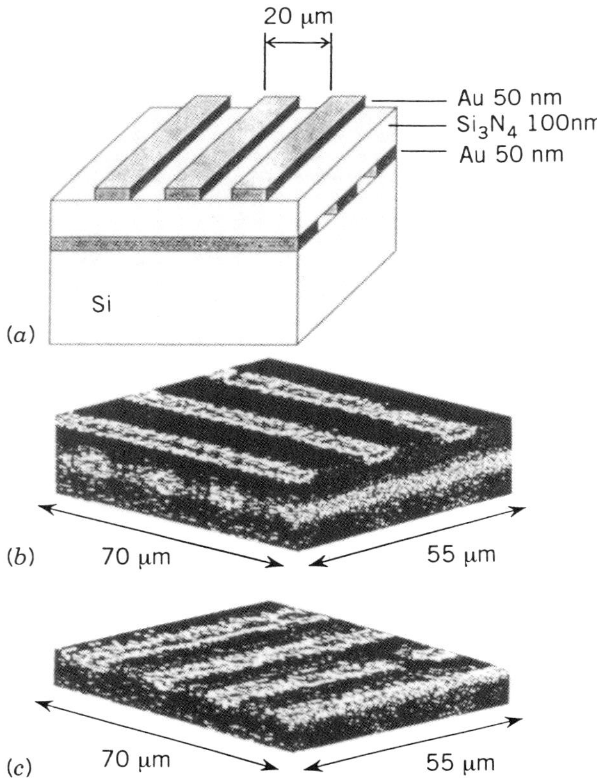

Figure 7.34. (a) A multilayer wiring test structure. (b) Three-dimensional backscattering image and (c) second layer of gold stripes after removing the top layer by software processing. Reprinted from Ref. 70 with kind permission from Elsevier Science B.V., Amsterdam, The Netherlands.

to SEUs using SEU imaging. Device layer structures and compositions can be quantified and imaged using a combination of backscattering spectrometry, PIXE, and STIM. None of these focused MeV ion beam techniques requires the surface layers to be removed, because of the long range and low lateral scattering of MeV light ions. Also, STIM can be used to locate micron-sized device areas for backscattering spectrometry.

REFERENCES

1. S.M. Sze, ed., *VLSI Technology*, McGraw–Hill, Singapore (1983).

2. D.J. Elliot, *Integrated Circuit Fabrication Technology*. McGraw–Hill, New York (1982).

3. D.F. Horne, *Microcircuit Production Technology*. Adam Hilger, Bristol (1986).

4. J.F. Ziegler, ed., *Ion Implantation, Science and Technology*, Academic, New York (1984).

5. M.J. Howes and D.V. Morgan, *Reliability and Degradation*. Wiley, Chichester (1981).

6. B.P. Richards and P.K. Footner, *GEC J. Res.* **1:**74 (1983).

7. B.P. Richards and P.K. Footner, *GEC J. Res.* **5:**1 (1987).

8. O.C. Wells. *Scanning Electron Microscopy*. McGraw–Hill, New York (1974).

9. L. Reiner, *Scanning Electron Microscopy*, Springer–Verlag, New York (1985).

10. J.I. Goldstein and H. Yakowitz, *Practical Scanning Electron Microscopy*. Plenum, New York (1975).

11. B.P. Richards and P.K. Falkner, *Microelectronics J.* **15:**5 (1984).

12. P.K. Footner, B.P. Richards, C.E. Stephens, and C.T. Amos, *24th Annual Conference on Reliability Physics* **24:**102 (1986).

13. P.B. Hirsch, A. Howie, R.B. Nicholson, D.W. Pashley, and M.J. Whelan, *Electron Microscopy of Thin Films*. Krieger, New York (1977).

14. G. Thomas and M.J. Goringe, *Transmission Electron Microscopy of Materials*. Wiley, New York (1979).

15. N.G. Chew and A.G. Cullis, *Ultramicroscopy* **23:**175 (1987).

16. A. Benninghoven, *Secondary Ion Mass Spectrometry*. Springer–Verlag, Berlin (1979).

17. A. van Oostrom, *Surf. Sci.* **89:**615 (1979).

18. M. Dellith, G.R. Booker, B.O. Kolbesen, W. Bergholz, and F. Gelsdorf, *Inst. Phys. Conf. Ser.* **134:**235 (1993).

19. H. Ahmed, *Microelectr. Eng.* **8:**235 (1988).

20. E. Druet and P.–H. Albarede, *Inst. Phys. Conf. Ser.* **134:**617 (1993).

21. R. Levi Seti, J.M. Chabala, Y. Wang, P. Hallegot, and C. Girod–Hallegot, In D.B. Williams, A.R. Pelton, and R. Gronsky, eds., *Images of Materials*. Oxford University Press, New York (1991).

22. G.A.D. Briggs and M. Hope, In D.B. Williams, A.R. Pelton, and R. Gronsky, eds., *Images of Materials*. Oxford University Press, New York (1991).

23. H.J. Leamy, *J. Appl. Phys.* **53:**R51 (1982).

24. C.J.R. Sheppard, *Scanning Microsc.* **3:**15 (1989).

25. C.J. Wu and D.B. Wittry, *J. Appl. Phys.* **49:**2827 (1978).

26. J.C. Chi and H.C. Gatos, *J. Appl. Phys.* **50:**3433 (1979).

27. K.F. Galloway, K.O. Leedy, and W.J. Keery, *IEEE Trans. Parts Hybrids Packag.* **12:**231 (1976).

28. D.C. Newbury, D.C. Joy, P. Echlin, C.E. Fiori, and J.I. Goldstein, *Advanced Scanning Electron Microscopy and X–ray Microanalysis.* Plenum, New York (1986).

29. L.C. Feldman and J.W. Mayer, *Fundamentals of Surface and Thin Film Analysis,* North–Holland, New York (1986).

30. C.R.M. Grovenor, *Microelectronic Materials.* Adam Hilger, Bristol (1989).

31. M.B.H. Breese, P.J.C. King, G.W. Grime, and F. Watt, *J. Appl. Phys.* **72**(6):2097 (1992).

32. M.B.H. Breese, G.W. Grime, and F. Watt, *Nucl. Instr. Meth.* **B77:**301 (1993).

33. M.B.H. Breese, C.H. Sow, D.N. Jamieson, and F. Watt, *Nucl. Instr. Meth.* **B85:**790 (1994).

34. M.B.H. Breese, J.S. Laird, G.R. Moloney, A. Saint, and D.N. Jamieson, *Appl. Phys. Lett.* **64**(15):1962 (1994).

35. G. Bertolini and A. Coche, eds., *Semiconductor Detectors,* North–Holland, Amsterdam (1968).

36. M.B.H. Breese, A. Saint, F.W. Sexton, H.A. Schöne, K.M. Horn, B.L. Doyle, J.S. Laird, and G.J.F. Legge, *J. Appl. Phys.* **77**(8):3734 (1995).

37. A. Saint, M.B.H. Breese, and G.J.F. Legge, unpublished.

38. M.B.H. Breese, G.W. Grime, F. Watt, and R.J. Blaikie, *Vacuum* **44:**175 (1993).

39. R.J. Blaikie, J.R.A. Cleaver, H. Ahmed, and K. Nakazato, *Appl. Phys. Lett.* **60:**1618 (1992).

40. T.C. May and M.H. Woods, *IEEE Trans. Electron Devices* **26:**2 (1979).

41. A.R. Knudson and A.B. Campbell, *Nucl. Instr. Meth.* **218:**625 (1983).

42. P.J. McNulty, W.G. Abdel–Kader, A.B. Campbell, A.R. Knudson, P. Shapiro, F. Eisen, and S. Roosild, *IEEE Trans. Nuc. Sci.* **31:**1128 (1984).

43. P.J. McNulty, *Nucl. Tracks Rad. Meas.* **16:**197 (1989).

44. S. Buchner, K. Kang, W.J. Stapor, A.B. Campbell, A.R. Knudson, P. MacDonald, and S. Rivet, *IEEE Trans. Nuc. Sci.* **37:**1825 (1990).

45. L.D. Flesner and R. Zuleeg, *Microelectron. Eng.* **12:**163 (1990).

46. K.M. Horn, B.L. Doyle, D.S. Walsh, and F.W. Sexton, *Scanning Microscopy* **5:**969 (1992).

47. K.M. Horn, B.L. Doyle, F.W. Sexton, J.S. Laird, A. Saint, M. Cholewa, and G.J.F. Legge, *Nucl. Instr. Meth.* **B77:**355 (1993).

48. F.W. Sexton, K.M. Horn, B.L. Doyle, J.S. Laird, M. Cholewa, A. Saint, and G.J.F. Legge, *Nucl. Instr. Meth.* **B79:**436 (1993).

49. S. Metzger, J. Dreute, W. Heinrich, H. Rocher, B.E. Fischer, R. Harboe–Sorensen, and L. Adams, *IEEE Conf. Records,* No. 93TH0616–3 (in press).

50. T. Matsukawa, K. Noritake, M. Koh, K. Hara, M. Goto, and I. Ohdomari, *Nucl. Instr. Meth.* **B77:**239 (1993).

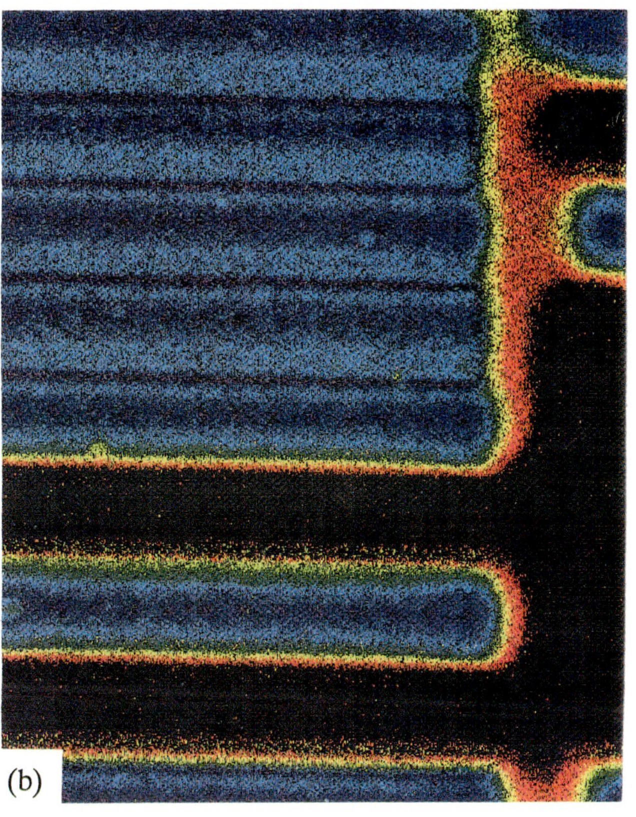

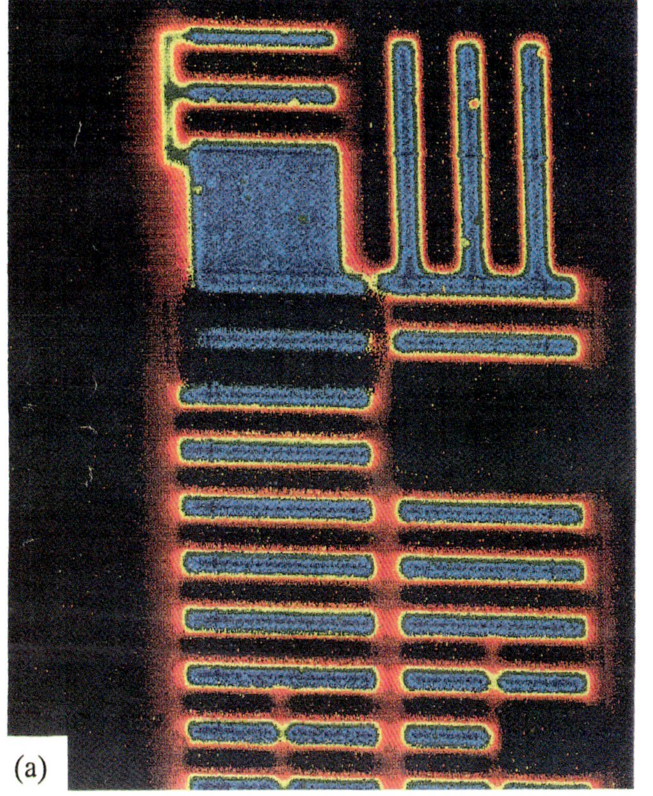

Color Plate 1. 2.3 MeV ^{4}He IBIC images of the EPROM device, with a scan size of (a) 200×200 μm^2 and (b) 60×60 μm^2. Charge pulses range from dark blue for the largest, to light blue, to yellow, to red for the smallest. See Section 7.1 for more details.

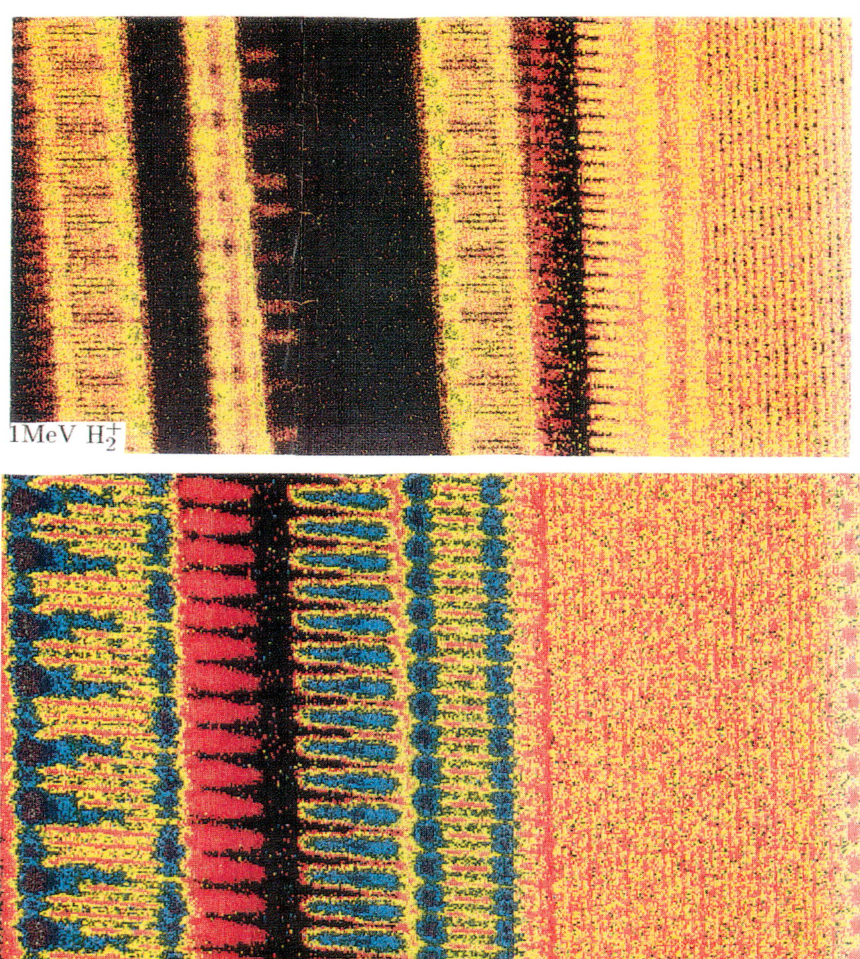

Color Plate 2. Two large area IBIC images of the central logic and the memory fields of the SA3002 device. (Top) 1 mm wide 1 MeV $^1H_2$ ion image and (bottom) 0.5 mm wide 2 MeV ^{4}He image from the lower right quadrant of (top). Charge pulses range from dark blue for the largest, to light blue, to yellow, to red for the smallest. See Section 7.1 for more details.

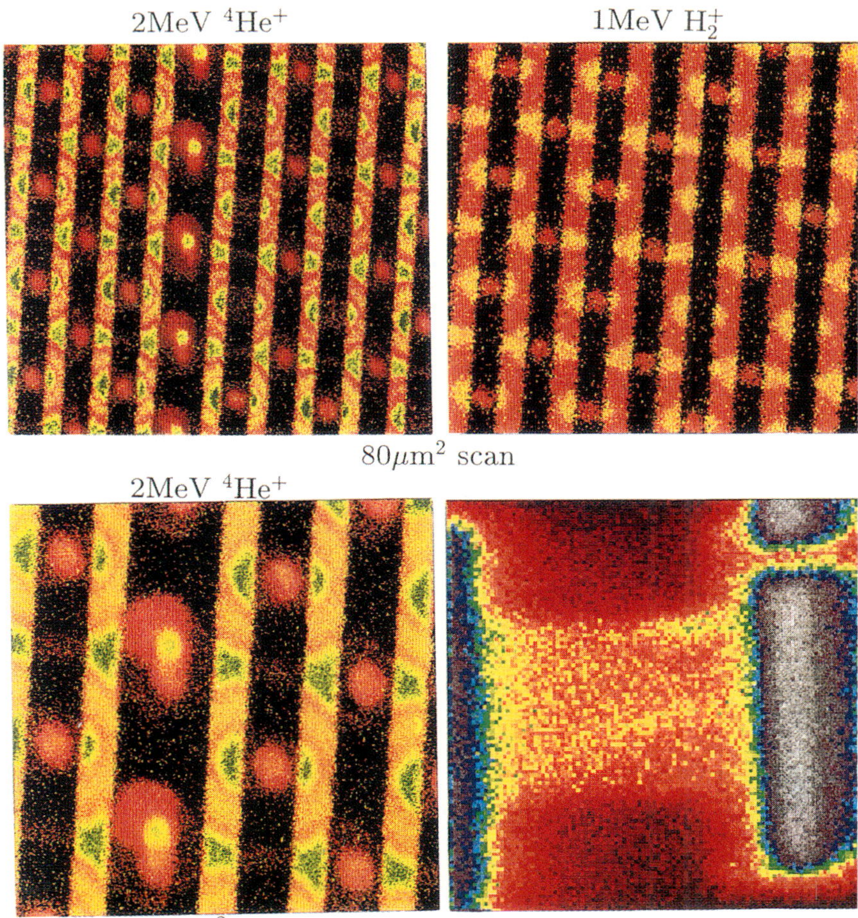

Color Plate 3. *Top row and bottom left:* IBIC images showing the average charge pulse height of a 80 × 80 μm² region of the device memory field for 2 MeV ^{4}He ions and 1 MeV ^{1}H$_2$ ions and a 40×40 μm² region for 2 MeV ^{4}He ions. See Section 7.1 for more details. *Bottom right:* A CCM image of a photodetector device fabricated in HgCdTe. Dechanneling from linear growth defects is visible. Note also the break in the gold metallization contact in the upper right corner. Scan size was 150×150 μm². See Section 9.3 for more details. Charge pulses range from dark blue for the largest, to light blue, to yellow, to red for the smallest.

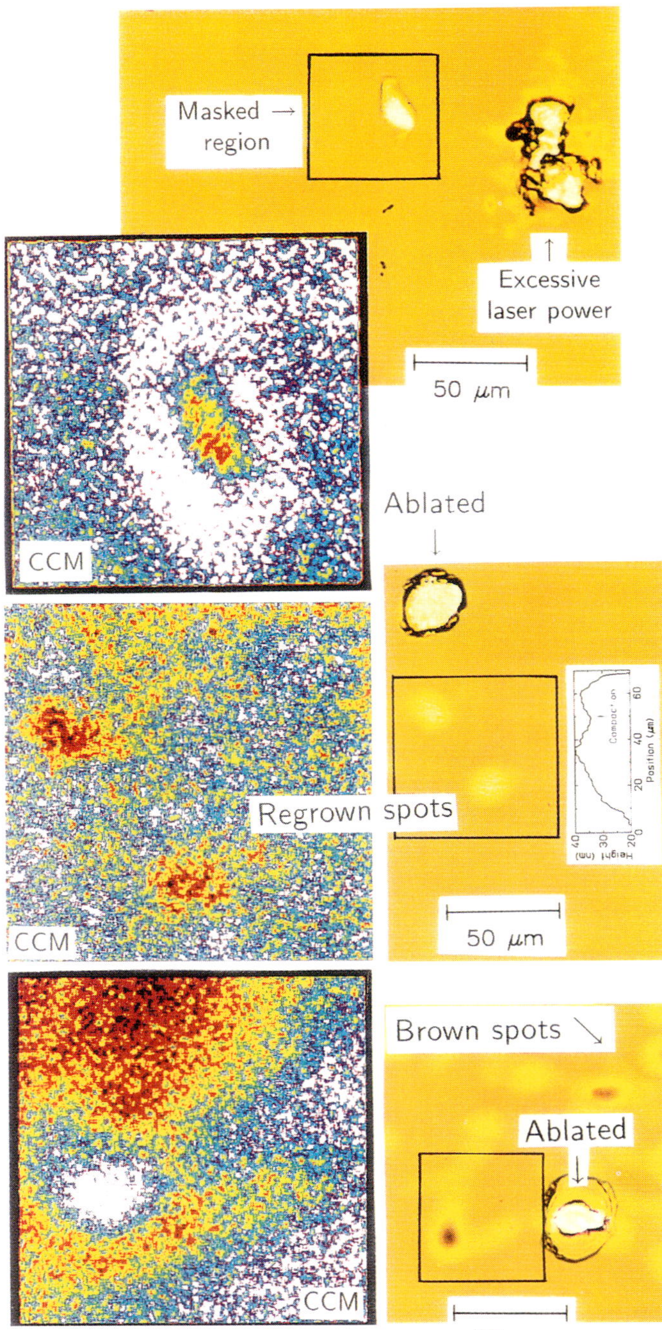

Color Plate 4. Laser-annealed ion-implanted (3×10^{15} carbon ions/cm^2 at 2.8 MeV) diamond single crystals, imaged with CCM (left) and optical microscopy (right). The CCM images are from regions indicated by the black boxes in the optical images. *Top:* CCM image of a masked region that shows the dechanneling from swollen edges. *Middle:* CCM image of laser-annealed regions. *Bottom:* CCM image of a region where excessive laser power has produced a buried, graphitic zone. In the CCM images, white/blue represent high yield (poor channeling) and red/black represent low yield (good channeling). More details are given in Section 9.2 and Reference 9 of Chapter 9.

51. K.H. Lee, J.C. Desko, R.A. Kohler, C.W. Lawrence, W.J. Nagy, J.A. Shimer, S.D. Steenwyk, R.E. Anderson, and J.S. Fu, *IEEE Trans. Nuc. Sci.* **34:**1460 (1987).

52. J.P. Kreskovsky and H.L. Grubin, *IEEE Trans. Nuc. Sci.* **32:**4140 (1985).

53. R.A. Bardos, G.F. Moorhead, G.N. Taylor, J.S. Laird, and A. Saint, *Nucl. Instr. Meth. B* (unpublished).

54. J.S. Williams, In F. Watt and G.W. Grime, eds., *Principles and Applications of High Energy Ion Microbeams.* Adam Hilger, Bristol (1987).

55. J.R. Bird and J.S. Williams, eds. *Ion Beams for Materials Analysis*, Academic Press, Orlando (1989).

56. J.S. Williams, J.C. McCallum, and R.A. Brown, *Nucl. Instr. Meth.* **B30:**480 (1988).

57. W.G. Morris, S. Fesseha, and H. Bakhru, *Nucl. Instr. Meth.* **B24/25:**635 (1987).

58. W.G. Morris, H. Bakhru, and A. Haberl, *Nucl. Instr. Meth.* **B15:**661 (1986).

59. H. Bakhru, W.G. Morris, and A. Haberl, *Nucl. Instr. Meth.* **B10/11:**697 (1985).

60. B.L. Doyle, Sandia National Laboratory, Rept. SAND–87–1138C (1987).

61. A. Kinomura, T. Lohner, Y. Katayama, M. Takai, H. Ryssel, R. Schork, A. Chayahara, Y. Horino, K. Fuji, and M. Satou, *Nucl. Instr. Meth.* **B77:**369 (1993).

62. M.B.H. Breese, F. Watt, G.W. Grime, and P.J.C. King, *Inst. Phys. Conf. Ser.* **117:**101 (1991).

63. M.B.H. Breese, J.P. Landsberg, P.J.C. King, G.W. Grime, and F. Watt, *Nucl. Instr. Meth.* **B64:**505 (1992).

64. M.B.H. Breese, G.W. Grime, and F. Watt, *Nucl. Instr. Meth.* **B75:**341 (1993).

65. M.B.H. Breese, J.A. Cookson, H.E. Bishop, and S. Greenwood, *Semiconductor Sci. Technol.* **6:**325 (1991).

66. D.N. Jamieson, G.W. Grime, F. Watt, and D.A. Williams, *Inst. Phys. Conf. Ser.* **100:**87 (1989).

67. Y. Horino, A. Chayahara, M. Kiuchi, K. Fujii, M. Satou, and M. Takai, *Nucl. Instr. Meth.* **B59/60:**139 (1991).

68. Y. Horino, A. Chayahara, M. Kiuchi, K. Fujii, M. Satou, and M. Fujimoto, *Jpn. J. Appl. Phys.* **29:**1230 (1990).

69. Y. Mokuno, Y. Horino, A. Chayahara, M. Kiuchi, K. Fujii, M. Satou, and M. Takai, *Nucl. Instr. Meth.* **B77:**373 (1993).

70. M. Takai, *Nucl. Instr. Meth.* **B85:**664 (1994).

71. C.L. Churms and R. Pretorius, *Nucl. Instr. Meth.* **B85:**699 (1994).

8

CRYSTAL DEFECT IMAGING WITH A NUCLEAR MICROPROBE

8.1. INTRODUCTION

This chapter is an account of observations of crystal defects (dislocations and stacking faults) made using the Oxford nuclear microprobe. The primary technique involved is CSTIM, but with a section also demonstrating IBIC images of dislocations.

The study of crystal defects is important for our understanding of the mechanical and electrical properties of crystalline materials. The presence of dislocations drastically influences the growth rate of crystals from supersaturated solution and the shear stress at which plastic deformation occurs [1]. The strain field around dislocations attracts impurity atoms, which can then form precipitate particles. In semiconducting crystals, these can increase the dislocation's conductivity, affecting the performance of *pn* junctions subsequently manufactured in the crystal [2]. Dislocations also result in new electronic states being introduced near the center of a semiconducting material's band gap, leading to locally enhanced recombination of charge carriers. Dislocations and stacking faults can also be used beneficially in semiconducting crystals. Defects deliberately introduced into non-vital areas of a crystal can attract impurity atoms away from regions where devices are to be made, a process called gettering [3].

This chapter presents CSTIM and IBIC studies of misfit dislocations and stacking faults in silicon-based semiconductors. As a precursor to this, Section 8.2 describes other analytical techniques commonly used for the observation of defects in semiconductor crystals. In Sections 8.3, 8.4, and 8.5, CSTIM investigations of $Si_{1-x}Ge_x/Si$ crystals are presented, and Section 8.6 demonstrates how CSTIM can be used to image stacking faults in a silicon crystal. Section 8.7 describes IBIC images of dislocations in a third $Si_{1-x}Ge_x/Si$ crystal. The chapter finishes by comparing CSTIM and IBIC with the defect imaging techniques introduced at the beginning of the chapter. As described in Section 5.5.3, all of the CSTIM images presented in this chapter are printed in grey scale with darker greys representing higher average transmitted energy loss.

The nuclear microprobe results described in this chapter have been taken from Refs. 4 to 10 for CSTIM investigations of $Si_{1-x}Ge_x/Si$ dislocations, Refs. 11 to 13 for CSTIM results from stacking faults, and Refs. 14 and 15 for the IBIC results.

8.2. CRYSTAL DEFECT IMAGING TECHNIQUES

In this section, the various techniques that can be used for the imaging of crystal defects, such as dislocations and stacking faults, are described. Each have their own strengths and weaknesses, determined by such factors as the spatial resolution of the technique, the amount of sample preparation required, the depth of defects that can be imaged and whether information on dislocation Burgers vectors (Section 5.2.1) and stacking fault translation vectors (Section 5.2.2) can be obtained.

8.2.1. Transmission Electron Microscopy

Probably the most powerful and widely-used technique for studying crystal defects is transmission electron microscopy (TEM) [16,17]. Conventional TEM employs a beam of 60 to 400 keV electrons. The beam is not scanned, but illuminates an area of the sample as in an optical microscope. The sample must be thin enough to be electron transparent, and the transmitted electrons are used to generate images with contrast caused by local variations in sample structure and composition.

Image contrast for defect analysis is caused by elastic scattering (diffraction) of the electron beam by the crystal lattice planes. Images can be produced either using the intensity of the undiffracted electrons (bright field imaging) or that of the electrons scattered into a particular diffraction spot (dark field imaging). For example, in a dark field image, regions of the crystal in which the lattice planes are oriented for diffraction show up brightly, whereas regions where the planes are not diffracting (such as where they are locally distorted by dislocations) are dark. Transmission electron microscopy is able to produce images that show variations in crystallographic orientation (different crystal grains, precipitate particles) or that reveal lattice strain (around dislocations or precipitate particles). Dislocation Burgers vectors and stacking fault translation vectors can be found by using different sets of lattice planes for diffraction; depending on its Burgers vector, a dislocation will cause a distortion of some sets of lattice planes but not others. Typical image widths for dislocations, determined by the extent of the strain field around the defect, are 15 to 20 nm, although the technique of weak beam imaging can produce image widths as low as 1.5 nm. Examples of plan-view and cross-sectional TEM images are given in Figure 8.16.

The relative phases of the diffracted and forward-scattered beams can also be used to produce images with TEM. Under the proper conditions, such phase contrast images can reveal the atomic structure of the material under investigation, and Figure 8.1 shows an atomic resolution image of a stacking fault in silicon from Ref. 18. Images with a resolution of less than 0.2 nm can be achieved with careful sample preparation and a microscope that is highly mechanically and electrically stable.

Samples for TEM are normally prepared in the form of 3 mm diameter discs. These have to be produced either by chemical or electrochemical thinning, or

Figure 8.1. High-resolution TEM image of an extrinsic fault in silicon. The extra half plane of atoms enters near the center of the top of the image. Note the shift in atomic rows on crossing the fault and the plane rotation surrounding the partial dislocation at the termination of the extra half plane. The closest separation between white dots is 0.32 nm. Reprinted with permission from Ref. 18.

by ion beam milling, to make a region for analysis that normally must be less than 1 μm thick. Care must be taken during the thinning process to minimize the likelihood of existing dislocations moving and new dislocations being introduced into the sample. The sample area that can be analyzed is limited to the region that has been thinned sufficiently and a problem for TEM is the small analytical volume.

8.2.2. Scanning Electron Microscopy Techniques

The electron beam analogue of the nuclear microprobe is the scanning electron microscope (SEM). A SEM provides a focused, scanned beam of electrons with energy in the 1 to 40 keV range for analytical investigations of materials using the various signals produced by the beam-sample interactions. Three SEM techniques can be used for crystal defect imaging.

8.2.2.1. *Electron Channeling Contrast Imaging (ECCI)*   TEM uses diffraction of the beam electrons to produce images of defects. Electron diffraction can also be used to produce contrast in an SEM by detection of electrons that have been backscattered from near the sample surface rather than those transmitted through the sample.

If the sample is oriented so that a set of lattice planes are just away from the exact position for diffraction (Bragg position), the backscattered electron (BSE) intensity is either increased or decreased (depending on which side of the Bragg position the beam is positioned) owing to anomalous absorption or transmission of electrons [19]. These changes in the intensity of BSE emission are used to produce images of crystal defects. A crystal is oriented so that a particular set of planes is at the exact Bragg position and a region of the sample is scanned by the electron beam. Any local distortion of the lattice planes away from the Bragg position is detected as an increase or decrease in the BSE yield. Images showing the locations of dislocations and other crystal defects can be generated. Dislocation images with a width of the order of 10 nm can be produced, and Burgers vector analysis is possible [20].

Bunches of misfit dislocations in a $Si_{1-x}Ge_x/Si$ crystal have been imaged by ECCI [21]. The dislocations were 0.7 μm below the sample surface, and the widths of the bunches in the images were approximately 1 μm. The $Si_{0.85}Ge_{0.15}/Si$ mesa sample, of which CSTIM analysis is described in Section 8.4 below, has also been studied using ECCI [22]. An ECCI image of a rectangular mesa, 20 μm wide along its shorter side, is shown in Figure 8.2. In this study, electron diffraction was also used to measure the amount of lattice plane rotation caused by elastic relaxation of the $Si_{0.85}Ge_{0.15}$ layer on the top of a 10 μm wide mesa [22].

Electron channeling contrast imaging has the advantages of high spatial resolution and the ability to analyze bulk samples with no preparation. However, the information obtained comes only from the top 200 nm or so of the sample surface [23] as only those backscattered electrons that have lost little energy

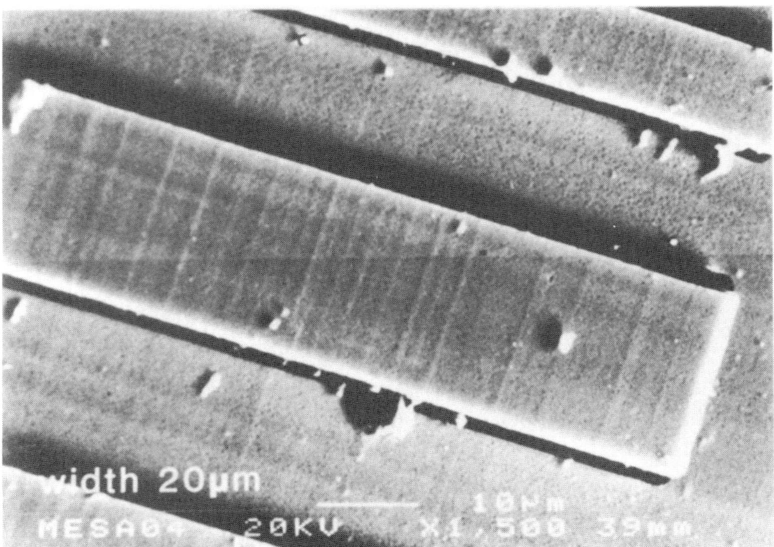

Figure 8.2. ECCI image of dislocations in a 20 μm wide mesa on a portion of the $Si_{0.85}Ge_{0.15}$/Si wafer, CSTIM studies of which are presented in Section 8.4. Image taken in 10–15 sec with a few nanoamps of 20 keV electrons at the Bragg position of the (220) planes. The sample was tilted by approximately 50° about the (220) plane normal and so is foreshortened in the horizontal direction. Reprinted with permission from Dr. A. J. Wilkinson, Department of Materials, University of Oxford.

contribute to the image contrast. For the dislocation bunches 0.7 μm below the sample surface [21], the authors believed that they were rendered visible by their long-range strain field which was present in the top 100 nm of the sample. By varying the electron beam energy, depth-resolved information from the near-surface region of a sample can be obtained.

8.2.2.2. Electron Beam Induced Current (EBIC)

When an electron beam is incident on a semiconducting crystal, electron-hole (*eh*) pairs are generated that can diffuse away from their generation site. If they enter a region of electric field, the electrons and holes move in opposite directions under the field's influence, producing a measurable current. The electric field can be supplied by either a *pn* junction within the sample or by the fabrication of a Schottky barrier on the sample surface. The EBIC technique uses the size of the induced current to produce an image as the sample is scanned by the focused electron beam [24]. Defects in the crystal can locally change the magnitude of the current. For example, dislocations are sites of enhanced charge carrier recombination, so cause fewer electrons and holes to diffuse to the field region, locally reducing the size of the measured current.

The EBIC technique is able to image electrically active defects that are approximately 1 μm below the sample surface (depending on the electron energy and semiconducting material) with a spatial resolution of about 1 μm, which is primarily governed by the electron beam spreading in the material. The technique provides information on the electrical properties of crystal defects as well as producing images of them, but it does not give direct information on the Burgers vector of dislocations. The only sample preparation necessary is the fabrication of a Schottky barrier on the sample surface, and the lateral area of the sample that can be analyzed is limited only by the barrier size. By its nature, EBIC can only be used to examine semiconducting crystals. Aspects of EBIC are compared with IBIC, the analogous ion beam technique, in Chapter 6.

8.2.2.3. *Cathodoluminescence (CL) Imaging*

Cathodoluminescence (CL) Imaging Cathodoluminescence imaging [25] also involves the generation of charge carriers in a semiconducting crystal by an electron beam. When charge carriers recombine in some materials, the energy released can be given out as a photon with wavelength in the visible or near-visible range. These photons are detected and their intensity mapped as the sample is scanned by the beam to produce an image.

Crystal defects, such as point defects and dislocations, locally change the CL signal intensity. For example, dislocations can cause *eh* pairs to recombine without the emission of a photon, locally lowering the CL intensity. Images of dislocations, in which the dislocations appear dark compared with their surroundings, can therefore be produced as the beam is scanned over the sample.

By selecting appropriate wavelengths of light in the CL spectrum, images based on one particular type of defect, or from one particular layer in a multilayer structure, can be produced. By varying the electron beam energy, and therefore the electron penetration depth, depth-resolved images from the top few microns of a sample can be produced. The spatial resolution in CL imaging is primarily dependent on the spreading of the electron beam within the sample and is therefore of the order of 1 μm. However, improvements on this figure can be obtained by thinning the sample to a thickness of a few hundred nm. No sample preparation is necessary for CL imaging, and large areas of a sample can be imaged. The technique is only appropriate for direct band gap semiconducting crystals. The analogous ion beam technique is IBIL which is described in Section 4.9.

8.2.3. X–Ray Topography

X–ray topography uses the diffraction of X–rays by crystal lattice planes to produce images of defects [26,27]. The crystal is illuminated by a collimated beam of X–rays and images based on the intensity of X–rays scattered into a particular diffraction spot are produced on photographic film. The topograph can be recorded in reflection and transmission. Local lattice strain produced by dislocations or precipitate particles changes the diffracted X–ray intensity revealing the defect in the image. By using different sets of lattice planes for diffraction, dislocation Burgers vectors can be determined. The technique is highly sensitive to lattice strain, resulting in dislocation image widths of the order of a few micrometers. X–ray topography can be performed without thinning of the sample and over a large sample area.

8.2.4. Infrared Microscopy

Semiconductors are transparent to infrared (IR) radiation if the photon energy is smaller than the electronic band gap of the material. However, absorption of the radiation can occur at precipitate particles and some crystal defects. Mapping of the intensity of transmitted IR radiation enables the distributions of such defects to be determined. This technique has been used to study the distribution of precipitate particles formed by the diffusion of copper atoms into silicon crystals [28]. Infrared radiation transmitted through a 1 mm thick silicon crystal was detected using a microscope fitted with an infrared image tube. Linear aggregates of the particles revealed the lines of dislocations running through the samples. The distribution of EL2 defects in gallium arsenide, believed to be caused by arsenic atoms on gallium sites, has also been studied [29]. Whole gallium arsenide wafers were scanned through a narrow beam of IR radiation to obtain images of the defects with a spatial resolution of approximately 0.2 mm in the scan direction.

The technique of scanning infrared microscopy (SIRM) has also been developed [30]. In this case, a beam of IR radiation from a laser is brought to a focus within the crystal and the intensity of transmitted or scattered radiation is measured. Particles within the crystal cause scattering of the beam, locally reducing the transmitted radiation intensity (and increasing the scattered intensity). Dislocations can be imaged either by detection of the particles decorating them or by imaging the strain field itself. Use of polarized light can enable information on dislocation Burgers vectors to be obtained. Depth distributions of defects can be determined by changing the position of the beam focus within the sample, and the depth of field (depth resolution) can be as low as 7 μm. The spatial resolution, determined by the beam spot size, is approximately 1 μm when the technique is used in confocal mode, although particles smaller than this can be detected. Bulk samples up to 3 mm in thickness and 150 mm in diameter can be analyzed, with the only requirement for good images being that the beam entrance and exit surfaces are flat and damage free.

8.2.5. Optical Microscopy

Optical microscopy can be used to reveal near-surface crystal defects. In particular, surface slip steps caused by dislocations can be revealed by Nomarski interference contrast [31,32], which is sensitive to local variations in the surface height (examples of Nomarski contrast images are given in Figures 8.4 and 8.15). Nomarski microscopy can reveal the sign of the slope of slip steps and, therefore, differentiate between dislocations of different Burgers vectors.

The points at which dislocations meet the surface of a crystal can be revealed optically if the sample is placed in a chemical (etch) that removes atoms from the sample surface [33]. The lattice distortion, the geometry of the lattice planes, or the presence of impurity atoms around a dislocation can all locally change the etch rate, leaving an etch pit (etch pits can be seen in the photograph shown in Figure 8.4) or hill marking the dislocation location. The size of the etch pits, normally greater than 1 μm, means that etching is unsuitable for crystals with high defect densities.

8.3. CHANNELING SCANNING TRANSMISSION ION MICROSCOPY IMAGES FROM A $Si_{0.95}Ge_{0.05}$/Si CRYSTAL

The following sections demonstrate how ion channeling can be used to produce images of individual crystal defects using the CSTIM technique. This section and the next two sections are concerned with CSTIM images of dislocations (CCM images of bunches of dislocations are also given in Section 8.5). Dislocations cause local distortions of the crystal lattice planes, which can lead to channeled ions being dechanneled, as was described in Section 5.2.1.

8.3.1. Description of the Sample

Silicon-germanium alloys have been the focus of much investigation because of the potential they offer for the production of fast transistors and novel electronic devices that cannot be produced in pure silicon [34,35] but that can be made using existing silicon device fabrication technology. The lattice parameters of pure silicon and germanium are 5.431 and 5.658 Å, respectively, so that an alloy of silicon with germanium (written $Si_{1-x}Ge_x$, x being the mole fraction of germanium) has a larger lattice parameter than pure silicon. If a thin layer of $Si_{1-x}Ge_x$ is grown on to a silicon crystal (by molecular beam epitaxy, MBE [26], for example), the layer atoms initially register exactly with the silicon substrate atoms in the plane of the interface. This results in the unit cell of the alloy layer being compressed along directions within the alloy–silicon interface plane and elongated perpendicular to this plane. The strain in the alloy layer produced by this tetragonal distortion (Figure 8.3a) modifies the electronic band gap of the layer, with the band gap falling as the germanium fraction is increased.

As the alloy is grown thicker, the strain in the layer rises so that, above a certain layer thickness, which depends on the alloy composition, it is no longer energetically favorable for the layer atoms to be in registry with those of the silicon substrate. Instead, the layer relaxes and dislocations, called misfit dislocations, are produced at the layer–substrate interface (Figure 8.3b). Above a certain thickness it is energetically possible for misfit dislocations to form as a result of the extension of existing threading dislocations in the silicon substrate [36,37], but significant relaxation of the layer does not occur until the layer is thick enough (thickness h_c) for the strain energy to allow new dislocations

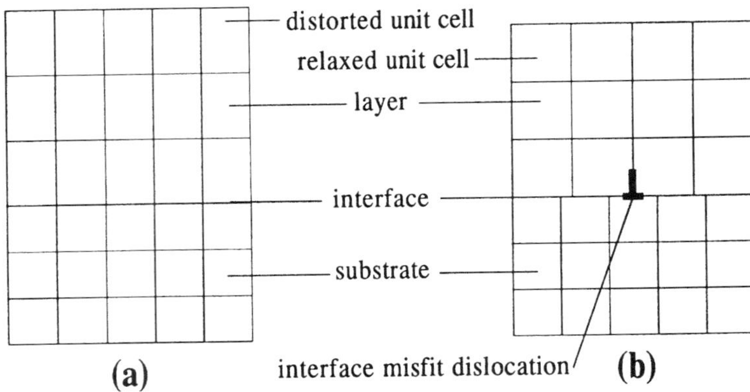

Figure 8.3. Epitaxial layer growth. (a) Registered layer, tetragonally distorted. (b) Relaxed layer showing interface dislocation.

to be produced at the sites of existing dislocations. For example, for $x = 0.05$ misfit dislocations can be produced for layers thicker than approximately 0.07 μm, whereas h_c is approximately 4 μm [38].

This section focuses on CSTIM investigations of a crystal consisting of a 1.8 μm thick layer of $Si_{0.95}Ge_{0.05}$ grown by MBE on to an (001) silicon crystal. The alloy layer was, therefore, between the two thicknesses given above, and it was found that misfit dislocations had formed at the layer–substrate interface in isolated regions of the sample.

Figure 8.4 shows an optical image of a portion of the sample taken with Nomarski interference contrast (Section 8.2.5). The sample had been given a chemical etch (a Schimmel etch [39] for 75 s) which preferentially attacked regions of the sample where lattice strain was present. Near the center of the image, a cross can be seen consisting of two bright lines on the grey background, one running along the [110] sample direction and the other along the [$\bar{1}$10] direction. These two lines are small steps on the sample surface caused by groups of misfit dislocations at the interface below. Such cross-patterns could be seen at intervals over the whole of the sample surface, and Figure 8.5 shows schematically one arm of such a cross.

The cross-patterns were the result of the relaxation of localized regions of the alloy layer owing to the stress produced by the layer-substrate lattice mismatch. Existing defects in the layer acted as nucleation sites to trigger misfit dislocation formation [32,37], and the misfit dislocations ran out from such sites along the [110] and [$\bar{1}$10] directions, or along the orthogonal [$\bar{1}$10] and [1$\bar{1}$0] directions, to form a cross. Several misfit dislocations could be produced at each nucleation site. Figure 8.5 shows a cross arm consisting of three misfit dislocations that have run out from a central defect (not shown) along a [110] direction. At the end of each misfit dislocation, another dislocation, called a threading dislocation, runs up to the sample surface, so that six of these threading seg-

Figure 8.4. Nomarski interference contrast photograph of an etched misfit dislocation cross pattern in the $Si_{0.95}Ge_{0.05}$/Si sample. Reprinted from Ref. 5 with kind permission from Elsevier Science B.V., Amsterdam, The Netherlands.

ments are shown. Careful inspection of the dislocation cross photograph (Figure 8.4) reveals that periodically black dots are visible along the cross arms; these dots are etch pits marking the points at which the threading dislocations from the misfit dislocations at the interface met the sample surface. By counting the dots, it was therefore possible to determine the number of misfit dislocations

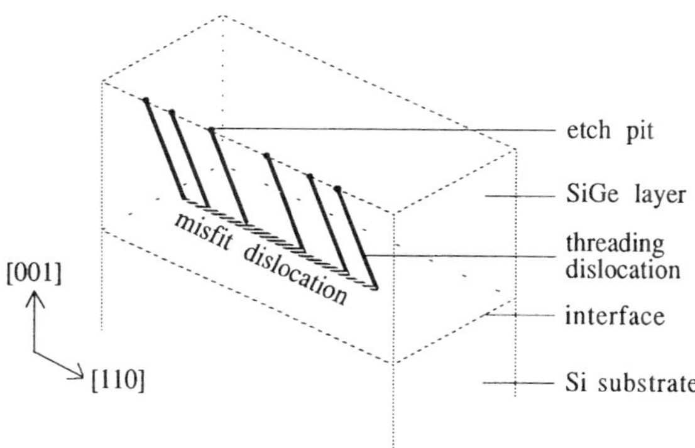

Figure 8.5. Schematic of the arrangement of misfit dislocations in an arm of the cross patterns seen in the $Si_{0.95}Ge_{0.05}$/Si sample. Reprinted from Ref. 5 with kind permission of Elsevier Science B.V., Amsterdam, The Netherlands.

in each arm of the cross shown in Figure 8.4: the top, bottom, left, and right arms consisted of 7, 9, 15, and 14 dislocations, respectively.

The majority of CSTIM images of the $Si_{0.95}Ge_{0.05}/Si$ crystal shown in this chapter were taken from etched portions of the crystal in order that the number of misfit dislocations present in the imaged region could be determined. The authors believe that the effect of etching on the CSTIM contrast was negligible: unetched cross pattern features were imaged with CSTIM and produced similar information as described in Section 8.3.2 for an etched sample.

8.3.2. Channeling Scanning Transmission Ion Microscopy Results

Figure 8.6 is a CSTIM average energy loss image of the cross-pattern shown in Figure 8.4 taken using 2 MeV ^{1}H ions. The sample was thinned to approximately 25 μm, and the image shown was taken with the beam just off ($\sim$0.1°) channeling alignment with the [001] axis of the crystal for reasons that will be

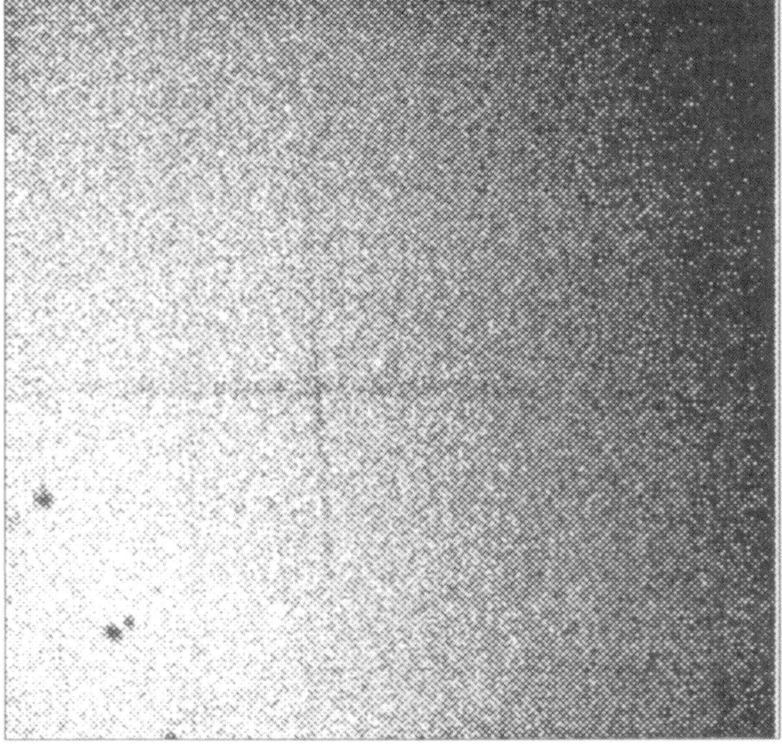

Figure 8.6. 200 μm wide CSTIM average energy loss image of the etched cross pattern feature of Figure 8.4, taken with a 2 MeV ^{1}H ion beam and with the sample tilted approximately 0.1° about both goniometer axes from the [001] channeling axis.

given below. The image was obtained using an uncollimated detector, and took 30 min to collect at an average count rate of 1,780 ^{1}H ions/s. In the center of the image, a cross can be seen corresponding to the cross pattern of misfit dislocations revealed in the optical image. The cross is dark with respect to the background, meaning that the ^{1}H ions passing through the cross region had a higher average energy loss than those passing through surrounding virgin crystal. The dislocations at the interface were therefore causing dechanneling of the beam. The vertical arms of the cross are visible in regions containing five dislocations, whereas the horizontal arms are visible in regions containing approximately eight dislocations. Gross contrast changes consisting of the image background getting darker toward the image edges can also be seen. These changes occurred because the strain in the alloy layer caused the thinned crystal to bend. This meant that only a relatively small area of the crystal, of the order of $100 \times 100 \ \mu m^2$, was at exact channeling alignment at any one time, producing changes in the average transmitted ^{1}H ion energy loss across the image.

8.3.2.1. *Contrast Changes on Tilting Off the Channeling Direction* Figure 8.7 shows eight images of the central region of this cross. These were all taken with a 2 MeV ^{1}H ion beam channeled in the $(\bar{1}10)$ planes of the sample, but with the sample tilted so that the beam made slightly different angles to (110) alignment. Each image was obtained by collecting the transmitted ^{1}H ions with an uncollimated detector at a rate of approximately 1,300 ^{1}H ions/s for half an hour (except the image taken at 0.0° to the (110) planes which was collected for 113 min). The following observations can be made from this set of images:

1. The cross arm running parallel to the [110] crystal direction remains visible in all the images as a line that is dark on its upper half and bright on its lower half. It changes little in contrast over the range of tilt angles shown.

2. The cross arm running parallel to the $[\bar{1}10]$ direction is invisible in the images at −0.3° and +0.4°.

3. This arm becomes visible as a bright line in the −0.2° image. It remains a bright line in the −0.1° image, but is bright on one side and dark on the other in the image taken closest to the channeling direction. On the other side of the channeling direction (+0.1°, +0.2°, +0.3°), it appears as a dark line. In other words, the $[\bar{1}10]$ cross arm changes contrast on going through the (110) channeling direction.

Figure 8.8a shows the average energy loss of the ^{1}H ions as a function of tilt angle from (110) channeling taken from the images of Figure 8.7 for a region of the $[\bar{1}10]$ cross arm just below the cross center, and also for the same region of the sample if the cross arm had not been there. Both sets of point have been fitted with Gaussians, assuming that the background level was the same for

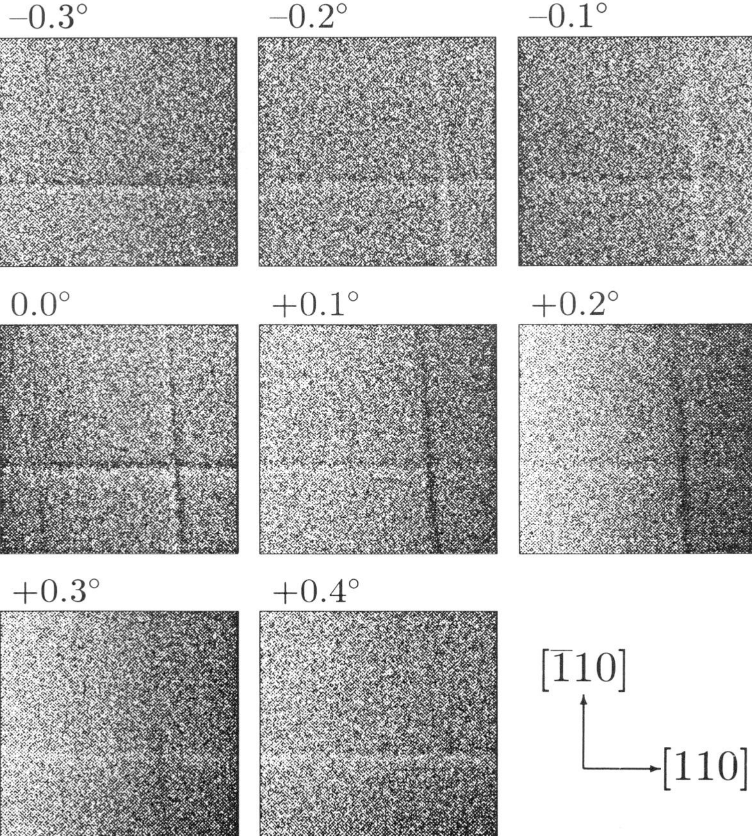

Figure 8.7. 60 μm wide images of the cross pattern of Figure 8.4, taken with a 2 MeV ^{1}H ion beam. For all the images, the beam was channeled in the ($\bar{1}$10) planes, but the sample was tilted so that the beam angle to the (110) planes was changed. The angle to (110) channeling is given by each image. Reprinted from Ref. 5 with kind permission of Elsevier Science B.V., Amsterdam, The Netherlands.

both curves. The position of the minimum of the curve for perfect crystal has been placed at 0.0°. The curve for the dislocation cross arm shows a slightly shallower minimum, which is shifted in angle by $0.04 \pm 0.01°$ from the curve for perfect crystal.

The visibility of the cross arm in CSTIM images is the difference between the two curves of Figure 8.8a, and this difference curve is shown in Figure 8.8b. The surprising feature of this curve is that the maximum contrast exhibited by the cross arm occurs at 0.1° from the best channeling position of the virgin (i.e., perfect crystal) curve in Figure 8.8a, on the side of the channeling direction when the cross arm is showing dark contrast. This is the reason that an image

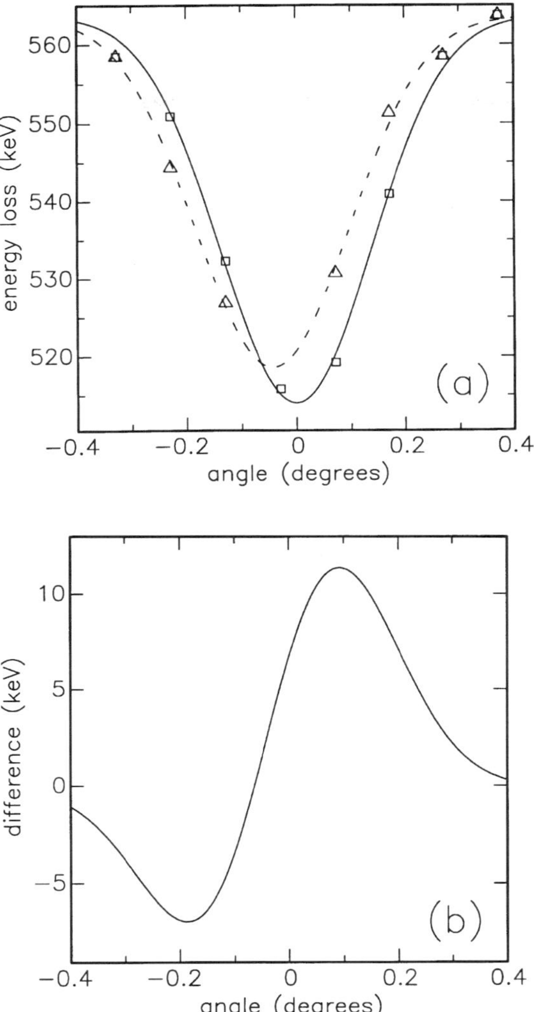

Figure 8.8. (a) Plot of the average transmitted ^{1}H ion energy versus angle from (110) channeling for the vertical cross arm of Figure 8.7 (dashed curve) and dislocation free material (solid curve). (b) The difference between the two curves of (a) showing fault contrast versus tilt angle. Reprinted from Ref. 5 with kind permission of Elsevier Science B.V., Amsterdam, The Netherlands.

of the cross arm taken just away from the channeling direction was selected for Figure 8.6.

To understand the behavior of the dislocation cross arm described above, the nature of the misfit dislocations must be considered. Misfit dislocations in Si$_{1-x}$Ge$_x$/Si crystals are known to be mixed dislocations (Section 5.2.1); they are in fact 60° dislocations [27,40], where pure edge and pure screw dislocations are described as 90° and 0°, respectively. One way to determine the effects on the crystal of such dislocations is to split them into components that are either pure edge or pure screw in nature. It was found that the contrast changes observed in CSTIM images on tilting about the channeling direction could be explained by reference primarily to the largest component of the dislocations which was an edge component.

Figure 8.9 shows a diagram of the effects of several of these components on the (110) planes of the sample. The case shown is for a group of misfit dislocations running parallel to the [$\bar{1}$10] crystal direction, as was the case for the cross arm displaying contrast changes in Figure 8.7. The dislocation component considered consists of an extra half plane of atoms, inserted parallel to the layer–substrate interface plane, and the diagram shows four of these extra planes. The effect of the extra half planes is to locally tilt the (110) planes in the alloy layer, and the effect can be considered to be similar to a low angle grain boundary [33]. This tilting of the channeling planes interferes with the channeling process, and produces the contrast seen in the CSTIM images.

The [$\bar{1}$10] cross arm could not be seen in images of the sample taken with

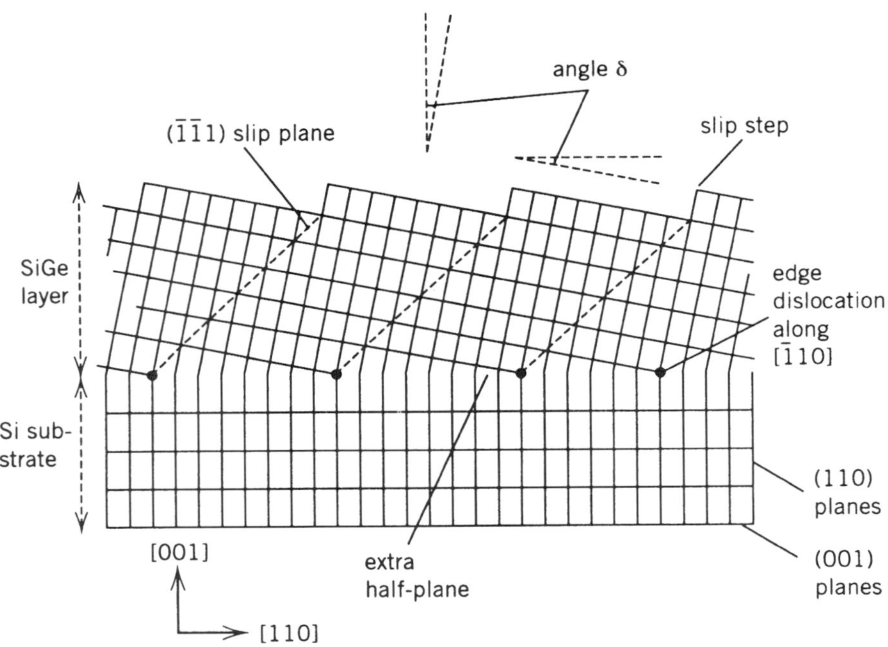

Figure 8.9. The effect of several of the largest components of the 60° misfit dislocations on the (110) planes in the SiGe layer of the $Si_{0.95}Ge_{0.05}/Si$ sample. The planes are locally rotated through an angle δ. Reprinted from Ref. 4 with permission (©1993 American Institute of Physics, Woodbury, NY).

the beam channeled in the $(\bar{1}10)$ planes but away from the (110) channeling direction (Figure 8.7). It is therefore concluded that the group of dislocations along the $[\bar{1}10]$ direction were not significantly affecting the $(\bar{1}10)$ planes, and indeed the extra half planes shown in Figure 8.9 would cause a tilting of the (110) planes but not the $(\bar{1}10)$ planes.

As the (110) channeling direction was approached in the images of Figure 8.7, the $[\bar{1}10]$ cross arm firstly showed bright contrast, indicating that it was enhancing channeling over that produced by virgin crystal. This can be understood from Figure 8.9 as occurring when the beam was aligned with the tilted (110) lattice planes produced by the extra half planes of the misfit dislocations. On the other side of the channeling direction, the $[\bar{1}10]$ cross arm produced dark contrast, meaning that it was causing dechanneling compared with virgin crystal. This can be understood to have occurred when the beam was misaligned with the tilted lattice planes. These two conditions are shown in Figure 8.10.

Monte Carlo computer simulations made using the computer code FLUX3 [41] of the passage of channeled ions through a crystal with lattice planes in the surface layer rotated were performed to test the above explanation of the CSTIM contrast changes [9]. The passage of 3 MeV ^{1}H ions through a 20 μm thick silicon crystal, which had the lattice planes in the top 1 μm rotated by a small amount, was simulated. Figure 8.11 shows the simulated average trans-

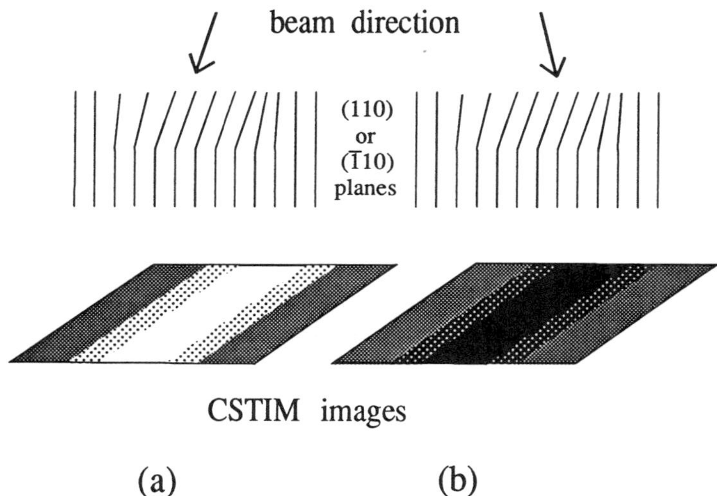

Figure 8.10. The effects of small rotations of the channeling planes on CSTIM image contrast. (a) Beam aligned with the rotated planes, which locally allow better channeling than surrounding material. The lower transmitted energy loss of the ^{1}H ions passing through this region produces a bright line in the CSTIM image. (b) Beam on the opposite side of the channeling direction to (a): rotated planes disrupt the channeling, producing transmitted ^{1}H ions with a higher energy loss than surrounding material and a dark line in CSTIM images.

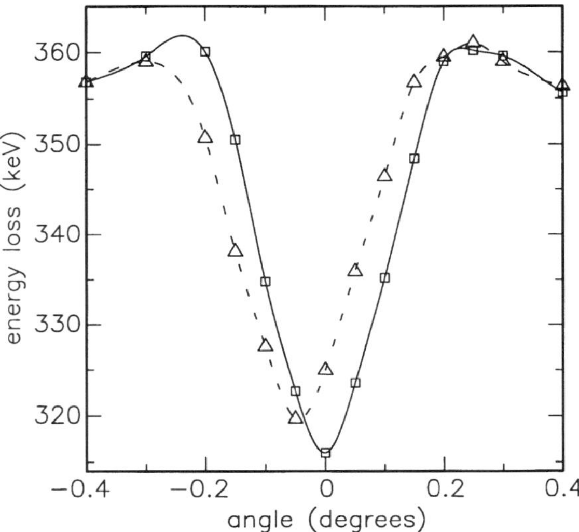

Figure 8.11. Average energy loss values for 3 MeV ^{1}H ions passing through virgin crystal (solid line) and crystal with the lattice planes rotated by 0.05° in the top 1 μm, produced by computer simulation. The points are joined by smooth curves. Reprinted from Ref. 10 with kind permission of Elsevier Science B.V., Amsterdam, The Netherlands.

mitted energy loss of ^{1}H ions passing through virgin crystal and through crystal with planes rotated by 0.05°, versus angle from the (110) channeling direction. This plot can be compared with that of Figure 8.8a. The simulated curve for the crystal with rotated lattice planes is slightly displaced in angle from that of the virgin crystal. The simulations showed that if the plane rotation was less than the channeling critical angle, the amount of shift is a measure of the plane rotation angle. It was shown above (Figure 8.8a) that the position of best channeling at the dislocation cross arm was shifted from that of the surrounding good crystal by 0.04 ± 0.01°. This is, therefore, a direct measurement of the amount of lattice plane rotation produced by the dislocations in the arm (and can be compared with the half-width-at-half-maximum of the energy loss curve from virgin crystal, Figure 8.8a, which was 0.16° and is a measurement of the channeling critical angle).

The bright-dark appearance of the [$\bar{1}$10] cross arm close to the channeling direction is harder to explain on the basis of the extra half plane edge component of the misfit dislocations alone, and it is believed that the other components of the dislocation, in particular an edge component with extra half plane perpendicular to the interface plane, must be considered, together with the possible effects of the curvature of the crystal owing to the thinning process.

Finally, the observation that the cross arm running parallel to the [110] sample direction changes little in contrast on tilting about the (110) channeling direction can be explained again on the basis of the edge component of the dislocations with the extra half plane parallel to the interface plane. For the [110] dislocations, this component would have primarily affected the ($\bar{1}$10) planes and not the (110) planes. The [110] cross arm, therefore, remains visible throughout all of the images of Figure 8.7, which were all taken with the beam channeled in the ($\bar{1}$10) planes. Its contrast is largely unaffected by the change in alignment with respect to the (110) planes.

8.3.2.2. *Channeling Scanning Transmission Ion Microscopy Contrast Versus Dislocation Number*

At the beginning of this section, it was described how the number of misfit dislocations in an arm of the cross-pattern feature could be determined using chemical etching of the sample and optical microscopy. Etching produced small pits that marked the points at which threading dislocations from the misfit segments met the sample surface. The number of misfit dislocations at any point in an arm could therefore be determined by counting the etch pits.

Figures 8.12a and 8.12b are two further CSTIM images of the dislocation cross shown in Figure 8.4. A 3 MeV ^{1}H ion beam was used, and the images were collected using an uncollimated detector for 1 hr, 45 min each at a count rate of just less than 2,000 ^{1}H ions/s. The image of Figure 8.12a was taken with the beam to one side (~0.15°) of the (110) planar channeling direction with the [$\bar{1}$10] cross arm showing bright contrast, and the image of Figure 8.12b was taken with the beam close to the (110) channeling direction with the cross arm showing dark contrast. Marked between the two images are the positions of the etch pits as measured from the optical image, and the number of dislocations in each part of the arm is given between the etch pit positions. Figure 8.12c is an optical image of another portion of the sample, and the most prominent feature visible in the image is of a group of 13 dislocations running along the [$\bar{1}$10] direction of the sample. A CSTIM image of this region of the sample is shown in Figure 8.12d, taken again with a 3 MeV ^{1}H ion beam channeled in the (110) planes of the sample. The bunch of 13 dislocations can clearly be seen in the image and shows predominantly dark contrast.

Figure 8.13 is a plot of the contrast of the [$\bar{1}$10] dislocation cross arms in Figures 8.12a and 8.12b and of the group of 13 dislocations in Figure 8.12d versus dislocation number. The contrast is defined as the difference in average transmitted ^{1}H ion energy loss between the cross arm and the surrounding virgin crystal. The contrast in all three cases increases fairly linearly with dislocation number, with rates of increase of between 1.0 keV and 1.6 keV per dislocation.

The optical image in Figure 8.12c also shows very faintly the slip trace of a single misfit dislocation just to the left of the group of 13 dislocations. In the CSTIM image of Figure 8.12d, this single dislocation is visible as a thin black line. The single dislocation feature in the CSTIM image is 0.6 ± 0.2 μm wide, a combination of the beam spot size and the width of the region around the dislocation within which ^{1}H ions were dechanneled. The height of the feature above the background is 4.2±0.2 keV, and the standard deviation of the energy-

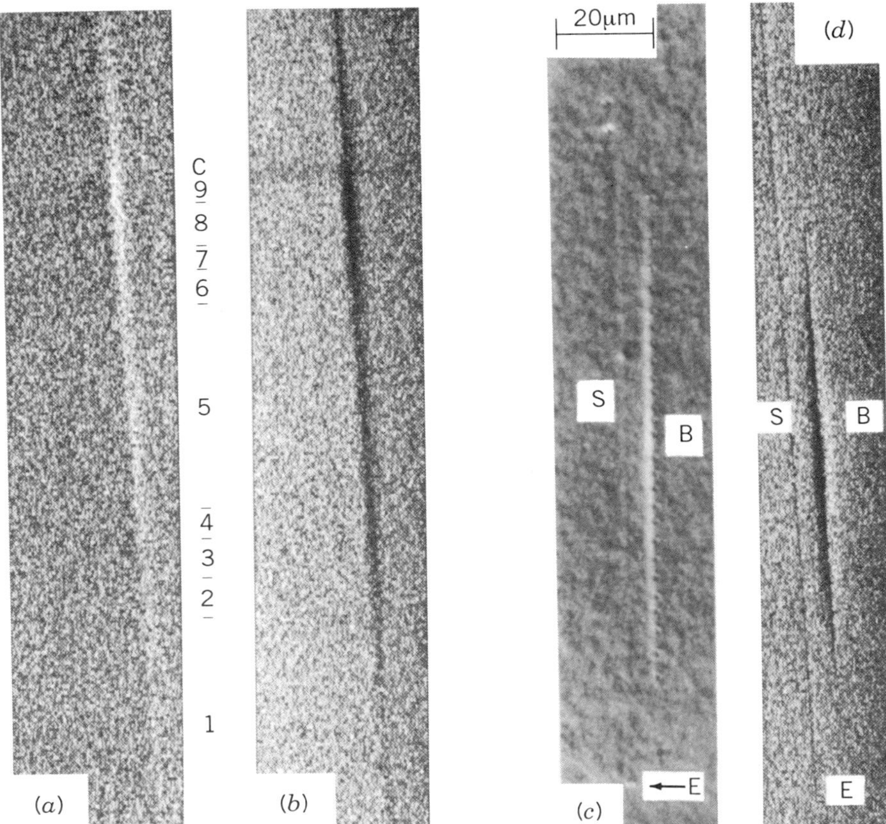

Figure 8.12. (a) and (b) 30×150 μm^2 (110) planar channeled images of the vertical arm of the cross shown in Figure 8.4. (a) Beam ~0.15° from the channeling direction. (b) Beam close to the channeling direction. Marked between these two images are the positions of the etch pits as measured from optical images of the sample (dashes) with the number of dislocations making up the arm given between each dash. C marks the cross center. (c) Nomarski optical image showing the slip trace of a single misfit dislocation (S) ending at an etch pit (E) and a group of thirteen misfit dislocations (B). (d) 30×200 μm^2 CSTIM image of the portion of the sample shown in (c).

loss values in the image (the noise on the image) is 4.3 ± 0.2 keV. The image was collected using an uncollimated detector for 149 min at an average count rate of just under 1,500 ^{1}H ions per second.

8.3.2.3. Data Collection Time for Dislocation Imaging

This section uses the above images as examples to describe the relationship between counting statistics, image noise, and minimum resolvable contrast. The noise level on a CSTIM average energy loss image, and hence the ability to detect low contrast

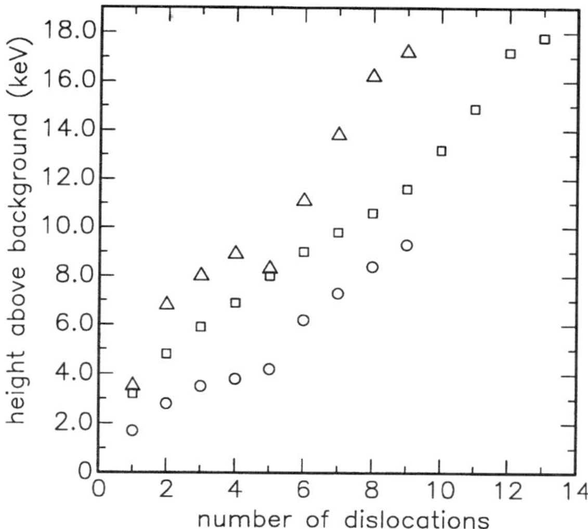

Figure 8.13. Plot of the dislocation contrast versus dislocation number for the vertical cross arms shown in Figure 8.12a (circles), Figure 8.12b (triangles), and Figure 8.12d (squares). These latter points are averages from the two sides of the arm of thirteen dislocations. The error on each of the points is approximately 1 keV.

features, depends on the length of the data acquisition time, for a given ^{1}H ion count rate. The greater the number of ion energy loss values recorded at each pixel, the smaller the error in the average ion energy loss calculated from these values and the lower the noise level on the CSTIM average energy loss image.

It is assumed that at a given pixel, N ion energy loss values are recorded, and that they form a distribution with standard deviation σ_p (it is assumed that the distribution is Gaussian, although the distribution can be very non-Gaussian for a channeled beam). From these values, the average ion energy loss is calculated. The noise in the CSTIM average energy loss image occurs owing to the difference between the calculated average ion energy loss and the true average ion energy loss at each pixel. The error in the average energy loss values σ_m [42] is given approximately by

$$\sigma_m^2 = \frac{\sigma_p^2}{N} \tag{8.1}$$

If a count rate of 2,000 ions/s and an image consisting of 256×256 pixels is assumed, then the number of ion energy loss values at each pixel is

$$N = \frac{2000 \times t}{256 \times 256} \tag{8.2}$$

where t the data acquisition time in seconds. Rearranging these two equations gives the data collection time required to produce an image with noise level σ_m when the ion distribution at each pixel has standard deviation σ_p

$$t(s) = \left(\frac{\sigma_p}{\sigma_m} \right)^2 \times \frac{256 \times 256}{2000} \tag{8.3}$$

To estimate the data acquisition time to image a group of n dislocations for imaging conditions similar to those used for the CSTIM images of Figures 8.12a and 8.12b, the value of σ_p for the image of Figure 8.12b is used. This image was collected from a sample approximately 25 μm thick with the beam channeled in the (110) planes and using an uncollimated detector, and σ_p is approximately 70 keV. Using the contrast values given in Figure 8.13 taken from the image of Figure 8.12b, it is estimated that the minimum data acquisition times to image groups of one, two, four, six, and eight dislocations at a count rate of 2,000 ions/s are 80 min, 21 min, 12 min, 8 min, and 4 min, respectively. This assumes that a dislocation group becomes visible when the noise level on the CSTIM image is 1.6 times the contrast produced by the group. This large value of the noise level compared with the group contrast is used because the image of the group consists of rows of pixels rather than isolated pixels. In the image of Figure 8.12b, the [$\bar{1}$10] cross arm is just visible in a region of the arm consisting of a single dislocation. The noise level on the image, $\sigma_m = 5.5 \pm 0.3$ keV, is approximately 1.6 times the contrast of the single dislocation region of the cross arm, 3.5 ± 1 keV.

Changing the experimental conditions of course alters the times given in the previous paragraph. Obviously, increasing the ion count rate will reduce the data acquisition time. Using a restricted acceptance angle detector (collimated detector) also changes the time as the contrast of features in CSTIM images can be increased. Changing the sample thickness and the channeling direction used will also affect the time as these change both feature contrast and σ_p.

8.4. CHANNELING SCANNING TRANSMISSION ION MICROSCOPY IMAGES FROM A $Si_{0.85}Ge_{0.15}$/Si CRYSTAL WITH MESAS

Because of the detrimental effect of misfit dislocations on electronic device performance, attempts have been made to grow $Si_{1-x}Ge_x$ layers that are free from such defects. One method used is to grow the layer on to a silicon substrate that has been patterned to produce raised, isolated islands (mesas). It was shown theoretically that growth of $Si_{1-x}Ge_x$ on a small enough area allows the layer to relax elastically along directions in the plane of interface [43]. This reduces the misfit strain to below the level at which misfit dislocations form.

A second test sample used for CSTIM analysis consisted of an (001) silicon wafer that was patterned by plasma etching [3] to produce mesas that were mainly square or rectangular, about 3 μm high and varying in width from 1 μm to over 1 mm. A 1 μm thick epitaxial layer of $Si_{0.85}Ge_{0.15}$ was then grown by MBE on to the surface of the wafer. A diagram of the cross-section through a mesa is given in Figure 8.14a. Figure 8.14b is a 1,220 μm wide, nonchanneled STIM image of the sample taken with 2 MeV ^{1}H ions (with the sample thinned to approximately 25 μm). The mesas can be seen as dark grey shapes owing to the fact that the sample was 3 μm thicker in these regions than in the non-mesa regions, so that the ^{1}H ions lost more energy at the mesas. This image demonstrates that relatively large areas of a sample can be thinned and easily imaged with STIM.

8.4.1. Optical and Electron Microscopy Results

The critical thickness for large-area growth of a layer containing 15% Ge is approximately 0.4 μm [38], so that interface dislocations were expected in this sample. Figure 8.15 is an optical image of a portion of the sample taken with Nomarski interference contrast. Lines of contrast on average 2 μm wide can be seen both on and off the mesas running along [110] and [$\bar{1}$10] directions. These lines are believed to be steps at the sample surface associated with misfit dislocations at the interface.

Plan-view TEM analysis of regions away from the mesas (Figure 8.16a) revealed an interface dislocation network. The dislocations were found to occur bunched together in some regions and more widely spaced in others. The bunched regions were approximately 0.25 to 0.5 μm wide and contained five

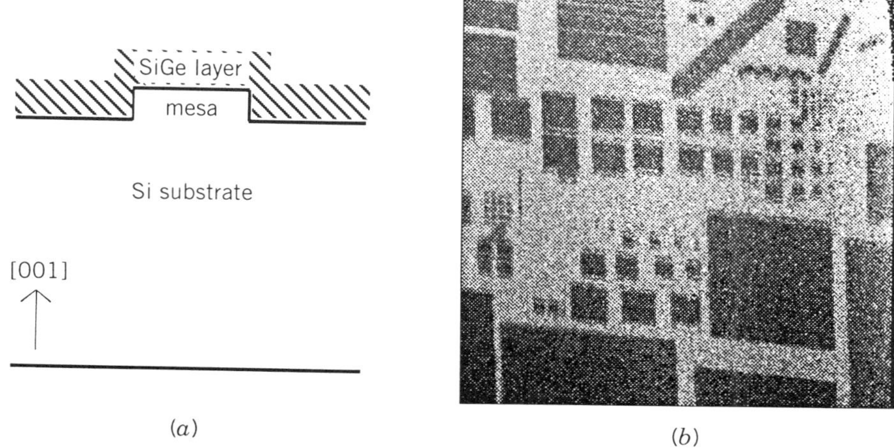

Figure 8.14. (a) Schematic diagram of the $Si_{0.85}Ge_{0.15}$/Si mesa sample. (b) 1,220 μm wide nonchanneled STIM image. The mesas are darker grey than surrounding regions.

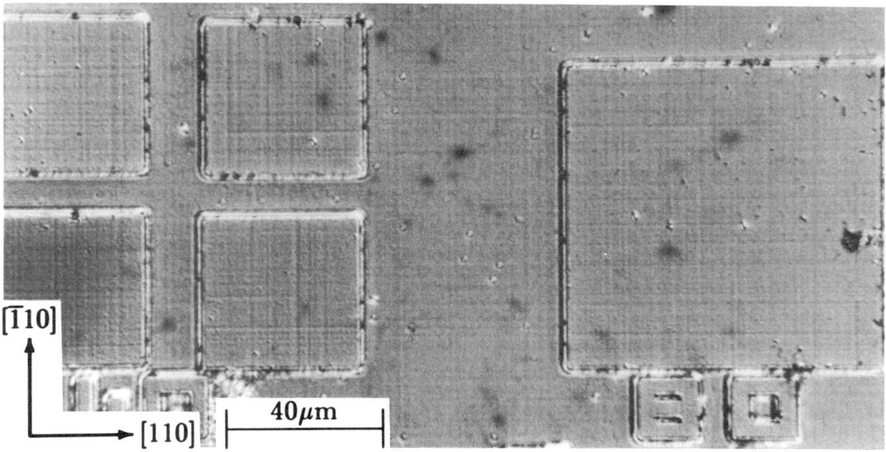

Figure 8.15. Nomarski contrast optical image of an 80 μm wide mesa and four 40 μm wide mesas. The bands of contrast running parallel to the [110] and [$\bar{1}$10] directions are steps at the sample surface produced by misfit dislocations at the interface below.

to ten parallel dislocations with an average dislocation spacing of approximately 50 nm. The regions of more widely spaced dislocations were about the same width, but with dislocations spaced on average 300 nm apart. The dislocations were found to be mainly of the 60° type, with the dislocations in any one band having predominantly the same Burgers vector.

Figure 8.16b is a cross-sectional TEM image taken through a 100 μm wide mesa. It reveals that around the base of the mesa was a moat where the $Si_{0.85}Ge_{0.15}$ layer was thinner and that the side face of the mesa was covered with a 0.3 μm thick $Si_{0.85}Ge_{0.15}$ layer. This side face of the mesa was found to contain a high density of crystal defects. It should be noted that plan-view TEM analysis of the small mesas was not possible as they would have fallen through the bottom of the substrate had it been thinned sufficiently.

8.4.2. Channeling Scanning Transmission Ion Microscopy Results

Figure 8.17a shows a CSTIM average energy loss image of a group of sixteen 10 μm wide mesas measured with a nonchanneled 2 MeV ^{1}H ion beam. The mesa and nonmesa regions show fairly uniform contrast. Figure 8.17b is a CSTIM image of the same region of the sample taken with the beam channeled along the [001] crystal axis. Bands of contrast can now be seen running along [110] and [$\bar{1}$10] directions. The bands have an average width of 1.5 to 3 μm and a spacing of 4 to 10 μm. They are 20 $\pm$ 5 keV above the background energy loss level, although this rises to 40 $\pm$ 5 keV above the background where a [110] band crosses a [$\bar{1}$10] band. The image took 16 min to collect at an average count rate of approximately 1,140 ^{1}H ions/s. The dark contrast bands are associated with channeling effects, as they are not visible in Figure 8.17a.

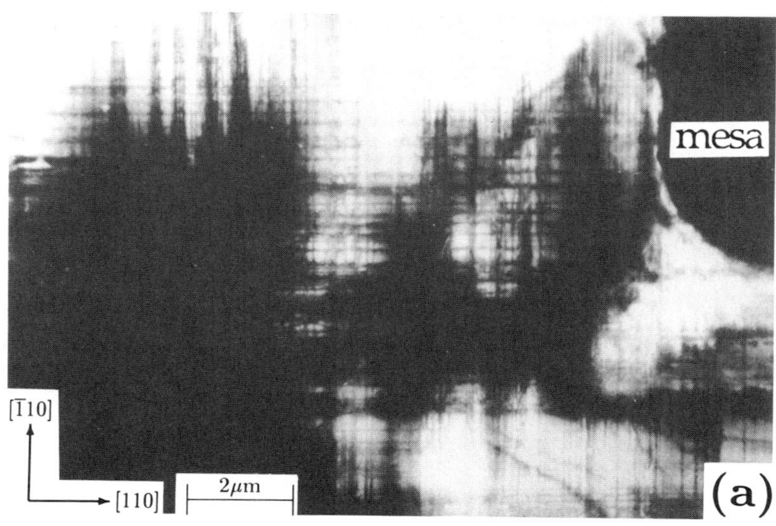

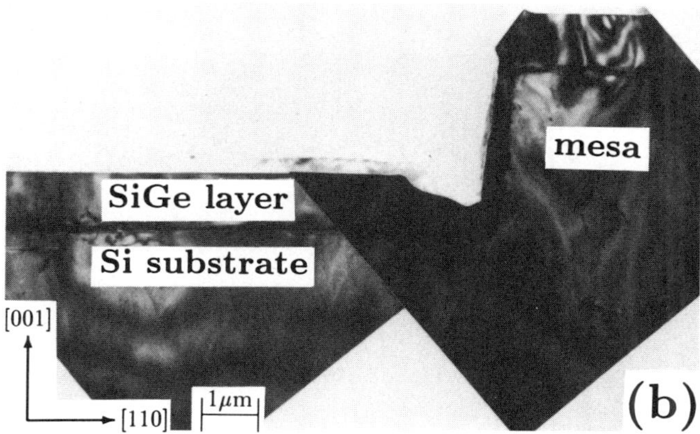

Figure 8.16. (a) Plan-view TEM image of the mesa sample showing the misfit dislocation network. (b) Cross-sectional TEM image taken through a 100 μm wide mesa. Reprinted from Ref. 4 with permission (©1994 American Institute of Physics, Woodbury, NY).

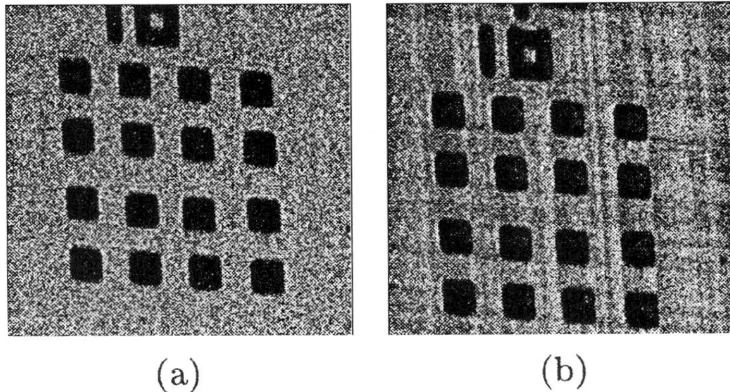

(a)   (b)

Figure 8.17. 120 μm wide CSTIM images of a group of sixteen 10 μm wide mesas. (a) Beam not channeled. (b) Beam aligned with the [001] axis of the sample. The [110] and [$\bar{1}$10] crystal directions run from left to right and from bottom to top, respectively. Reprinted from Ref. 4 with permission (©1994 American Institute of Physics, Woodbury, NY).

Such contrast bands were also visible in the larger mesas. Figures 8.18a and 8.18b are nonchanneled and [001] axially channeled CSTIM images of an 80 μm wide mesa. Again, the mesa shows uniform contrast in the nonchanneled image, but is crossed by bands of contrast when the beam was channeled. The bands of contrast seen in the CSTIM images of Figures 8.17b and 8.18b are the

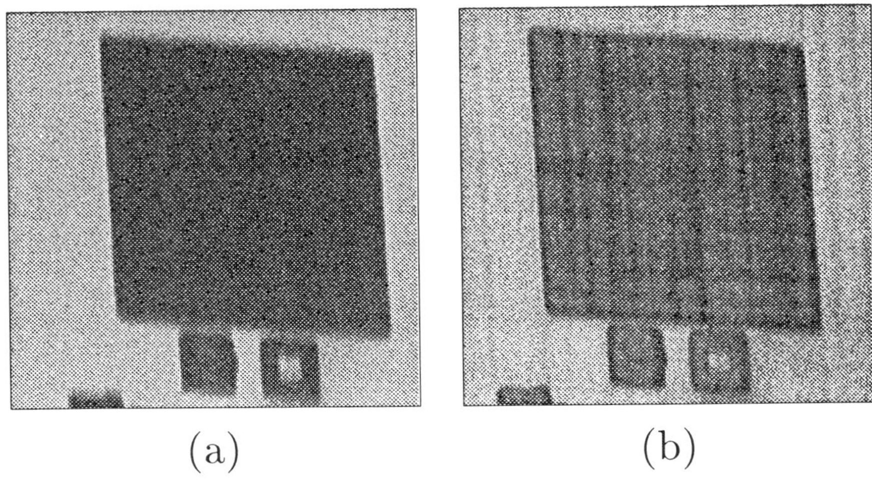

(a)   (b)

Figure 8.18. 120 μm wide CSTIM images of an 80 μm wide mesa. (a) Beam not channeled. (b) Beam channeled along the [001] axis. These images have been plotted over the same range of energy-loss values with no histogram equalization. Reprinted from Ref. 5 with kind permission of Elsevier Science, B.V., Amsterdam, The Netherlands.

results of the effects on channeling of the misfit dislocations present at the layer-substrate interface of the sample. The widths of the bands meant that they encompassed several of the bunches of closely spaced dislocations revealed by TEM.

8.4.2.1. Contrast Changes on Tilting off the Channeling Direction As for CSTIM images of the misfit dislocation cross patterns shown in Figure 8.7, contrast changes were observed when the mesa sample was tilted just off alignment for channeling. Figure 8.19 shows six CSTIM images of the central

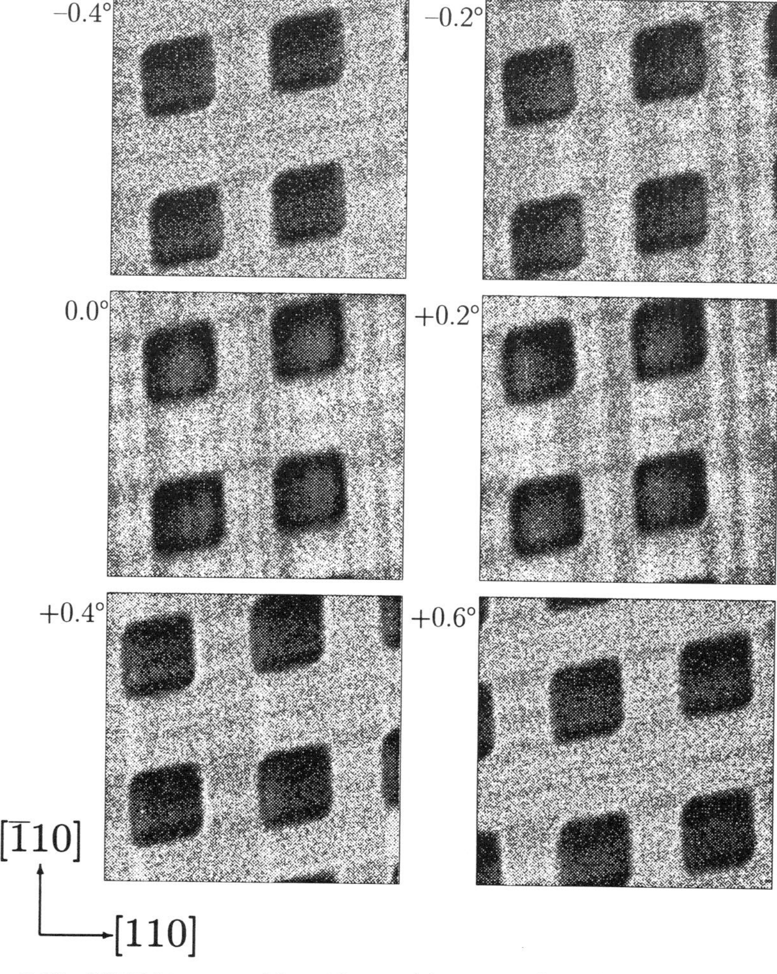

Figure 8.19. CSTIM images of four 10 μm wide mesas taken with the beam channeled in the ($\bar{1}$10) planes but with the beam angle to the (110) planes varied and given by each image. Reprinted from Ref. 4 with permission (©1993 American Institute of Physics, Woodbury, NY).

four 10 μm wide mesas shown in Figure 8.17. They show the effects of tilting the sample in 0.2° steps, so that the beam went through channeling in the (110) planes, while leaving the beam channeled in the ($\bar{1}$10) planes. The following observations can be made from the images:

1. The bands running along the [110] direction change little in contrast throughout the images.
2. The [$\bar{1}$10] bands are only faintly visible in the images taken 0.4° from the (110) channeling direction and are invisible in the image at +0.6°.
3. The contrast in the [$\bar{1}$10] bands has completely reversed on going from the image at −0.2° to that at +0.2°.
4. The [$\bar{1}$10] bands are narrower and more numerous in the image taken closest to the channeling direction than in those taken at ±0.2°.

Figure 8.20 shows the effect of tilting the sample to one side of the ($\bar{1}$10) channeling direction. In these images, the [110] bands of contrast disappear, whilst the [$\bar{1}$10] bands remain in contrast.

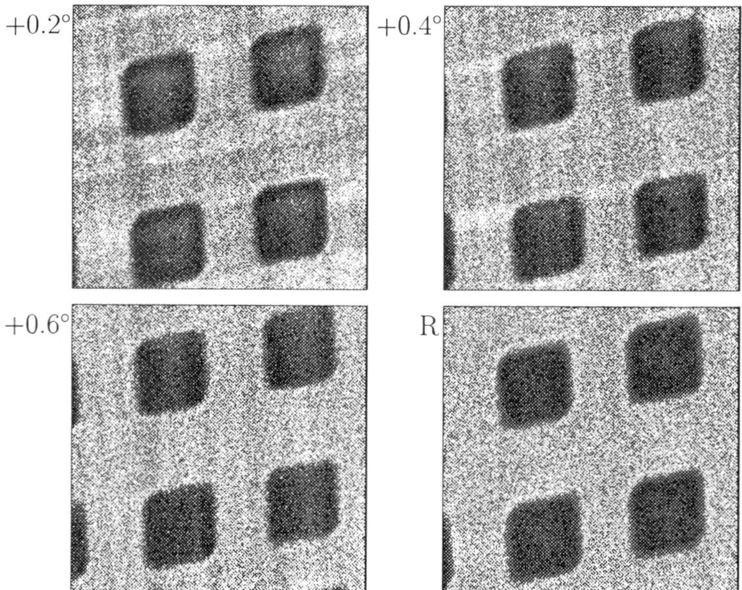

Figure 8.20. CSTIM images of the four 10 μm wide mesas at different tilt angles to the ($\bar{1}$10) channeling direction. For the first three, the beam was channeled in the (110) planes and the angle to ($\bar{1}$10) channeling is given at the top of each. The image labeled R was taken with the sample wound 1° about both goniometer axes from the [001] channeling direction, so it is a nonchanneled image of this region of the sample.

These contrast features are very similar to those exhibited by the misfit dislocation cross arm shown in Figure 8.7, and their explanation is also similar. The bands seen in the CSTIM images are believed to be caused by several of the bunches of misfit dislocations, with the dislocations within each band having predominantly the same Burgers vector. As with the misfit dislocations in the cross arms, the CSTIM image contrast can be primarily attributed to the largest component of the dislocations: the edge component with extra half plane parallel to the interface plane. This component caused the (110) planes (for the dislocations running along the [$\bar{1}$10] direction) in the layer to be locally rotated (and similarly the dislocations along [110] affected the ($\bar{1}$10) planes). In any one band, the dislocations with predominantly the same Burgers vector caused the planes to be rotated so as to cause the band to show bright contrast with the beam to one side of the channeling direction and dark contrast with the beam to the other. Neighboring bands that showed opposite contrast would have been comprised of dislocations that rotated the lattice planes in opposite senses.

8.4.2.2. Contrast Within Small Mesas

The contrast changes that occurred within the small mesas themselves were also of interest. In Figure 8.19, the region of the 10 μm wide mesas that allowed the best channeling in each image varied from the right-hand side of the mesas (image at $-0.2°$) to the left-hand side of the mesas (images at $+0.2°$ and $+0.4°$) for each of the mesas in the images. Figure 8.21 shows six images of a 10 μm wide mesa at various sample tilts to the (110) and ($\bar{1}$10) channeling directions. The region of the mesa allowing the best channeling (the lightest region of the mesa) varies in position within the mesa, and this suggests lattice plane rotation within the layer on the mesa as for regions off the mesa. From the images of Figure 8.21, the sense of the plane rotation can be determined, and it is found that, to cause the region of best channeling to move as it does, the (110) and ($\bar{1}$10) planes in the layer must have been rotated outward, away from the center of the mesa, as shown schematically in Figure 8.22. This is also true for the (110) planes in the layer on the four 10 μm mesas shown in Figure 8.19. Figure 8.23 shows a CSTIM image of several small mesas. In each, the position of best channeling is at the top right corner of the mesa, suggesting that the {110} planes in the layer on each of these small mesas were rotated in the same sense.

It is possible that the plane rotations within the layer on the tops of the small mesas were caused by dislocations in a similar fashion to the plane rotations in the layer away from the mesas. The evidence presented above suggests that elastic relaxation of the layer along directions in the interface plane, as proposed by Luryi and Suhir [43], was also responsible for the plane rotations. This is deduced from the fact that the planes in the 10 μm mesas were rotated outwards from the mesa center, the correct sense for a layer with lattice parameter greater than that of the substrate, and from

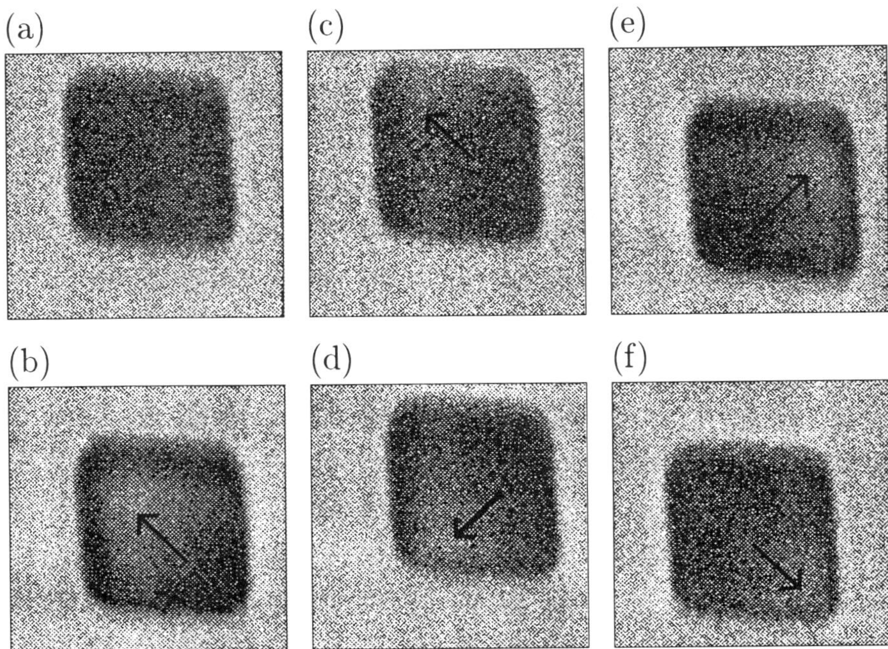

Figure 8.21. CSTIM images of a 10 μm wide mesa at different tilt angles to the [001] channeling direction: (a) x and y axes both $-1.0°$ from channeling. (b) x and y axes at channeling alignment. (c) x axis $+0.3°$, y axis channeled. (d) x axis $-0.3°$, y axis channeled. (e) x axis channeled, y axis $-0.3°$. (f) x and y axes both $-0.3°$. The arrow in each image points to the region of best channeling.

the suggestion that all of the small mesas had layer lattice planes that were rotated in the same sense. Evidence for the elastic relaxation of the layer on the small mesas has also come from optical microscopic and chemical etching studies [44] and from backscattered electron diffraction observations [22].

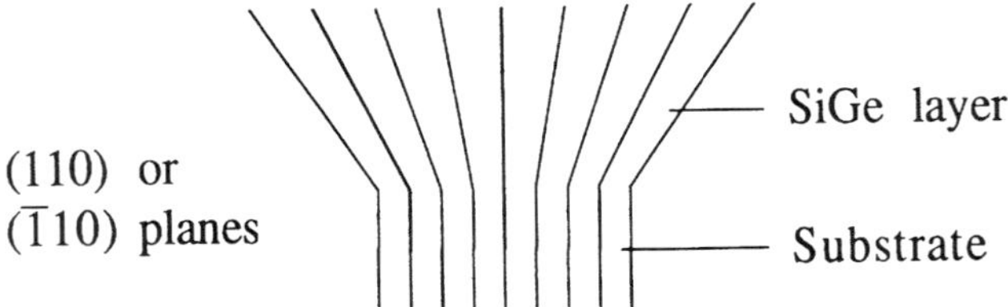

Figure 8.22. Schematic of the rotation of the (110) and ($\bar{1}$10) lattice planes in the layer on the small mesas.

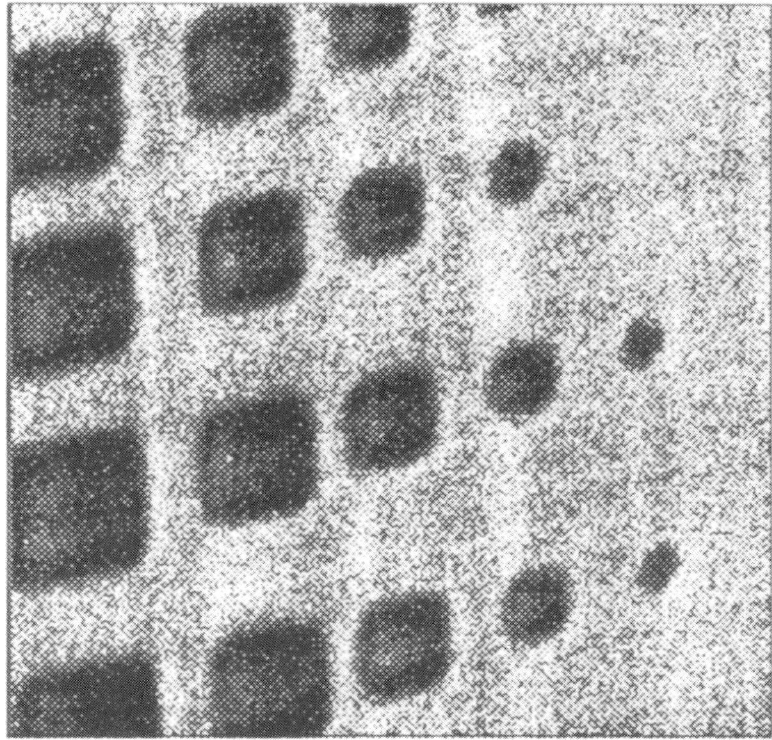

Figure 8.23. 60 μm wide image of a group of small mesas that range in size from approximately 2.5 μm to 10 μm wide. The position of best channeling in each mesa is toward its lower left corner.

8.5. CHANNELING SCANNING TRANSMISSION ION MICROSCOPY AND CHANNELING CONTRAST MICROSCOPY IMAGES FROM A $Si_{0.875}Ge_{0.125}/Si$ CRYSTAL WITH A HIGH DISLOCATION DENSITY

Both CCM and CSTIM images were produced from a sample consisting of a nominally 4 μm thick layer of $Si_{0.875}Ge_{0.125}$ deposited by MBE on to a (001) silicon substrate. The critical layer thickness for the formation of misfit dislocations for a layer containing 12.5% Ge is ~0.6 μm, so this sample was expected to contain significant numbers of dislocations.

Figure 8.24 is an optical image of the sample taken with Nomarski interference contrast to reveal the surface topography. Bands of contrast can be seen running approximately parallel to the [110] and [$\bar{1}$10] directions of the sample. The bands are 1 to 4 μm wide and are believed to be surface steps caused by misfit dislocations present at the layer–substrate interface. Such bands were

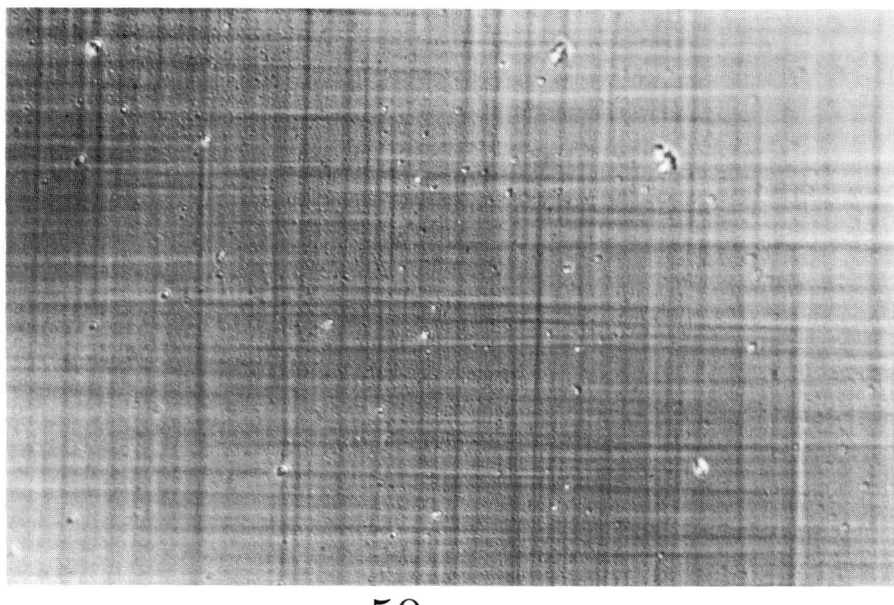

Figure 8.24. Optical image of the highly dislocated $Si_{0.875}Ge_{0.125}$ sample taken with Nomarski interference contrast. Reprinted from Ref. 14 with permission (©1993 American Institute of Physics, Woodbury, NY).

observed over the whole of the sample surface and their density suggests that large numbers of interface dislocations were present in the sample.

8.5.1. Channeling Contrast Microscopy and Channeling Scanning Transmission Ion Microscopy Results

Figure 8.25a shows a 100 μm wide CSTIM average energy loss image of a 30±5 μm thick portion of the sample taken with a 2 MeV ^{1}H ion beam channeled along the [001] crystal axis. Channeling contrast from dislocations can be seen. As a general observation from this and previous CSTIM images, there is less resolvable contrast from samples with a higher dislocation density and also the resolvable regions of contrast are larger. The image took 18 min to collect and resulted in the irradiated region receiving a dose of 1.7×10^{10} ^{1}H ions/cm^2.

Figure 8.26 shows channeled and nonchanneled backscattered ion energy spectra also taken from the sample using 2 MeV ^{1}H ions. The channeled spectrum (B) shows that the backscattered ion yield was significantly affected by channeling, although the minimum yield of 20% from the germanium atoms at the sample surface is an order of magnitude greater than would be expected from a defect-free crystal.

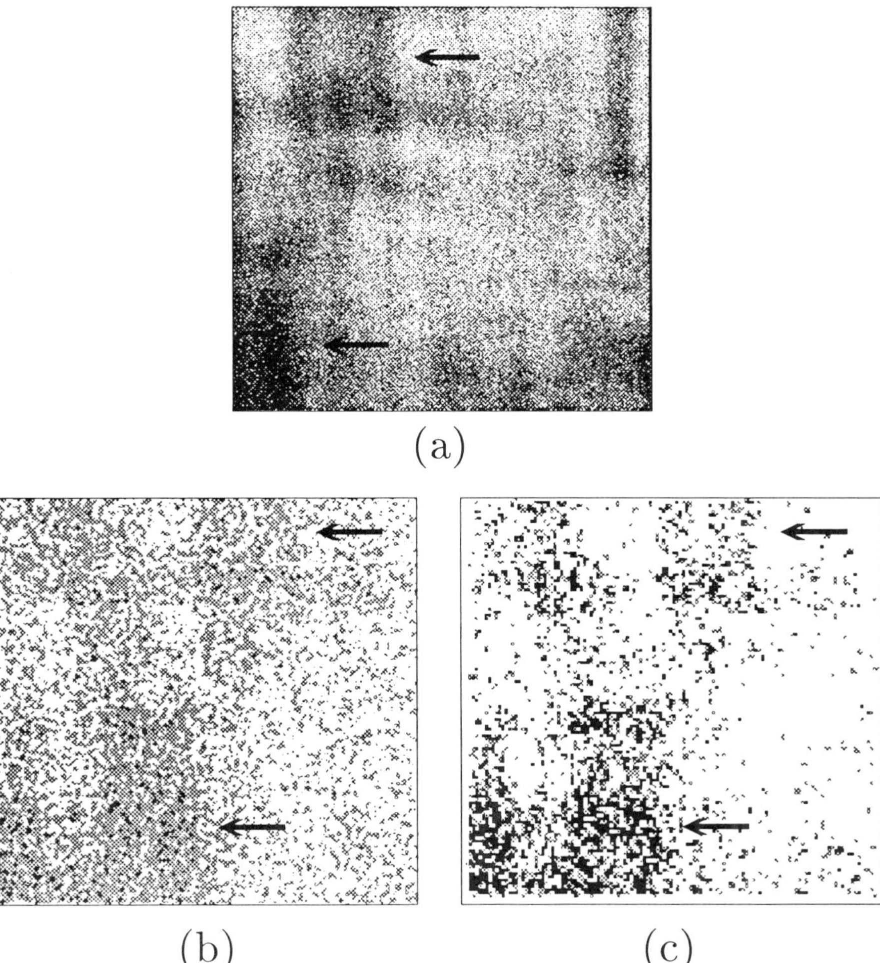

Figure 8.25. (a) 100 μm wide CSTIM energy loss image of the higher dislocated Si$_{0.875}$Ge$_{0.125}$/Si sample. (b) and (c) CCM images of the same sample region as (a), except for a small lateral shift. (b) shows the local variations in Si K X–ray counts and (c) shows the local variations in backscattered ion yield from the layer (using the energy window marked on Figure 8.26). Corresponding regions between the three images are arrowed.

Figures 8.25b and 8.25c are two CCM images from the same sample region as the CSTIM image shown in Figure 8.25a (there is a lateral shift of about 20 μm however). Figure 8.25b shows the variation in intensity of Si K X–rays. Figure 8.25c shows the variation in intensity produced from a window set on the backscattering spectrum shown in Figure 8.26 between energies corresponding to scattering from silicon atoms between the sample surface and 2 μm deep and

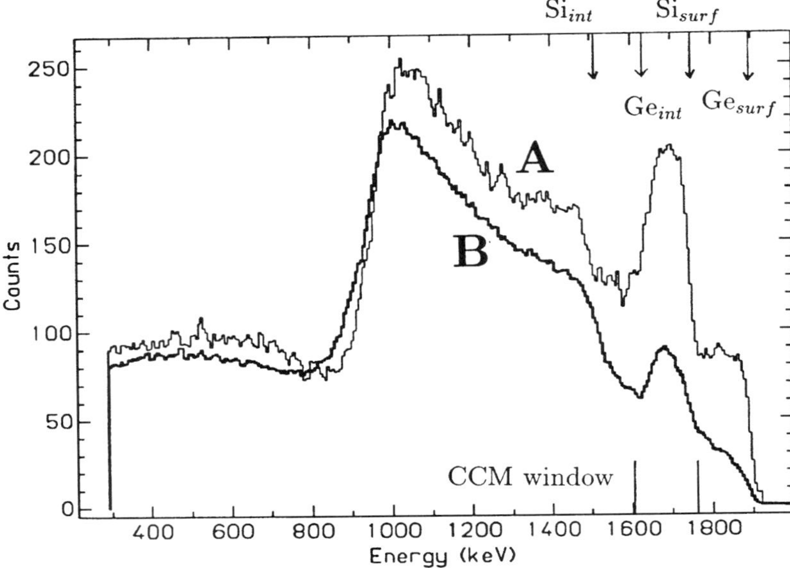

Figure 8.26. Nonchanneled (A) and channeled (B) backscattered ion energy spectra taken from the highly dislocated Si$_{0.875}$Ge$_{0.125}$/Si sample. Si$_{surf}$ and Ge$_{surf}$ are the energies of ^{1}H ions backscattered from surface silicon and germanium atoms, and Si$_{int}$ and Ge$_{int}$ are the energies of ^{1}H ions from silicon and germanium atoms at the interface. Both spectra were taken with the beam scanning a 100×100 μm^2 region of the sample (the region shown in Figure 8.25), and they have been normalized to the same incident beam dose.

from germanium atoms between 2 μm and 4 μm (i.e., the window was within the surface layer of the sample). Where dislocations caused dechanneling of the beam, the backscattered ion and Si X–ray counts were locally increased. The two arrows on these images are directed at such regions, and the arrows on the CSTIM image in Figure 8.25a point to the same regions. The CCM images have been smoothed from the originals, but, even so, show poor statistics. They took just over 30 min to collect at a beam current of approximately 100 pA and the scanned region received 1.30×10^{16} ^{1}H ions/cm^2.

The advantages and disadvantages of using CSTIM and CCM analysis were discussed in Section 5.3.4, and are well illustrated by Figure 8.25. Whereas CCM requires no sample preparation and can produce depth-resolved information, care must be taken to avoid damaging the sample by the beam. The CCM images shown above required a ^{1}H ion dose a million times greater than that used to produce the CSTIM energy loss image of the same region. The CSTIM energy loss image has much better statistics than those produced by CCM, because virtually every incident ^{1}H ion was recorded and used in its production. The CCM images shown had on average fewer than one X–ray or backscattered ion count per pixel before smoothing, whereas an average of

28 ^{1}H ion energy values were recorded at each pixel and used to produce the CSTIM average energy loss image.

Channeled and nonchanneled IBIC results from this same material are described in Section 8.7.

8.6. CHANNELING SCANNING TRANSMISSION ION MICROSCOPY IMAGES OF OXIDATION INDUCED STACKING FAULTS

8.6.1. Description of the Sample

Stacking faults were produced at the top surface of a silicon wafer by damaging the wafer surface and then annealing the wafer in an atmosphere containing oxygen. Such oxidation induced stacking faults (OISFs) [45] are extrinsic faults on {111} planes. Figure 8.27a shows the effects of adding an extra {111} plane

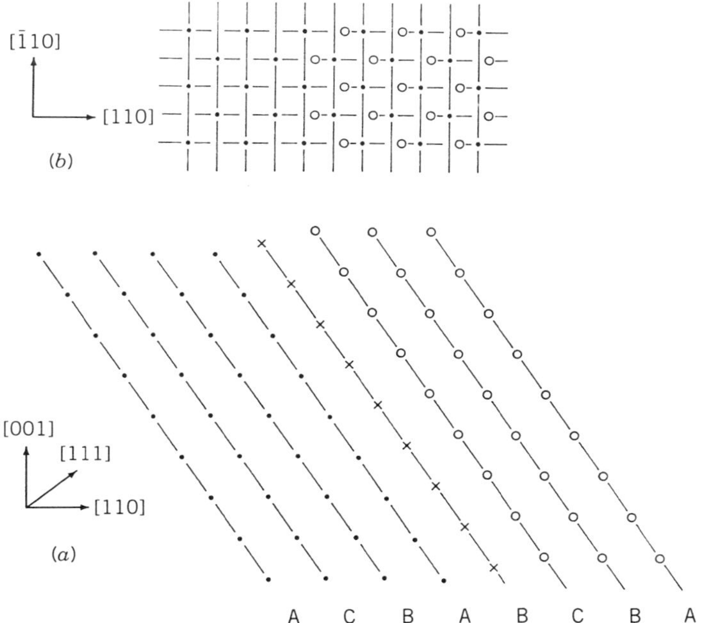

Figure 8.27. The effect of an extrinsic stacking fault on the (111) planes of the silicon lattice. (a) View of the lattice projected on to the ($\bar{1}$10) plane. For clarity, atoms on only one of the two sublattices have been shown. Atoms below the fault are filled circles, those above it are open circles, and those in the fault plane are crosses. (b) View of the fault looking along the [00$\bar{1}$] direction. In the fault region, rows of atoms below the fault are moved into the channels formed by those above the fault. A similar diagram for an intrinsic fault, showing views along ⟨110⟩ and ⟨111⟩ axes as well, is given in Ref. 46. Reprinted from Ref. 11 with permission (© 1995 American Physical Society, Woodbury, NY).

into the lattice of an f.c.c. crystal [a similar figure for an intrinsic fault is given in Ref. 46]. The effect of such a fault can be described as a translation of the top part of the crystal through a distance and direction given by the vector $\mathbf{R} = (a/3)\langle 111 \rangle$, where a is the silicon unit cell dimension, so that the fault vector is along the fault plane normal direction. Figure 8.27b shows a view of a fault along the [001] sample normal direction. The effect of the translation between the two halves of the crystal can be seen: the (110) planes below the fault are moved into the channels created by those above the fault by an amount equal to one-third of the interplanar distance.

The crystal used for CSTIM investigations in this section had the [001] crystal direction as its surface normal, so that the four different {111} planes on which the faults could lie were inclined at 54.7° to the surface plane. Figure 8.28 shows one of these planes in the crystal unit cell together with a schematic diagram of a fault. The appearance of this fault when viewed along the crystal surface normal direction is also shown. The faults intersected the sample surface along [110] or [$\bar{1}$10] directions and ran down into the crystal to a depth of a few microns. Their lower edge was curved so that the faults were D-shaped, and they were bounded on the curved edge by a partial dislocation (Section 5.2.1). Figure 8.29 shows an optical image of an etched portion of the sample. The faults appear dumbbell shaped in this photograph, as the etch preferentially attacked the sample where the bounding partial dislocation met the surface at each end of the fault.

8.6.2. Channeling Scanning Transmission Ion Microscopy Results

The CSTIM images from the wafer described above are presented in this section, taken with the sample thinned to a thickness of approximately 40 μm. Figure 8.30a is a 100 μm wide CSTIM average energy loss image taken with a 3 MeV ^{1}H ion beam channeled along the [001] axis of the sample. The image was generated using an uncollimated detector over a 90-min period with an an average count rate of 1,640 ^{1}H ions/s. D-shaped stacking faults lying on each of the four {111} planes can be seen; the fault shape in the image is the projection of the fault on to the (001) surface plane. The faults are darker than the image background. This means that the ^{1}H ions transmitted through the faults had a higher average energy loss than those transmitted through the surrounding material, which is to be expected for faults causing dechanneling. The faults vary in length from 13 μm to 24 μm and in width from 2.5 μm to 3.7 μm.

8.6.2.1. *Invisibility Criterion for Planar Channeling*   (110) and ($\bar{1}$10) planar channeled images of this portion of the sample were obtained by tilting just off the [001] axis; these are shown in Figures 8.30b and 8.30c. In the (110) channeled image, the faults that intersected the sample surface along the [110] direction are no longer visible. These faults were the ones lying on the ($\bar{1}$11) and (1$\bar{1}$1) planes, with translation vectors $\mathbf{R} = (a/3)[\bar{1}11]$ and $\mathbf{R} = (a/3)[1\bar{1}1]$, respectively. Both of these vectors lie within the (110) channeling planes, so the shift of the portion of the crystal above the fault with respect to that below it had

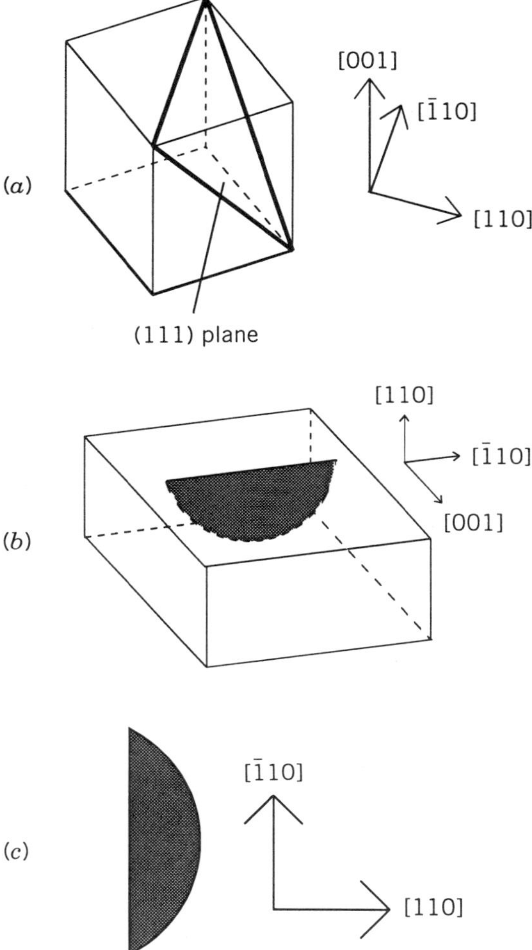

Figure 8.28. (a) Diagram showing the position of the (111) plane in the cubic unit cell. (b) Schematic diagram of a fault on a (111) plane. (c) Projected appearance of a fault on a (111) plane as viewed along the sample surface normal direction.

no effect on these planes. Faults on these two types of {111} planes, therefore, caused no dechanneling of the beam, and so they do not appear in the image of Figure 8.30b. Similarly, in the image taken with the beam channeled in the $(\bar{1}10)$ planes, those faults that intersected the sample surface along the $[\bar{1}10]$ direction are invisible. These faults had translation vectors $\mathbf{R} = (a/3)[111]$ and $\mathbf{R} = (a/3)[\bar{1}\,\bar{1}1]$, which lie within the $(\bar{1}10)$ planes.

It is possible to produce a general criterion for the invisibility of stacking faults in ion channeling images analogous to that which exists for transmission

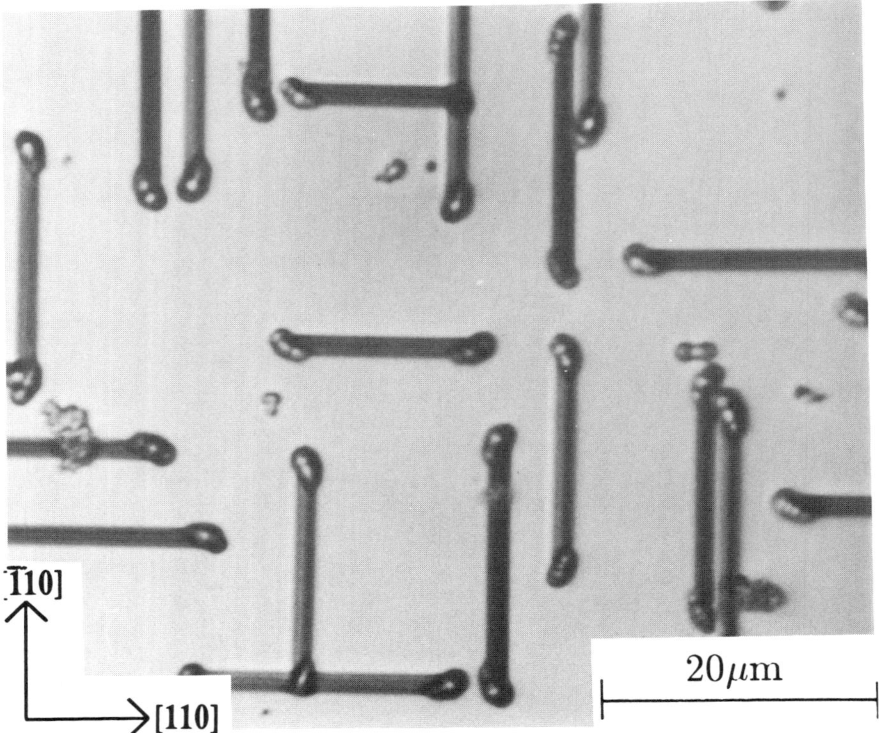

Figure 8.29. Optical image of stacking faults made visible by chemical etching of the wafer. Reproduced with permission from M. deCoteau, Department of Materials, University of Oxford.

electron microscopy (TEM). For TEM, faults are invisible in images if $\mathbf{g} \cdot \mathbf{R} = 0, \pm 1, \pm 2$, etc., where $\mathbf{g}$ is the reciprocal lattice vector corresponding to the planes used for diffraction and $\mathbf{R}$ is the fault vector [45]. $\mathbf{g} \cdot \mathbf{R} = 0$ implies that the fault vector is contained within the reflecting planes and so no contrast is observed (i.e., the fault causes no shift to the planes). Invisibility for $\mathbf{g} \cdot \mathbf{R} = \pm 1, \pm 2$, etc. arises because addition of a lattice vector to the fault vector does not affect the contrast and $\mathbf{R}$ dotted with a lattice vector is an integer. In this case, the fault causes a shift to the channeling planes that is equal to exactly one or more interplanar distances so that the channeled ions see no disruption to the planes. For example, a translation vector $\mathbf{R} = (a/3)[111]$ is equivalent (as far as displacement of rows into channels) to a vector $\mathbf{R} = (a/6)[2\bar{1}\,\bar{1}]$ because

$$\frac{a}{3}[111] = \frac{a}{6}[2\bar{1}\,\bar{1}] + \frac{a}{2}[011]$$

and $(a/2)[011]$ is a lattice vector.

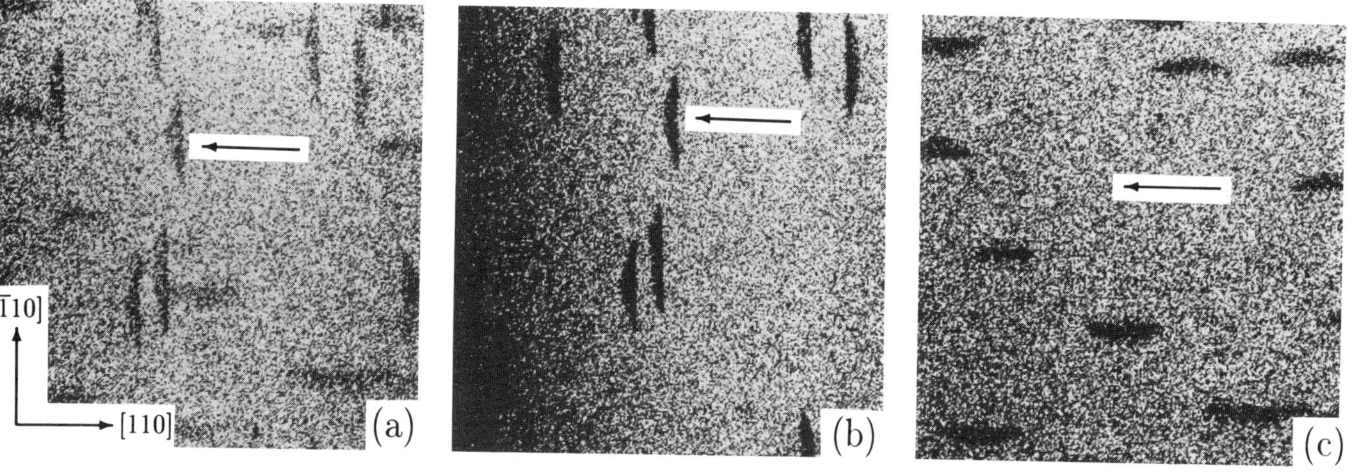

Figure 8.30. 100 μm wide CSTIM images of stacking faults. (a) Beam channeled along the [001] crystal axis. (b) and (c) Beam channeled in the (110) and ($\bar{1}$10) planes, respectively. The same fault is arrowed in (a) and (b), and its position in (c), had it been visible, is also arrowed. Reprinted from Ref. 13 with the kind permission of Elsevier Science, B.V., Amsterdam, The Netherlands.

For ion channeling, the same $\mathbf{g} \cdot \mathbf{R}$ criterion can be used to predict the faults that will be invisible for channeling in a given set of planes, although care must be taken to use the correct $\mathbf{g}$ vector. The channeling planes used to produce the images of Figures 8.30b and 8.30c were described as (110) and ($\bar{1}$10), as the indices used to describe planes are reduced to the lowest form possible. However, the reciprocal lattice vectors for these planes are $\mathbf{g}$ = [220] and $\mathbf{g}$ = [$\bar{2}$20], and it is these vectors that must be used when calculating $\mathbf{g} \cdot \mathbf{R}$. This is because the use of [220] and [$\bar{2}$20] correctly takes into account the interplanar spacing. For example, in Figure 8.30c, the faults on the (111) planes, with fault vector $\mathbf{R} = (a/3)[111]$, are invisible. In this case $\mathbf{g} \cdot \mathbf{R}$ is calculated from $[\bar{2}20] \cdot (a/3)[111] = 0$. Equivalently, with the fault vector represented by $(a/6)[2\bar{1}\,\bar{1}]$, $\mathbf{g} \cdot \mathbf{R} = 1$. This demonstrates that, as in TEM, $\mathbf{g} \cdot \mathbf{R}$ equal to zero or an integer produces invisibility; the disappearance of faults on the ($\bar{1}$11), (1$\bar{1}$1), and ($\bar{1}\,\bar{1}$1) planes in the images of Figures 8.30b and 8.30c can be similarly explained. The direction of the translation vector of a particular fault can be determined by finding two or more planar channeling directions of the incident beam that cause the fault to be invisible in the resulting CSTIM images.

Care must be taken, however, to ensure that the invisibility of a given fault in a CSTIM image is caused by the $\mathbf{g} \cdot \mathbf{R}$ criterion. Figure 8.31 shows two images of a portion of the sample, both taken with the beam channeled in the (111) planes. In the image of Figure 8.31a, which was taken with the beam close to the [$\bar{1}\,\bar{1}$2] axis, faults on the ($\bar{1}$11) planes are visible and two are arrowed. In Figure 8.31b, taken with the beam close to the [0$\bar{1}$1] axis, these two faults cannot be seen. The reason for this is not that the faults had $\mathbf{g} \cdot \mathbf{R} = 0$ (or an integer) in

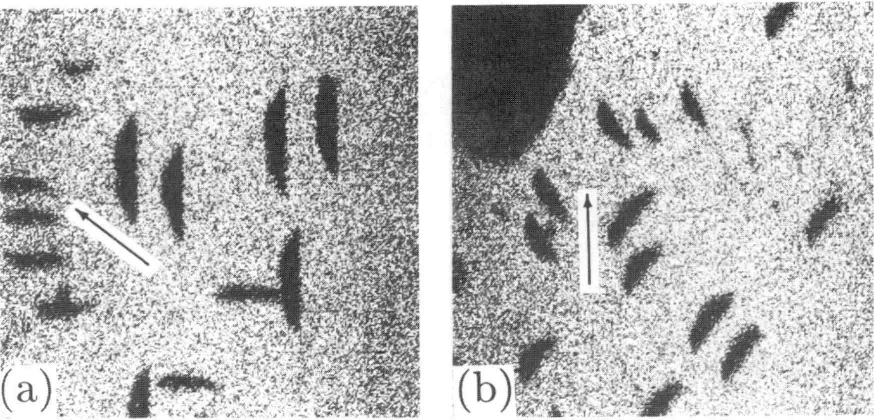

Figure 8.31. CSTIM images of stacking faults, beam channeled in the (111) planes for both. (a) 100 μm wide, beam near the [$\bar{1}\,\bar{1}$2] axis. (b) 125 μm wide, beam near the [0$\bar{1}$1] axis. The two faults near the head of the arrow in (a) are invisible in (b) because the ($\bar{1}$11) planes are close to end-on in (b) and not in (a).

the second image (**g** and **R** are the same for these faults in both images). Instead, when the beam was close to the [0$\bar{1}$1] axis (where the ($\bar{1}$11) and (111) planes intersect), the ($\bar{1}$11) planes were almost end-on. Faults on these planes were, therefore, effectively invisible in this image owing to the very narrow width of the image of an end-on fault. Faults on the (111) planes are invisible in both images of Figure 8.31 for the same reason. End-on faults are visible in low noise or small scan images as thin dark lines (see Figure 8.35a).

8.6.2.2. Contrast Variations Between Axial and Planar Channeling

Planar channeling can be used to give information on fault translation vectors that axial channeling cannot give. It was found that (110) and ($\bar{1}$10) planar channeling also caused the OISFs to be visible with stronger contrast than [001] axial channeling. Figure 8.32 shows two images of a fault on a ($\bar{1}\,\bar{1}$1) plane (the same fault as was arrowed in Figure 8.30). The first image was taken with the beam channeled along the [001] axis of the crystal, the second with the beam channeled in the (110) planes. The images have been printed over the same range of energy loss values so that the fault contrast can be directly compared between them. The fault shows much stronger contrast in the planar channeled image: the difference between the average ^{1}H ion energy loss at the fault and that away from it is approximately 22 keV for the planar channeled image compared with 12 keV for the axially channeled image. The reason for this is that, assuming the fault can be treated as a new crystal surface, approximately 30% of the channeled ^{1}H ions would have been dechanneled by the fault for planar channeling, whereas only 2% to 3% would have been dechanneled by the fault when the beam was axially channeled.

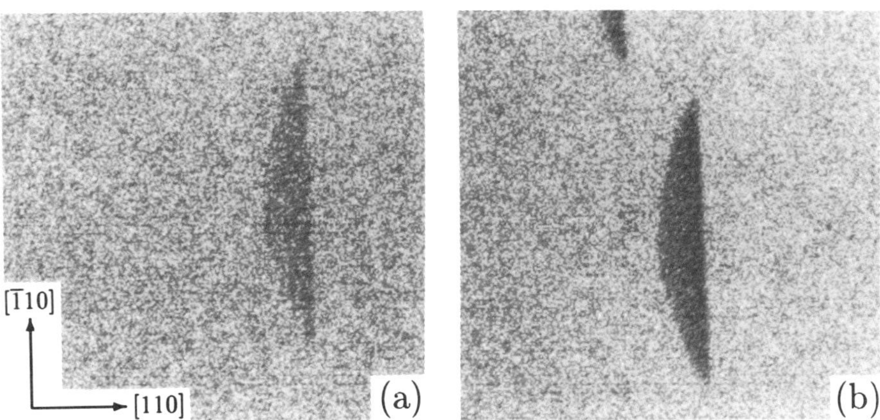

Figure 8.32. 30 μm wide CSTIM images of the fault arrowed in Figure 8.30. (a) Beam channeled along the [001] axis. (b) Beam channeled in the (110) planes. Reprinted from Ref. 11 with permission (© 1995 American Physical Society, Woodbury, NY).

8.6.2.3. Depth of the Deepest Fault that Can Be Imaged with Channeling Scanning Transmission Ion Microscopy

Figure 8.33a shows the difference in average ^{1}H ion energy loss between the stacking fault and the image background as a function of distance across the fault for the fault arrowed in Figure 8.30. The linescan is across the center of the fault in Figure 8.32b and was taken with a 3 MeV ^{1}H ion beam using an uncollimated detector with the beam channeled in the (110) planes close to the sample normal direction. The right-hand edge of the fault corresponds to where it intersected the sample surface, and the left-hand edge to where it ended inside the crystal. From this linescan, the depth of the fault at its deepest point can be determined. The width of the fault from Figure 8.33a is 3.0 ± 0.2 μm, so that it was $3.0 \times \tan 54.7°$ μm below the sample surface at its deepest, i.e., 4.2 ± 0.3 μm.

On going from the right hand side of the fault to the left, that is, on going from the sample surface to a depth of 4.2 μm, the height of the fault above the background dropped. This was caused by the 'natural' dechanneling of the ^{1}H ions as they penetrated into the sample which meant that the fraction of ^{1}H ions remaining channeled decreased with depth into the sample. Because the channeled ^{1}H ions are responsible for producing CSTIM image contrast, the contrast across the fault fell with fault depth. Those ^{1}H ions that are dechanneled at a depth of 4 μm traveled this distance with the reduced energy loss rate, also lowering the contrast from deeper parts of the fault. From the rate of decrease of the fault contrast in Figure 8.33a, it is expected that faults of the order of 10 μm below the sample surface of silicon crystals would be detectable by CSTIM when 3 MeV ^{1}H ions channeled in $\{110\}$ planes and an uncollimated detector are used.

However, the actual depth of the deepest fault that can be imaged is greater than 10 μm. Figure 8.33b shows a similar plot of fault contrast versus distance across the fault taken with the 3 MeV ^{1}H ion beam channeled in the ($\bar{1}\bar{1}1$) planes of the crystal. The image was taken using a detector collimated to an acceptance half-angle of 1° (see Section 5.4.7). In this case, the fault on the (111) plane was closest to the sample surface on the left hand side of the image, and was 5.0 ± 0.3 μm below the sample surface at the right-hand, deepest side. There was no noticeable fall off in contrast across the fault, and this resulted from a combination of using a more collimated detector to exclude many of the ^{1}H ions that were initially not channeled (or which dechanneled close to the sample surface), and of using $\{111\}$ planes for channeling, which have a greater dechanneling half distance than $\{110\}$ planes for ^{1}H ions in silicon [47,48]. The fault contrast of ~75 keV was greater than that of the fault in Figure 8.33a (22 keV). This again was because of the use of a collimated detector and $\{111\}$ channeling planes.

A portion of the stacking fault sample was thinned, so that it was wedge-shaped, with a thickness that varied from 5 μm upward. The wedge was mounted so that the faults were on the side of the crystal away from the beam. A detector with an acceptance half-angle of 0.4° was used to capture the transmitted ^{1}H ions, and the incident beam was channeled in a set of the $\{111\}$ planes

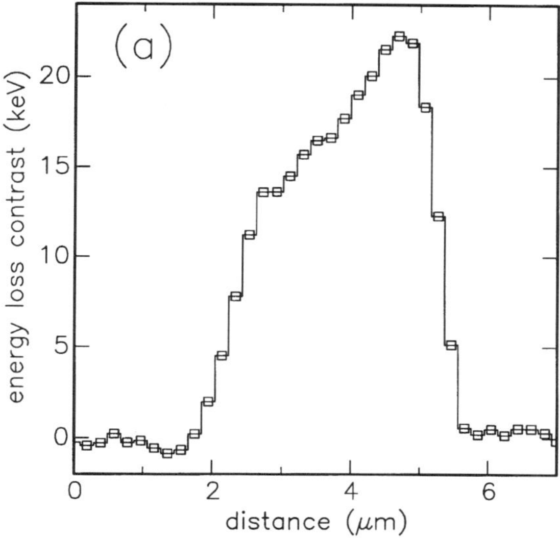

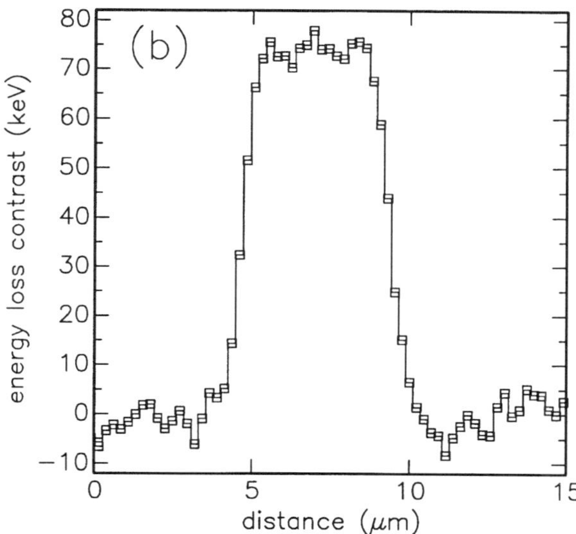

Figure 8.33. (a) Contrast of the fault shown in Figure 8.32 versus distance across the fault. Data taken using an uncollimated detector, beam channeled in the (110) planes. Each point has been smoothed with those ±0.2 μm on either side. The right-hand side of the fault was closest to the sample surface. (b) Contrast versus distance across a second fault (labeled A in Figure 8.35a), taken with the beam $(\bar{1}\,\bar{1}1)$ channeled, detector collimated to an acceptance half-angle of 1°. For this fault, the left-hand side was closest to the sample surface. Reprinted from Ref. 11 with permission (© 1995 American Physical Society, Woodbury, NY).

of the sample. Shown in Figure 8.34 is a montage of CSTIM images of the wedge, with the wedge thickness varying from 15 μm near the bottom of the montage to 50 μm near the top. Stacking faults that were on the back surface of the wedge as viewed by the beam can clearly be seen over the whole range of wedge thicknesses shown. This demonstrates that by use of {111} planes for channeling, sufficient numbers of ^{1}H ions were channeled up to 50 μm into the sample, enabling the faults to be detected. It was found that faults on the back surface of the wedge were not revealed in images taken for 1 hr using a detector with a larger acceptance angle ($\sim 4°$ half-angle), or using a collimated detector but with the incident beam channeled in {110} planes.

8.6.2.4. Contrast Changes on Tilting off the Channeling Direction   As in the case of CSTIM images of misfit dislocations discussed earlier in this chapter, changes in stacking fault contrast were also observed in CSTIM images on tilting the crystal so that the beam was just away from a planar channeling direction. However, in the stacking fault case, the contrast changes occurred symmetrically on both sides of the channeling direction, requiring a different explanation to that used for the dislocations.

Figure 8.35 shows four images of a stacking fault (labeled A in Figure 35a) on a (111) plane of the crystal. The images were taken with the sample tilted so that the beam angle to the $(\bar{1}\,\bar{1}1)$ channeling direction was slightly different in each image, and the tilt angle from exact channeling alignment is given. In the image taken closest to the channeling direction, the fault is uniformly dark. It was therefore causing dechanneling of the beam. On tilting away from the channeling direction by $+0.11°$ (an amount close to the channeling critical angle) the near-surface part of the fault changed in contrast to be bright, whereas the deeper parts of the fault remained dark. This meant that the portion of the fault close to the sample surface was actually allowing the ^{1}H ions to be transmitted with a lower average energy loss than surrounding virgin crystal. On tilting further from the channeling direction (image at $\pm 0.17°$), the whole fault became bright in contrast, and so was causing the ^{1}H ions to suffer a lower average energy loss than surrounding crystal. However, it can be seen that this effect was happening most strongly at the part of the fault closest to the sample surface.

Figure 8.34. CSTIM image montage of the stacking fault sample wedge, with the faults on the back surface of the sample. Image taken with the incident beam channeled in {111} planes of the sample and a collimated detector. The sample thickness varies from 15 μm at the bottom of the image to 50 μm at the top of the image. The images have each had the sloping background (caused by the variation in sample thickness) removed and have been smoothed and processed with histogram equalization to increase the fault contrast.

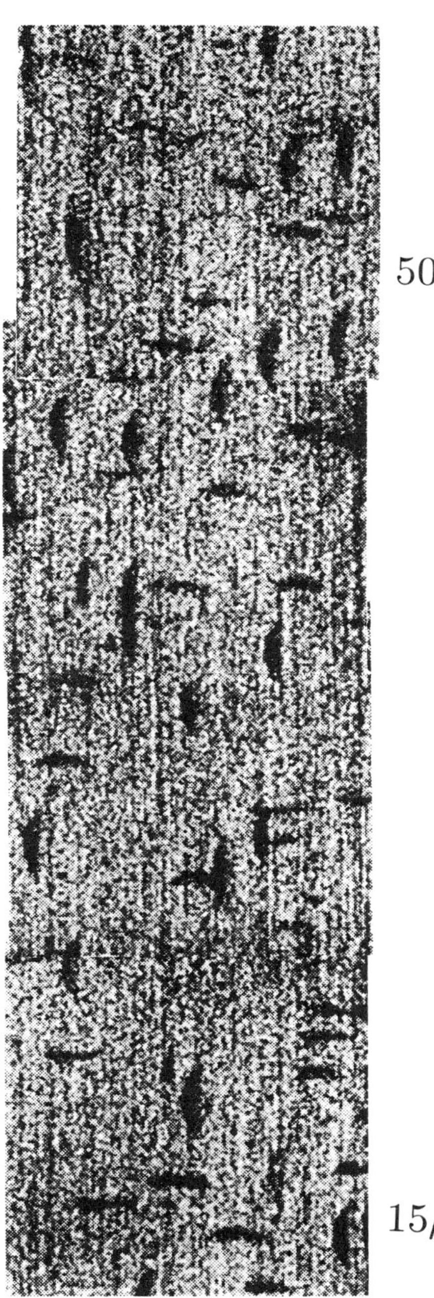
50μm
15μm

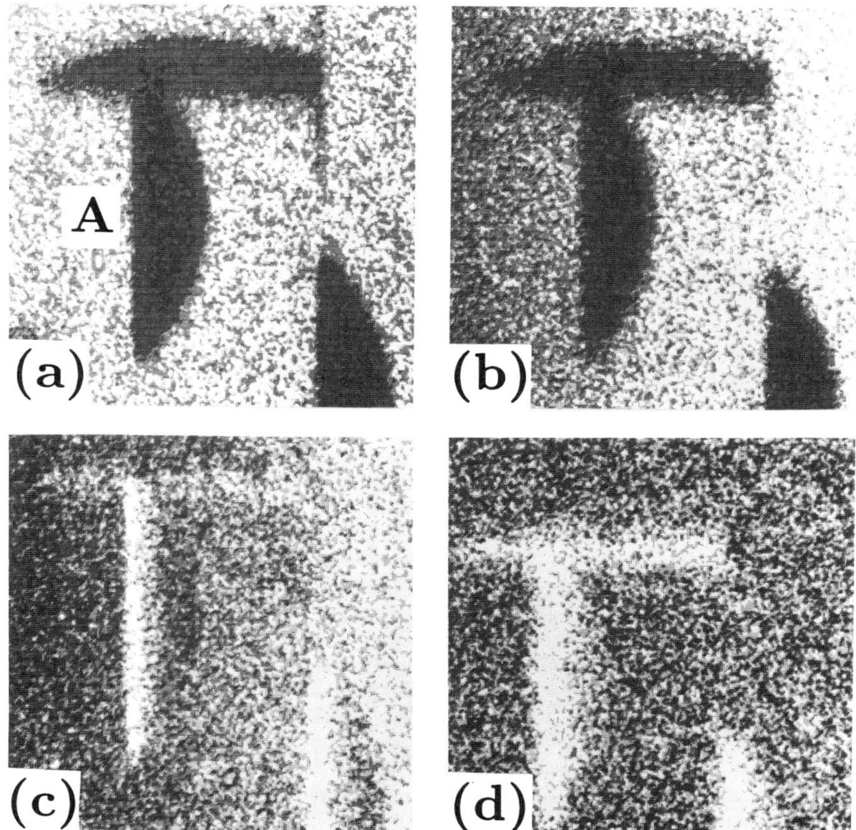

Figure 8.35. CSTIM images of a fault on a (111) plane, taken with the beam angle to the ($\bar{1}\,\bar{1}1$) channeling direction varied. (a) Beam channeled. (b), (c), (d), Beam +0.05°, +0.11°, +0.17° from ($\bar{1}\,\bar{1}1$) channeling, respectively. The thin, vertical line near the top right corner of (a) is an end-on fault on a ($\bar{1}\,\bar{1}1$) plane. Images smoothed and processed with histogram equalization. Reprinted from Ref. 12 with permission (© 1995 American Physical Society, Woodbury, NY).

Figure 8.36 shows this effect even more clearly. Figure 8.36a shows the average energy of ^{1}H ions transmitted through a part of a (different) fault where it was close to the sample surface, and through virgin crystal, versus angle to the ($\bar{1}\,\bar{1}1$) planes. In Figure 8.36b, the difference between these two curves is plotted. This second graph is effectively a plot of fault contrast versus tilt angle, with positive values meaning that the fault would have been dark in CSTIM images and negative values meaning that it would have been bright. For angles

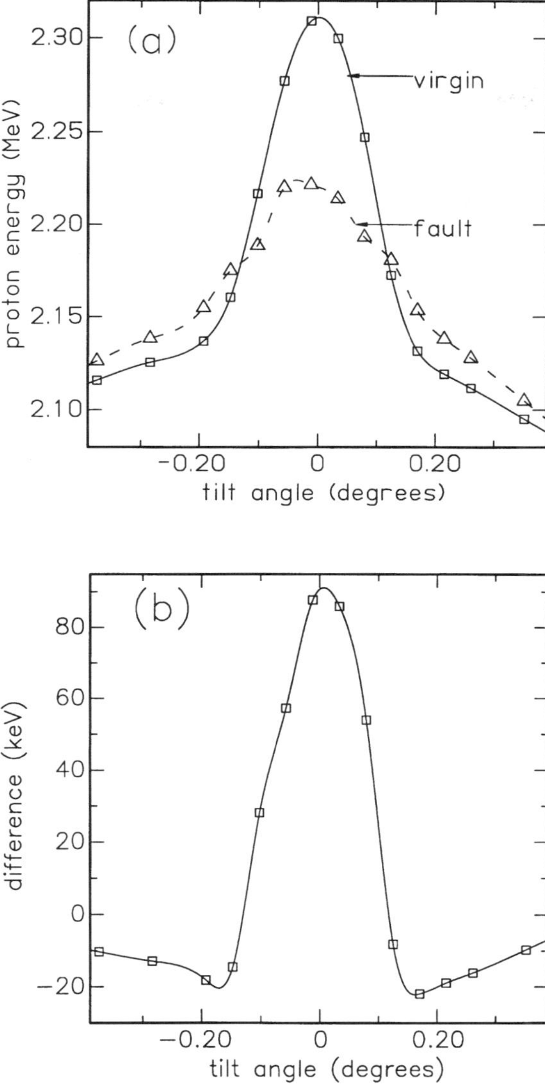

Figure 8.36. (a) Average transmitted energy versus tilt angle from $(\bar{1}\,\bar{1}1)$ channeling for a region of a stacking fault close to the sample surface and a near-by virgin region of the sample. (b) Difference between the two curves of (a) showing fault contrast versus tilt angle. A negative value indicates the fault would have appeared bright in a CSTIM image, a positive value that it would have appeared dark (the points at $+0.03°$ and $+0.08°$ on the curve for the fault in (a) were taken from a slightly deeper part of the fault than the other points and so are likely to be slightly lower in energy than if they had been taken from a part close to the sample surface). Reprinted from Ref. 12 with permission (© 1995 American Physical Society, Woodbury, NY).

within approximately $0.12°$ of the channeling direction, the net effect of the fault was to cause dechanneling, so that the average transmitted energy at the fault was lower than that away from the fault. Further from the channeling direction, however, the fault transmitted ^1H ions with a higher average energy than the surrounding crystal, and it can be seen (Figure 8.36b) that this effect was greatest when the beam was approximately $0.2°$ from the channeling direction. The fault contrast is symmetrical about the channeling direction. The two plots in Figure 8.36 can be compared with those of Figure 8.8 which show analogous graphs for the bunch of five misfit dislocations in the $Si_{0.95}Ge_{0.05}/Si$ crystal.

The ability of stacking faults to cause ^1H ions to be transmitted with a lower energy loss than surrounding material when the incident beam was just away from exact channeling alignment cannot be due to rotated lattice planes, as was found for the case of misfit dislocations. This is because stacking faults are not associated with rotated planes (except at the bounding partial dislocation at the fault edge inside the crystal).

To understand this phenomenon, Monte Carlo computer simulations of the passage of ions through a stacking fault were performed [41]. It was found that stacking faults could cause initially nonchanneled ions to become channeled. Figure 8.37a shows simulated trajectories of 100 ions for virgin crystal at exact planar alignment and at $0.17°$ to alignment and for crystal containing a fault with the beam at $0.17°$ to alignment. With the crystal tilted just off planar alignment, a significant fraction of the ions become channeled on passing through the fault plane (trajectories parallel to the horizontal axis in Figure 8.37c). Figure 8.38 shows four individual ^1H ions trajectories, also produced by simulation, two of which become channeled at the fault and two of which do not.

This phenomenon can be understood by consideration of ions with blocked trajectories (described in Section 1.4). At a tilt angle of $0.17°$ to planar alignment (for the case of the 3 MeV ^1H ions described here), a large fraction of the ^1H ion trajectories fall initially into this category. Such ^1H ions have a transverse energy that is sufficient to take them through the atomic planes, but they still feel the steering effects of the continuum potential for some of their path so that they spend a greater than average time in the vicinity of the planes. When such an ion is close to a plane, it has a low transverse kinetic energy and high transverse potential energy (assuming that its total transverse energy is a constant and swaps between kinetic and potential depending on the ion's distance from a plane). If it then passes through a stacking fault, it experiences a sudden shift in the positions of the atomic planes which can result in a lowering of its potential energy. By this process, its total transverse energy is lowered, in some cases enough for the ion to become channeled. At beam angles to the channeling direction just greater than the channeling critical angle, it is therefore possible for a stacking fault to locally enhance channeling by converting ions with blocked trajectories into those that are channeled. The average transmitted energy loss of ions passing through the fault is then lower than that of those passing through surrounding virgin crystal.

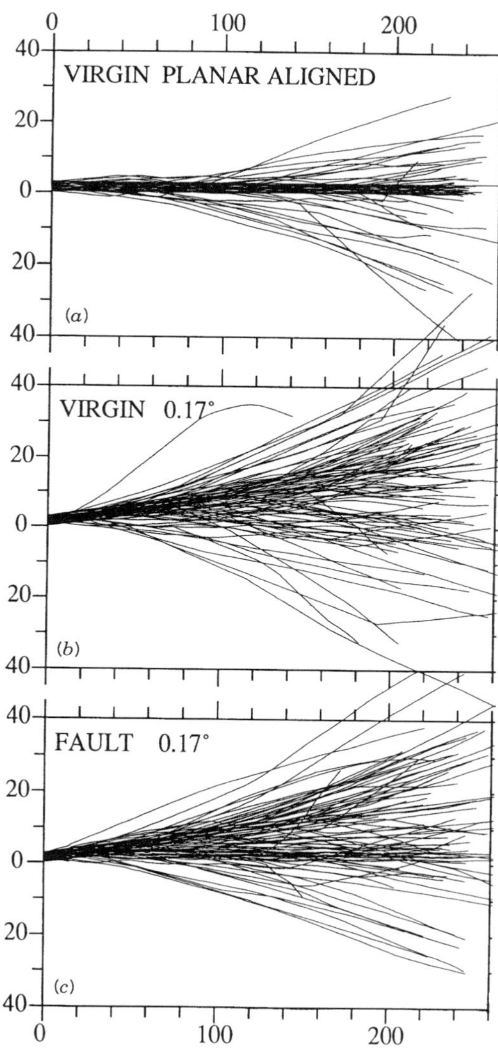

Figure 8.37. Simulated trajectories of 100 ions incident at or close to ($\bar{1}\,\bar{1}1$) channeling. The incident ion path was 1.7° from the [112] crystal axis, and the projections of the ion paths on to the plane normal to this axis is shown. The ion path length, projected on to the [112] axis, was 1,500 nm. The units along the axes are interplanar distances, with ($\bar{1}10$) planes at integer x values and ($\bar{1}\,\bar{1}1$) planes at $y = 0.75$, 1.0, 1.75, and 2.0. (a) Virgin crystal, beam aligned with the ($\bar{1}\,\bar{1}1$) planes. (b) Virgin crystal, beam at 0.17° to alignment. (c) Crystal with a fault at 100 nm depth, beam at 0.17° to alignment. Notice in (c) a considerable channeling fraction (paths parallel to the x axis) which is absent in (b). Reprinted from Ref. 12 with permission (© 1995 American Physical Society, Woodbury, NY).

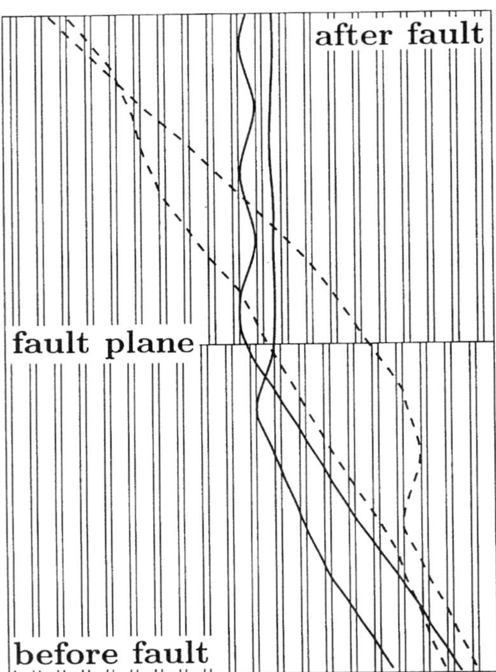

Figure 8.38. Simulated trajectories of individual ^{1}H ions passing through a stacking fault. Projection is the same as for Figure 8.34. Vertical lines are {111} planes. The ^{1}H ions' paths start in the bottom right, and two (solid lines) become channeled at the fault while two (broken lines) do not. Reprinted from Ref. 12 with permission (© 1995 American Physical Society, Woodbury, NY).

The transition from blocking to channeling can be seen in the energy spectra of the transmitted ^{1}H ions. Figure 8.39 shows two such spectra extracted from the data used to produce Figure 8.35d. Spectrum A is from a region of virgin crystal and spectrum B from a region where the stacking fault was approximately 1 μm deep and showing bright contrast. Spectrum B shows fewer ^{1}H ions transmitted with a low energy but a larger number with a high energy characteristic of channeled ^{1}H ions. This is the reverse of the situation that occurred when the beam was aligned with the channeling direction (Figure 8.35a, spectra on and off the fault shown in Figure 5.8) when the fraction of ^{1}H ions in the high-energy part of the spectrum at the fault was smaller than that of the spectrum from virgin crystal.

The observation of this blocking to channeling transition at stacking faults demonstrates that images of crystal defects produced by ion channeling can be used for studies of the physics of the interaction of channeled ions with crystal defects as well as for investigations of the defects themselves.

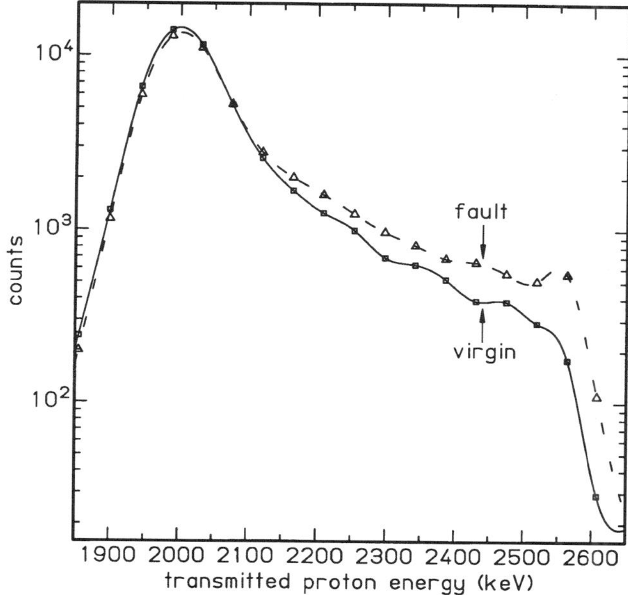

Figure 8.39. Transmitted ^{1}H ion energy spectra from the data used to produce Figure 8.35d. The spectrum shown in a solid line is from a region away from the fault. The spectrum shown as a broken line is from a region where the fault was approximately 1 μm deep. Note the presence of more high-energy counts measured in the spectrum from the fault. Reprinted from Ref. 12 with permission (© 1995 American Physical Society, Woodbury, NY).

8.6.3. Large Area Imaging and Imaging under Thin Surface Layers

The final image of stacking faults (Figure 8.40) is a montage of several CSTIM images taken to show that a large region of a sample can be mapped with this technique. This sample was similar to the one described above, except that copper had been diffused into the sample and formed colonies of small precipitate particles nucleated at the dislocations bounding the faults. With the beam channeled in the {111} planes of the sample, 200 μm wide CSTIM images were taken at 150 μm wide intervals. The sample had to be tilted very slightly between each image so that the beam remained channeled owing to the curvature of the crystal. The area of the sample that has been imaged is 735 × 325 μm^2, and the images have been plotted each over the same range of energy loss values and with no histogram equalization.

There are a number of contrast features that are visible in the montage. Over the whole image, dark grey D shapes can be seen that are stacking faults. The image gets lighter toward the right hand side, and this is caused by a decrease in the sample thickness across the imaged region. Finally, there is a darker,

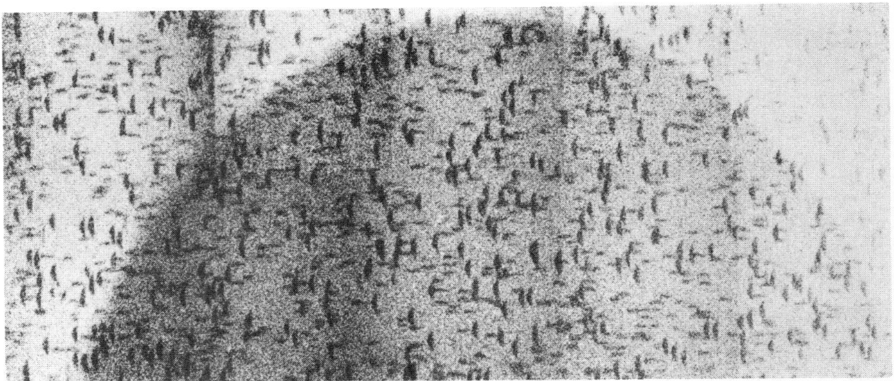

Figure 8.40. Montage of CSTIM images of a sample containing stacking faults. The sample area shown is approximately $735 \times 325 \ \mu m^2$. The darker grey semicircular region is the part of the sample covered with a thin metal layer.

semicircular region covering most of the center of the figure. This marks the location of a thin metal layer (a few tens of nanometres in thickness, composed of gold and palladium) that had been deposited on the surface of the sample to form a Schottky barrier in order to examine the faults with the IBIC and EBIC techniques. The layer caused some scattering of the incident beam, so that fewer ^{1}H ions were channeled on passing through the layer region than through surrounding crystal. The layer region therefore appears darker than regions where it is absent. However, stacking faults can clearly be seen in the sample where the layer is present, demonstrating that CSTIM can detect defects under thin surface layers when used with a MeV ^{1}H ion beam.

8.7. DISLOCATION IMAGING USING ION BEAM INDUCED CHARGE MICROSCOPY

IBIC microscopy has also been used to image bands of misfit dislocations in a 4 μm thick epitaxial layer of $Si_{0.875}Ge_{0.125}$ grown on a silicon substrate (CSTIM and CCM images from this sample were discussed in Section 5.5). The sample contained a high dislocation density, as revealed in the Nomarski optical image shown in Figure 8.24.

A gold layer, approximately 100 nm thick, was deposited on the front surface to form a 2 mm diameter Schottky barrier, and a gold wire was connected to this using silver loaded paint to form the front connection to the sample. The back face of the sample was attached to the microprobe sample holder using silver loaded paint to make the ohmic contact. Capacitance–voltage measurements showed that the n-type substrate had a doping concentration of approximately $2 \times 10^{15} \ cm^{-3}$, and the depletion layer thickness with no external bias voltage was approximately 0.7 μm. The network of dislocations at the $Si_{0.875}Ge_{0.125}/Si$ interface was thus outside the depletion region. At room temperature, the maximum measured charge pulse height at the preamplifier output was 230 keV and the noise level was 140 keV, with most of the charge pulse height spectrum not resolved from the noise level because of the large barrier area fabricated on this sample. Thus, the IBIC images here show the variation in the measured number of counts in a single window from 140 to 230 keV in the IBIC pulse height spectrum. No bias voltage was applied to the sample as the consequent increase in leakage current was found to ruin the IBIC image contrast.

Three IBIC images of this sample are shown in Figure 8.41 [14], from different regions of the sample to avoid ion induced damage effects. Line scans extracted from across the middle of these three images are shown in Figure 8.42. For these images the sample was tilted away from any major crystal axis to prevent channeling effects altering the image contrast. On these images, there are bands visible running along the [110] and [$\bar{1}$10] directions; dark areas represent a high number of measured counts, and light areas represent a low number of measured counts. The interpretation of the image contrast is that the dark areas represent regions of low dislocation density, whereas the light regions represent regions of high dislocation density. The lowest number of counts per pixel is limited by the noise level, and the maximum number of counts is limited by the dislocation density.

In the 50×50 μm^2 IBIC image there are only approximately five ^{1}H ions used at each pixel superimposed on a noise level of four counts, with an ion dose of ~130 ^{1}H ions/μm^2. The standard deviation in the average number of counts in each pixel in the measured image was $\sigma = 2.5$, and this was reduced to $\sigma = 0.8$ when ten adjacent vertical pixels (pixels 120 to 130, i.e., a 2 μm wide section) from across the middle of the image were combined to a single horizontal line scan. This was further reduced to $\sigma = 0.5$ by smoothing the combined horizontal data set in groups of ±1 pixel as shown in Figure 8.42c.

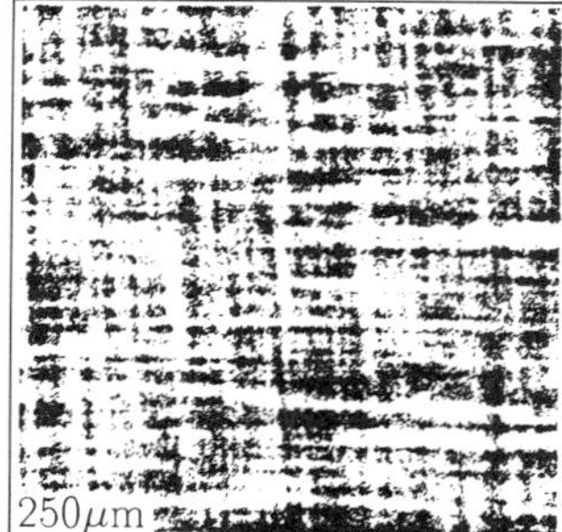

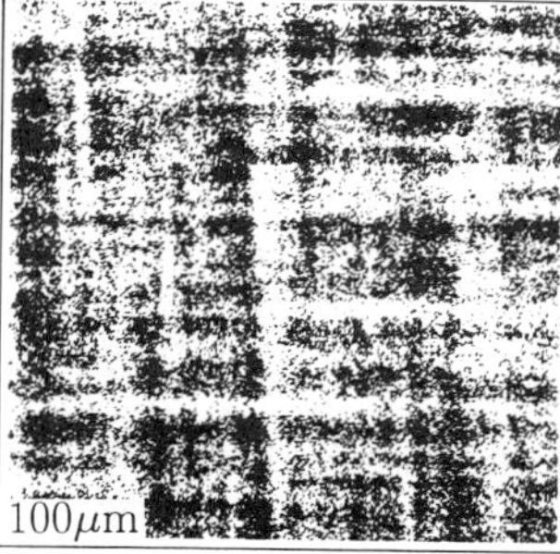

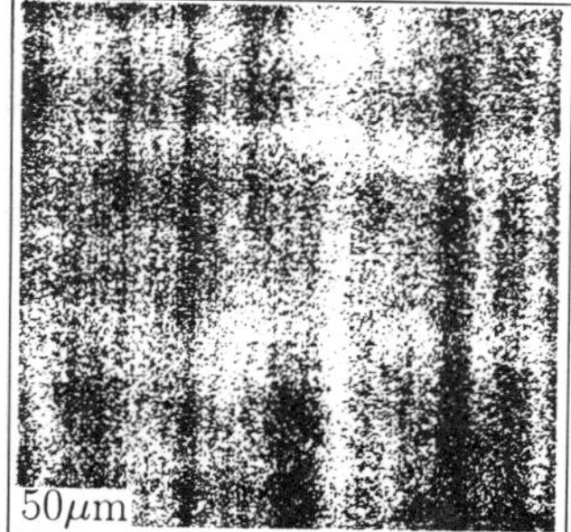

Figure 8.41. Three IBIC images of different areas of the $Si_{0.875}Ge_{0.125}$ sample, with the scan sizes shown in the bottom left corner of each image. Reprinted from Ref. 14 with permission (©1993 American Institute of Physics, Woodbury, NY).

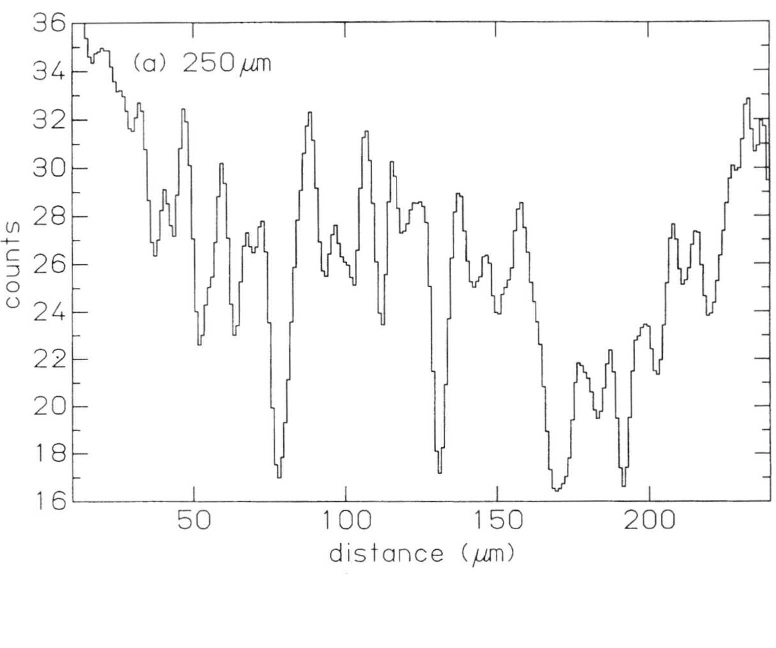

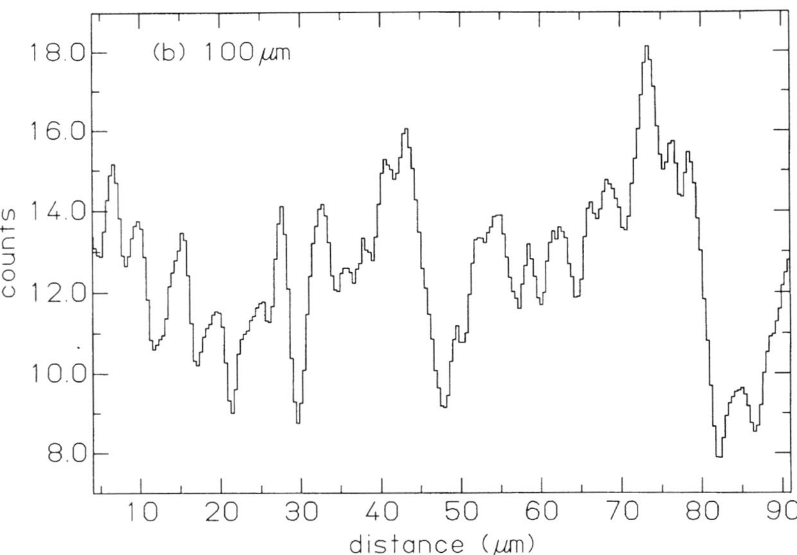

Figure 8.42. Horizontal line scans extracted from the ten central adjacent vertical pixels from (a) the 250×250 μm^2, (b) 100×100 μm^2, and (c) 50×50 μm^2 IBIC images shown in Figure 8.41, and smoothed in groups of ± 1 horizontal pixels. The minimum resolvable band width of 0.8 μm is arrowed in (c), which also shows a horizontal line scan from another set of ten vertical pixels, offset by +3 counts for clarity. Reprinted from Ref. 14 with permission (©1993 American Institute of Physics, Woodbury, NY).

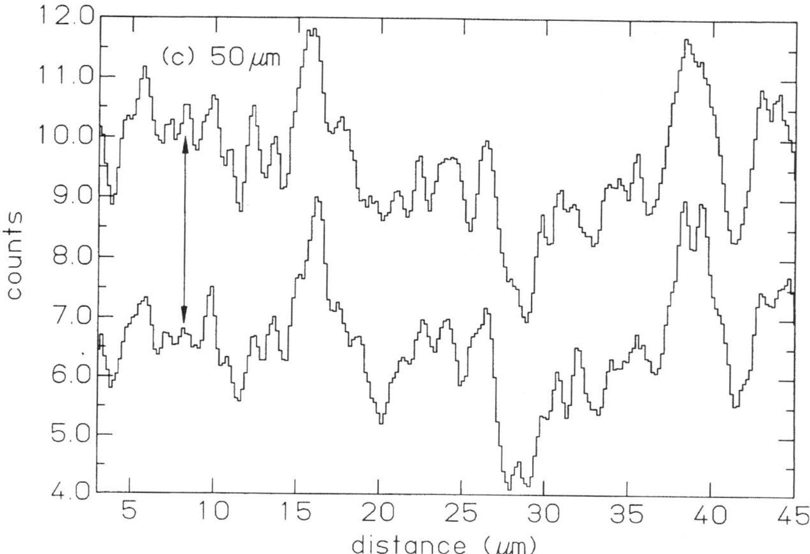

Figure 8.42. (*Continued*)

The minimum resolvable band width in the smoothed 50 μm horizontal line scan shown in Figure 8.42c was 0.8 μm (i.e., four pixels wide), and this band is arrowed. Averaging over three horizontal pixels decreases the horizontal resolution to three pixels, which represents a distance of 0.6 μm in Figure 8.41. The minimum resolvable band width of four pixels in Figure 8.42c is thus likely to be limited by this horizontal smoothing process. To further show that this structure is real, and not an artifact of the low signal to noise level and the smoothing process, another horizontal line scan from the 50×50 μm^2 IBIC image is also shown in Figure 8.42c. This is from another consecutive group of ten adjacent vertical pixels combined together (i.e., a 2 μm wide section), and has been combined and smoothed as for the other line scans, as described above. This second line scan has been offset vertically by three counts from the lower line scan for clarity, and the same band is arrowed here.

The maximum measured IBIC pulse height of 230 keV is equivalent to the number of charge carriers generated by the beam passing through approximately 11 μm of silicon. Because the depletion region was less than 1 μm thick, most of the carriers measured in the charge pulses were from outside the depletion region. Since the Si$_{0.875}$Ge$_{0.125}$ layer is only 4 μm thick, then some carriers were measured from below the dislocation network. The most likely mechanism of observed dislocation contrast is as follows. Where the ion beam passed through a region of low dislocation density at the Si$_{0.875}$Ge$_{0.125}$/Si interface, carriers from below the interface were measured as they could diffuse to the depletion region without being trapped at the interface. However, where the ion beam passed through a region of high dislocation density at the $Si_{0.875}Ge_{0.125}/Si$ interface, carriers from below the interface were not detected because they were trapped and recombined at the interface, resulting in a smaller measured charge pulse. This smaller charge pulse was not resolvable from the noise level, so only large pulses from regions of lower dislocation density were detected.

The maximum number of ^1H ions per pixel in the 50×50 μm^2 IBIC image was limited by the maximum number of 130 ^1H ions/μm^2 that could be used before ion induced damage prevented any more pulses from being measured. It was thus ion induced damage that prevented the use of smaller scan sizes and so limited the minimum resolvable band width.

Following from this, further work reported evidence that the size of MeV ion induced charge pulses measured from this same material depended on both the crystallographic and electrical properties of the $60°$ misfit dislocations present [15]. The results were correlated with both backscattered and transmission ion channeling analysis. With the sample in nonchanneled alignment, the measured ion induced charge pulses depended on the number of charge carriers that recombine at the dislocations. With the sample in [001] alignment, the rotated (110) and ($\bar{1}$10) planes around the $60°$ dislocations affected the local rate of carrier generation and so altered the size of the measured ion induced charge pulses.

An example of this behavior is given in Figure 8.43, which shows two 250 μm linescans extracted from the same 10 μm wide strip of channeled and non-

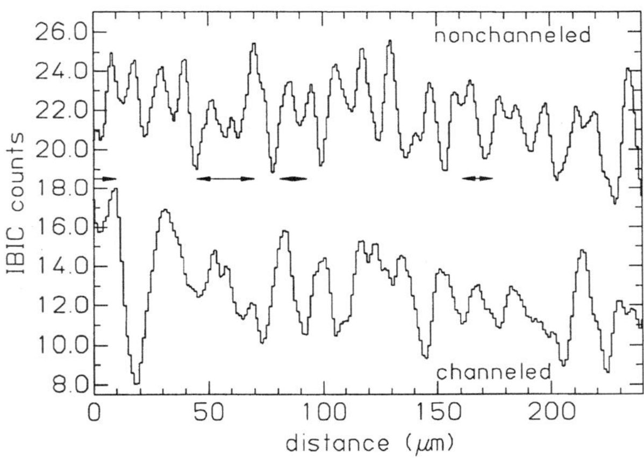

Figure 8.43. Channeled and nonchanneled 250 μm line scans extracted from a 10 μm wide strip from 250×250 μm^2 IBIC images of the sample. Each pixel is smoothed by averaging with each pixel on either side of it, and the nonchanneled line scan is offset vertically by twelve counts for clarity. Zones where the spatial variation of maxima and minima coincide in the offset line scans are indicated by arrows between them. Regions where there is no agreement between the locations of the maxima and minima are not indicated by arrows. Reprinted from Ref. 15 with permission (©1995 American Institute of Physics, Woodbury, NY).

channeled IBIC images of the same area. Most regions of the channeled IBIC linescans are totally different from the nonchanneled IBIC line scans, which demonstrates that the ion induced charge pulses are indeed affected by both the electrical and crystallographic nature of the 60° dislocations. A possible explanation of this behavior is given in Ref. 15.

Because the signal strength measured with EBIC microscopy [24] and with OBIC microscopy [49] does not depend on the crystallographic orientation of the sample, neither can determine the Burgers vector of dislocations. This work therefore demonstrates the important observation that the measured charge pulses with MeV ions are sensitive to both the crystallographic and electrical nature of dislocations, so it should be possible to determine the Burgers vector of dislocations with IBIC microscopy.

8.8. COMPARISON OF CRYSTAL DEFECT IMAGING TECHNIQUES

This chapter began with a description of established techniques for crystal defect imaging, and has introduced the new techniques of CSTIM, CCM, and IBIC. Table 8.1 compares the abilities and requirements of all of these methods.

The newest of the techniques are CSTIM and IBIC, so it is of interest to consider what role these techniques have alongside all of those described in Section 8.2. CSTIM is capable of imaging dislocations and stacking faults that are several microns below the surface of crystals that are tens of microns thick with a spatial resolution of a few hundred nanometers. In all of these respects, it falls neatly between the commonly used techniques of TEM and X–ray topography: TEM can image defects throughout the thickness of samples thinned to approximately 1 μm with a spatial resolution of a few nanometers, and X–ray topography can image defects throughout wafers that are hundreds of microns thick with a spatial resolution of approximately 1 μm. Because CSTIM has a strain sensitivity of the order of 0.05%, it again falls between the TEM and X–ray topography whose sensitivities are of the order of 0.5% and 0.01%, respectively. The information in ECCI images comes only from very close to the sample surface; CSTIM information comes from much deeper, so that defects further into the sample can be observed, but CSTIM information is averaged over its analytical depth. Imaging by EBIC and CL can use bulk samples and provide information on the electrical properties of defects, but CSTIM has a better spatial resolution and can see deeper defects.

The CSTIM seems to find its niche in being able to image defects several microns deep with sub-micron spatial resolution and with fault vector information available. The technique also offers the potential of enabling the interaction of channeled ions with crystal defects to be studied in more detail to increase our understanding of the channeling process. IBIC offers the potential of being able to give both structural and electronic information on crystal defects when combined with ion channeling, while EBIC and electron channeling cannot be

TABLE 8.1. Comparison of Semiconductor Crystal Defect Imaging Techniques[a]

Technique	Measured Signal	Spatial Resolution	Max. Defect Depth	b Analysis?	Sample Preparation	Comments
CSTIM	Channeled ion energy loss	0.3 μm (beam spot size)	>10 μm	Yes	Thinned to a few 10s of microns	Analysis under thin surface layers
CCM	Backscattered ion or X–ray yield	1 μm (beam spot size)	~10 μm	Yes?	None	Depth information, needs high beam current
IBIC	Electron-hole charge pulses	~0.3 μm (beam spot size)	~10 μm	Yes?	Schottky barrier fabrication	Semiconductors only; analysis under layers
TEM	Transmitted electron diffraction	1.5 nm (dislocation strain field)	1 μm	Yes	Thinned to <1 μm	Lattice imaging with high resolution TEM
ECCI	Backscattered electron diffraction	10 nm (dislocation strain field)	200 nm	Yes	None	
EBIC	Electron-hole current	1 μm (beam spreading)	5 μm	No	Schottky barrier fabrication	Semiconductors only
CL	Light from electron-hole recombination	1 μm (beam spreading)	A few microns	No	None	Semiconductors only
X–ray topography	X–ray diffraction	A few microns (dislocation strain field)	mm	Yes	None	Images through whole wafer
SIRM	Transmitted or scattered IR light	1 μm (beam spot size)	mm	Yes	Damage-free surfaces	Depth information through whole wafer
Optical microscopy	Amplitude/phase of visible light	0.5 μm (diffraction limit)	Surface	No	Chemical etching	

[a]In the spatial resolution column, the comment in brackets is the factor limiting the spatial resolution to the value quoted. "b analysis?" is whether or not Burgers vector analysis is possible with the technique. A question mark means the capability is unproved.

combined together as electron channeling is prohibited by the beam scattering in the gold layer required for the Schottky barrier.

REFERENCES

1. J.P. Hirth and J. Lothe, *Theory of Dislocations*. McGraw-Hill, New York (1968).

2. H.J. Queisser, *Mat. Res. Soc. Symp. Proc.* **14**:323 (1983).

3. S.M. Sze, ed., *VLSI Technology*. McGraw-Hill, New York (1983).

4. M.B.H. Breese, P.J.C. King, J. Whitehurst, G.R. Booker, G.W. Grime, F. Watt, L.T. Romano, and E.H.C. Parker, *J. Appl. Phys.*, **73**(6):2640 (1993).

5. P.J.C. King, M.B.H. Breese, G.R. Booker, J. Whitehurst, P.R. Wilshaw, G.W. Grime, F. Watt, and M.J. Goringe, *Nucl. Instr. Meth.*, **B77**:320 (1993).

6. P.J.C. King, M.B.H. Breese, P.R. Wilshaw, G.R. Booker, G.W. Grime, F. Watt, and M.J. Goringe, In: A.G. Cullis, A.E. Staton-Bevan, and J.L. Hutchinson, eds., *Microscopy of Semiconducting Materials*, Conference Series No. 134. Institute of Physics, Bristol (1993), p. 153.

7. P.J.C. King, PhD thesis, University of Oxford (1993).

8. P.J.C. King, M.B.H. Breese, P.R. Wilshaw, G.R. Booker, and G.W. Grime, *Ann. Chim. Fr.* **19**:257 (1994).

9. M.B.H. Breese, P.J.C. King, P.J.M. Smulders, and G.W. Grime, *Phys. Rev.* **B51**(5):2742 (1995).

10. P.J.C. King, M.B.H. Breese, P.R. Wilshaw, P.J.M. Smulders, and G.W. Grime, *Nucl. Instr. Meth.* (in press).

11. P.J.C. King, M.B.H. Breese, P.R. Wilshaw, and G.W. Grime, *Phys. Rev.* **B51**(5):2732 (1995).

12. P.J.C. King, M.B.H. Breese, P.J.M. Smulders, P.R. Wilshaw, and G.W. Grime, *Phys. Rev. Lett.* **74**(3):411 (1995).

13. P.J.C. King, M.B.H. Breese, P.R. Wilshaw, and G.W. Grime, *Nucl. Instr. Meth.* (in press).

14. M.B.H. Breese, P.J.C. King, G.W. Grime, and P.R. Wilshaw, *Appl. Phys. Lett.* **62**(25):3309 (1993).

15. M.B.H. Breese, P.J.C. King, and G.W. Grime, *Appl. Phys. Lett.* **65**(25):3227 (1994).

16. P.B. Hirsch, A. Howie, R. Nicholson, D.W. Pashley, and M. J. Whelan, *Electron Microscopy of Thin Crystals*. Krieger, Malabar, Florida (1977).

17. G. Thomas and M.J. Goringe, *Transmission Electron Microscopy of Materials*. Wiley-Interscience, New York (1979).

18. D.B. Williams, A.R. Pelton, and R. Gronsky, eds., *Images of Materials*. Oxford University Press, New York (1991).

19. G.R. Booker, In: S. Amelinckx, ed., *Modern Diffraction and Imaging Techniques in Materials Science*. North Holland, Amsterdam (1971), pp. 613–653.

20. G.R. Booker. In *Analytical Techniques for Semiconducting Materials*, Edited by B.O. Kolbesen, D.V. McCaughan, and W. Vanderworst, *Electrochem. Soc. Proc.* **90**(11) Electrochemical Society, NJ (1990), p.107.

21. A.J. Wilkinson, G.R. Anstis, J.T. Czernuska, N.J. Long, and P.B. Hirsch, *Phil. Mag.* **A68**(1):59 (1993).

22. A.J. Wilkinson, *Proceedings of the 13th International Conference on Electron Microscopy*, (1994), p. 97.

23. J.T. Czernuska, N.J. Long, E.D. Boyes, and P.B. Hirsch, *Mat. Res. Soc. Symp. Proc.* **209**:289 (1991).

24. H.J. Leamy, *J. Appl. Phys.* **53**:R51 (1982).

25. K. Löhnert and E. Kubalek, In: *Microscopy of Semiconducting Materials*, Conference Series No. 67. Institute of Physics, Bristol (1983), p. 303.

26. C.R.M. Grovenor, *Microelectronic Materials*. Institute of Physics, Bristol (1992).

27. M.A. Capano, L. Hart, D.K. Bowen, D. Gordon-Smith, C.R. Thomas, C.J. Gibbings, M.A.G. Halliwell, and L.W. Hobbs, *J. Cryst. Growth* **116**:260 (1992).

28. W.C. Dash, *J. Appl. Phys.* **27**(10):1193 (1956).

29. M.R. Brozel, I. Grant, R.M. Ware, and D.J. Stirland, *Appl. Phys. Lett.* **42**(7):610 (1983).

30. G.R. Booker, Z. Laczik, and P. Kidd, *Semicond. Sci. Technol.* **7**:A110 (1992).

31. V.A. Phillips, *Modern Metallographic Techniques and their Applications*. Wiley-Interscience, New York (1971).

32. C.J. Gibbings, C.G. Tuppen, and M. Hockly, *Appl. Phys. Lett.* **54**(2):148 (1989).

33. D. Hull, *Introduction to Dislocations*. Pergamon Press, Oxford (1975).

34. R. People, *IEEE J. Quant. Elec.* **22**(9):1696 (1986).

35. B.S. Meyerson, High-speed silicon-germanium electronics. *Sci. Am.* **270**(3):42 (1994).

36. J.W. Matthews and A.E. Blakeslee, *J. Cryst. Growth* **27**:118 (1974).

37. C.G. Tuppen, C.J. Gibbings, and M. Hockly, *J. Crystal Growth* **94**:392 (1989).

38. J.C. Bean, *J. Vac. Sci. Technol.* **B4**(6):1427 (1986).

39. D.G. Schimmel, *J. Electrochem. Soc.* **126**(3):479 (1979).

40. P.M.J. Marrée, J.C. Barbour, J.F. van der Veen, K.L. Kavanagh, C.W.T. Bulle-Lieuwma, and M.P.A. Viegers, *J. Appl. Phys.* **62**(11):4413 (1987).

41. P.J.M. Smulders and D.O. Boerma, *Nucl. Instr. Meth.* **B29**:471 (1987).

42. P.R. Bevington, *Data Reduction and Error Analysis for the Physical Sciences*. McGraw–Hill, New York (1969).

43. S. Luryi and E. Suhir, *Appl. Phys. Lett.* **49**(3):140 (1986).

44. A.R. Powell, R.A. Kubiak, T.E. Whall, E.H.C. Parker, and D.K. Bowen, *Mat. Res. Soc. Symp. Proc.* **220**:277 (1991).

45. G.R. Booker and W.J. Tunstall, *Phil. Mag.* **13**:71 (1966).

46. S.U. Campisano, G. Foti, E. Rimini, and S.T. Picraux, *Nucl. Instr. Meth.* **149**:371 (1978).

47. J.A. Davies, J. Denhartog, and J.L. Whitton, *Phys. Rev.* **165**(2):345 (1968).

48. L.C. Feldman and B.R. Appleton, *Phys. Rev.* **B8**(3):935 (1973).

49. C.J.R. Sheppard, *Scanning Micros.* **3**:15 (1989).

9

OTHER MATERIALS ANALYSIS AND MODIFICATION

This chapter describes further applications of the nuclear microprobe for materials analysis of crystalline and metallurgical samples. Many such applications are given in the nuclear microprobe conference series listed in Appendix 1. Section 9.1 describes the use of channeling contrast microscopy (CCM) to image the mosaic spread present in single YBaCuO high-T_C (high critical temperature) superconductor crystals. It is then shown how a nuclear microprobe can be used to assess the substitutionality of iron in YBa(Cu–Fe)O single crystals using a method of combined backscattering and PIXE channeling. The use of a nuclear microprobe enables the effect of any mosaic spread in the crystal to be isolated so that it does not affect the channeling results.

Section 9.2 describes the use of CCM to image regrowth regions in laser-annealed diamond. Here, the ability of CCM to analyze the material without the requirement of any sample preparation was very important owing to the difficulty of thinning diamond for transmission electron microscopy (TEM) analysis, for example.

Section 9.3 describes channeling contrast analysis of mercury–cadmium–telluride photodetectors, in which different types of defect structures are identified. Section 9.4 describes the use of a nuclear microprobe in metallurgical analysis. An AlPb alloy was analyzed to show the distribution of lead precipitates present within the aluminum matrix. Images from STIM, PIXE, and EPMA of the same area are compared, and it is shown how depth-resolved information using STIM can be obtained by tilting the sample about the beam axis.

A different subject is covered in Section 9.5, where the use of the nuclear microprobe for materials modification and fabrication is described. Firstly it is shown how high-energy, heavy ions, such as xenon and chromium ions, can be used for microlithography using a nuclear microprobe to create structures such as acoustic microscope lenses. It is then shown how narrow, high aspect ratio channels can be created in resist materials such as polymethyl methacrylate (PMMA) for potential applications in X–ray lithography, using focused MeV ^{1}H ions. Finally, it is described how irradiation of glass using a nuclear microprobe can create optical waveguides using MeV ^{1}H ions and ^{4}He ions.

9.1. MOSAIC SPREAD IN HIGH-T_C SUPERCONDUCTOR CRYSTALS

This section describes the use of the Oxford nuclear microprobe to give information on the amount of strain, misorientation, and crystalline order from three different high-T_C YBaCuO crystals with CCM using 2 MeV ^{1}H ions, based on work described elsewhere [1]. The relationship with the observed surface structure using optical microscopy and channeling contrast is described. Three crystals studied here are (a) with a smooth surface, (b) exhibiting surface defects, and (c) a twinned crystal with a smooth face, discussed in more detail below.

Figure 9.1 shows a [001] aligned CCM image of the measured variation in the number of 2 MeV ^{1}H ions backscattered from barium up to a depth of 2 μm below the surface. Barium was chosen, because it is the heaviest element

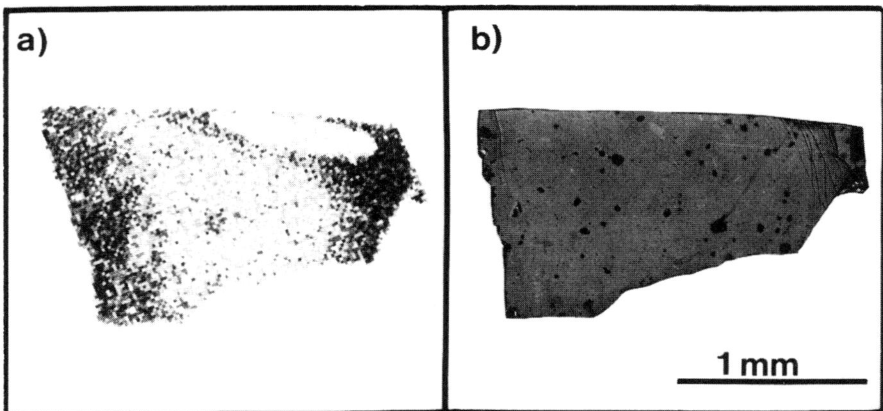

Figure 9.1. (a) [001] CCM image for barium in YBaCuO from a depth of 2 μm. (b) Corresponding optical image of the YBaCuO crystal showing the surface to be free of steps and smooth, although some pitting is observed. Reprinted from Ref. 1 with kind permission from Elsevier Science B.V., Amsterdam, The Netherlands.

present, so the backscattered signal was most clearly resolved. Figure 9.1b is an optical image of this crystal showing a smooth surface, apart from a few pits. The measured minimum yield of 3% from the central region of this crystal, together with the uniform channeling contrast, indicated good crystal quality. The higher backscattered yield observed at the left- and right-hand edges of the crystal is caused by deflection of the beam away from the channeling axis because of the large scanned area in this case. This experimental limitation was only a problem at this very large scan size and does not affect any of the work described below. It should also be noted that this effect has only been observed with MeV ^{1}H ions, where the axial channeling critical angle is ~0.5°, because the larger channeling critical angle for heavier ions makes this effect insignificant.

Figure 9.2a shows a [001] aligned backscattering CCM image from a crystal exhibiting surface defects, and Figure 9.2b shows a corresponding optical image. The CCM image shows uniform channeling in the central region corresponding to the smooth area of the crystal in Figure 9.2b, and the measured minimum yield from the central region was 10%. In comparison, the upper region exhibits a higher yield indicating either poorer crystal quality or a difference in the location of the [001] channeling axis. The deeper steps at the bottom of the image show a large contrast, and further scans at different tilt angles indicated a continuous variation in the tilt angle for the best channeling of up to 0.5° away from the [001] axis over the steps. PIXE images of the top region of the crystal showed that the major constituent of the chipped and pitted region was barium.

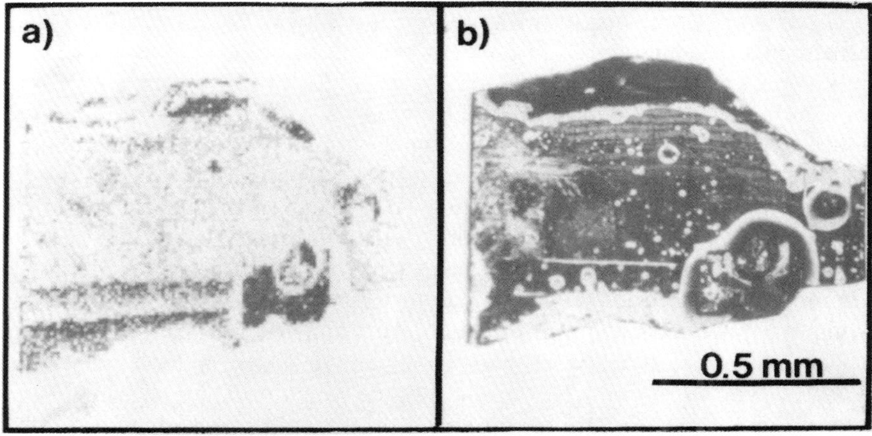

Figure 9.2. (a) [001] CCM image for barium in YBaCuO from a depth of 2 μm. (b) Corresponding optical image of the YBaCuO crystal showing surface steps and inclusions, such as a chipped top edge, shallow, narrow steps on the top part, deeper (~5 μm), wider steps on the bottom part and a pitted defect in the lower right corner. Reprinted from Ref. 1 with kind permission from Elsevier Science B.V., Amsterdam, The Netherlands.

Figure 9.3. Polarized light image of a twinned YBaCuO single crystal with no surface defects. Reprinted from Ref. 1 with kind permission from Elsevier Science B.V., Amsterdam, The Netherlands.

The third crystal, shown in Figure 9.3, had a twin structure and a smooth surface. Figure 9.4 shows CCM images from this crystal for different tilt angles with respect to the [001] axis for backscattered 2 MeV ^{1}H ions from barium at a near-surface (B1) and in a deeper region (B2), barium L X–rays, and ion induced electrons emitted from the sample surface. The measured minimum yield from this crystal for ^{1}H ions backscattered from barium close to the surface was approximately 5%, indicating good local crystal quality again. All these images show regions of different yields indicating nonuniform channeling across the sample, with similar changes in contrast observed for each of the different signals used. There is more contrast in window B1 than in window B2 owing to natural dechanneling of the initially channeled ^{1}H ions with increasing penetration depth. The variation of the channeling contrast in these images does not correspond to any optically visible surface structure. The images taken at a tilt angle of $(-1°, 0°)$ with respect to the [001] axis (i.e., planar channeled) show uniform contrast owing to the small ion fraction that was initially channeled at this orientation. The channeling contrast is thought to be caused by strain in the crystal caused by the twinned structure, which has been previously suggested in TEM experiments [2].

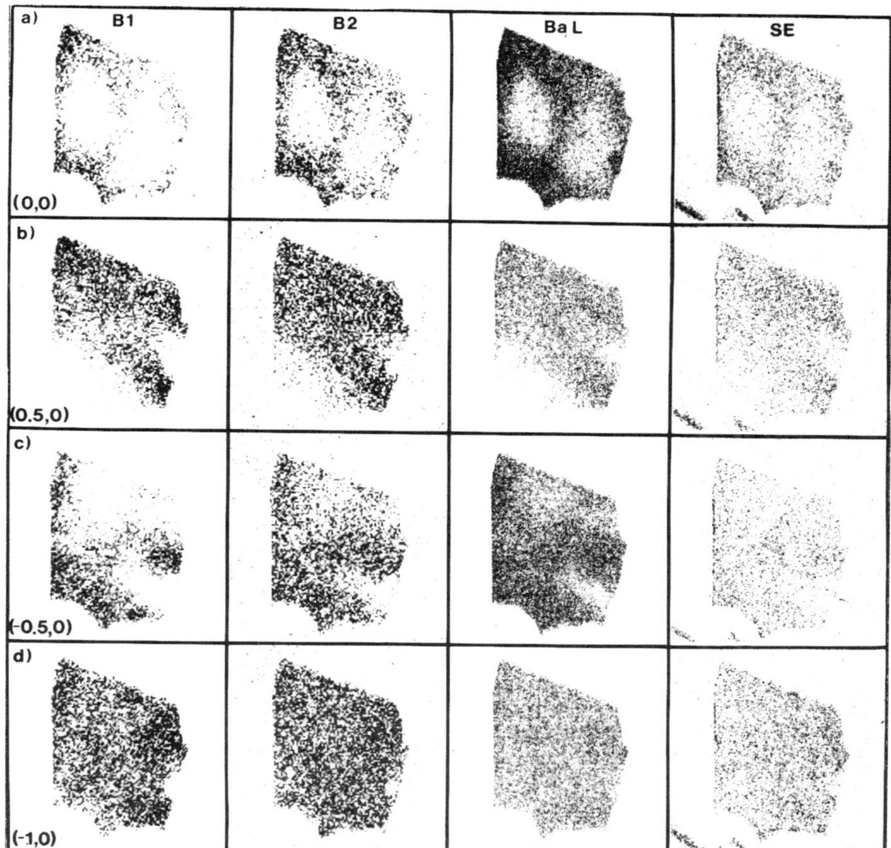

Figure 9.4. CCM images for barium from close to the surface (B1), barium from deeper in the crystal (B2), Ba L X–rays (Ba L), and ion induced electrons (SE). The tilt angles with respect to the [001] axis are shown for each set of four horizontally running images. Reprinted from Ref. 1 with kind permission from Elsevier Science B.V., Amsterdam, The Netherlands.

This work demonstrates the ability of the nuclear microprobe to give unique channeling information on crystalline quality in localized areas and to image the mosaic spread in small crystals. The sensitivity of the contrast in the ion induced electron images to channeling effects had not been previously observed. This is an important result, because the channeling contrast observed in the ion induced electron images in Figure 9.4 arises from variations in the crystallographic orientation in just the top few nanometers of the sample. More work should be carried out using this approach to give channeling information in the near surface region.

9.1.2. Lattice Position of Iron in $YBa_2(Cu_{1-x}Fe_x)_3O_{7-y}$ Single Crystals

In this work, the lattice position of iron in single crystal $YBa_2(Cu_{1-x}Fe_x)_3O_{7-y}$, for $x \leq 0.05$, was investigated using combined backscattering and PIXE channeling [3]. This analysis was only possible with a nuclear microprobe, firstly because of the small size of the single crystals and secondly because it allowed any variation of the minimum yield across the surface of the crystals to be imaged, as described above. This enabled channeling information to be measured free from uncertainties as to whether the analyzed region had a variation in channeling angle across its surface, which would render the channeling information meaningless.

With channeling measurements in conjunction with backscattering spectrometry the mass resolution is usually not sufficient to distinguish between each element present in the YBa(Cu–Fe)O crystal, the structure of which is shown in Figure 9.5. Figure 9.6 shows backscattering spectra for 2 MeV ^{1}H ions channeled along the (a) [001] and (b) [110] axis, and also the nonchanneled spectra in both cases. These backscattering energy spectra show the locations of the backscattering edges for the major elements present, and because iron and copper have a similar mass they cannot be distinguished because their surface energies cannot be resolved. However, each element present generates a different energy X–ray, so iron can be distinguished from the other elements present, allowing the substitutionality to be investigated using PIXE channeling. PIXE channeling does not directly give depth resolved information. The measured PIXE minimum yields for the different elements present were high owing to the contribution to the measured number of X–rays from the beam fraction that has dechanneled at a depth of several microns beneath the surface.

The approach developed to extract lattice location information from the backscattering and PIXE channeling information has been described in [3,4]. It involves using the channeled and nonchanneled backscattering spectra to determine the rate at which the initially channeled ^{1}H ion beam dechanneled with increasing depth into the crystal, along both the [001] and [011] channeling axes. This measured rate of dechanneling was then used to calculate the expected ratio of PIXE minimum yields measured from the different elements present, assuming that they were substitutional. Figure 9.7 shows the measured backscattering minimum yield from barium, for iron fractions of $x = 0.0$, 0.025, and 0.05 and also the measured and calculated PIXE minimum yields for iron concentrations of $x = 0.025$ and 0.05, assuming iron substitutionality. There is a large discrepancy between the measured and calculated minimum yields for iron along the [001] axis. The calculated PIXE minimum yield, assuming substitutionality, is typically 20% lower than the measured iron K X–ray minimum yield. Conversely, along the [011] direction, this was not the case and there is reasonable agreement between the measured and calculated minimum yield for iron, assuming substitutionality. It should be noted that this analysis only allowed the determination of whether an element is or is not substitutional;

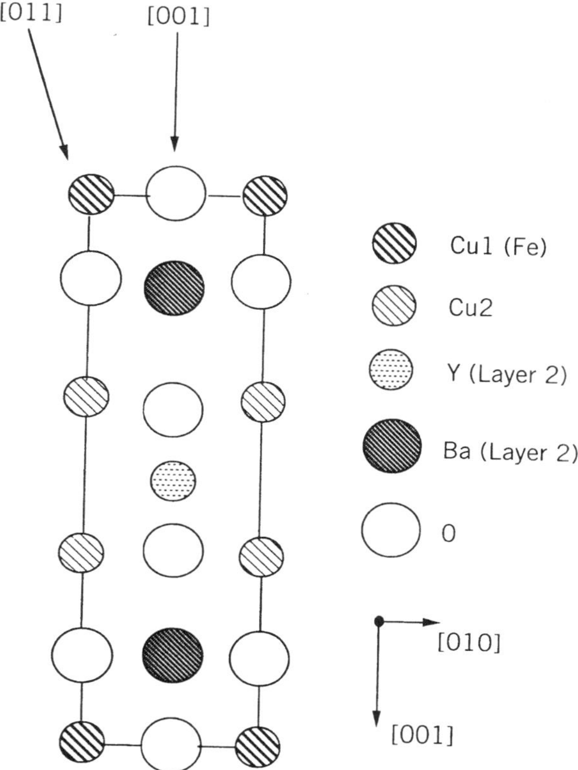

Figure 9.5. [100] projection of YBa(Cu–Fe)O showing the directions used for ion channeling analysis. Reprinted from Ref. 3 with permission (© 1991 American Physical Society, Woodbury, NY).

it does not allow the exact location of the interstitial atoms to be determined, as is the case using channeling in conjunction with backscattering spectrometry [5].

These channeling results were interpreted as indicating that the iron was displaced into postulated positions, as shown in Figure 9.8. The strain induced by the displacement in the presence of extra oxygen may enhance iron clustering along the [110] direction and the attraction of further oxygen as nearest neighbors, causing a change in twin orientation and promoting the growth of microtwins as observed in TEM analysis. Possible ordering of iron along the [110] axis has also been suggested [6].

Further nuclear microprobe analysis of quinternary superconductors is a matter of active research.

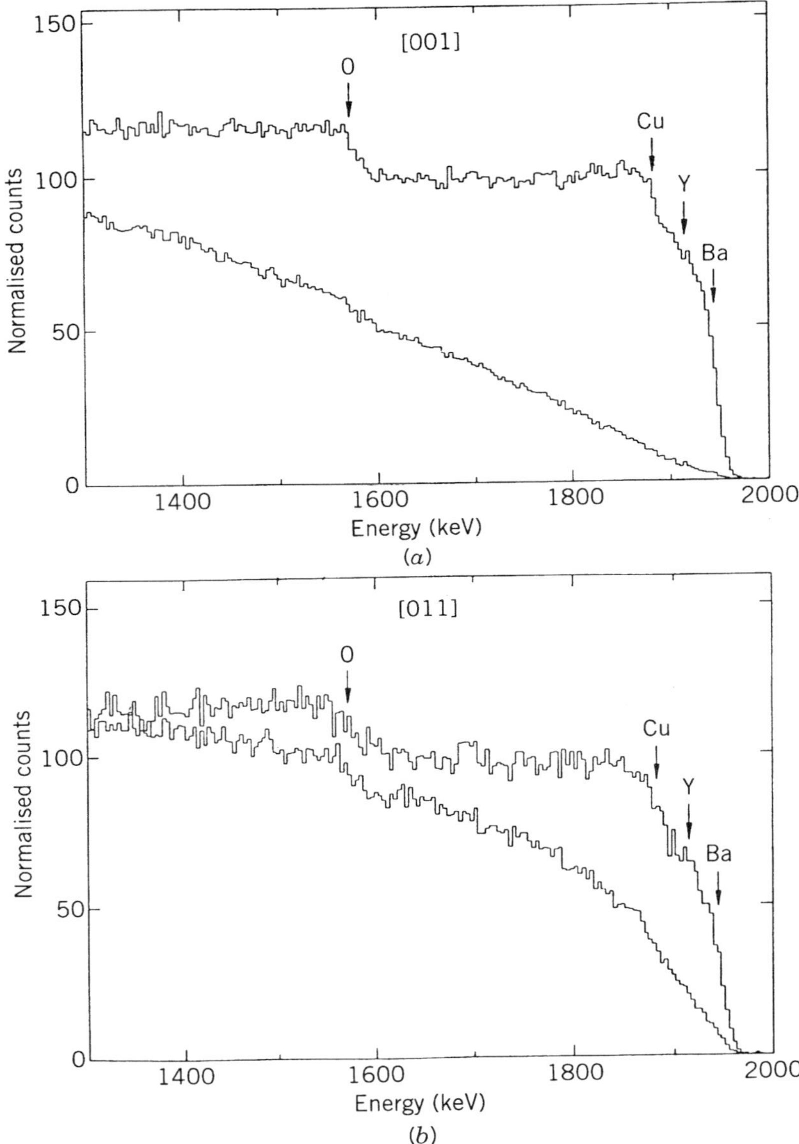

Figure 9.6. Random (i.e., nonchanneled) and channeled backscattering spectra from YBa(Cu–Fe)O for $x = 0.05$ with 2 MeV ^{1}H ions along the (a) [001] axis and (b) [011] axis, for the same beam dose. Reprinted from Ref. 3 with permission (© 1991 American Physical Society, Woodbury, NY).

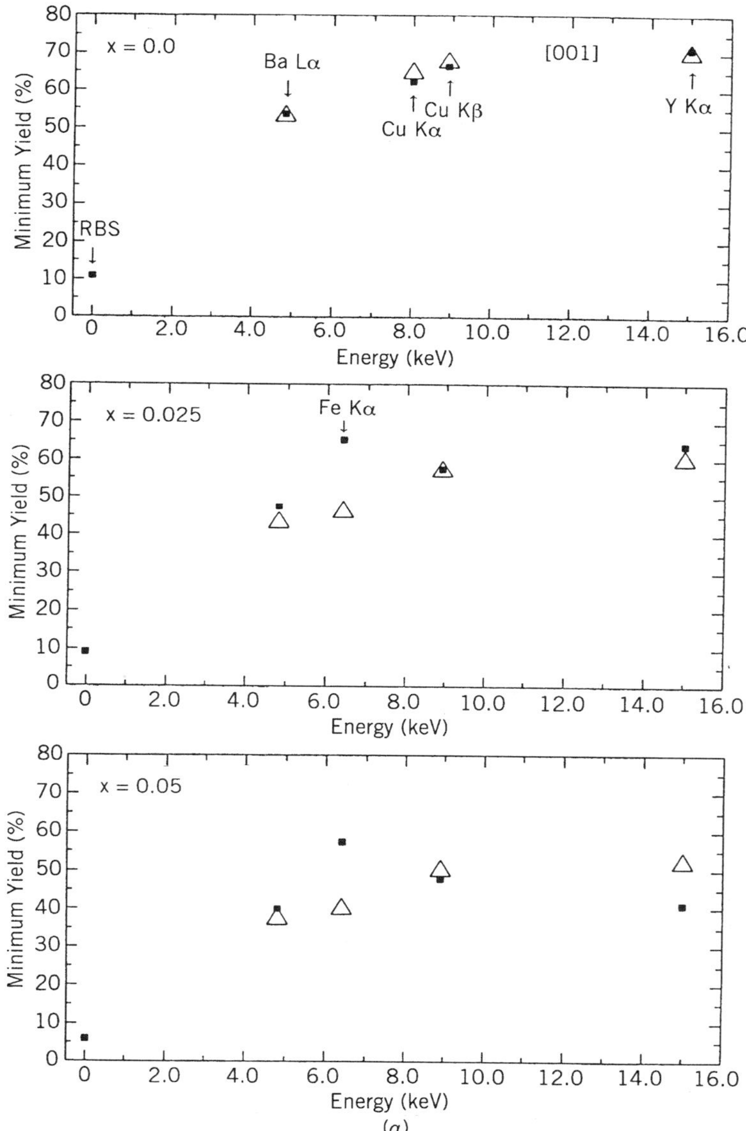

Figure 9.7. Measured backscattering minimum yield from barium (shown at 0 keV), and PIXE minimum yields (solid squares) and calculated PIXE minimum yields (triangles) along (a) the [001] axis and (b) the [011] axis, for iron concentrations of x = 0.0, 0.025, and 0.05. Reprinted from Ref. 3 with permission (© American Physical Society, Woodbury, NY).

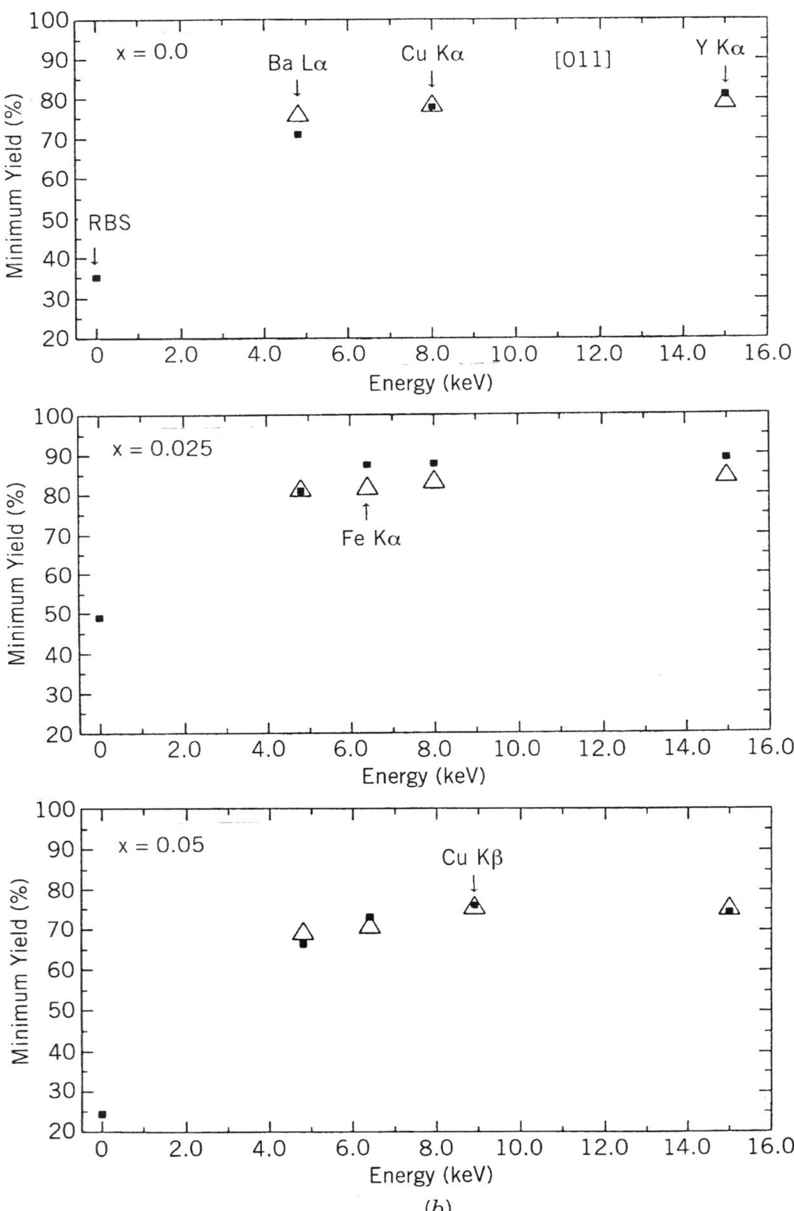

Figure 9.7. (*Continued*)

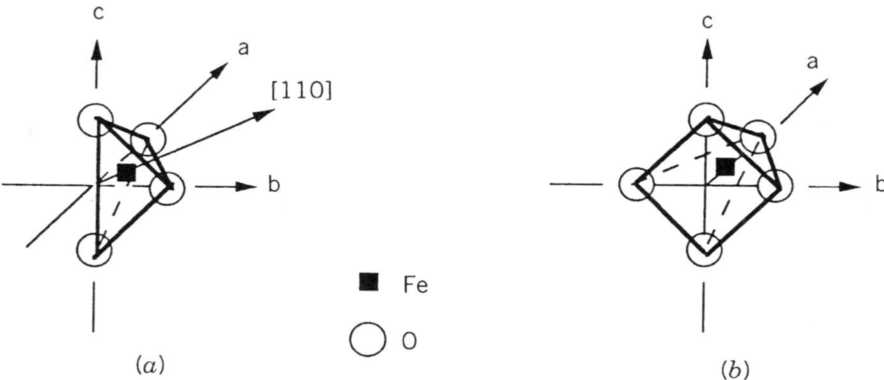

Figure 9.8. Models for the possible positions of iron substituting for copper in (a) a tetrahedral site along the [110] direction and (b) a pyramidal site along the [110] direction. Reprinted from Ref. 3 with permission (© American Physical Society, Woodbury, NY).

9.2. ANALYSIS OF LASER-ANNEALED DIAMOND

A potentially excellent material for semiconductor devices is diamond. This is because of its superior heat conducting and electrical properties, which hold promise for the use of diamond in high-temperature, radiation-hard, and fast microelectronic devices and detectors. However, diamond-based semiconductor devices have proved to be an elusive goal, mainly because of the difficulties of producing n-type diamond. One approach to the study of this problem makes use of natural diamond crystals that are modified by ion implantation. In this work, the nuclear microprobe has an important role to play.

The Melbourne group has devised a scheme of deep, MeV ion implantation followed by focused laser annealing which may have potential to produce n-type diamond. Discussed here is the first step of the scheme. The nuclear microprobe is used to measure the efficacy of the annealing process.

To study the annealing process, type IIa optically polished <110> natural diamond windows were first implanted with MeV carbon or phosphorus ions to a depth of 1.4 μm. This produced a buried layer of heavily damaged diamond, which must be annealed. The diamond also turned brown and swelled approximately 30 nm for a dose of 1×10^{15} ions/cm^2. The buried damage layer was covered by an essentially undamaged surface cap, typically 1 μm thick. Laser annealing was then performed with a Q-switched, frequency-doubled (532 nm) Nd-YAG laser focused through a conventional optical microscope objective to a spot approximately 15 μm in diameter [7]. The cap layer can prevent graphitization of the damaged layer when it was heated by the selectively absorbed laser pulse. Regrowth, graphitization, and melting followed by ablation were observed for increasing power single laser pulses [8,9]. Optical and CCM images showing this phenomena are given in Color Plate 4 (see color insert). The use of multiple laser pulses of ramped-up power on the same spot produced the best regrowth [10]. CCM was used to assess lattice perfection inside the annealed volume of the diamond.

The analysis of diamond with ion beams requires some care because of swelling induced by the analysis beam, which is a severe problem with ^{4}He ions. Furthermore, the low atomic number of carbon means that the scattering cross-section for ^{4}He ions is low compared with silicon, for example. For these two reasons, it is preferable to use ^{1}H ions for analysis. With ^{1}H ions, swelling is less of a problem and, at 1 to 2 MeV, the scattering cross-section is enhanced by more than a factor of five over the Rutherford cross-section.

The ramped-up power multiple-pulse annealing scheme produced good annealing of the implanted diamond, as is shown by the CCM image in Figure 9.9

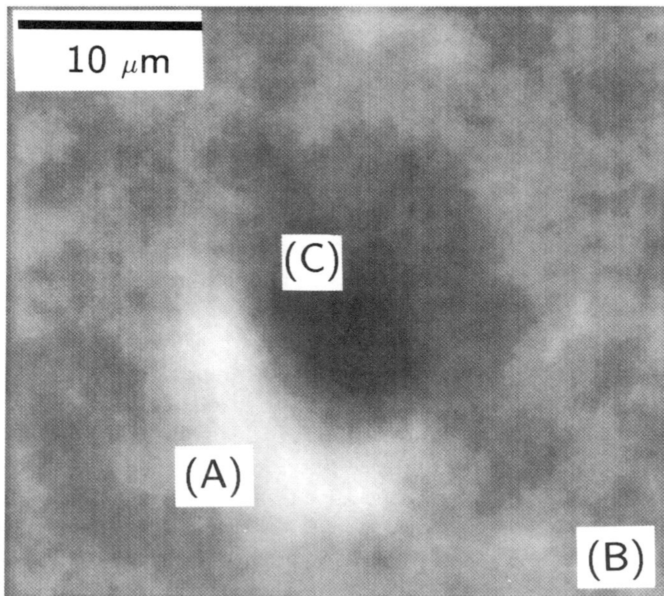

Figure 9.9. CCM image of a multiple-pulse laser-annealed diamond originally implanted to a dose of 1×10^{15} 4 MeV ^{31}P ions/cm^2. A dose of ~0.1 μC of 1.4 MeV ^{1}H ions, measured with a particle detector at a scattering angle of 150° were used to make this image. The image depicts the yield of backscattered ions with energies 700 to 1,400 keV. Black corresponds to low yield and hence good channeling, white corresponds to high yield and hence poor channeling. The incident beam was aligned with the <110> axis. The different regions indicated are: A, swelling; B, as implanted; and C, laser annealed.

[for more information, see 10]. The CCM image revealed several interesting phenomena. Most obvious was the bright rim of the laser spot. This was a result of the compaction of the annealed region, which was caused by the return to a high-quality, dense, single crystal from the less dense damaged material. The crystal planes are tilted around the compacted area and hence dechannel the incident analysis beam. In the center of the laser pulse, the yield is low, and, hence, the crystallinity is good. A spectrum extracted from the laser annealed region is compared with a spectrum from an unimplanted diamond in Figure 9.10. These spectra show that the regrowth has indeed been near perfect, at least as measured by CCM. Further measurements with other techniques, such as Raman spectroscopy, can reveal the presence of residual defects. Very challenging further work is required to determine the lattice site of the implanted phosphorus ions as well as to perform electrical measurements on the annealed regions.

The use of natural diamond crystals in experiments can be expensive! This is particularly true if a wide variety of experiments must be performed to map out the behavior of the material as a function of one, or more, of the experimental parameters. The nuclear microprobe, teamed with a focused laser beam, offers many economies in this area.

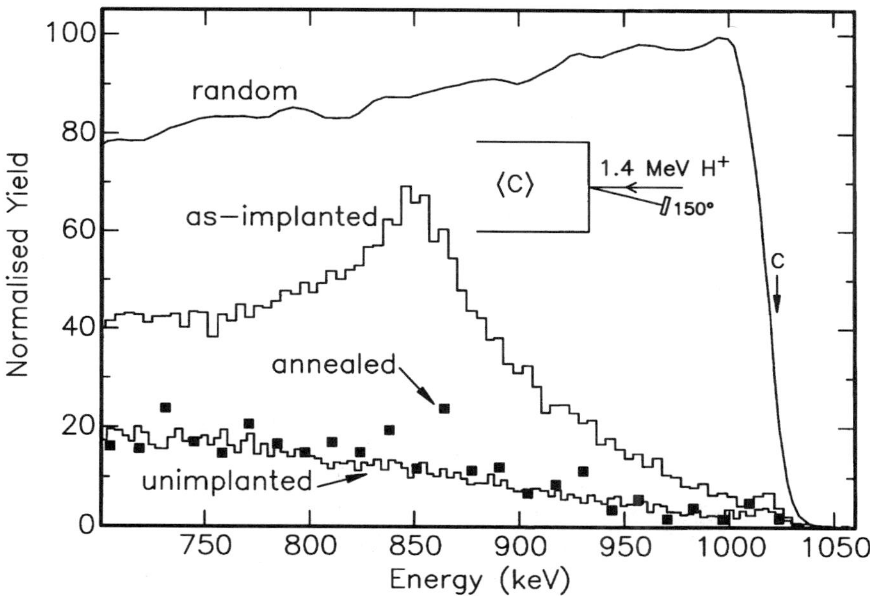

Figure 9.10. Backscattering spectra extracted from the same data set as was used to make the CCM image in Figure 9.9. Also shown are spectra from an unimplanted diamond and from a randomly oriented diamond. Adapted from Ref. 10.

9.3. ANALYSIS OF TERNARY ALLOY SEMICONDUCTORS

One of the most important applications of CCM is the study of materials used for microelectronic applications. In Melbourne, extensive studies have been performed on materials used to fabricate infrared detectors for optical fiber communication systems. These studies included investigations of the substrate material [11], novel etching techniques [12], and complete devices [13,14].

The devices are fabricated in epitaxial layers of $Hg_{1-x}Cd_xTe$ ($x = 0.62$ to 0.70) grown on gallium arsenide substrates. The epitaxial layers contain pyramidal hillocks, as revealed by SEM images. These defects were believed to have nucleated on etch pits in the substrate. The CCM images, shown in Figure 9.11a,

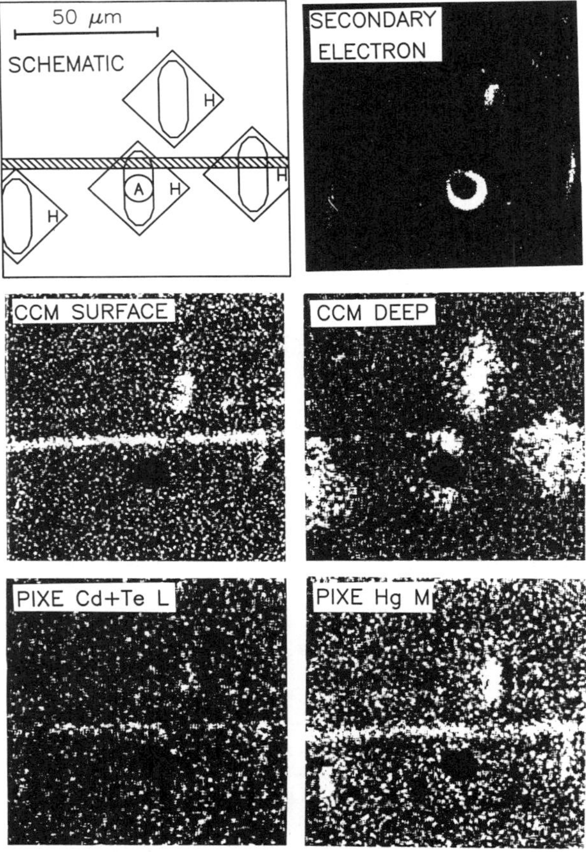

Figure 9.11. Images of pyramidal hillocks and linear growth defects in epitaxial $Hg_{1-x}Cd_xTe$ on <100> gallium arsenide, using 2 MeV ^{4}He ions. In the schematic, H designates a hillock, A designates a high hillock, and the hatched region represents a linear growth defect. All images were obtained with the beam aligned with the <100> axis. White represents high yield and black low yield.

of the hillocks revealed a $\chi_{min} = 13\%$ that was larger than that of the high-quality surrounding material where $\chi_{min} = 9\%$. However, PIXE analysis showed evidence for an increased concentration of mercury in the hillocks, suggesting that mercury was preferentially displaced from lattice sites, in accordance with other studies [11].

The CCM images in Figure 9.11a of the $Hg_{1-x}Cd_xTe$ films also show linear growth defects, with a shape suggesting the nucleation of defective or misaligned material on scratches in the substrate. Indeed, further detailed studies on functioning devices showed linear growth defects bridging the electrical contacts on some photodetector devices, as shown in Color plate 3. It was found that two out of three devices on some wafers suffered from these linear defects [13]. No other microscopy method showed any contrast from images of the surface of these faulty devices. It is likely that these linear growth defects accounted for the erratic electrical properties from different devices on the same wafer [14].

9.4. ANALYSIS OF AN ALUMINUM–LEAD METAL ALLOY

All previous nuclear microprobe metallurgical studies have used NRA to determine light element concentrations, as described in Section 4.3. However, the work presented here describes the use of PIXE, backscattering spectrometry and STIM to determine the distribution and concentration of lead present in a chill cast aluminium ingot of a bearing alloy [15] that had a nominal composition of Al 5wt%Pb. However, gravity segregation can cause a nonuniform dispersion of lead precipitates depending on the processing conditions [16]. This material was analyzed previously using optical microscopy, scanning and transmission electron microscopy, and EPMA [16], but none of these techniques can analyze a depth of greater than 1 μm. Since the range of 3 MeV ^{1}H ions in aluminum is 82 μm, the sample composition and distribution was analyzed through a considerably greater depth using nuclear microprobe techniques than with bulk electron beam techniques.

The aluminum–lead sample was prepared for analysis with a nuclear microprobe by cutting and polishing a 40 μm thick slice. Backscattering spectrometry was used to determine the sample stoichiometry, and Figure 9.12a shows the backscattering spectrum from a $75 \times 75 \ \mu m^2$ area of this sample. Figure 9.12b shows the high-energy part of the backscattering spectrum of Figure 9.12a, fitted with a simulation to determine the average lead content within the scanned area. The nominal sample composition was Al 5wt%Pb, but both EPMA results [16] and these backscattering results show large local variations present in the measured lead concentration.

EPMA, PIXE, and STIM images were collected on the same area of this material. Figure 9.13a shows a $100 \times 100 \ \mu m^2$ EPMA image generated using 10 nA of 15 keV electrons and wavelength dispersive X–ray analysis with the lead $M\alpha$ line selected. The white regions correspond to the lead precipitates, which are between 0.5 μm and 5 μm in diameter. Smaller precipitates formed a cellular

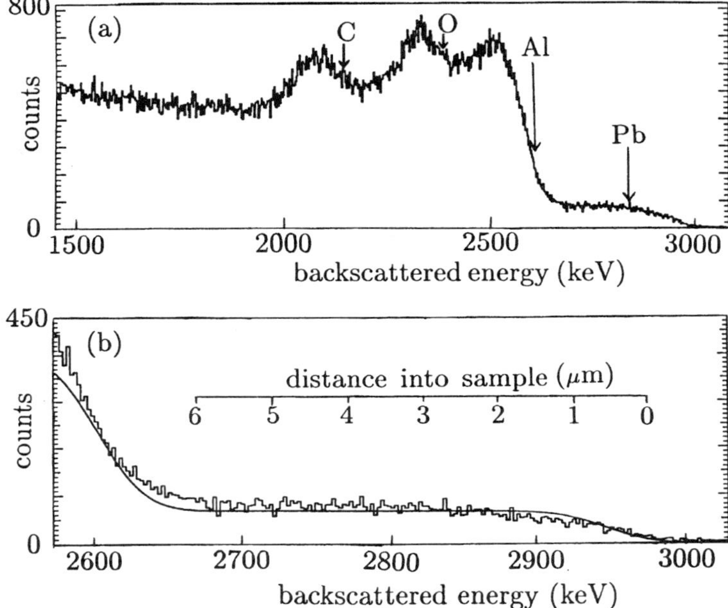

Figure 9.12. (a) Backscattering spectrum from the AlPb sample, measured using 3 MeV ^{1}H ions, showing the positions of the elements aluminum, lead, carbon, and oxygen. (b) The high energy part of the same spectrum showing a simulated composition of aluminum 2.1wt% Pb. Reprinted from Ref. 15 with permission (© 1992 Materials Research Society).

structure, which has been reported elsewhere for this material [16]. Figure 9.13b shows a $60 \times 60 \ \mu m^2$ PIXE lead M image, and Figure 9.13c shows the PIXE lead L image centered about the same position on the sample surface as for the EPMA image. Figure 9.13b shows the same three large lead precipitates as in the EPMA image, and closer inspection shows that a similar cellular structure is just visible in the background. However the smaller precipitates visible in the EPMA image cannot be distinguished in Figure 9.13b, in which the smallest resolvable precipitates are approximately 2 μm in diameter. This is because the average X–ray production depth for lead Mα X–rays generated with 15 keV electrons is approximately 1 μm, whereas it is 3 to 4 μm using 3 MeV ^{1}H ions. The smaller precipitates are not resolvable because of image overlap from those below. To look at the large precipitates through a greater sample depth than 3 to 4 μm, the PIXE lead L X–rays were used, as shown in Figure 9.13c. The large precipitates in Figure 9.13a,b are still seen in Figure 9.13c, but with overlying precipitates since the average X–ray production depth was now approximately 20 μm.

Figure 9.14 shows three STIM images centered about the same sample area

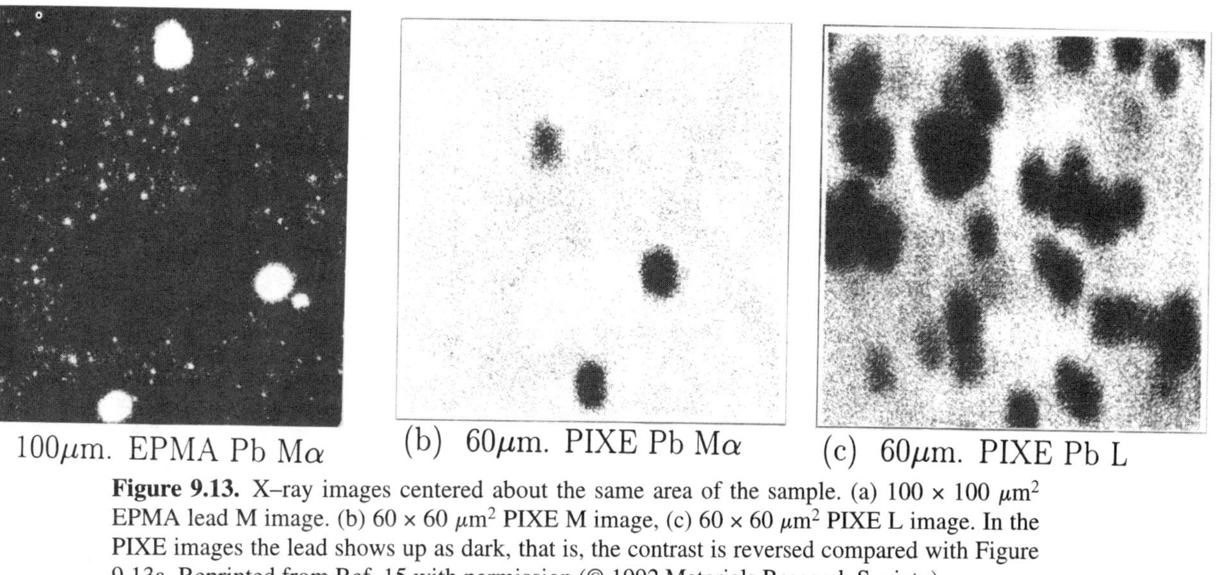

Figure 9.13. X–ray images centered about the same area of the sample. (a) $100 \times 100\ \mu m^2$ EPMA lead M image. (b) $60 \times 60\ \mu m^2$ PIXE M image, (c) $60 \times 60\ \mu m^2$ PIXE L image. In the PIXE images the lead shows up as dark, that is, the contrast is reversed compared with Figure 9.13a. Reprinted from Ref. 15 with permission (© 1992 Materials Research Society).

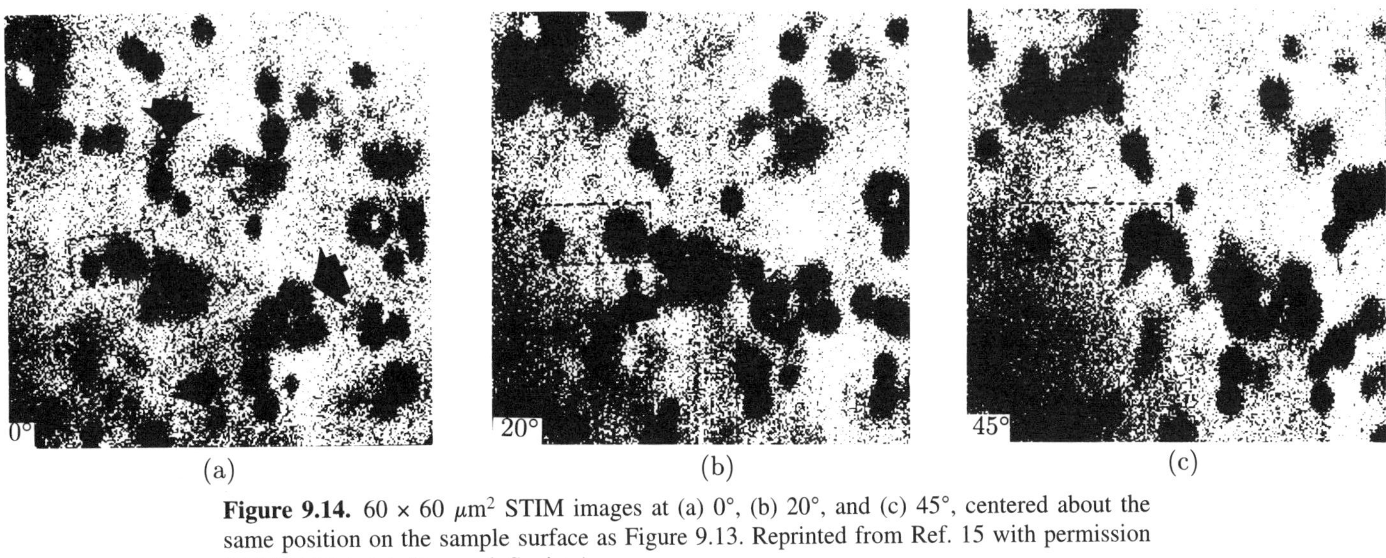

Figure 9.14. 60×60 μm^2 STIM images at (a) 0°, (b) 20°, and (c) 45°, centered about the same position on the sample surface as Figure 9.13. Reprinted from Ref. 15 with permission (© 1992 Materials Research Society).

shown in Figure 9.13 with different sample tilt orientations. The scan size was fixed at 60×60 μm^2 in each case and was not changed to compensate for the larger surface area scanned at larger tilt angles. Regions of high energy loss by the transmitted ^{1}H ion beam are indicated by the dark areas and correspond to the lead precipitates, since they are of higher density compared with the aluminium matrix. In Figure 9.14a the sample was perpendicular (0°) to the beam, and the same three precipitates that are shown in Figure 9.14a,b are arrowed. All the lead precipitates in the PIXE L X–ray image of Figure 9.14c were imaged in addition to others, because of the greater analytical depth of STIM. Using 3 MeV ^{1}H ions, the minimum resolvable precipitate size using STIM was approximately 2 μm in this case. This was considerably larger than in the EPMA image in Figure 9.13a, and similar to the minimum resolvable size in the PIXE images.

Figures 9.14b,c are STIM images measured with the sample tilted by 20° and 45°, respectively about the vertical axis. The lead precipitates all appear round from any orientation, indicating that they are spherical within the resolution of the method (in fact the precipitates are faceted on {111} planes, as determined by TEM). The relative position of most of the precipitates moved consistently with increasing tilt angle. Some pairs of precipitates showed very little relative displacement, which indicated that the precipitate pair was nearly at the same depth. The precipitate pair enclosed by the dashed box in Figure 9.14 is separated by 7 μm at 0° tilt, whereas at 20° tilt it is separated by 14 μm, which indicates that these two precipitates are separated by approximately 20 μm in depth.

Precipitates that lie beneath one another were resolved by imaging the same area at different angles, as in Figure 9.14. For example, the precipitate indicated with the arrow in Figure 9.14b is now resolved from those lying to the right, which hid it in Figure 9.14a. The number of precipitates with a given diameter is plotted in Figure 9.15, from which it can be calculated that 0.5% of the sample volume consisted of lead precipitates with a diameter larger than 2 μm in this region, assuming the precipitates to be spherical.

9.5. MATERIALS MODIFICATION AND FABRICATION

This section describes the use of a nuclear microprobe as a tool for materials modification and fabrication, whereas the rest of this book has only discussed materials analysis. The use of nuclear microprobes to create structures has been developed since the early 1980s and several groups are now working in this area.

9.5.1. MeV Heavy Ion Microlithography

The work by Fischer at GSI Darmstadt [17–21] was the first serious attempt to use focused beams of MeV ions for microlithography to create novel structures and for materials analysis.

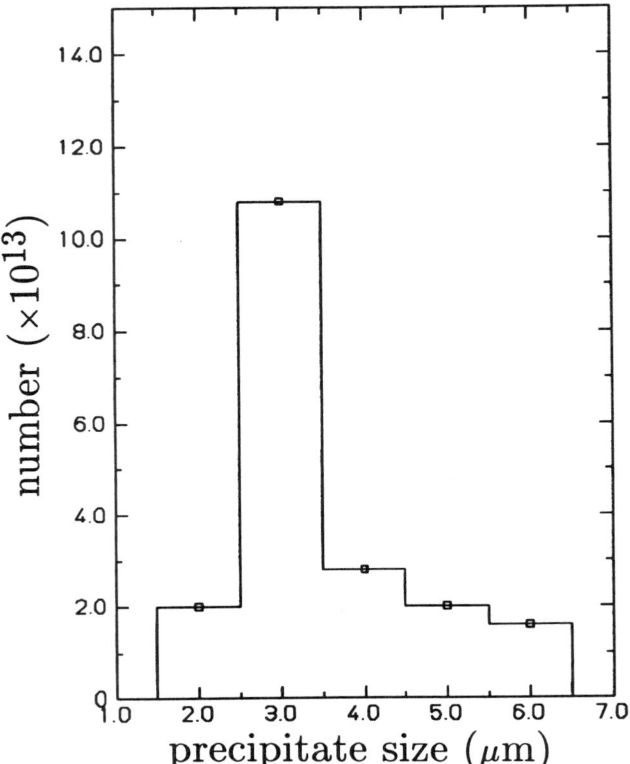

Figure 9.15. Number of lead precipitates per cubic meter of a given diameter. Reprinted from Ref. 15 with permission (© 1992 Materials Research Society).

Figure 9.16 shows the first method used to generate microlithographic areal density images with submicron resolution of thin structures, which gave similar information as STIM images described in Section 4.7. Here the thin sample was placed on a radiation-sensitive material such as mica. A shadow of the sample was cast into the mica by the passage of high-energy heavy ions through the sample, as shown in Figure 9.16a. Variations in the areal density of the sample resulted in variations in the amount of ion energy deposited in the underlying mica. The more dense regions were produced as raised portions in the relief structure, because less energy was deposited in the mica at these locations. After developing the mica for 30 min in 40% hydrofluoric acid, a relief-like replica of the internal structure of the sample was produced, as shown in Figure 9.16b. Figure 9.16c shows a SEM image of a sample relief, which was a part of a small insect, produced by this method.

Figure 9.17 shows a different example of heavy ion microlithography whereby a pattern was directly written onto a piece of CR–39 plastic using

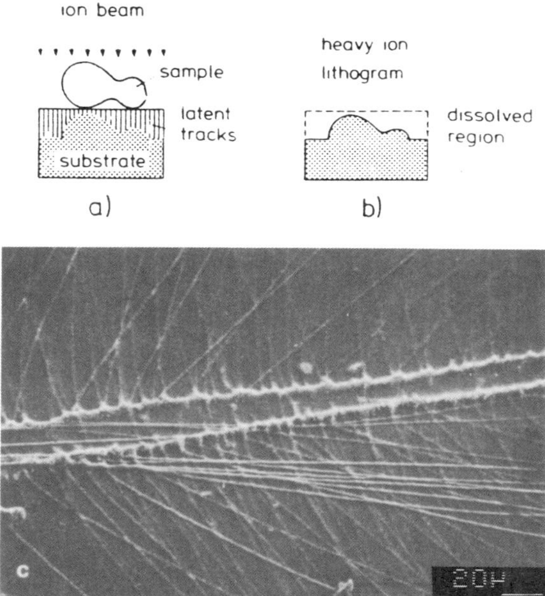

Figure 9.16. (a) Schematic of the sample geometry used to create different length ion tracks in the resist material underneath. (b) After developing the resist a relief-like replica of the sample is produced. (c) SEM image of an example of this process, called a heavy ion microlithogram, produced using a dose of approximately 10^{11} 1.4 MeV/u ^{131}Xe ions/cm^2 on mica. The sample is a small insect and the antenna are shown here with a lateral resolution of approximately 0.5 μm. Reprinted from Ref. 18 with kind permission from Elsevier Science B.V., Amsterdam, The Netherlands.

focused ^{16}O ions from a nuclear microprobe, with no mask used. The complex pattern was produced by loading the pattern into the computer-controlled scan generator. The sample was developed in a solution of NaOH for 30 min and then imaged in a scanning electron microscope. This work showed that the beam dose delivered to different regions of the irradiated areas could be controlled well enough to produce very intricate patterns.

It has been known for many years that spherical surfaces can be produced by irradiation of nuclear tracks. This principle has been used by Fischer [20] to produce lenses that have a radius of 10 to 100 μm for the scanning acoustic microscope to focus ultrasonic sound waves onto the sample. This work investigated fabricating these lenses using heavy ion microlithography as a cheaper and more precise alternative method to more conventional means. Figure 9.18a shows a spherical concave lens produced by developing a track formed by a single ion in fused silica. The requirements for producing a microprobe capable of delivering single ions to required locations are described in Refs. 19 and 20 and form an interesting technical challenge. This general process has also

Figure 9.17. SEM image of an artificial pattern written into a piece of CR–39 plastic using a focused ^{16}O beam from a nuclear microprobe. Reprinted from Ref. 19 with kind permission from Elsevier Science B.V., Amsterdam, The Netherlands.

been named *single ion micromechanics*. Figure 9.18b shows an array of such concave lenses produced in soda glass using approximately 2 min of ion irradiation. Every etch pit is the result of a single ion directed to its intended position by the scan controller. The few pits produced that are not part of the regular pattern are from problems associated with trying to control the destination of every ion in the microprobe, but this work demonstrates the feasibility of the fabrication process and it is hoped that further applications will arise for this novel technique.

9.5.2. MeV ^{1}H Ion Beam Lithography

One limitation of X–ray lithography mask contrast is the thickness of the patterned absorber layer. This is limited by the need to use thin resist layers of material such as polymethyl methacrylate (PMMA) to maintain the high spatial resolution when fabricated using electron-beam lithography, because electron scattering severely limits the minimum achievable linewidth in thick samples. Much effort has been devoted to developing methods to circumvent this serious problem of low absorber thickness for high spatial resolution X–ray masks [22–24].

The 3 MeV ^{1}H ions suffer little lateral straggle and have a range of approximately 125 μm in PMMA; the possibility of their use for fabricating high-

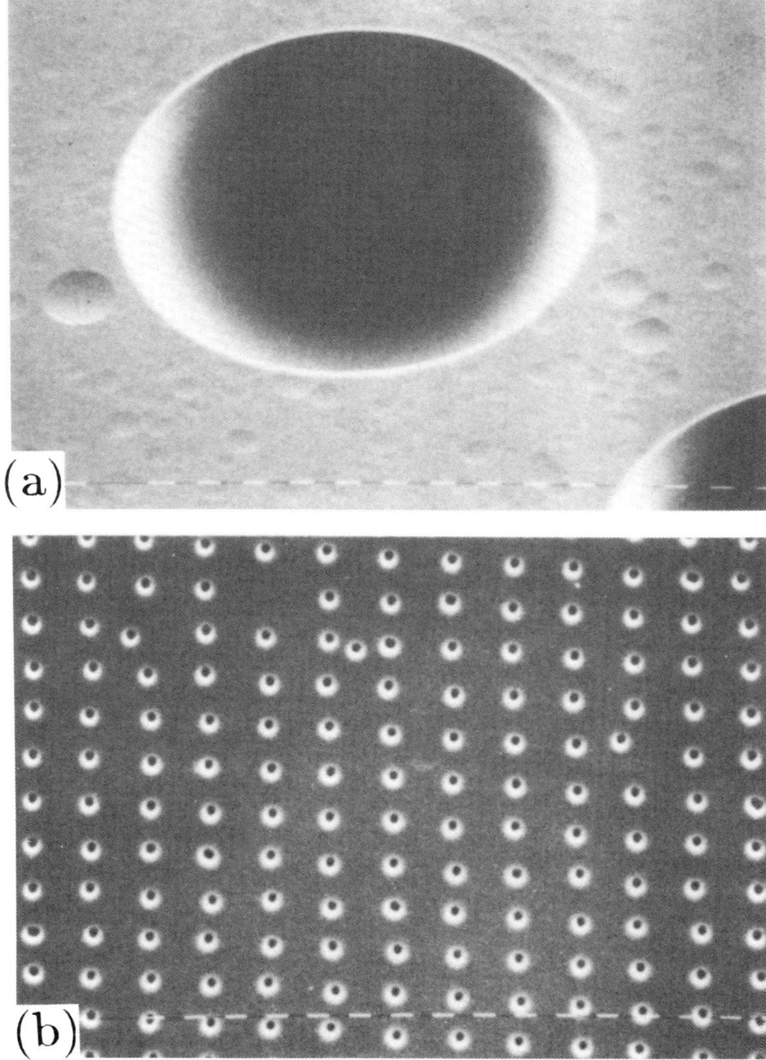

Figure 9.18. (a) SEM image of spherical concave lens produced by track etching in fused silica. Scale bars are 10 μm. (b). SEM image of a regular array of etch pits produced by the impact of single 1.4 MeV/u ^{52}Cr ions onto soda glass. Scale bars are 10 μm. Reprinted from Ref. 20 with kind permission from Elsevier Science B.V., Amsterdam, The Netherlands.

contrast X–ray masks is described in a preliminary study using the Oxford nuclear microprobe [25]. This paper presented simulations demonstrating that the minimum attainable linewidths for 3 MeV ^{1}H ions are approximately 120 nm through 10 μm of PMMA.

Figure 9.19 shows an optical image of a side view of eight etched channels caused by different doses of 3 MeV ^{1}H ions scanning a line over the edge of a 1 mm thick square of PMMA. The beam current was 20 pA and the PMMA was developed using a solution of 80% methyl isobutyl ketone and 20% isopropyl alcohol for 2 min, and then rinsing with isopropyl alcohol for 30 sec [according to 26]. The developing solution preferentially dissolved the ion irradiated areas of the positive resist PMMA. The channels created using a dose of 1 to 12 pC/μm^2 are approximately 125 μm long. In general the channels created using a higher dose have etched wider than the lines created using a low dose. Figure 9.20 shows an SEM image of the four channels in Figure 9.19 created with doses of 0.2 to 1.6 pC/μm^2, which have been imaged such that both the channel along the surface of the PMMA and the channel profile over the edge are visible.

Figure 9.21 shows an optical image of the same PMMA sample with three channel profiles etched with doses of 0.3 to 1.0 pC/μm^2. The beam current of 3 pA was scanned over the edge of the sample, as for the channels shown in Figure 9.19. These channels are considerably less than 125 μm long, indicating that they have not etched along the full ion path. Figure 9.22 shows an SEM image of the bottom of the channel with a dose of 0.7 pC/μm^2 that is shown in

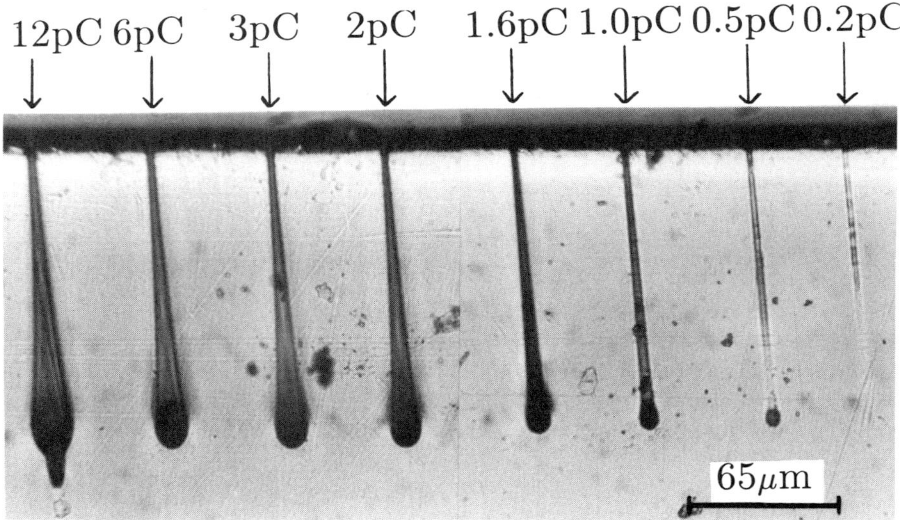

Figure 9.19. Optical image of eight channels created in PMMA using a current of 20 pA. The dose used for each channel is shown in units of pC/μm^2. Reprinted from Ref. 14 with kind permission from Elsevier Science B.V., Amsterdam, The Netherlands.

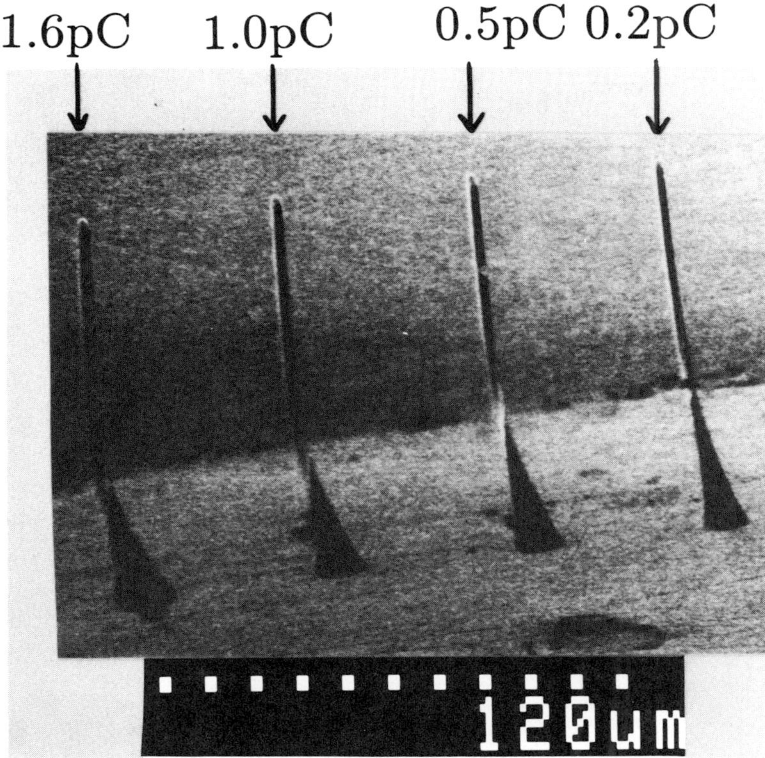

Figure 9.20. SEM image of some of the channels shown in Figure 9.16. The upper part of the figure is the top face of the PMMA and the bottom part shows the channel profile from the side of the PMMA. Reprinted from Ref. 25 with kind permission from Elsevier Science B.V., Amsterdam, The Netherlands.

Figure 9.21. The channel width at the surface and at this depth of approximately 30 μm is approximately 2.2 μm in both cases. This indicates the minimum dose of 0.2 pC/μm^2, determined on the basis of whether a channel could be detected using optical microscopy, was probably too high [25].

Here the channel half-width at the PMMA surface of approximately 1.0 μm is considerably greater than the beam resolution from over-etching in this case. However, this work does demonstrate that focused MeV ^{1}H ion beams can create tracks through tens of microns in PMMA and that the track width is not limited by lateral scattering.

9.5.3. Optical Waveguide Fabrication

With unfocused keV or MeV ion beams, graded index planar optical waveguides have been fabricated by implantation into a wide variety of optical mate-

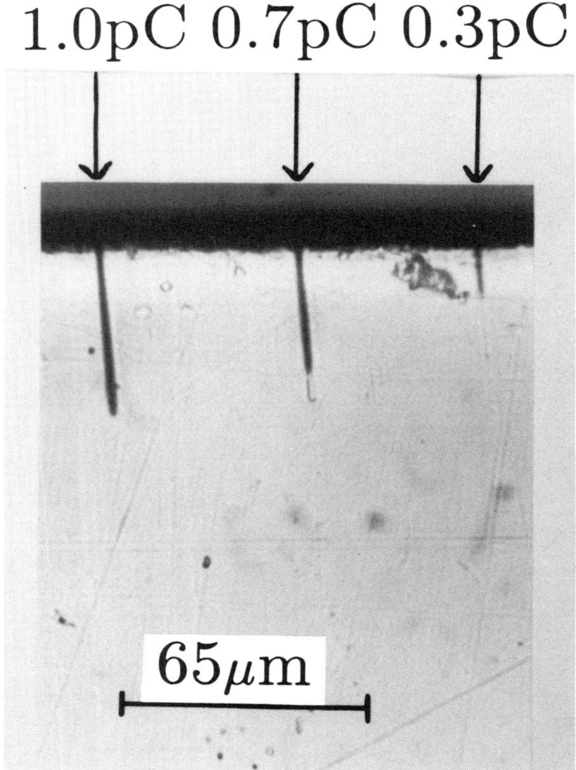

Figure 9.21. Optical image of three channels created using a current of 3 pA. The ion dose used for each channel is shown above in units of $pC/\mu m^2$. Reprinted from Ref. 25 with kind permission from Elsevier Science B.V., Amsterdam, The Netherlands.

rials. Channel waveguides are normally produced by placing a mask over the sample and allowing a selected area to be irradiated [27,28]. Although effective for keV ions used to produce shallow guides, the use of masks at higher incident energies to produce deep guides is difficult, since the masks must be relatively thick (tens of microns) and have a high aspect ratio to adequately stop the MeV ions. These problems can be overcome by the use of a focused MeV ion beam to direct-write microstructures.

In Melbourne, focused MeV ion beams have been used to fabricate channel guides in fused silica. The waveguides were fabricated with ^{1}H ions with an energy in the range of 2 to 3 MeV, focused to a spot size of a few microns. The silica sample was mounted on a translation stage and moved with respect to the fixed beam spot to trace out the desired pattern. This procedure was done in preference to magnetic or electrostatic scanning, which

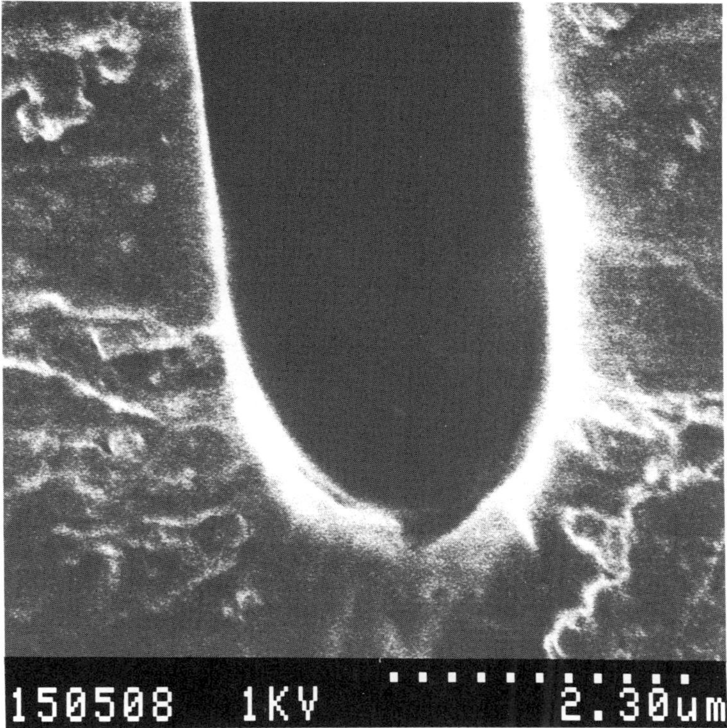

Figure 9.22. SEM image of the base of the channel created with a dose of 0.7 pC/μm^2, which is shown in Figure 9.21. Reprinted from Ref. 25 with kind permission from Elsevier Science B.V., Amsterdam, The Netherlands.

would otherwise degrade the resolution of the probe owing to the large scan sizes typically required. Both planar and channel waveguides have been fabricated using this technique at proton doses ranging between 1 and 100 pC/μm^2.

The most promising guides have been fabricated using doses in the range of 1 to 10 pC/μm^2. An analysis of the guides using Raman spectroscopy confirms the assumption that the refractive index increase at the ion end-of-range is caused by an increase in molecular defects. The defect distribution at the end-of-range is in agreement with TRIM calculations showing the nuclear energy deposition. The guides have been used to transmit focused white light, and light from a helium–neon laser, butt-coupled via a single mode fiber into the guide, and a typical example is shown in Figure 9.23. Techniques are currently being investigated to recover the refractive index distribution from the output intensity distribution of the guides.

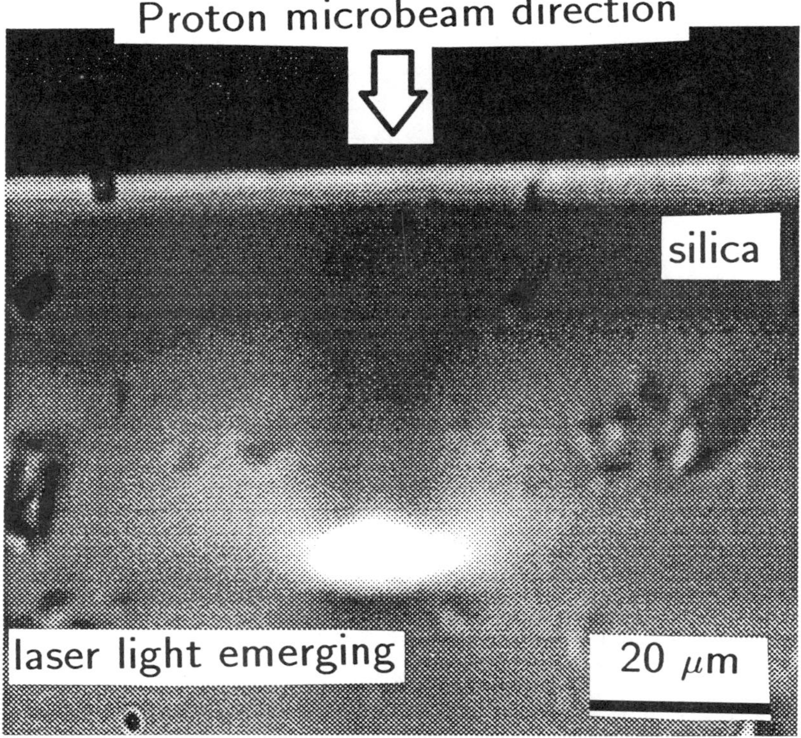

Figure 9.23. Light transmitted through a channel waveguide fabricated in silica with a focused 2 MeV ^{1}H ion beam. The ion dose was 10 μC/cm^2 and the waveguide was approximately 5 mm long.

REFERENCES

1. L.T. Romano, M.B.H. Breese, G.W. Grime, and F. Watt, *Nucl. Instr. Meth.* **B54:**234 (1991).

2. A.F. Moodie and H.J. Whitfield, *Ultramicroscopy* **24:**329 (1988).

3. L.T. Romano, M.B.H. Breese, D.N. Jamieson, C. Chen, G.W. Grime, and F. Watt, *Phys. Rev.* **B44**(13):6927 (1991).

4. M.B.H. Breese, P.J.C. King, L.T. Romano, G.W. Grime, and F. Watt, *Nucl. Instr. Meth.* **B62:**133 (1991).

5. L.C. Feldman, J.W. Mayer, and S.T. Picraux, *Materials Analysis by Ion Channeling*, Academic Press, New York (1982).

6. P. Bordet, J.L. Hodeau, P. Strobel, M. Marezio, and A. Santoro, *Sol. Stat. Comm.* **66:**435 (1988).

7. S. Prawer, D.N. Jamieson, S.P. Dooley, and R. Kalish, *Mat. Res. Soc. Symp. Proc.* **235:**431 (1992).

8. D.N. Jamieson, S. Prawer, S.P. Dooley, and R. Kalish, *Nucl. Instr. Meth.* **B77:**457 (1993).

9. S. Prawer, D.N. Jamieson, and R. Kalish, *Phys. Rev. Lett.* **69:**2991 (1992).

10. M.G. Allen, S. Prawer, D.N. Jamieson, and R. Kalish, *Appl. Phys. Lett.* **63:**2062 (1993).

11. D.N. Jamieson, S.P. Dooley, G.N. Pain, and S.P. Russo, *Mat. Res. Soc. Symp. Proc.* **235:**253 (1992).

12. A. Semu, L. Montelius, P. Leech, D.N. Jamieson, and P. Silverberg, *Appl. Phys. Lett.* **59:**1752 (1991).

13. P.W. Leech, P.J. Gwynn, G.N. Pain, N. Petkovic, J. Thompson, and D.N. Jamieson, *M.R.S. Symp. Proc.* **216:**11 (1990).

14. P.W. Leech, L.A. Witham, S.P. Dooley, and D.N. Jamieson, *J. Vac. Sci. Technol.* **A13:**21 (1995).

15. M.B.H. Breese, L.T. Romano, C.J. Salter, G.W. Grime, and F. Watt, *J. Mater. Res.* **7**(9):2373 (1992).

16. K.I. Moore, D.L. Zhang, and B. Cantor, *Acta Metall. Mater.* **38:**1327 (1990).

17. B.E. Fischer and R. Spohr, *Microtooling and Microscopy with Heavy Ions*, GSI Darmstadt, Germany preprint 80–24 (1980).

18. B.E. Fischer and R. Spohr, *Nucl. Instr. Meth.* **168:**241 (1980).

19. B.E. Fischer, *Nucl. Instr. Meth.* **B10/11:**693 (1985).

20. B.E. Fischer, *Nucl. Instr. Meth.* **B30:**284 (1988).

21. B.E. Fischer, *Nucl. Instr. Meth.* **B54:**401 (1991).

22. M. Hatzakis, D. Hofer, and T.H.P. Chang, *J. Vac. Sci. Technol.* **16:**1631 (1979).

23. P. Parrens, E. Tabouret, and M.C. Tacussel, *J. Vac. Sci. Technol.* **16:**1965 (1979).

24. Conference volumes *J. Vac. Sci. Technol.* **B7**(6), **B8**(6).

25. M.B.H. Breese, G.W. Grime, F. Watt, and D. Williams, *Nucl. Instr. Meth.* **B77:**169 (1993).

26. T.M. Hall, A. Wagner, and L.F. Thompson, *J. Vac. Sci. Technol.* **16:**1889 (1979).

27. E.R. Schineller, R.P. Flam, and D.W. Wilmot, *J. Opt. Soc. Am.* **58**(9):1171 (1968).

28. P.D. Townsend, *Nucl. Instr. Meth.* **B46:**18 (1990).

APPENDIX 1

RELEVANT CONFERENCE PROCEEDINGS

APPLICATIONS OF ACCELERATORS IN RESEARCH AND INDUSTRY

6th Conference. *IEEE Trans. Nuc. Sci.* **28** (1989), J.L. Duggan, I.L. Morgan, and J.A. Martin, eds.

8th Conference. *Nucl. Instr. Meth.* **B10/11** (1985), J.L. Duggan, I.L. Morgan, and J.A. Martin, eds.

9th Conference. *Nucl. Instr. Meth.* **B24/25** (1987), J.L. Duggan and I.L. Morgan, eds.

10th Conference. *Nucl. Instr. Meth.* **B40/41** (1989), J.L. Duggan and I.L. Morgan, eds.

11th Conference. *Nucl. Instr. Meth.* **B56/57** (1991), J.L. Duggan and I.L. Morgan, eds.

12th Conference. *Nucl. Instr. Meth.* **B79** (1993), J.L. Duggan and I.L. Morgan, eds.

13th Conference. *Nucl. Instr. Meth.* **B99** (1995), J.L. Duggan and I.L. Morgan, eds.

ATOMIC COLLISIONS IN SOLIDS

10th Conference. *Nucl. Instr. Meth.* **230** (1984), H.E. Roosendaal, H.O. Lutz, W. Heiland, and H.J. Andrä, eds.

11th Conference. *Nucl. Instr. Meth.* **B13** (1986), T.M. Buck, N. Cue, T.E. Madey, N.H. Tolk, and P.A. Treado, eds.

12th Conference. *Nucl. Instr. Meth.* **B33** (1988), F. Fujimoto, ed.

13th Conference. *Nucl. Instr. Meth.* **B48** (1990), F. Besenbacher, J.U. Andersen, and A.H. Sørensen, eds.

14th Conference. *Nucl. Instr. Meth.* **B67** (1992), J.A. van den Berg, P.C. Zalm, and G.A. Stephens, eds.

EUROPEAN CONFERENCE ON ACCELERATOR APPLICATIONS IN RESEARCH AND TECHNOLOGY (ECAART)

1st Conference. *Nucl. Instr. Meth.* **B50** (1990), K. Bethge, F. Rauch, and P. Misaelides, eds.

2nd Conference. *Nucl. Instr. Meth.* **B68** (1992), K. Bethge, F. Rauch, J.D. Meyer, and P. Misaelides, eds.

3rd Conference. *Nucl. Instr. Meth.* **B89** (1994), J.L. Debrun, K. Bethge, J.D. Meyer, and P. Misaelides, eds.

ION BEAM ANALYSIS (IBA)

4th Conference. *Nucl. Instr. Meth.* **168** (1980), H.H. Andersen, J. Bøttiger, and H. Knudsen, eds.

5th Conference. *Nucl. Instr. Meth.* **191** (1981), J.A. Bird and G.J. Clark, eds.

6th Conference. *Nucl. Instr. Meth.* **218** (1983), W.A. Lanford, I.S.T. Tsong, and P. Williams, eds.

7th Conference. *Nucl. Instr. Meth.* **B15** (1986), J.P. Biersack and K. Wittmaack, eds.

8th Conference. *Nucl. Instr. Meth.* **B35**, (1988), H.J. Annegarn, T.E. Derry, R.W. Fearick, J.P.F. Sellschop, J.I.W. Watterson, and J.K. Basson, eds.

9th Conference. *Nucl. Instr. Meth.* **B45** (1990), J.F. Ziegler, P.J. Scanlon, W.A. Lanford, and J.L. Duggan, eds.

10th Conference. *Nucl. Instr. Meth.* **B64** (1992), H.H. Brongersma and M.J.A. de Voigt, eds.

11th Conference. *Nucl. Instr. Meth.* **B85** (1994), J. Gyulai, F. Pászti, T. Lohner, and G. Battistig, eds.

ION BEAM MODIFICATION OF MATERIALS (IBMM)

3rd Conference. *Nucl. Instr. Meth.* **209/210** (1983), B. Biasse, G. Destefanis, J.P. Gailliard, eds.

4th Conference. *Nucl. Instr. Meth.* **B7/8** (1985), B.M. Ullrich, ed.

5th Conference. *Nucl. Instr. Meth.* **B19/20** (1987), S.U. Campisano, G. Foti, P. Mazzoldi, and E. Rimini, eds.

6th Conference. *Nucl. Instr. Meth.* **B39** (1989), S. Namba, N. Itoh, and M. Iwaki, eds.

7th Conference. *Nucl. Instr. Meth.* **B59/60** (1991), S.P. Withrow and D.B. Poker, eds.

8th Conference. *Nucl. Instr. Meth.* **B80/81** (1993), S. Kalbitzer, O. Meyer, and G.K. Wolf, eds.

ION IMPLANTATION TECHNOLOGY

7th Conference. *Nucl. Instr. Meth.* **B37/38** (1989), T. Takagi, ed.

8th Conference. *Nucl. Instr. Meth.* **B55** (1991), K.G. Stephens, P.L.F. Hemment, K.J. Reeson, B.J. Sealy, J.S. Colligon, D.W. Palmer, and M.F. Jackson, eds.

NUCLEAR MICROPROBE TECHNOLOGY AND APPLICATIONS

1st Conference. *Nucl. Instr. Meth.* **B30** (1988), G.W. Grime and F. Watt, eds.

2nd Conference. *Nucl. Instr. Meth.* **B54** (1991), G.J.F. Legge and D.N. Jamieson, eds.

3rd Conference. *Nucl. Instr. Meth.* **B77** (1993), U. Lindh, ed.

4th Conference. *Nucl. Instr. Meth.* **B104** (1995), F. Yang, ed.

PARTICLE INDUCED X–RAY EMISSION AND ITS ANALYTICAL APPLICATION

2nd Conference. *Nucl. Instr. Meth.* **181** (1981), S.A.E. Johansson, ed.

3rd Conference. *Nucl. Instr. Meth.* **B3**, (1984), B. Martin, ed.

4th Conference. *Nucl. Instr. Meth.* **B22** (1987), H. van Rinsvelt, S. Bauman, J.W. Nelson, and J.W. Winchester, eds.

5th Conference. *Nucl. Instr. Meth.* **B49** (1990), R.D. Vis, ed.

6th Conference. *Nucl. Instr. Meth.* **B75** (1993), M. Uda, ed.

2nd Conference on Chemical Analysis. Microanalysis using Charged Particle Accelerators. *Nucl. Instr. Meth.* **197** (1982), G. Demortier, ed.

APPENDIX 2

STANDARD STEREOGRAPHIC PROJECTIONS FOR CUBIC CRYSTALS

Reprinted with permission from B.R. Appleton and G. Foti, eds. *Ion Beam Handbook for Materials Analysis.* Academic Press, New York (1977).

$\langle 100 \rangle$

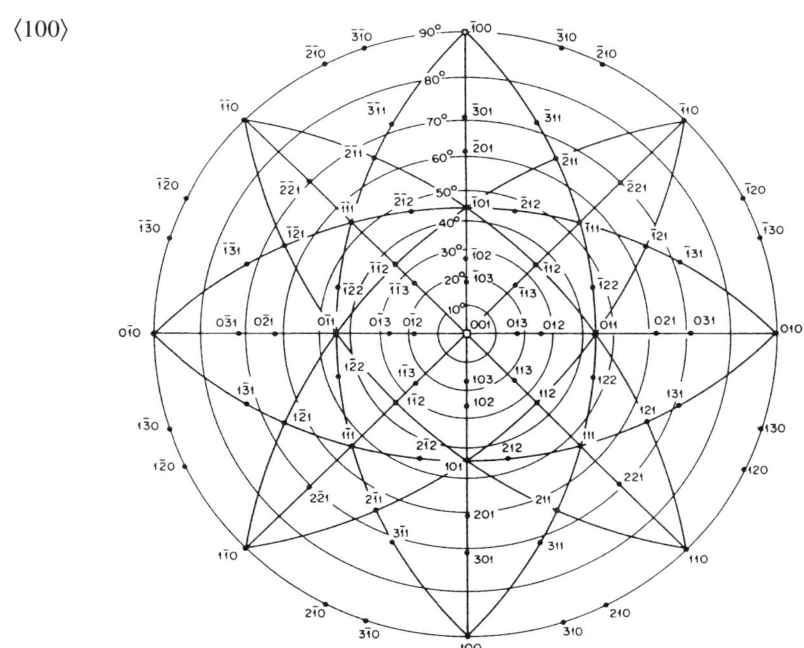

⟨110⟩

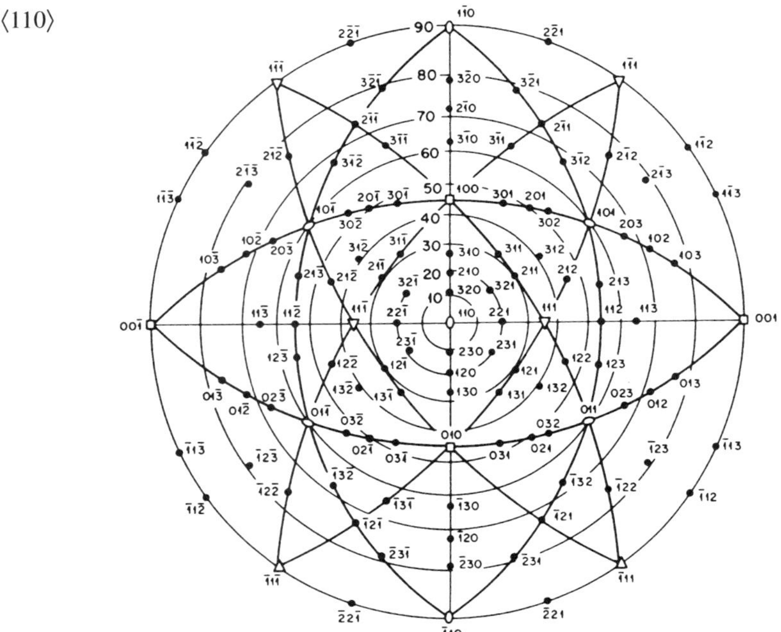

⟨111⟩

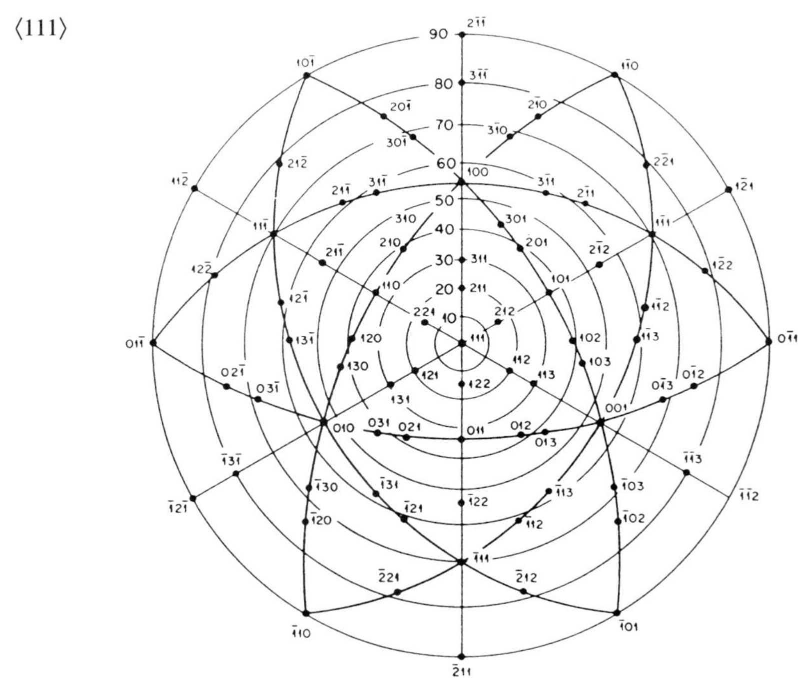

INDEX